Engineering Geology of Waste Disposal

Geological Society Engineering Geology Special Publications
Series Editor M. EDDLESTON

Geological Society Engineering Geology Special Publication No. 11

Engineering Geology of Waste Disposal

EDITED BY

Stephen P. Bentley
School of Engineering
University of Wales
Cardiff, UK

1996
Published by
The Geological Society
London

THE GEOLOGICAL SOCIETY

The Society was founded in 1807 as The Geological Society of London and is the oldest geological society in the world. It received its Royal Charter in 1825 for the purpose of 'investigating the mineral structure of the Earth'. The Society is Britain's national society for geology with a membership of around 7500. It has countrywide coverage and approximately 1000 members reside overseas. The Society is responsible for all aspects of the geological sciences including professional matters. The Society has its own publishing house, which produces the Society's international journals, books and maps, and which acts as the European distributor for publications of the American Association of Petroleum Geologists, SEPM and the Geological Society of America.

Fellowship is open to those holding a recognized honours degree in geology or cognate subject and who have at least two years' relevant postgraduate experience, or who have not less than six years' relevant experience in geology or a cognate subject. A Fellow who has not less than five years' relevant postgraduate experience in the practice of geology may apply for validation and, subject to approval, may be able to use the designatory letters C Geol (Chartered Geologist).

Further information about the Society is available from the Membership Manager, The Geological Society, Burlington House, Piccadilly, London W1V 0JU, UK. The Society is a Registered Charity, No. 210161.

Published by The Geological Society from: The Geological Society Publishing House, Unit 7, Brassmill Enterprise Centre, Brassmill Lane, Bath BA1 3JN, UK (*Orders*: Tel. 01225 445046; Fax 01225 442836)

First published 1996

British Library Cataloguing in Publication Data

A catalogue record for this book is available from the British Library.

ISBN 1-897799-46-2

Typeset by Aarontype Ltd, Unit 47, Easton Business Centre, Felix Road, Bristol BS5 0HE, UK.

Printed by The Alden Press, Osney Mead, Oxford, UK

Distributors

USA
AAPG Bookstore
PO Box 979
Tulsa
OK 74101-0979
USA
(*Orders*: Tel. (918) 584-2555
Fax (918) 560-2632)

Australia
Australian Mineral Foundation
63 Conyngham Street
Glenside
South Australia 5065
Australia
(*Orders*: Tel. (08) 379-0444
Fax (08) 379-4634)

India
Affiliated East-West Press PVT Ltd
G-1/16 Ansari Road
New Delhi 110 002
India
(*Orders:* Tel. (11) 327-9113
Fax (11) 326-0538)

Japan
Kanda Book Trading Co.
Tanikawa Building
3-2 Kanda Surugadai
Chiyoda-Ku
Tokyo 101
Japan
(*Orders*: Tel. (03) 3255-3497
Fax (03) 3255-3495)

Contents

Section 3: Geotechnics of underground repositories

Section 4: Containment properties of natural clays

Section 5: Standards of landfill engineering

Preface

This book is based on papers presented to the 29th Annual Conference of the Engineering Group of the Geological Society, which was held at the School of Engineering, University of Wales, Cardiff, between 6 and 9 September 1993. The theme of the conference was 'Engineering Geology of Waste Disposal' and covered the whole spectrum of activities associated with domestic waste disposal and nuclear waste disposal.

Forty-nine papers are included in this book; all have been refereed and revised since they were submitted to the conference.

Most of the papers presented at the conference are represented in this volume, together with other papers that, because of time constraints, could not be presented orally. Additionally, there are several keynote papers that were not included in the conference preprint volume.

The papers represent the combined experience of a wide cross-section of the engineering geology, geotechnical engineering and environmental science professionals, not only in Britain but from around the world. Many of the authors have international reputations for their expertise in this discipline; for many others time will show them to be equally distinguished. I sincerely hope that each book becomes a well-thumbed source of reference for many years to come.

Stephen P. Bentley

Cardiff, Wales, October 1995

Acknowledgements

The editor gratefully acknowledges the work of David Ogden of the Geological Society Publishing House in preparing this volume and the conference preprints.

I should also like to thank all authors, keynote speakers and discussion contributors for making this volume possible.

Thanks are also due to my colleagues in the School of Engineering, Cardiff, for their contributions.

SECTION 1

INVESTIGATION, HAZARD ASSESSMENT AND REMEDIATION OF EXISTING LANDFILLS

Investigation, hazard assessment and remediation of existing landfills

A. H. Marsh & A. Garnham

STATS Geotechnical Ltd, Porterswood House, Porters Wood, St Albans, Herts AL3 6PQ, UK

Disposing of waste to land is an activity as old as the human race itself. Any ancient deposits still in existence are now principally of archaeological interest. However, with the advent of the industrial revolution the volume of man's waste started to rapidly increase and with it the problems.

In Britain and other similar developed countries the nature and composition of waste has evolved over the decades, reflecting contemporary industrial and domestic practices. Perhaps the most significant such trend concerns the gradual change from the relatively high-density, low putrescible content of Victorian and early to mid-twentieth century waste, when it was the practice to burn domestic waste and dispose of the ashes, to modern unburnt, low-density, highly putrescible domestic waste. The 1950s and 1960s witnessed the most rapid period of change in this respect.

The manner in which waste has been disposed of to land has similarly undergone a process of gradual evolution, but with three key events taking place in the UK in modern times. Historically, waste has been disposed of either heaped on the ground, with the natural strata beneath remaining essentially undisturbed, or in convenient holes previously excavated for other purposes, generally quarries of one sort or another. In the latter case the waste abuts natural strata in the side walls and, often different, strata or the groundwater table beneath.

Any attempts to engineer the containment of waste and control the spread of pollution from it were the exception not the rule until the Control of Pollution Act (1974), COPA, which came into effect on 1st January 1976. This date, with a few enlightened older exceptions, represents the start of modern landfill science and engineering practice in Britain.

The second notable event was the explosion of landfill gases in March 1986 which destroyed a house at Loscoe adjacent to a landfill in Derbyshire (Williams & Aitkenhead 1989). This brought landfills very much to the attention of central government, local authority Environmental Health Departments and planners.

The third major event was the passing of the Environmental Protection Act (1990), EPA, which in general terms introduced the concept in law of 'the polluter must pay' principle, followed by the proposed 'Landfill Tax' introduced in the 1994 budget. The EPA contains specific requirements relating to landfills, some of which have yet to come into force.

In addition to these headline events, regulations on water quality, particularly the EC Groundwater Directive (80/68/EEC) emanating from Europe, have helped to transform the standards pertaining to acceptable landfilling practice.

In the investigation and assessment of existing landfills a working understanding of this historical background can be helpful. These studies are almost always complex, requiring the application of multi-disciplinary skills in science and engineering. A clear understanding of the risks involved in waste disposal by landfilling should be sought by owners, operators, regulators and their advisers and contractors. Jefferies *et al.* in their key note paper on risk assessment provide a wide-ranging review of this vital topic.

Investigation

In the study of existing closed landfills it is often helpful to remember that they are essentially geological deposits, albeit very recent and of a rather specialized nature. From this starting point the geologist will seek to understand the source of the sediment in the deposit, i.e. the waste, and the manner in which it was laid down—not by the normal agents of water and wind but the less predictable hand of man. The geologist will then expect that deposit to change with time, both physically by settlement and possibly slumping, and chemically by degradation processes, leaching and so on. These changes will go on certainly for decades, if not centuries, and lead to complex four-dimensional situations.

The investigation team can usually only take a relatively short exposure snapshot of the characteristics of a landfill. Reynolds & Taylor provide a useful list of potential unknowns for any landfill site at the start of an investigation. All these variables add up to low predictability and emphasize the need for the careful design of investigations.

A significant number of ground investigations carried out each year for the construction industry have proved inadequate for one or more reasons. Such investigations typically deal with 'normal' geology with well-established principles which make for reasonable predictions. Working with landfill is less predictable. It therefore cannot be stressed enough that the Site Investigation in Construction (Anon 1993) initiative led by Professor

From Bentley, S. P. (ed.) *Engineering Geology of Waste Disposal*,
Geological Society Engineering Geology Special Publication No. 11, pp 3–7

Stuart Littlejohn to improve standards in site investigation applies doubly so to landfill work. This initiative promotes the view that 'in site investigation the greatest scope for misjudgements leading to unsatisfactory service is in the conceptual and planning stages'. The main stages of a typical landfill investigation are set out in Fig. 1.

Forth & Beaumond in their paper on coal-carbonation sites provide a review of some of the early steps, emphasizing the benefits of the desk study and discussing sampling strategies. ICRCL Guidance Note 17/78 (1990) sets out many of the factors to be taken into account in the investigation of landfills.

Because of the relatively high cost of chemical analysis, it has become apparent in the last few years that adopting a massive regular sampling and testing exercise along the lines advocated in BS DD175 (Anon 1988) is by no means always appropriate, particularly in the first instance.

In the search to home in on problem areas and obtain rapid general characterizations of landfills, geophysical techniques have very much come to the fore. This is reflected in the papers in this volume. Reynolds & Taylor provide a sound overview of geophysical techniques and Maurenbrecken describes shallow over-water seismic reflection technique from the Netherlands.

The importance of investigating the geological and hydrogeological character of the host ground around a landfill is brought out in many of the papers, particularly the case histories by Bell *et al.* from South Africa, by Coppola *et al.* from Italy and by McShane & Gregory working in Northern Ireland.

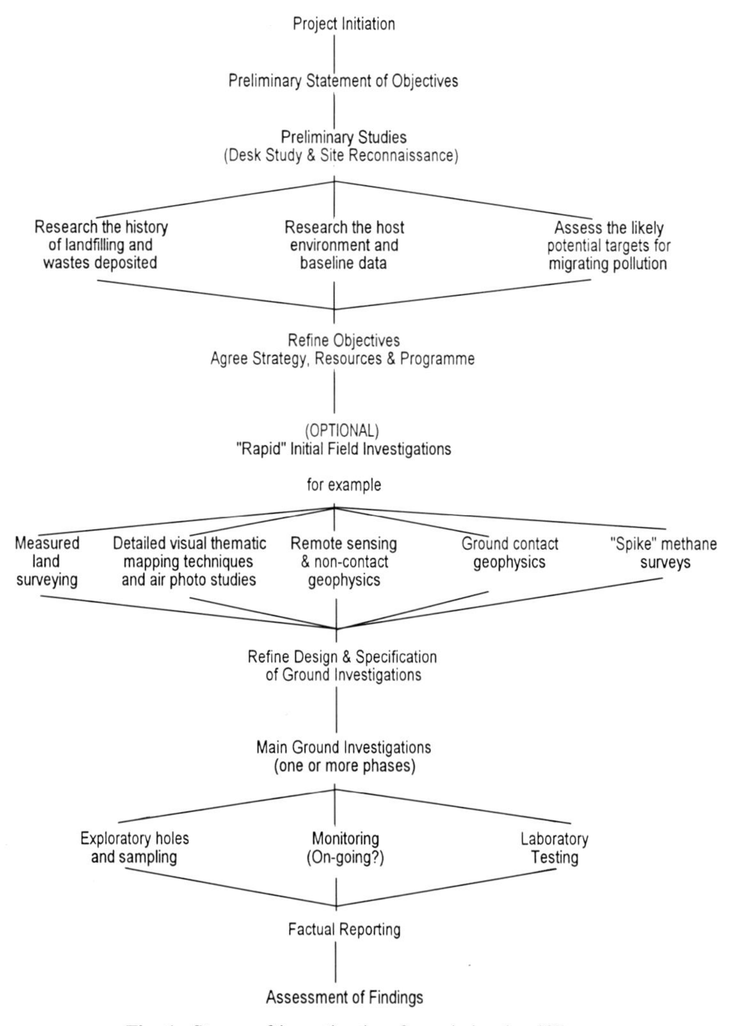

Fig. 1. Stages of investigation for existing landfills.

There are many omissions in the subject matter covered by the papers on the topic of investigations. Landfill gas is barely mentioned, a tribute perhaps to documents such as Waste Management Paper No. 27 (1991) on landfill gas, which is the standard reference on this topic. Problems with leachates clearly stand out as the area of greatest current concern.

Leachate investigations in particular can generate quantities of data. Data management is therefore a key issue. Bentley *et al.* describe one commercially available software package, Strata 3, designed to aid the presentation and interpretation of information.

This volume also contains three papers on rather special investigations. Di Stefano presents stimulating data on settlement measurements within landfills and their implications for long-term settlement prediction, challenging elements of current practice. Mosley & Crozier describe a technique for leak detection in newly laid geomembrane landfill liners. Mahmoud & Morley describe the engineering properties of two types of pfa in a waste disposal situation.

Assessment

There are always at least two main elements to assessment: the physical side of landfill behaviour and the chemistry. Neither area is comprehensively covered here. The British Government's published guidelines on many aspects of assessment are summarized in ICRCL Guidance Note 17/78 (1990) and this provides a useful first point of reference. Where it is proposed to build over a closed landfill, reference should also be made to Leach & Goodger (1991).

Coppola *et al.*'s paper is principally concerned with physical aspects of stability and makes the important point that landfills in relatively dry terrain can profoundly change the groundwater regime to the detriment of stability, in their case by supporting a shallow perched water-table beneath the waste. Bell *et al.* discuss the equally important question of water balance within a landfill.

The assessment of landfill chemistry is, by any measure, a complex task in which chemists should work with geologists. Matter exists in three physical states, namely gas, liquid and solid, and all landfills comprise a delicate and shifting balance between the three states. Any assessment of the state of a landfill and its environs must embrace consideration of the substances present in the landfill, their mobility now and in the future, the potential pathways along which pollutants can travel and the targets potentially at risk from the substances involved.

In terms of published guidelines for data assessment, landfill gases are probably best covered, with Waste Management Papers No. 26 (1986) and 27 (1991), Crowhurst & Manchester (1993) and BRE Report BR212 (1991) providing clear guidance on methods of measurement and on acceptable levels of landfill gases leaving landfill sites and beneath buildings.

Published guidelines for soils and other solid forms of contaminated ground are incomplete and in part contradictory. Waste arriving at a UK landfill site for disposal is generally 'controlled waste' and could be one of several categories; namely, household, industrial, commercial or clinical waste. These terms describe the origin of the waste but not its toxicity. The latter is normally determined to be either non-hazardous, hazardous or special waste. Further guidance on the legal definitions of these is given by Attewell (1993). For industrial waste the distinction between non-hazardous and hazardous wastes can be crucial to the cost of its disposal. Although there is no single set of government-approved values or finite list of chemical determinands, it is often the case that 'hazardous' waste is classified as such by the presence of one or more substances at concentrations in the 'heavily contaminated', or worst, category formulated by the former Greater London Council and published by Kelly (1980).

For the redevelopment of closed landfill sites reference is made in ICRCL Guidance Note 17/78 (1990) to the 'threshold and action trigger' concentrations published in ICRCL Guidance Note 59/83 (1987). These trigger concentrations only cover a relatively limited range of determinands, albeit including some of the more common substances involved. The interpretation of Tables 3 and 4 in ICRCL Guidance Note 59/83 (1987) is however rather subjective, with the majority of data often falling into the area between the threshold and action values where Fig. 1 therein states that the 'significance of risk depends on intended use and form of development, USE PROFESSIONAL JUDGEMENT TO DECIDE WHETHER ACTION IS NEEDED'. This provides both designers and regulators with plenty of scope for disagreement.

The third set of guidelines commonly referred to are those published by the Dutch Government (Keuzzenkamp 1990). These employ the definition of three categories (A to C) of contamination by a rather longer list of determinands than ICRCL Guidance Note 59/83 (1987), where A is the reference value below which soils are probably uncontaminated, B is the value above which there is a need for further investigation, and C is the value above which a clean-up is indicated. Recently amendments have been made to these guidelines (Denneman 1993) introducing a simpler system of 'target values' and 'intervention values'.

For liquids, assessment standards are crucially dependent upon the target to be affected by mobile contaminants. In the UK there are standards specifically for surface water courses (EC Directive 75/440/EEC) and for groundwater (EC Directive 80/68/EEC) to be abstracted for potable water supplies but little general guidance. Many people again rely on Dutch Government assessment standards for groundwater. Some South African water-quality standards are also presented in Bell *et al.* paper.

Forth & Beaumont make some mention of data interpretation and in particular the need to interpret data with the end-use of the site borne in mind.

Remediation

The remediation of 'failed' landfills can be achieved by a variety of methods, with the principal options given in Table 1. The most appropriate method for any particular landfill will depend upon many factors covering technical issues, planning considerations and economics. At the feasibility stage of remediation as wide a view as possible should be taken of the options. It is probable that the economic, planning and licensing conditions that pertained when the landfill was started have significantly changed by the time the failure has to be dealt with. One likely scenario in the future is that extra new landfill space may be available to operators at a particular site but tied in with a requirement to redeposit unsatisfactorily contained old waste. Reprocessing the old waste in transit, by such techniques as accelerated composting of degradable matter and solids reclamation of metals and inert rubble, could result in a major net gain in the final new landfill space once the original landfill space has been relined to contemporary standards of containment.

The principles of containment and control techniques are given in several standard texts including ICRCL Guidance Note 17/78 (1990) and Waste Management Papers No. 26 (1986) and No. 27 (1991). A bibliography of landfill gas-related remediation is given by Hartless (1992). The literature on leachate containment and control is rather fragmented but reference can usefully be made to Fetter (1992) and Jefferies (1990), amongst many.

There are three papers in the conference giving useful case histories of remediation projects. Leachate containment by drainage and cement: bentonite cut-off wall techniques is the main subject of the case histories reported from South Africa by Bell *et al.* At the peat-bog landfill site in Northern Ireland described by McShane & Gregory, an HDPE lined bund was the preferred solution. Drainage measures, essentially to reduce pore water pressures but with the added advantage of controlling pollution are described by Coppolla *et al.* from a site in Italy.

Confidentiality is certainly a major restriction at present to the free dissemination of data and experience in landfill investigation and remediation. A useful source of references specifically on methane and associated hazards including remediation schemes is contained in Hartless (1992).

Table 1. *Options for remediation*

Removal off-site	(i)	Bulk of selective excavation
Redeposition on-site	(i)	Bulk transfer to new containment cells
	(ii)	Process for volume reduction and place residue in new cells
Containment and control		
Passive systems	(i)	Create flow paths, e.g. venting trenches for gases and drains for leachates
	(ii)	Create barriers, e.g. vertical and horizontal
Active systems	(i)	Pumped collector systems

Concluding remarks

The selection of topics contained in papers submitted to the conference has reflected the evolving state of the art with emphasis being placed on investigating leachate problems at the expense of gas problems. A very wide range of investigative and monitoring techniques is now available and many useful data have been presented supporting their uses. The authors endorse the current trend towards tipping the balance of resources a little more in favour of initial rapid 'mass characterization' techniques, principally by geophysics, compared with more traditional sampling and testing programmes.

The current situation in the UK concerning criteria for the assessment of the condition of an existing landfill and its host environment is unsatisfactory. In general, criteria are only qualitative and not consistently applied across administrative boundaries. With particular reference to landfill chemistry, only the assessment of landfill gas has been reasonably satisfactorily set out in government-approved publications. Guidelines for assessing leachates and solid residues remain incomplete and often potentially contradictory.

Remedial works on existing landfills have been extensively carried out over the last decade or so, mainly to control the migration of gas. This technology is now relatively well proven. Leachate control has proved, and is likely to remain, more challenging to the industry. This arises from the chemical aggressiveness of many leachates combined with the long timescales in engineering terms during which systems will be required to operate.

References

ANON. 1988. *Draft for Development: Code of Practice for the Identification and Investigation of Potentially Contaminated Land.* British Standard DD175, British Standards Institution, HMSO, London.

—— 1993. *Site Investigation in Construction, Parts 1 to 4.* Site Investigation Steering Group, Thomas Telford, London.

ATTEWELL, P. B. 1993. *Ground pollution* E & F N Spon, London.

BRE 1991. *Construction of new buildings on gas-contaminated land.* BRE Report BR212, Building Research Establishment, Watford.

CROWHURST, D. & MANCHESTER, S. J. 1993. *The measurement of methane and other gases from the ground.* CIRIA Report 131, CIRIA, London.

DENNEMAN, C. 1993. *What is ecological risk?—risk based environmental quality objectives.* Dept of Soil Protections, Ministry of Housing, The Hague, the Netherlands.

FETTER, C. W. 1992. *Contaminant hydrogeology.* Macmillan, New York.

HARTLESS, R. 1992. *Methane and associated hazards to construction: a bibliography.* Ciria Special Publication 79, CIRIA, London.

INTERDEPARTMENTAL COMMITTEE ON THE REDEVELOPMENT OF CONTAMINATED LAND 1987. *Guidance on the assessment and redevelopment of contaminated land.* Guidance Note 59/83, UK Department of Environment, London.

—— 1990. *Notes on the development and after-use of landfill site.* Guidance Note 17/78, UK Department of Environment, London.

JEFFERIES, S. A. 1990. Cut-off walls: methods, materials and specifications. *In*: FORDE, M. C. (ed.) *Polluted and Marginal Land—90.* Proc. Int. Conf. on Polluted and Marginal Land, 28/29 June 1990, Brunel University, London. Engineering Geotechnics Press, Edinburgh, 177–225.

KELLY, R. T. 1980. Material problems and site investigation. *In*: *Reclamation of Contaminated Land.* Soc. Chemical Ind., London.

KEUZZENKAMP, K. 1990. Dutch policy on clean-up of contaminated soil. *Chemistry and Industry*, **1990**(3), 63–64.

LEACH, B. A. & GOODGER, H. K. 1991. *Building on derelict land.* CIRIA Special Publication 76, London.

WASTE MANAGEMENT PAPER NO. 26. 1986. *Landfilling Wastes, A Technical Memorandum for the Disposal of Wastes on Landfill Sites.* HMSO, London.

WASTE MANAGEMENT PAPER NO. 27. 1991. *Landfill Gas.* HMSO, London.

WILLIAMS, G. M. & AITKENHEAD, N. 1989. Case study—the gas explosion at Loscoe, Derbyshire, *In: Methane: Facing the Problems Symposium*, 26–28 September 1989, East Midlands Conference Centre, Nottingham.

Application of a geomembrane leak location survey at a UK waste disposal facility

N. G. Mosley & F. Crozier

Entec UK Ltd, 160–162 Abbey Foregate, Shrewsbury, Shropshire SY2 6AL, UK

Abstract. Electrical leak location surveys have been undertaken in the US for several years. 1993 has seen the first application of this technology on a newly constructed soil-covered HDPE geomembrane at a waste disposal site in the UK. The technology exploits the electrical insulating properties of geomembranes. A voltage placed across the geomembrane is surveyed to identify anomalies where electrical current is flowing through leaks in the liner. In its first UK application both 'pin hole' leaks and larger cuts in the geomembrane were located and repaired prior to commissioning the waste disposal facility.

The first UK application of an electrical leak location survey was undertaken at Antrim Borough Council's Craigmore Landfill Site, Randalstown, Northern Ireland. Phase 1 of the Craigmore Landfill was con-structed in 1988 utilizing a single HDPE geomembrane. Subsequently problems were encountered with the Phase 1 leachate containment system due to a failure in the integrity of the geomembrane liner. With the construction of Phase 2 of the Craigmore Landfill, additional measures were investigated and applied to ensure the containment properties of the geomembrane liner.

Craigmore Landfill Phase 2 liner design

Phase 2 of Craigmore Landfill was designed by Entec as a full containment landfill with a composite lining system. Due to a lack of suitable sources of natural clay at Craigmore, a bentonite-based liner was selected. Various options of bentonite-based liners were considered. Bentonite-enhanced sands were discounted due to the anticipated construction difficulties likely to be encountered as a result of the high levels of precipitation and wind-exposed setting of the site.

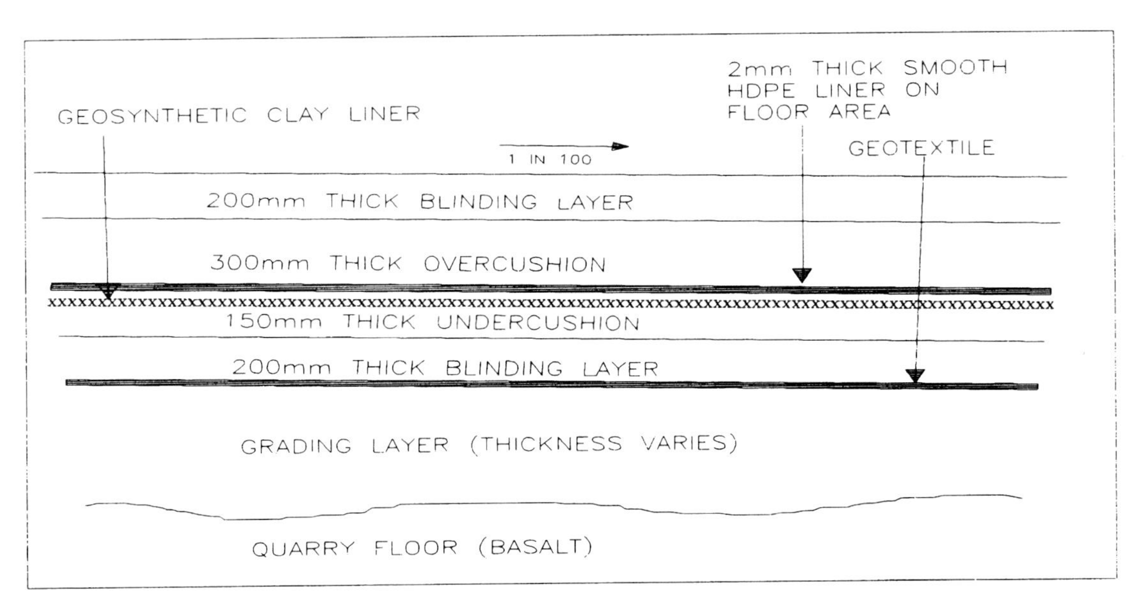

Fig. 1. Craigmore Landfill Phase 2. Section through the liner system.

From BENTLEY, S. P. (ed.) *Engineering Geology of Waste Disposal*,
Geological Society Engineering Geology Special Publication No. 11, pp 9–14

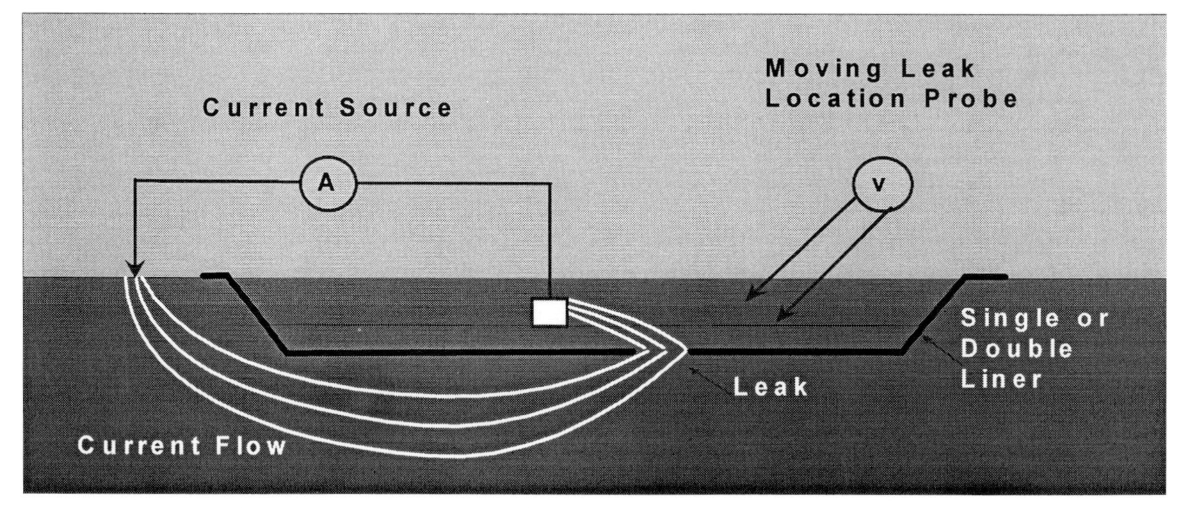

Fig. 2. Diagram of electrical leak location method on soil covered geomembrane.

A geosynthetic clay liner (GCL) was selected due to the ease of deployment, quality control and economics when compared to alternative systems.

The final composite liner design for Phase 2 at Craigmore comprised a primary liner of 2 mm HDPE geomembrane overlying a secondary liner of 'Bentomat'. 'Bentomat' is the trade name for Volclay Ltd's geosynthetic clay liner product. A typical section through the liner system installed in Phase 2 of Craigmore is shown in Fig. 1.

Leak location survey technique

Following a review of available techniques, an electrical leak location method was chosen to test the newly constructed geomembrane liner prior to commissioning the next phase of the landfill. The electrical leak location technique is illustrated in Fig. 2.

The method involves connecting an electrical power supply to electrodes placed above and below the liner. Probes are used to detect areas of localized electrical current flow through leaks in the otherwise insulating liner. If no leaks are present, the voltage impressed across the liner produces a very low current flow and a relatively uniform voltage distribution in the material above the liner. If the liner has a leak, electrical current flows through the leak causing a localized anomaly in the potential gradient. Leaks are located by measuring potential gradients in the material above the liner and by searching for localized areas of relatively high electrical potential gradients. The electrical leak location method can be used on liquid impoundments, and for pre-service inspection of solid-waste landfills with soil cover.

With the proper implementation of equipment and survey procedures, the electrical leak location method is a very sensitive and accurate technique. The measured amplitude of the leak signal is proportional to the amount of electrical current flowing through the leak, such that leak location surveys are conducted with the maximum practical safe impressed voltage. To increase the leak detection reliability the detector electronics are optimized for maximum sensitivity. In addition, extraneous electrical conductors such as metal pipes and pump wiring that provide an electrical conduction path through or around the geomembrane liner need to be eliminated or insulated.

The technique is dependent on electrical current transmitted through soils via the soil moisture. The surrounding soils (above and below the geomembrane) must be moist but not necessarily saturated. The moisture content required to achieve adequate compaction is sufficient to transmit an electrical current.

The non-hydrated bentonite in geosynthetic clay liners contains a residual moisture content of some 15% to 17%, making it highly conductive. These conditions were considered sufficient to transmit an electrical current. This was confirmed by undertaking laboratory trials on non-hydrated samples of the selected geosynthetic clay liner. The use of geosynthetic clay liners in the field, results in exposure to soil moisture which causes the bentonite to hydrate, so enhancing the conductivity.

Electrically leak location survey

Following the construction of the geomembrane liner in Phase 2 of Craigmore, a 300 mm layer of sand was

Fig. 3. Leak location survey. Data collection on soil covered geomembrane.

placed above the HDPE as a protective cover. The construction of the geomembrane liner and placement of the soil cover was carried out under a strict programme of construction quality assurance.

The electrical leak location survey was performed on the 300 mm sand layer covering the basal area of Phase 2. The survey lines were extended 5–10 m up the side slopes of the landfill containment.

Phase 2 at Craigmore is divided into four cells by means of a HDPE partition welded to the floor of the primary liner. The HDPE partitions extend approximately 1 m up the side slope. The sand located on the side slope directly above the partition was removed to electrically isolate each of the cells. In addition, the sand on the top of the side slope was removed around the full perimeter of the landfill to electrically isolate the sand from the soil substrate outside of the landfill.

The survey lines were spaced 1 m apart with potential data collected on 50 cm centres along each survey line. Survey data being collected on the soil-covered geomembrane are shown in Fig. 3. A total of 197 survey lines oriented in a north–south direction and varying in length from approximately 30 m to 80 m were surveyed in the landfill. Survey data were plotted and analysed for data quality and completeness in the field. Leaks indicated in the data were investigated and if verified the sand cover was removed to expose the leak in the HDPE liner. Representative plots of the data collected during the survey are shown in Fig. 4.

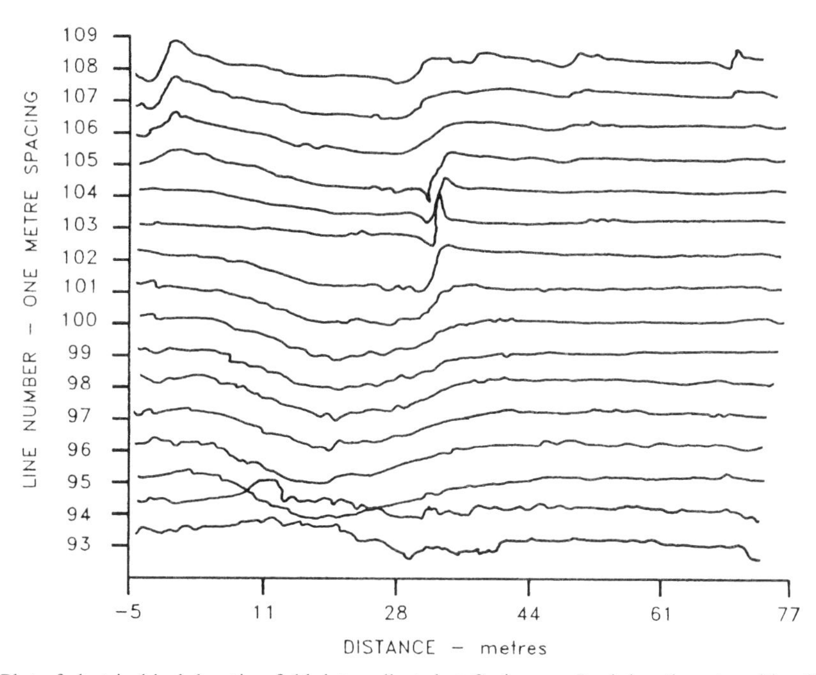

Fig. 4. Plot of electrical leak location field data collected at Craigmore. Leak location at position 30 in line 103.

Results

The electrical leak location survey at Craigmore detected and verified ten leaks in the primary HDPE geomembrane. The locations of the leaks on the CQA 'as built' panel layout drawing are shown in Fig. 5. All the leaks were uncovered, repaired and retested prior to recovering with sand. The following notes describe the leaks located in the geomembrane:

- Leak 1: located in extrusion fillet weld on pipe boot;
- Leak 2: 2 cm cut in geomembrane panel approximately 15 cm northeast of patch R2;
- Leak 3: leak on fillet weld between panels N2 and B2 approximately 1 m south of panel N1;
- Leak 4: leak in tee weld between panels B14 and S15/S16; this leak was visually confirmed with water flowing from the leak;
- Leak 5: leak in hand fillet weld on north side of patch R31/D5; water was observed flowing from the leak;
- Leak 6: leak located on north side of patch R44 in hand fillet weld;
- Leak 7: leak located in fillet weld on panel S24 near toe seam;
- Leak 8: leak located on the north side of patch R53;
- Leak 9: tee weld of panels S31/S32 and B33/B34;
- Leak 10: 15 cm cut in liner located in panel B39 near toe of side slope.

In addition to locating the leaks in the primary liner, the electrical current injected into each of the landfill cells was measured before and after the liner was repaired. Because the HDPE liner is an excellent electrical insulator, the electrical current injected into each cell is a measurement of current flow paths around or through the liner material. This is a relative measurement and is not used to quantify the number or size of leaks in the primary liner. If there are no leaks in the liner material and the soil located above the liner is not in contact with the soil located below the liner then with an input of 350 V the resultant current flow should be less than approximately 1 mA. This assumes a liner floor area of approximately 10 000 m^2 with a fairly high soil or water resistivity.

Results of other leak location surveys

Electrical leak location surveys have been available as a commercial service since 1985. During this time 169 surface impoundments and 17 landfill sites have been investigated. The method is used to locate leaks in the geomembrane liners of single-lined and double-lined facilities covered with water or a protective soil cover. The surveyed facilities have ranged in size from 9 m diameter above ground tanks to surface impoundments of approximately 6 ha in size.

To date, the majority of electrical leak location surveys have been conducted in the United States. However, surveys have been conducted in Japan, Saudi Arabia, Germany, Italy and Canada. Figure 6 shows a

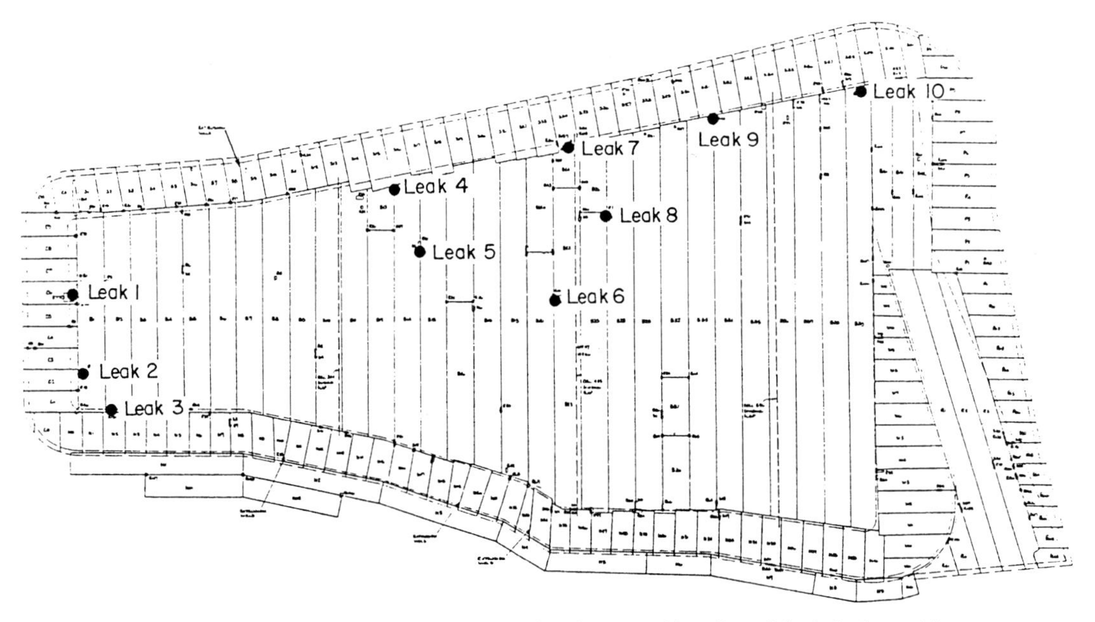

Fig. 5. Craigmore Landfill Phase 2 panel layout showing the general location of the leaks located in the geomembrane.

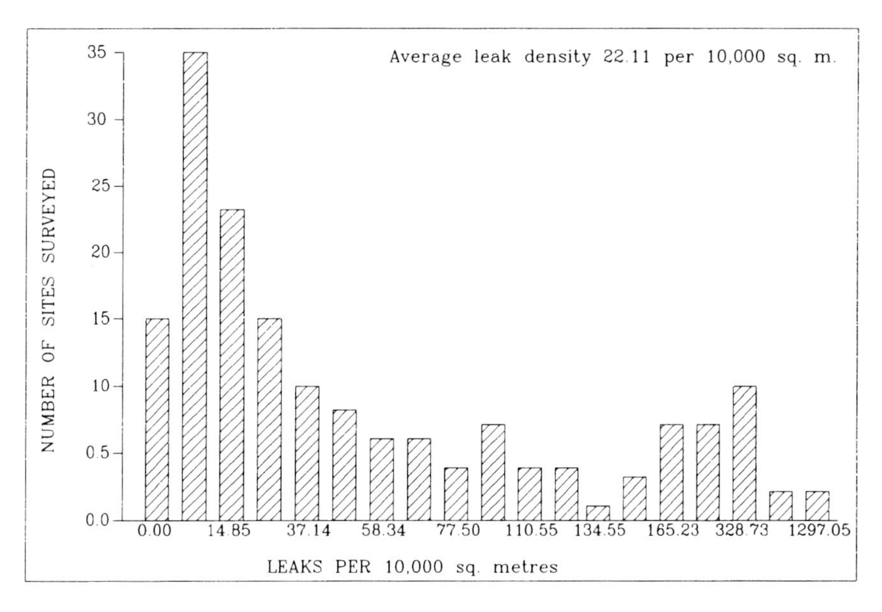

Fig. 6. Electrical leak location survey data for geomembranes covered with water.

histogram of the survey data for 169 sites where the geomembrane was covered with water. These data indicate that an average of 22 leaks per 10 000 m^2 of geomembrane tested were detected when an electrical leak location survey was conducted. In many cases the leak location survey was conducted on geomembrane liners that had been installed under strict construction quality assurance supervision.

Leak location survey results for 17 landfill sites where the geomembrane was covered with up to 75 cm of protective soil are shown in Fig. 7. The results indicate an average of 14 leaks per 10 000 m^2. The method is particularly valid on soil-covered landfill liners as the geomembrane is tested under load and after the liner has been exposed to potential damage during construction.

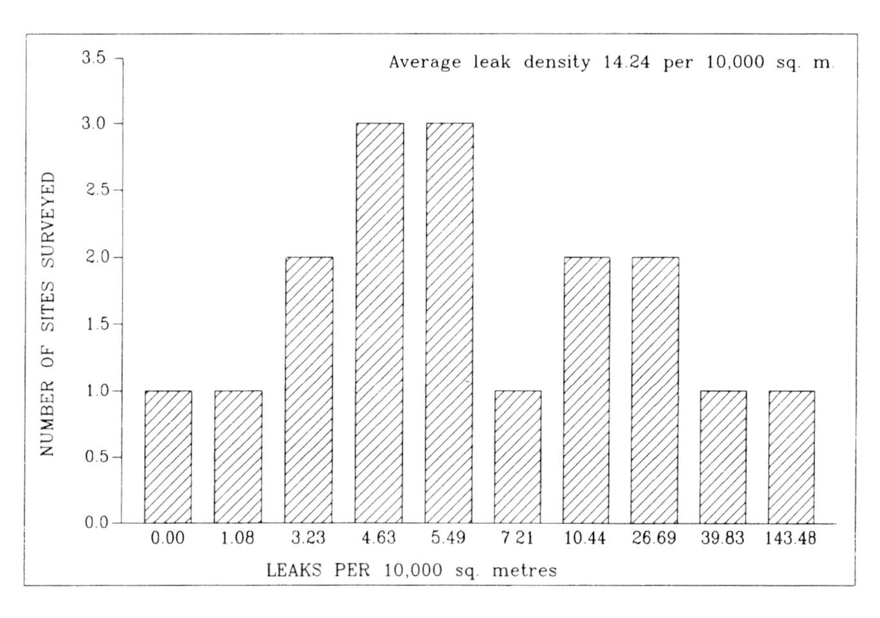

Fig. 7. Electrical leak location survey data for geomembranes covered with a protective soil.

Conclusions

For several years the application of electrical leak location surveys has proved effective in locating defects in geomembrane liners prior to commissioning the facilities. In its first UK application the technology proved very effective in pinpointing a variety of imperfections in a soil-covered geomembrane liner.

The leak location survey at Craigmore yielded similar results to other surveys using the same technology, in that most of the leaks resulted from imperfections in seams or welded joints. However, significant holes were also identified in geomembrane panels which would have had the potential for serious leachate escapes. The holes in the geomembrane panels are considered to be a result of damage caused during deployment of the protective soil overcushion layer.

The electrical leak location method described in this paper provides a quick, efficient and very effective technique for locating defects in geomembranes. The technique has proved highly effective when applied to both surface water impoundments and soil-covered landfill liners. Experience from leak location surveys using this technology indicates that despite strict construction quality assurance programmes during construction, most geomembrane liners leak.

Given the significant cost involved in constructing geomembrane-lined facilities, and the potential environmental impact that may occur if such liners fail, then a leak location survey using the technique described above is considered to be an additional tool to ensure the integrity and quality of geomembrane installations.

Acknowledgements. Mr John Quinn, Chief Technical Officer with Antrim Borough Council is gratefully acknowledged for allowing this case study to be published. In addition, Environment Service, DOE Northern Ireland are acknowledged for providing financial support, via the European Regional Development Fund under the Intereg Program, which enabled the survey to be conducted.

References

DARILEK, G. T., LAINE, D. L. & PARRA, J. O. 1989. *The electrical leak location method for geomembrane liners—development and applications.* Industrial Fabrics Association International Geosynthetics '89 Conference, San Diego, CA.

LAINE, D. L. & DARILEK, G. T. 1993. *Locating Leaks in Geomembrane Liners of Landfills Covered with a Protective Soil.* Industrial Fabrics Association International Geosynthetics '93 Conference, Vancouver, British Columbia, Canada.

—— & MIKLAS, M.P. 1989. *Detection and Location of Leaks in Geomembrane Liners using an Electrical Method: Case Histories.* Superfund '89, 10th National Conference, Washington, DC.

Landfills and associated leachate in the greater Durban area: two case histories

F. G. Bell, A. J. Sillito & C. A. Jermy

Department of Geology and Applied Geology, University of Natal, Durban, South Africa

Abstract. Two waste disposal sites in the greater Durban area used for codisposal of solid and liquid wastes are considered. One site is still operational and the other has been closed for a number of years.

An investigation of Site 1 revealed that much of the area is covered by Berea Red Sand. However, the clay content of much of this formation means that it generally has a low permeability. The sandstones and tillite which occur beneath the sand also are of low permeability or are virtually impermeable. This is necessary when the character of the leachate produced is considered. The very high values, for example, of chemical oxygen demand are attributable to a large extent to the disposal of hop waste. Accordingly the leachate is conveyed from a sump at the toe of the landfill to a sewerage works.

At site 2 very little has been done since its closure to ensure that leakage from the landfill does not pollute the surrounding ground or surface waters. It was suspected that this was occurring. Small non-perennial and perennial streams occur in the area of the site. The site itself is underlain by sandstones and tillite. These rocks have been weathered to form a mantle of residual soil of variable thickness which is overlain by colluvium on the valley slopes and alluvium in the valley floor. An investigation was undertaken to determine the extent of the pollution and revealed that a plume extended some 300 m downstream of the leachate pond. Two main remedial measures were proposed. First, a subsoil drain should be constructed immediately downstream of the landfill site. This should be located in bedrock, its purpose being to intercept seepage from the landfill. The drain should lead into a pump chamber that would remove polluted water which would then be conveyed to the sewer system. Secondly, to make doubly sure that the pollution was contained, a cement–bentonite cut-off trench, founded in bedrock, should be constructed 3 m downstream of the subsoil drain.

One of the major problems associated with waste disposal is the generation of leachate. Leachate is contaminated water which is produced when the rainwater or groundwater flows through a landfill dissolving the soluble fraction of the waste. This occurs once the absorbent characteristics of the refuse are exceeded. The composition of leachate depends on the materials present in the landfill and the environmental conditions existing at the site. It also varies with time. Domestic refuse is a heterogeneous collection of many things, much of which is capable of reacting with water to give a liquid rich in organic matter, mineral salts and bacteria. At many landfill sites where control measures are either inadequate or non-existent, leachate has moved into the soil, groundwater or surface water, and this can cause pollution. Such leachate problems may occur in relation to either operational or completed landfill sites. Ideally leachate production should be kept to a minimum and monitored.

Generally the largest contributor to leachate generation is precipitation, which percolates through the waste, reacting with it during the process. In codisposal operations (where both liquids and solids are disposed of together), leachate generation is enhanced by the addition of liquid wastes. An estimate of the volume of leachate that will be produced can be obtained from a water-balance determination for a landfill which consists of an evaluation of liquid input and output from the site. The first part of a water-balance calculation involves an assessment of the quantity of water entering a site, either as rainfall or by codisposal. The second part involves an evaluation of liquid retained in a landfill, as well as liquid lost due to evaporation and leachate flow. Water-balance calculations, however, are complex. One of the problems involved concerns the determination of the absorptive capacity of the waste material. Absorptive capacity is influenced by the composition of the waste, the method of waste treatment prior to disposal and the method of disposal.

The climate of the greater Durban area can be described as hot humid summers with mild winters. The average annual rainfall is 1013 mm, while the average annual evaporation is 1298 mm. In other words, on average, evaporation exceeds precipitation

From Bentley, S. P. (ed.) *Engineering Geology of Waste Disposal*,
Geological Society Engineering Geology Special Publication No. 11, pp 15–35

Table 1. *Average monthly figures for precipitation, evaporation and wind speed*

Month	*Precipitation (mm)*	*Evaporation (mm)*	*Wind speed ($m s^{-1}$)*
January	129	150	3.7
February	121	130	3.6
March	113	125	3.5
April	96	95	2.9
May	73	84	2.2
June	30	64	2.1
July	37	72	2.4
August	46	88	3.5
September	63	105	4.2
October	86	110	4.3
November	103	125	3.9
December	116	150	3.9

in every month except April (Table 1). The prevailing wind direction is predominantly northeasterly or southwesterly on a 50% basis. Wind speeds vary between $2.1 m s^{-1}$ and $4.3 m s^{-1}$, averaging $3.3 m s^{-1}$.

Case history 1

The site occurs within a southerly facing valley with elevations varying from 106 m above msl at the top of the crest at the northern edge of the site, to 45 m above msl at the valley invert at the southern boundary of the site (see Fig. 3). The valley side slopes vary from 1 in 4 in the northern area to 1 in 2 in the south. The steep valley slopes presumably facilitate the movement of leachate towards the toe of the landfill and into the sump.

Codisposal operations started in June 1986 and the site is still operational. The minimum available airspace for the landfill site is $750\,000 m^3$. Assuming an average annual volume of uncompacted waste of some $250\,000 m^3$ and a compaction ratio of 3:1, the site probably has a potential life of at least nine years.

The site is protected against stormwater runoff by cutoff drainage ditches around the perimeter. Any drainage to the site which existed prior to the start of operations was redirected around it. When the final cover is placed over the landfill the area will be graded to cross falls of not less than 1 in 20 to promote runoff of stormwater to the cutoff drains. Surface runoff will not be allowed to run over the face of the berms constructed to contain the waste.

Geology and investigation of the site

A site investigation was carried out prior to the formation of the landfill in order to determine the site geology, the position of the water-table and the permeability of the rocks and soils concerned. The position of the water-table and permeability of the rocks and soils is particularly important in relation to possible pollution of groundwater by leachate.

Most of the site is underlain by pinkish-grey quartzitic sandstones which, in places are arkosic. They belong to the Natal Group (Fig. 1). These are generally medium- to coarse-grained hard sandstones, except when weathered. Individual beds of sandstone vary in thickness from around 0.3 to 3 m and lenticular horizons of shale are present. The latter are often highly micaceous and their thickness rarely exceeds 0.5 m. Generally the sandstone dips in a northeasterly direction at around 15°.

A joint analysis carried out on the sandstone revealed two major joint sets orientated at 120° and 30°. The jointing is medium to widely spaced with spacing between joint sets ranging between 150 and 400 mm. Secondary random jointing was also developed in a number of the exposures examined. The joints vary from discontinuous and closed, to continuous and widely separated containing clay gouge. These discontinuities, together with the bedding planes, are primarily responsible for the permeability of the sandstone, which decreases with increasing depth.

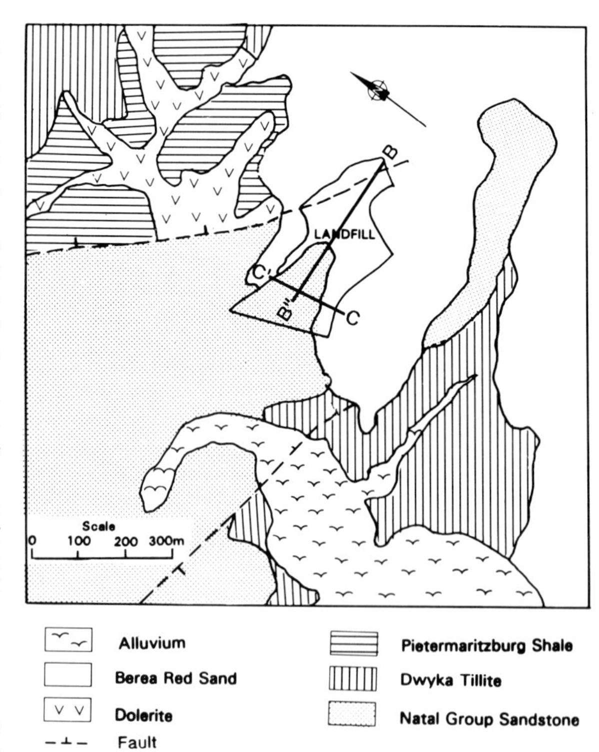

Fig. 1. Geology of Site 1.

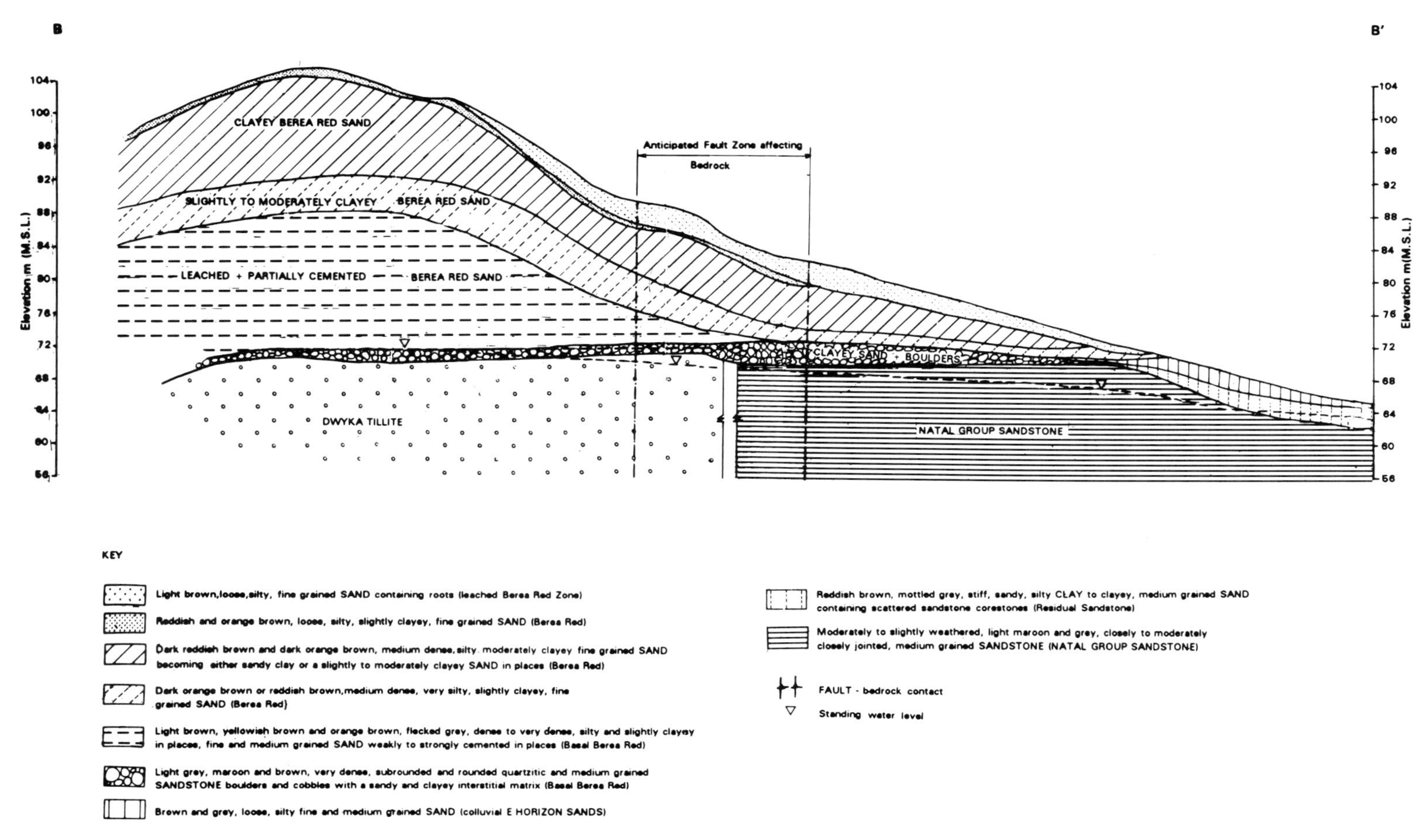

Fig. 2. Geological cross-sections of Site 1 (for location, see Fig.1).

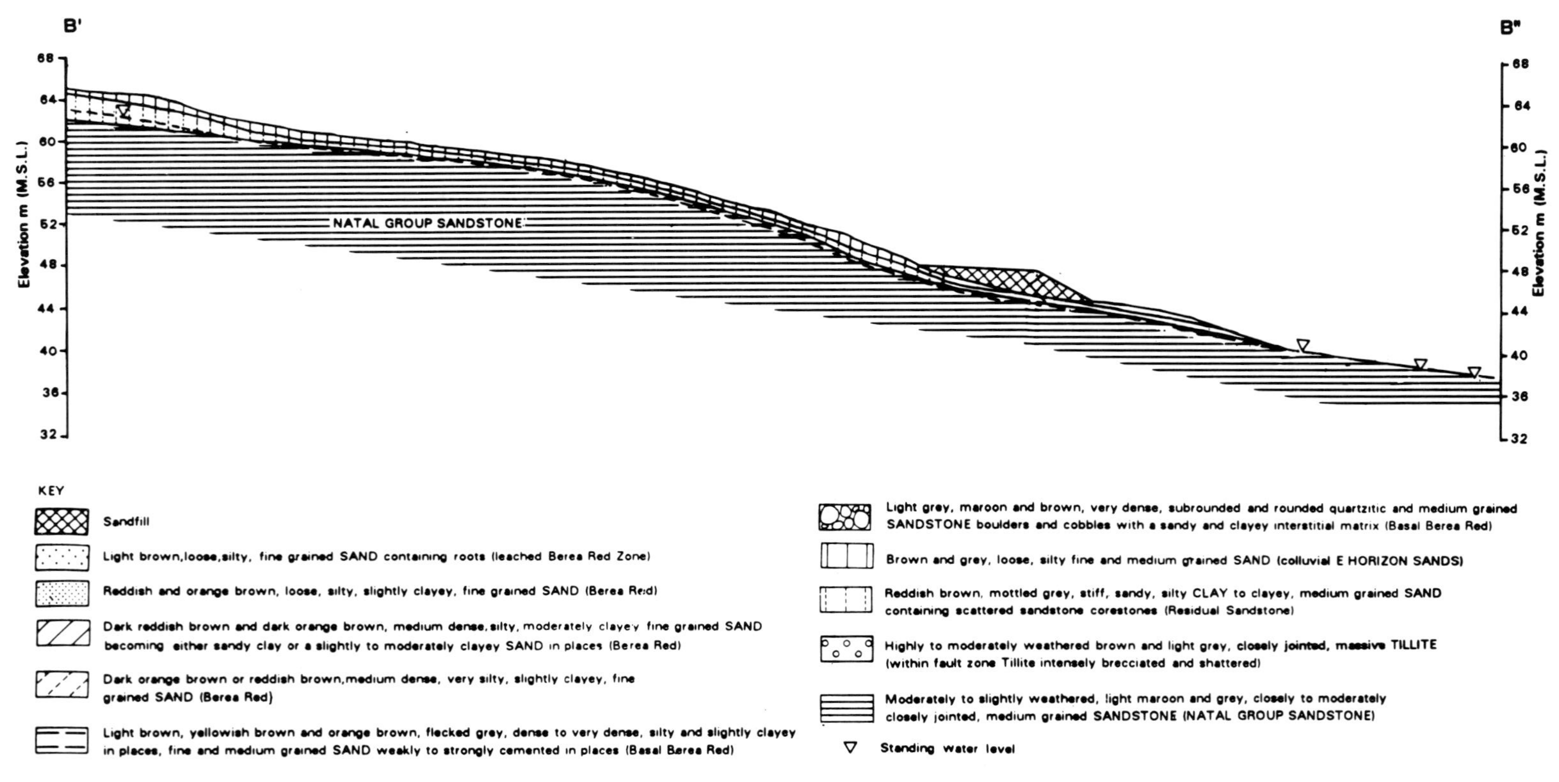

Fig. 2. Continued.

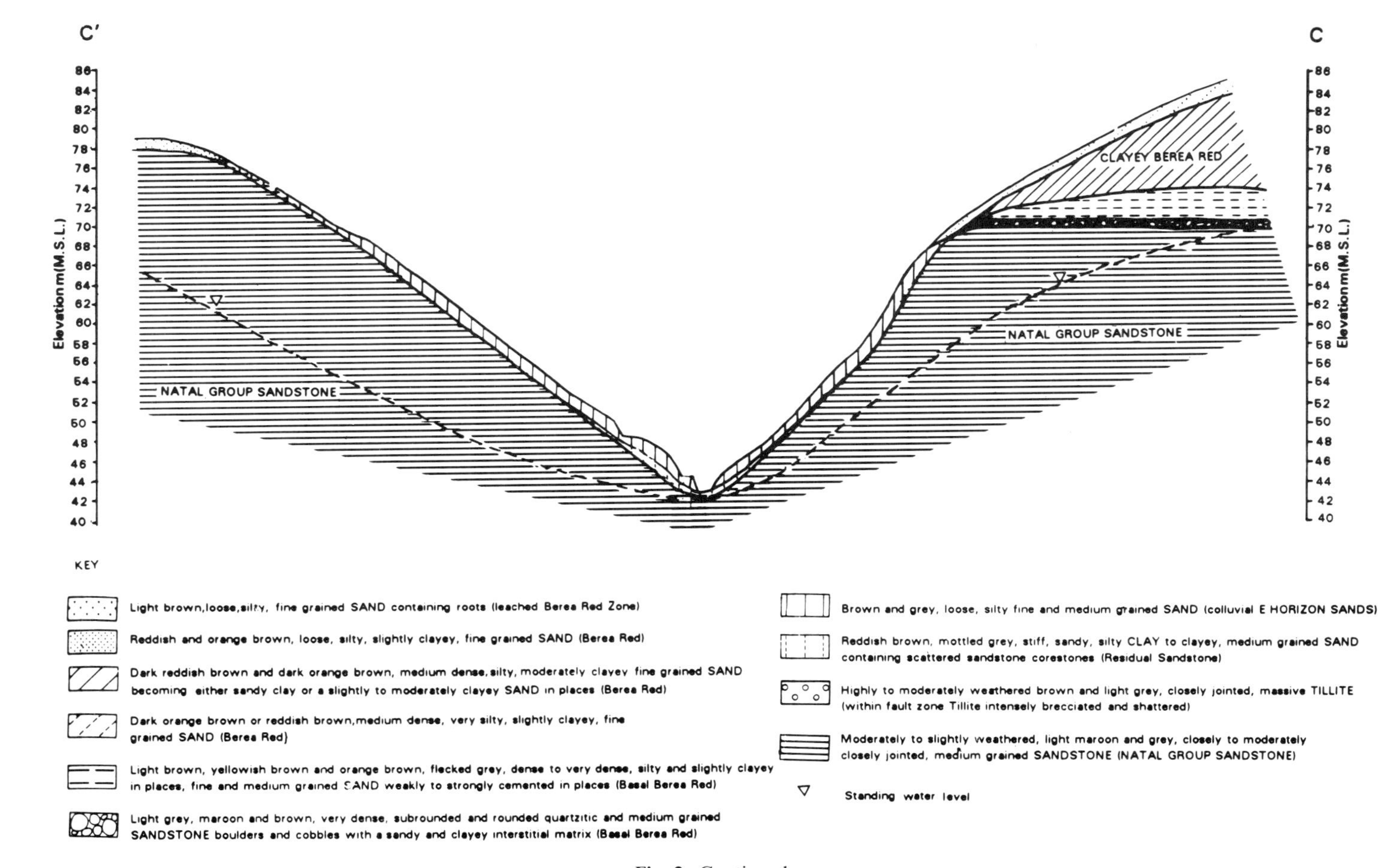

Fig. 2. Continued.

Drillholes which extended into the sandstone beneath the Berea Red Sands revealed a gently sloping, seaward-inclined surface indicative of a platform produced by marine abrasion. Over the southern half of the site, this marine-planed sandstone surface has been bisected by an incised drainage course leading down to a river to the south of the site. The nature, inclination and topography of this bedrock surface beneath the overburden is of major importance geohydrologically as it controls the flow and movement of groundwater within the bounds of the site. In other words, groundwater flow is directed towards the valley outlet at the lowermost part of the site.

Unweathered Dwyka tillite is an unstratified dark bluish-grey diamictite containing abundant inclusions of older rocks. It is so hard that it fractures through inclusions and matrix alike, its strength ranging from 120 to 250 MPa. The matrix is fine grained and consists of small angular fragments of quartz and rock material embedded in a fine base of rock floor. The inclusions vary in size up to several metres in diameter. The tillite is characterized by well-developed joints which are closely spaced with secondary silicification frequently evident. Preferential weathering develops along the joints and gives rise to corestones. The tillite on initial weathering is brown in colour and is still hard rock. It then grades into a softer yellowish-brown material and when completely weathered is a very soft yellowish-brown clay. The Dwyka tillite beneath the site exhibits an upper ferruginous capping layer (i.e. an immature iron pan) and passes downwards into a highly weathered soft medium-hard rock.

The Pietermaritzburg Shale consists of a uniform succession of dark bluish-black silty shales and mudstones. Harder indurated horizons are present within the shale and vary up to 0.3 m in thickness. The shales are well jointed and bedded. On weathering, the shale alters to a light brown or buff silty clay.

An appreciable thickness of Berea Red Sand overlies the Natal Group sandstones and the Dwyka tillite over much of the site (Figs. 1 and 2). The Berea Red Sand represents the *in situ* weathered product of a coastal dune deposit. It covers the more elevated northern, northwestern and eastern parts of the site, where as much as 32 m overlie the bedrock surface. It is absent from the southern and the southwestern parts of the site. This formation has been subdivided into three main horizons. The surface layer consists of light reddish to greyish brown, very loose sand which varies in thickness from 0.5 to 3.0 m, the maximum thickness occurring in the valley floor. The clay content of this layer generally varies from 0 to 5%.

Beneath this the sand is dark reddish-brown and clayey. The clay content of this horizon normally ranges between 20 and 30% (Fig. 3). It is mainly kaolinite and has been derived from the weathering of microcline. This material, because of its clay content, is considered very suitable for use as cover, as well as fill for those areas where the unsaturated zone is less than 2.0 m in thickness. The material can be obtained from the upper northern part of the site where it is planned to locate two football fields when site operations are terminated. Levelling this area to a reduced level of 98 m msl will yield approximately 55 000 m^3 of cover material. Additional good-quality material is available in a borrow pit in this clayey Berea Red Sand to the east of the site (Fig. 4). The borrow pit should provide about 67 000 m^3.

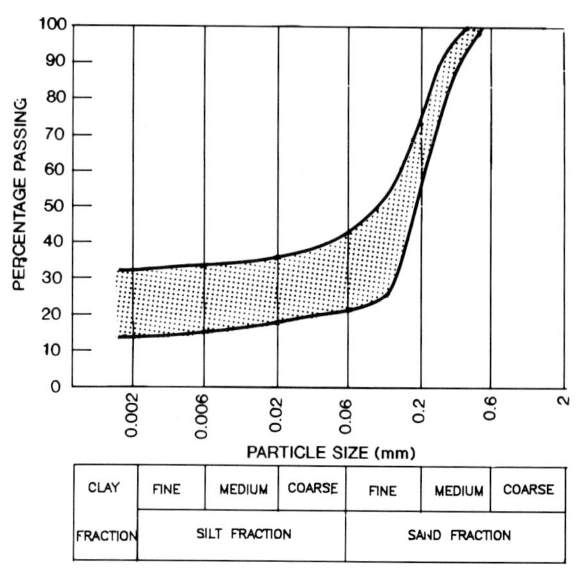

Fig. 3. Particle size distribution of Berea Red Sand.

A boulder bed of water-worn pebbles and boulders, in a clayey matrix, occurs at the base of the lowest horizon and is overlain by sands containing a low clay percentage. The basal horizon shows evidence of leaching and dissolution in some places, where water percolating through the sands has encountered the low-permeability bedrock of the underlying marine platform and has been forced to move laterally. This movement has given rise to bleached siliceous sand zones and local ferruginized bands which frequently occur within a zone immediately above the bedrock surface.

Generally, the upper sandy colluvium and underlying clayey residual soil are fairly extensive over the central parts of the site, attaining thicknesses of between approximately 4.5 and 5.0 m. The steeper side-slopes, especially those towards the lower end of the valley, possess only a thin mantle of silty sand overlying weathered sandstone. A stiff residual brown sandy clay overlies the Natal Group sandstones where they are not overlain by the Berea Red Sand in the southern part of the site, and is itself overlain by a thin cover of greyish-brown colluvial silty sand. The residual soils

Fig. 4. Clayey Berea Red Sand in the borrow pit area.

developed on the tillite comprise greyish-brown silty and clayey sands.

An east–west striking fault extends across the upper northwestern part of the site and has thrown Pietermaritzburg Shale and a dolerite sill, on its northern side, against the sandstones of the Natal Group. The throw is some 300 m. Off-site the fault zone is occupied by a hard fault breccia which has been silicified and so is virtually sealed. Furthermore, the fault zone within the site boundaries is overlain by Berea Red Sand which varies in thickness between 12 m and 32 m.

The investigation of the site for the landfill involved sinking a number of drillholes and a seismic refraction survey. This provided information on the variation in nature and thickness of the deeper sandy overburden, the nature of the bedrock formations, and the elevation and direction of groundwater flow.

Groundwater measurements were made with the aid of piezometers installed below the water-table. Determination of the position of the phreatic surface was also possible in a number of the inspection pits excavated at locations across the southern part of the site. Within the site boundaries the water-table generally lies below the upper surface of the bedrock formations (Fig. 5). The exception occurs along the northeastern boundary of the site where the water-table occurs in the leached basal Berea Red Sands. An intermittent water-table periodically occurs in the residual clayey soils in the valley axis over the southern part of the site where it forms a spring-line in the shallow erosion gully towards the southern boundary of the site.

Groundwater flow is confined within the valley defining the site. The main component of groundwater flow therefore is directed down the valley to the river south of the site. The steepening of the valley sides toward the southern part of the valley axis results in a rapid channelling of the groundwater into a very narrow neck located at the base of the valley close to the southern boundary of the site. Monitoring and control of potential pollutants could, therefore, be readily achieved at the exit of the drainage course from the site.

The permeability of the Natal Group sandstones and Dwyka tillite underlying the site is predominantly controlled by the discontinuities within the rock mass. The prevalence and pronounced nature of the jointing diminishes with increasing depth in the rock and is associated with a decrease in the degree of weathering of the rock masses.

Field permeability tests were carried out in the sandstones and tillite by means of single packer tests in NX sized drillholes. The packer tests comprised a series of five, 5–10 min runs at three increasing pressures followed by two decreasing pressures, which allowed assessment of the type of flow during testing. The packer tests provided an assessment of the nature of the permeabilities of both the sandstone and tillite bedrock with respect to potential flow of leachate through these formations. Permeabilities for the sandstones of the Natal Group vary from approximately 10^{-6} to $10^{-7}\,m\,s^{-1}$ whilst those for the Dwyka tillite range between 10^{-6} and $10^{-9}\,m\,s^{-1}$ (Table 2).

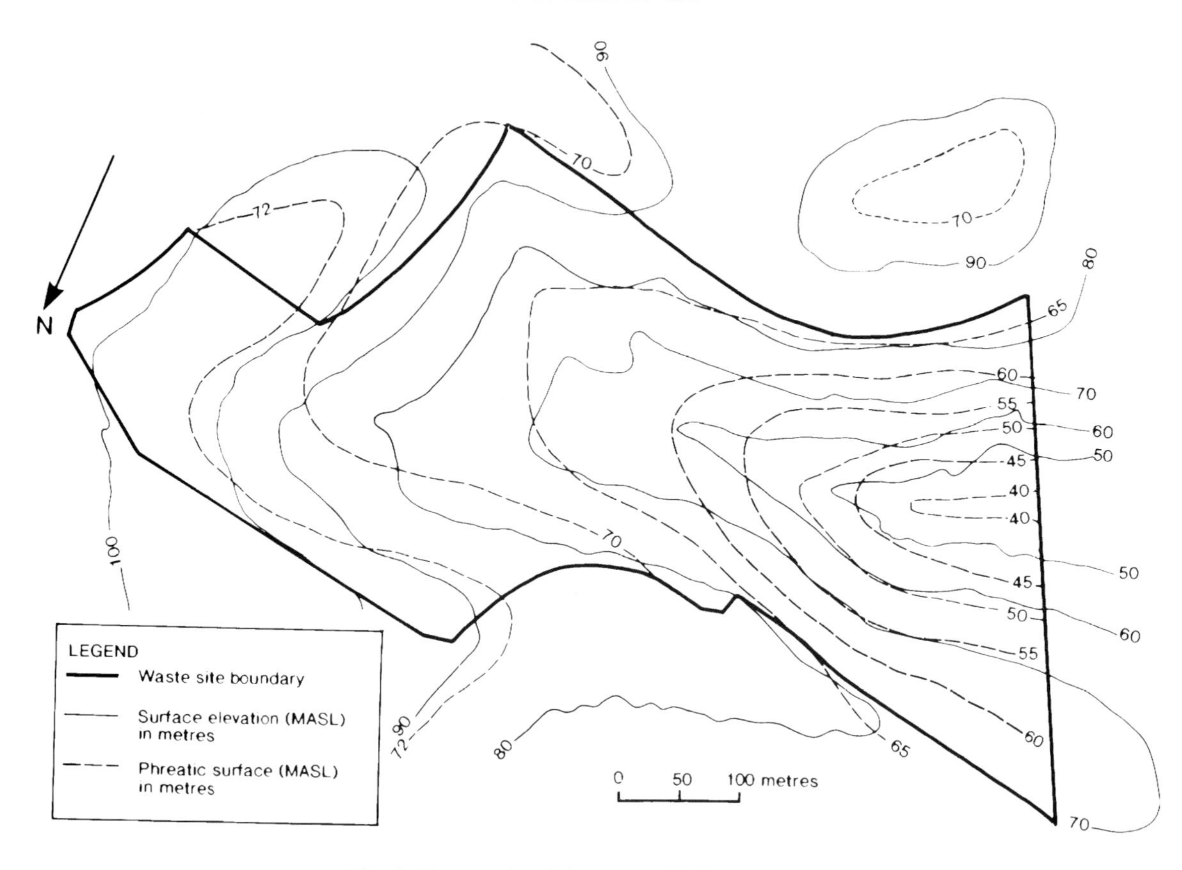

Fig. 5. Topography of the water-table beneath the site.

Falling head permeability tests were carried out on *in situ* Berea Red Sands in a number of boreholes put down for this purpose. The results of the *in situ* falling head permeability tests compared very favourably with the permeability tests performed on selected samples in the laboratory. The intermediate layer within the Berea Red Sand, because of the appreciable amount of clay present, has permeability values ranging from 10^{-7} to 10^{-10} m s^{-1}, with most being around 10^{-7} m s^{-1}.

Comparison of the coefficient of permeability at various compaction densities (85, 90 and 95% of modified AASHTO maximum dry density) indicated that it was only when the percentage of clay fraction exceeds 33% that there was any meaningful decrease in the coefficient of permeability with increased compaction. As mentioned above, the clay content of this part of the Berea Red Sand mainly varies between 20 and 30%.

Permeability tests were also performed on the colluvial sands and residual clays in the southern part of the site. The results indicated that they had coefficients of permeability of 10^{-5} and 10^{-11} m s^{-1} respectively.

Permeability testing carried out on the various rock types at the site indicated that they had coefficients of permeability in the range 10^{-6}–10^{-9} m s^{-1}. In other words, these rock masses either had low permeabilities or were virtually impermeable. Furthermore, the site generally was underlain by an unsaturated zone greater than 2.0 m in thickness. The material in this attenuation zone has a coefficient of permeability of 10^{-5} m s^{-1} or less. In the lowermost part of the valley the unsaturated zone can be less than 2 m thick as the intermittent water-table occasionally rises close to the ground surface. However, this area is of limited extent and the permeability here is generally around 10^{-7} m s^{-1}. This is somewhat lower than the accepted norm for a landfill. None the less, compacted fill was placed over this area as an added precaution. The fill consisted of clayey Berea Red Sand and had a maximum thickness of 2 m. It was considered that this would provide a satisfactory attenuation zone, especially in the area underlain by colluvial sands. In addition, a stone drain, surrounded by geofabric was formed along the exit to the site. As mentioned above, groundwater flow beneath the site is funnelled downstream towards the neck of the valley. The sump for the landfill is located in the neck of the valley (Fig. 6).

Table 2. *Results of field permeability tests in sandstone and tillite*

Borehole no.	*Depth of test section (m)*	*Packer test (kPa)*	*Coefficient of permeability* ($m\,s^{-1}$)	*Rock type*
1	21.0–23.5	100 180 250	2.24×10^{-6} 2.36×10^{-6} 1.96×10^{-6} Low	(Faulted) Dwyka tillite
2	12.5–15.7	200	2.97×10^{-6} Low	Highly weathered sandstone of Natal Group
2	22.4–25.4	100 180 250	1.61×10^{-6} 1.15×10^{-6} 1.90×10^{-6} Low	Moderately weathered sandstone of Natal Group; closely jointed
3	6.86–9.86	100 180 250	1.15×10^{-6} 2.29×10^{-6} 2.29×10^{-6} Low	Moderately to slightly weathered sandstone of Natal Group widely jointed
3	8.36–9.86	100 250	7.72×10^{-7} 5.82×10^{-7} Low	Moderately to slightly weathered sandstone of Natal Group; widely jointed
7	31.5–36.0	100 200 300	7.42×10^{-8} 2.72×10^{-8} 6.80×10^{-8} Very low to impermeable	Highly to moderately weathered Dwyka tillite; joints tight

Fig. 6. Sump for Site I.

Water balance and leachate generation

A water-balance calculation for this landfill site was carried out in an attempt to predict the amount of leachate that possibly could be generated. Data were available for quantities of solid and liquid/sludge entry. A compaction density of 650 $kg\,m^{-3}$ was used for the solid waste material. Campbell (1983) showed that such waste material has an absorptive capacity around 0.102 $m^3\,Mg^{-1}$, which represents 66.2 $l\,m^{-3}$ of waste (Fig. 7). The rainfall figures used were those obtained from Louis Botha Airport, the nearest recording station. The site covers an area of approximately 5.5 ha. A simplified water-balance equation was used to derive the water balance for the landfill and was as follows

$$L_g = P \times A + L_c - L_s$$

where L_g is the amount of leachate generated, P is the percentage percolation from the rainfall, A is the area of the landfill, L_c is the quantity of liquid codisposed and L_s is the quantity of liquid retained in storage which is primarily the absorptive capacity of the waste.

Table 3 gives the amounts of solids, their absorptive capacity, and the quantities of liquids associated with the landfill. The table indicates that a significantly larger quantity of liquid waste is disposed of into the landfill than any rainfall which ultimately percolates through it. It was assumed that 5% of the rainfall percolated into the landfill. This assumption was made on the basis that during the period 1980–1990 inclusive, the average annual rainfall recorded at Louis Botha Airport, Durban, was 1051.5 mm. The average annual evaporation calculated at the airport for the same period by use of the Symmons tank was 1365.1 mm. Evapotranspiration was therefore higher than rainfall so that this figure is probably an over-conservative one to choose. In fact, according to these figures, evaporation exceeds rainfall by about 30% in the greater Durban area. This obviously has a bearing on the disposal/codisposal ratios which can be used at waste sites. Liquid and sludges were brought to the site by bulk tanker from which they were decanted into trenches. Such disposal requires careful placement in order to maintain appropriate solid/liquid waste ratios. It is evident that the total liquid entry into the landfill appreciably exceeds the absorptive capacity of the solid waste and so a notable volume of liquid is available for the formation of leachate. The average monthly amount of liquid exceeding the absorptive capacity of the solids is 2 285 840 litres. However, when this quantity is compared with the average monthly amount of leachate as monitored by the Umgeni Water Board, it is almost three times higher (the approximate figure given by Umgeni Water is 1 000 000 litres per month). Unfortunately the figures for codisposed liquid and for codisposed sludge were not distinguished. Obviously only a part of the sludge would be available as liquid and some liquid would be held within voids whilst some would react with the solid materials present and some would be lost due to evaporation. Consequently this will reduce the amount of liquid available for leachate generation. If 20% of the liquid/sludge is unavailable due to the aforementioned reasons, then the amount of leachate generated per month is 1 013 763 l which compares favourably with the figure quoted by the Umgeni Water Board (Table 3).

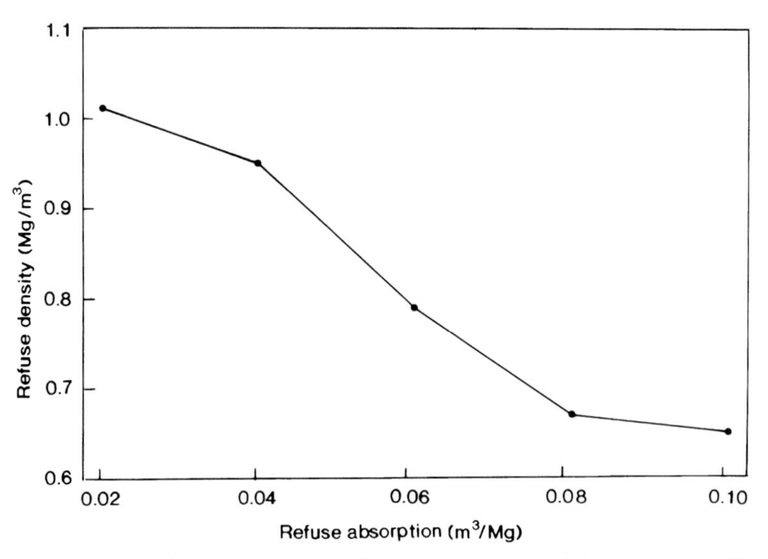

Fig. 7. Density of refuse in relation to its absorptive capacity (after Campbell 1983).

Table 3. *Solids, their absorption capacity, and liquids associated with Site 1*

Date	Solids (m^3)	Absorptive capacity (litres)	Liquid/sludge (litres) (a)	Liquid/sludge (litres) (b)	Precipitation 5% of total (litres)	Total liquid entry (litres) (a)	Total liquid entry (litres) (b)	Quantity of liquid exceeding absorptive capacity (litres) (a)	Quantity of liquid exceeding absorptive capacity (litres) (b)
Jan 1990	107 640	7 125 768	9 891 000	7 912 800	277 000	10 168 000	8 189 000	3 042 232	1 064 032
Feb 1990	66 130	4 277 806	6 609 000	5 287 200	376 000	6 986 000	5 663 200	2 707 194	1 385 394
Mar 1990	68 647	4 544 431	7 059 000	5 647 200	651 500	7 710 000	6 298 200	3 166 069	1 754 269
April 1990	57 841	3 829 074	5 951 000	4 760 800	108 000	6 059 000	4 868 800	2 229 926	1 039 726
May 1990	70 309	4 694 176	7 355 000	5 884 000	56 000	7 411 000	5 940 000	2 716 824	1 245 824
June 1990	67 095	4 441 689	6 769 000	5 415 200	28 000	6 797 000	5 443 200	2 355 311	1 001 511
July 1990	61 997	4 104 201	6 291 000	5 032 800	3 300	6 294 300	5 036 100	2 190 099	931 899
Aug 1990	63 890	4 229 518	5 854 000	4 683 200	41 650	5 896 350	4 725 550	1 667 132	496 332
Sept 1990	52 288	3 461 466	4 819 000	3 855 200	141 500	4 760 500	3 996 700	1 463 034	499 234
Oct 1990	62 604	4 144 385	5 334 000	4 267 200	338 500	3 672 500	4 605 700	1 528 115	461 315
Nov 1990	60 195	3 984 909	6 612 000	5 289 600	153 000	6 765 000	5 442 600	2 780 091	1 457 691
Dec 1990	50 911	3 370 308	5 047 000	4 037 600	320 000	5 367 000	4 357 600	2 176 692	1 167 292
Jan 1991	59 000	3 905 800	5 094 000	4 075 200	505 000	4 580 200	5 599 000	1 693 200	674 400
Average								2 285 840	1 013 763

Columns (a) take into consideration the total amounts of liquid/sludge disposed of, while in columns (b) the liquid/sludge values have been reduced by 20%.

During the formation of the landfill, chemical analyses were carried out on samples taken from the sump near the toe of the landfill over a five-year period from 1986 to 1991 (Table 4). The values in Table 4 can be compared with those of Table 5. The latter shows the results of chemical analyses carried out on samples of groundwater taken from the boreholes and the nearby river and two tributary streams. Table 4 shows some remarkably high values which presumably are associated with the type of waste disposed of. For instance, much of the sludge which goes into the landfill consists of hop waste from breweries. This accounts for the high levels of chemical oxygen demand, oxygen absorption, suspended solids, total dissolved solids, and conductivity.

In terms of pH value, the leachate was alkaline. It had an average pH of 7.7, ranging from 6.6 to 8.9. The values of COD and pH were analysed in terms of peaks and lows in rainfall. The highest COD and pH values tended to follow periods of low rainfall (Fig. 8). This is to be expected since the lower volume of liquid entry, no doubt, gives rise to a more concentrated leachate.

The potential volume of leachate generated depends on a number of factors which affect the overall water

Table 4. *Chemical analyses of leachate from sump at Site 1 taken over a five-year period from 1986 to 1991*

	pH	Suspended solids ($mg\,l^{-1}$)	Total dissolved solids ($mg\,l^{-1}$)	Conductivity ($mS\,m^{-1}$)	Oxygen absorption ($mg\,l^{-1}$)	Chemical oxygen demand ($mg\,l^{-1}$)
Max	8.9	6 044	45 041	8 080	6 200	70 900
Min	6.6	200	900	450	145	11.2
Mean	7.7	965	17 029	1 174	649	13 546
MPL	5.5–9.5		500	300	10	75
	Sodium ($mg\,l^{-1}$)	*Potassium ($mg\,l^{-1}$)*	*Calcium ($mg\,l^{-1}$)*	*Magnesium ($mg\,l^{-1}$)*	*Sulphate ($mg\,l^{-1}$)*	*Ammonium ($mg\,l^{-1}$)*
Max	8 249	2 646	1 236	759	2 237	3 530
Min	80	355	60	70	6.5	92
Mean	1 393	613	146	109	837	1 093
MPL	400	400	200	100	600	2

MPL, maximum permissible limit for insignificant risk (Anon 1993).

Table 5. *Analysis of ground and surface water from Site 1 area prior to the construction of the landfill*

	RS 1	*RS 2*	*TS 1*	*TS 2*	*BS 1*	*BS 2*	*MPL*
Alkalinity, as $CaCO_3$ ($\mathrm{mg\,l^{-1}}$)	126	206	52	52	64	38	300
Ammonia, as N ($\mathrm{mg\,l^{-1}}$)	9	11	0.8	0.5	1	0.6	2
Arsenic ($\mathrm{mg\,l^{-1}}$)	0.025	0.030	0.03	0.03	0.025	0.050	0.3
Cadmium ($\mathrm{mg\,l^{-1}}$)	<0.01	<0.01	<0.02	<0.01	<0.01	<0.01	0.02
Chemical oxygen demand ($\mathrm{mg\,l^{-1}}$)	87	75	10	491	77	40	75
Chloride ($\mathrm{mg\,l^{-1}}$)	92	85	57	64	206	170	600
Hexavalent chromium ($\mathrm{mg\,l^{-1}}$)	0.23	0.28	0.22	0.09	0.23	0.22	0.2
Conductivity ($\mathrm{mS\,m^{-1}}$)	68	71	40	60	93	73	300
Copper ($\mathrm{mg\,l^{-1}}$)	<0.01	<0.01	0.004	0.012	<0.01	<0.01	1
Cyanide ($\mathrm{mg\,l^{-1}}$)	N/D	N/D	N/D	N/D	N/D	N/D	0.3
Lead ($\mathrm{mg\,l^{-1}}$)	<0.05	<0.05	<0.05	<0.05	<0.05	<0.05	0.1
Manganese ($\mathrm{mg\,l^{-1}}$)	0.3	1.3	0.09	1.41	3.9	0.4	1
Mercury ($\mathrm{mg\,l^{-1}}$)	<0.001	<0.001	<0.002	<0.002	<0.001	<0.001	0.01
Nitrate ($\mathrm{mg\,l^{-1}}$)	0.7	0.6	4.0	2.5	0.1	2.3	10
Nitrate ($\mathrm{mg\,l^{-1}}$)	0.001	0.005	0.005	0.001	N/D	0.11	
pH at 25 °C	6.25	6.94			6.70	6.05	
Phenolic compounds, as phenol ($\mathrm{mg\,l^{-1}}$)	2.1	1.92	12	12	1.80	2.45	
Phosphates ($\mathrm{mg\,l^{-1}}$)	0.25	0.12	<0.001	<0.001	0.12	0.05	
Sodium ($\mathrm{mg\,l^{-1}}$)	75	83	53	72	124	107	400
Sulphate ($\mathrm{mg\,l^{-1}}$)	26	2.6	33	28	60	37	600
Sulphides ($\mathrm{mg\,l^{-1}}$)	N/D	N/D	N/D	N/D	N/D	N/D	
Suspended solids ($\mathrm{mg\,l^{-1}}$)	29	67	65	652	48 820	122	
Total organic carbon ($\mathrm{mg\,l^{-1}}$)	166	212	54	86	232	160	

RS1, RS2, river samples; TS1, TS2, tributary stream samples; BS1, BS2, borehole samples. MSL, maximum permissible limit (Anon 1993).

balance of a site. These include precipitation, evaporation, infiltration, surface runoff and volume of liquids codisposed at the site. A favourable factor with respect to potential leachate production is that the average annual evaporation exceeds the average annual precipitation in the greater Durban area. Hence with adequate stormwater cutoff drains around the disposal site, it generally should have a moisture deficiency and therefore a certain volume of codisposed liquids presumably are utilized in making good this deficiency. Potential leachate production consequently depends primarily on the volume of liquids codisposed over and above the volume required to make up any moisture deficiency at the site and should be controllable in the event of groundwater pollution being detected. Another factor influencing the potential for groundwater pollution is the relatively steep slopes on this site which tend to promote runoff of any liquids towards the leachate collection sump located at the toe of the disposal site rather than the major portion reaching the groundwater through the attenuation zone. Taking all factors into account, it is considered that only limited quantities of leachate reach the water-table and that any leachate which does recharge the groundwater is sufficiently attenuated as to fall within accepted norms. In addition, in order to try and avoid the problem of pollution, the leachate produced at this site is pumped directly from sump to sewer and conveyed to a sewerage plant for treatment. Lastly, no evidence of extraction of groundwater for human consumption or irrigation purposes occurs with a radius of 1 km of the site, either by means of pumping from the boreholes or extraction from springs.

Case history 2

A landfill site which had served the needs of an area within the greater Durban region was closed in 1989. Unfortunately, however, seepage occurred from the toe of the landfill and it became necessary to carry out an investigation to determine the extent of the seepage, whether or not it posed a problem and, if so, what remedial measures should be taken.

The area surrounding the landfill site is undeveloped and consists of low rounded hills and small valleys. Small non-perennial and perennial streams flow into one of the notable rivers of the region. The area downstream of the toe of the landfill slopes gently eastwards, with a valley located immediately to the southeast of the boundary of the landfill. A stream occupies this valley, commencing just downstream of the leachate pond and collection sump. A second stream occurs approximately 40 m to the northeast of the previously mentioned stream, and both flow into the river referred to.

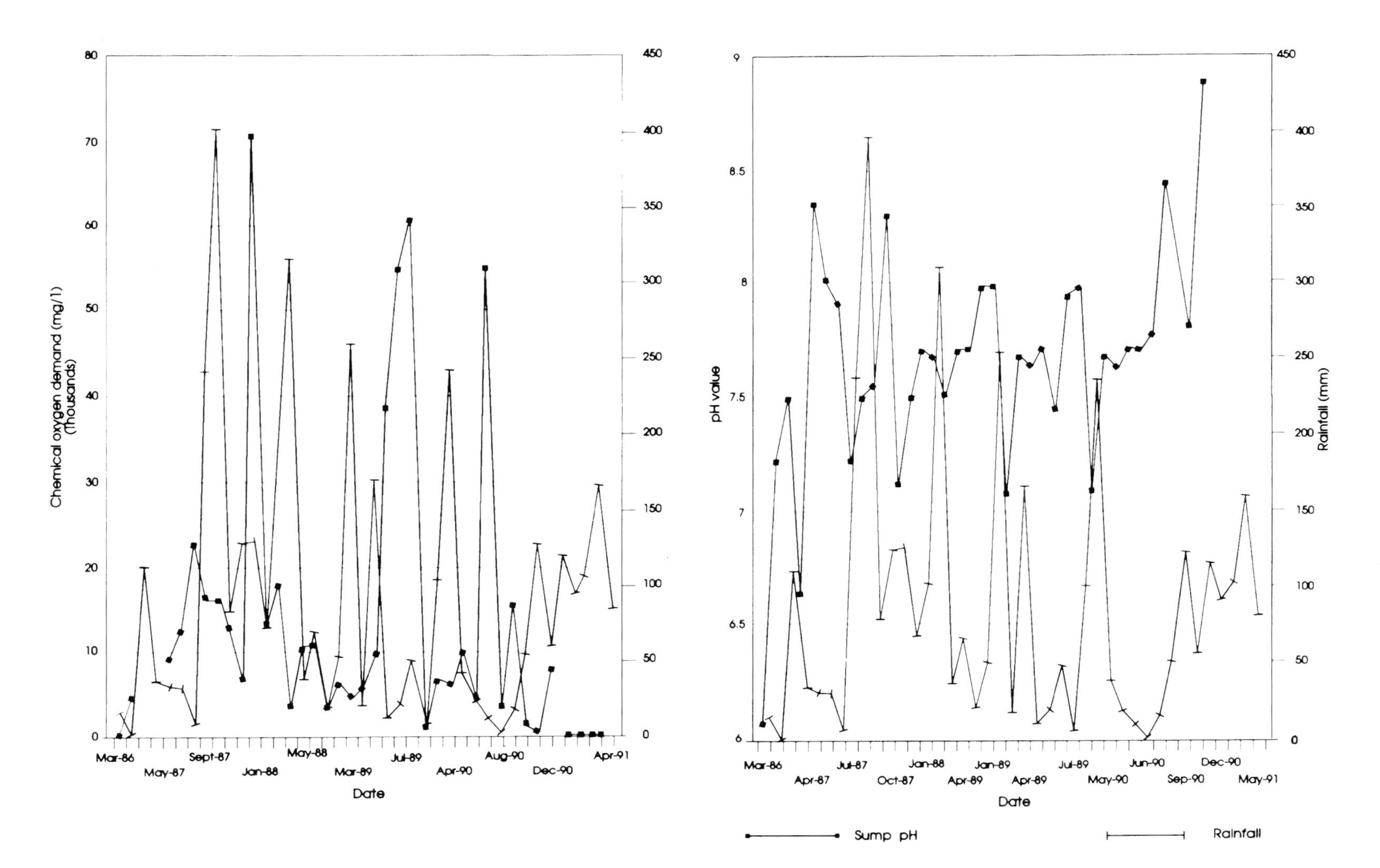

Fig. 8. Graphs of (a) COD against rainfall and (b) pH value against rainfall.

Fig. 9. Leachate pond for Site 2.

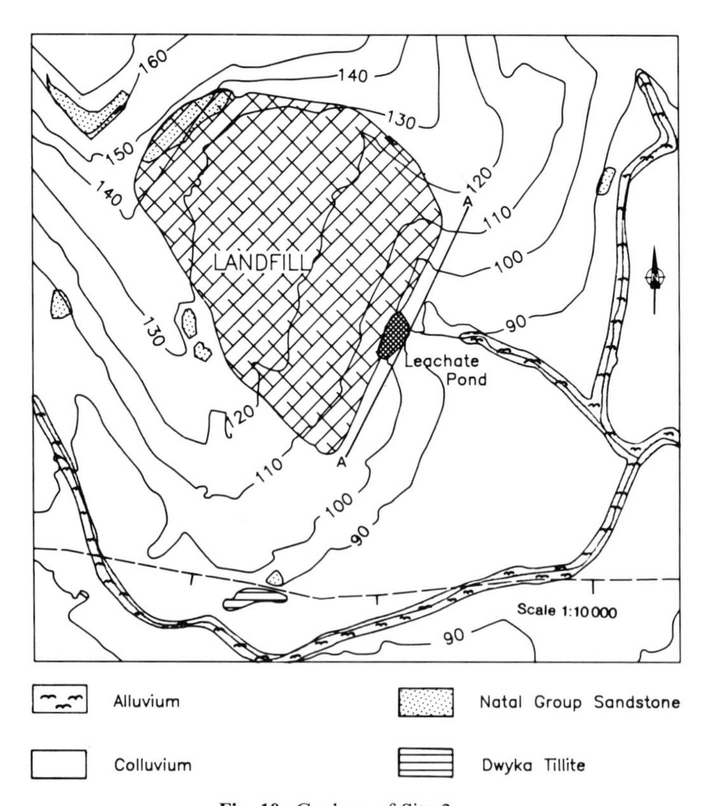

Fig. 10. Geology of Site 2.

The waste materials disposed of at the site consisted mainly of domestic and commercial wastes. The leachate generated by the waste was collected in a leachate pond situated at the toe of the landfill adjacent to the eastern boundary of the site (Fig. 9). A small bund surrounded the leachate pond on the downstream side. A trench around the perimeter of the landfill helped convey leachate seeping from the sides of the fill to the leachate pond. An attempt had been made at the head of the stream downstream of the leachate pond to form a leachate sump to contain leachate migrating beyond the pond and prevent pollution of the stream. Leachate was pumped from the sump periodically.

Geology and investigation of the site

The solid geology of the area around the landfill consists of sandstones of the Natal Group and tillite of the Dwyka Formation (Fig. 10). These are overlain by a mantle of residual soils of variable thickness. The residual soils are, in turn, overlain by colluvial soils on the hillslopes and by alluvial soils in the valley floors.

The sandstones of the Natal Group are exposed in scattered outcrops across the area. They vary from soft (where weathered) to medium hard rock and are closely jointed. These sandstones occupy most of the area except to the south where they are downthrown by a fault against Dwyka tillite. The sandstones have an overall dip to the south, with angles ranging from 12° to 28°.

Where the tillite of the Dwyka Formation is exposed it is highly to extremely weathered. Closely spaced joints are still present in the weathered tillite, which varies from very soft to medium-hard and grades from grey to light yellow in colour.

The residual soils associated with the sandstones are variable in character. They are generally dark reddish-brown micaceous silty sands and sandy clay of soft to firm consistency. They range in thickness from 1.6 to 2.1 m. However, in some places the residual soils are absent and here colluvial soils rest directly on sandstone.

The colluvial soils overlying the residual sands and clays commonly comprise dark brownish-grey, loose to medium dense, fine- to medium-grained silty sand. The thickness of the colluvial soils ranges from 0.3 to 1.8 m. A pebble marker horizon usually occurs at the base of the colluvial soils. It is brownish, loose to medium dense, ferruginized sandy gravel. Most of the gravel-sized particles are composed of sandstone.

It is not easy to distinguish between the residual and colluvial soils overlying the tillite. These soils consist of brown, soft to firm, slightly sandy clay, which range in thickness from 0.3 to 0.7 m. Again a pebble horizon occurs at the base of the soil. Patches of tillite gravel also overlie the tillite.

Alluvial soils primarily consist of light brown, very loose, fine- to medium-grained sand, clayey sands and sandy clay of soft to firm consistency. They occur along the valley bottoms occupied by the streams and main river, and reach a maximum thickness of 2.5 m. In places, alluvial soils are overlain by colluvium.

The area around the site was mapped geologically and exposures were described in detail. Twenty-two inspection pits were excavated (Fig. 11). The pits were positioned to give a general coverage of the area, and to enable a geological cross-section (Fig. 12) to be drawn of the subsoil conditions along the southeast boundary of the landfill where seepage was occurring. The pits were excavated by a mechanical excavator except where soft wet marshy conditions existed immediately downstream

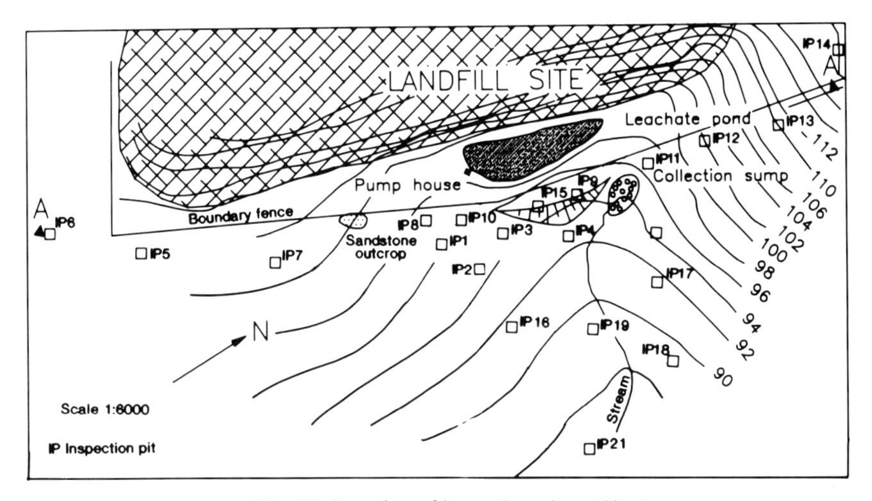

Fig. 11. Location of inspection pits at Site 2.

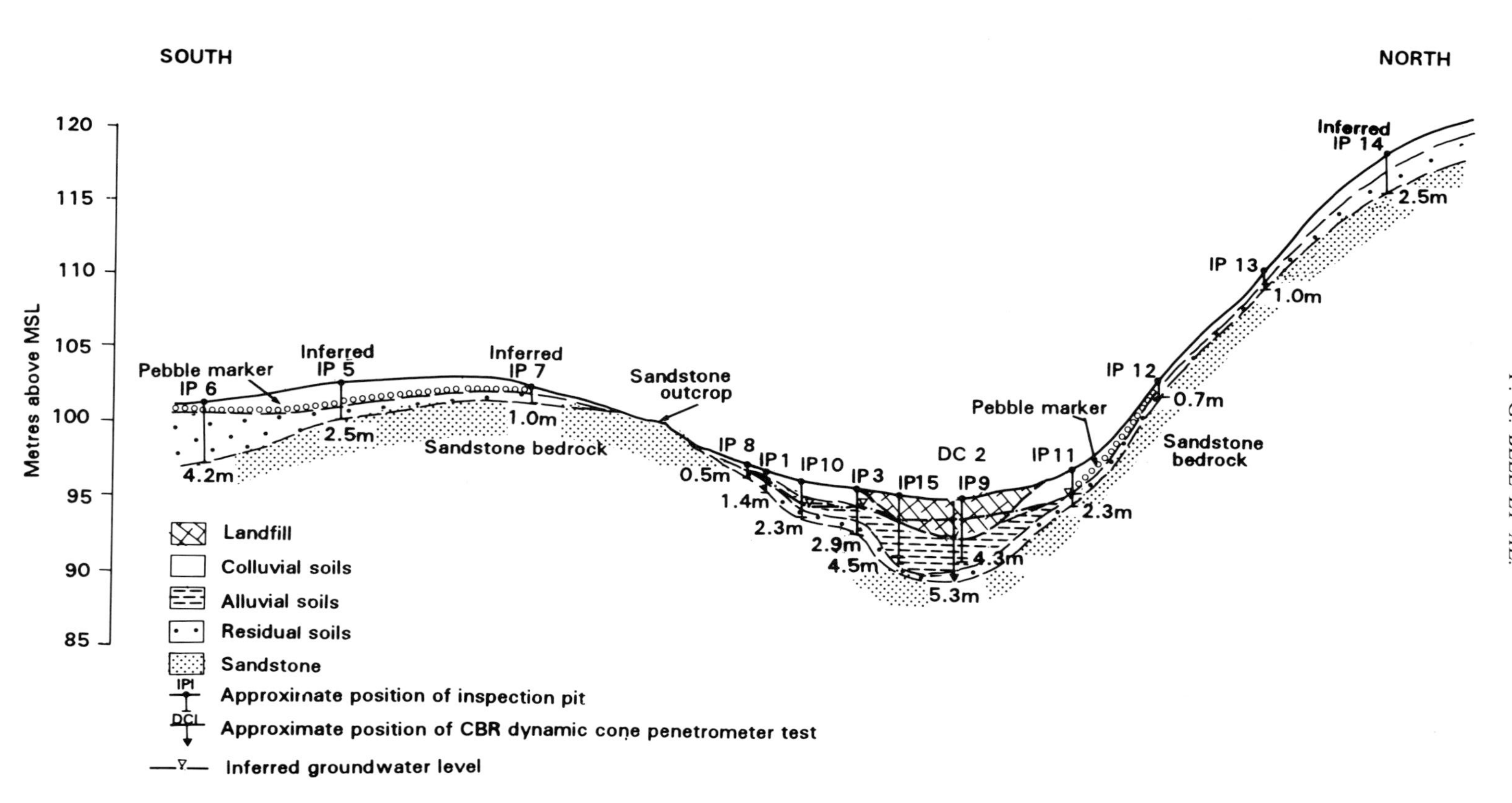

Fig. 12. Geological cross-section along the southeast boundary of Site 2.

of the toe of the landfill. Here the pits were excavated by hand. The latter were excavated to a depth of 1 m, whereas those which were mechanically excavated extended to around 4.5 m below ground surface. Profiles of the soils in each pit were made and soil samples were taken for testing. Water samples were also collected from the inspection pits. Dynamic cone penetrometer tests were carried out to assess the depth to bedrock and relative consistency of the subsoils in the valley floor immediately southeast of the landfill. Bedrock occurred at between 3.5 m and 5.0 m below ground level.

Groundwater seepage in the inspection pits varied over the site from very slight to appreciable. It tended to occur within the upper 2 m of soil, principally at the interface between the more permeable colluvial sandy soils (estimated permeability, 10^{-3}–10^{-5} m s^{-1}) and the underlying less permeable alluvial or residual soils (estimated permeability, 10^{-4}–10^{-7} m s^{-1}). The water samples taken from the inspection pits were analysed and examples of the results are given in Table 6.

The results of the investigation indicated that leachate was polluting the groundwater by seeping from the toe of the landfill and from the leachate pond, thereby demonstrating that the containment was inadequate. The concentration levels or isopleths of sulphides, chlorides, total dissolved solids, oxygen absorption, conductivity, chemical oxygen demand and pH downstream of the site are shown in Fig. 13(a)–(g). High levels of concentration of pollutants were found, which in most cases greatly exceeded the maximum permissible.

These pollution indicators show similar isopleth trends in that the plume originates from the leachate pond and extends downstream, the degree of pollution decreasing downstream of the landfill. However, Fig. 13 indicates that sulphides decrease in concentration until acceptable levels are encountered beyond approximately 310 m downstream from the boundary (i.e. beyond the 1.0 mg l^{-1} isopleth). On the other hand, acceptable concentrations of chlorides are encountered some 180 m downstream from the boundary (i.e. beyond the 600 mg l^{-1} isopleth). The pH value of the groundwater downstream of the site remains around neutral.

Remedial measures

In order to avoid disturbing the landfill and to prevent further migration of seepage from it, it was decided to construct a cutoff barrier immediately downstream of the leachate pond, together with a subsoil drain and pump chamber (Fig. 14). The maximum depth to bedrock at the proposed location of the cutoff barrier is around 5 m. Most of the soil present at this location is colluvial and residual sand, reaching a maximum thickness of 4.5 m in the valley immediately downstream of the landfill, with alluvium beneath. In the centre of the valley, groundwater is present at approximately 1.5 m below the ground surface.

It was suggested that the most suitable form of barrier would be a cement–bentonite cutoff wall (Bell 1993). The choice was economically attractive because its construction would involve limited use of materials and would be completed with a minimum sequence of

Table 6. *Examples of chemical analyses carried out on groundwater obtained from inspection pits, Site 2*

Water sample no. of inspection pit	*Depth (m)*	*pH*	*Chloride (mg l^{-1})*	*Sulphide (mg l^{-1})*	*Total dissolved solids (mg l^{-1})*	*Conductivity (mS m^{-1})*	*Oxygen absorption (mg l^{-1})*	*Chemical oxygen demand (mg l^{-1})*
IP 9	1.4	7.3	2 782	8.0	9 726	1 448.4	778	3 520
IP 10	1.7	6.9	3 938	1.4	11 047	1 397.4	74	656
IP 11	1.8	7.5	3 832	5.6	9 902	1 530.0	339	1 920
IP 17	0.4	6.9	76	6.1	557	85.7	1 577	6 240
IP 19	Surface	7.7	1 033	3.5	3 722	581.4	170	960
IP 20	0.6	6.9	504	0.48	2 166	300.9	40	272
IP 21	0.4	7.05	820	0.8	2 807	418.2	88	448
Spring S2	Surface	7.35	71	0.32	261	38.8	7	40
Recommended maximum limit*	—	6.0–8.0	250	—	—	10.0	—	—
Maximum permissible limit*	—	5.5–9.5	600	1.0	500	300	10	75

* After Anon (1993).

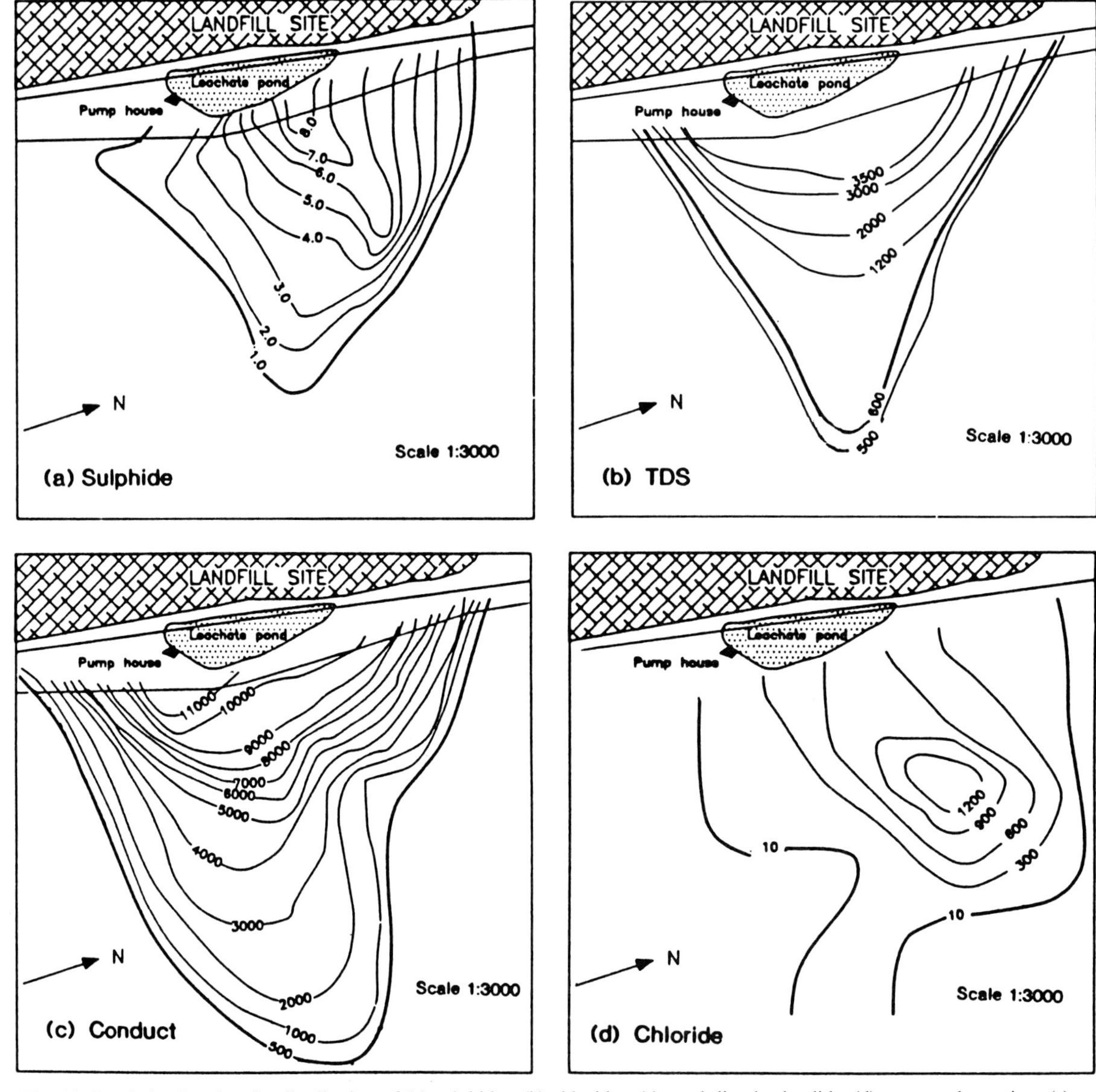

Fig. 13. Isopleths showing the distribution of (a) sulphides, (b) chlorides, (c) total dissolved solids, (d) oxygen absorption, (e) conductivity, (f) chemical oxygen demand, and (g) pH value in the groundwater. All in $mg\,l^{-1}$ except conductivity, which is in $mS\,m^{-1}$, and pH value.

operations. It was further suggested that if sand was used to replace some of the cement, about 50% by volume, then the cost of materials could be reduced significantly. A permeability of $10^{-9}\,m\,s^{-1}$ could still be achieved.

It was proposed that the cement–bentonite wall and associated subsoil drain be constructed along the entire southeastern boundary of the landfill to a point some 20 m south of the southern boundary and should be keyed into bedrock (i.e. the sandstones of the Natal Group). The addition of sand, obtained from the nearby stream, will mean that polymer-treated sodium bentonite, Culseal, which is especially designed for containment of polluted water, will have to be used to ensure the durability and permeability of the cutoff wall. As the pH values of the water samples tested ranged from 7 to 7.5, the cement in the backfill should not be affected adversely.

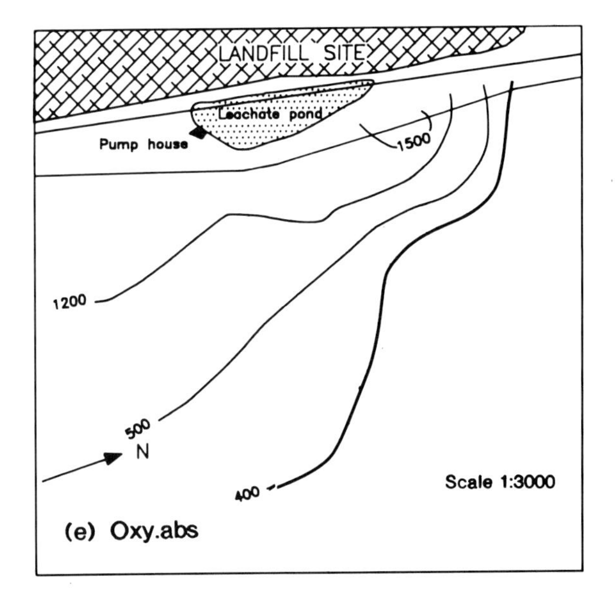

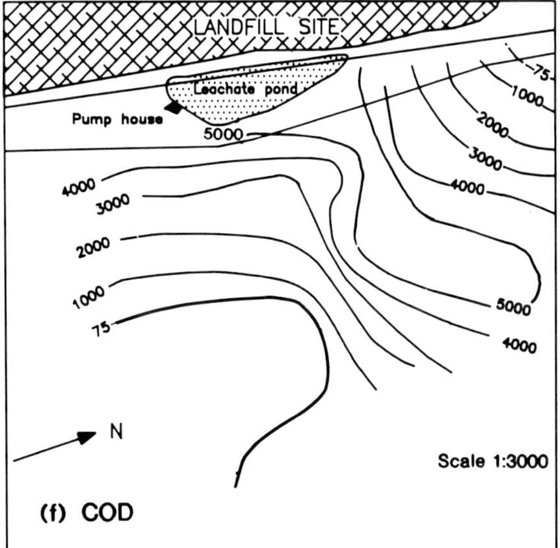

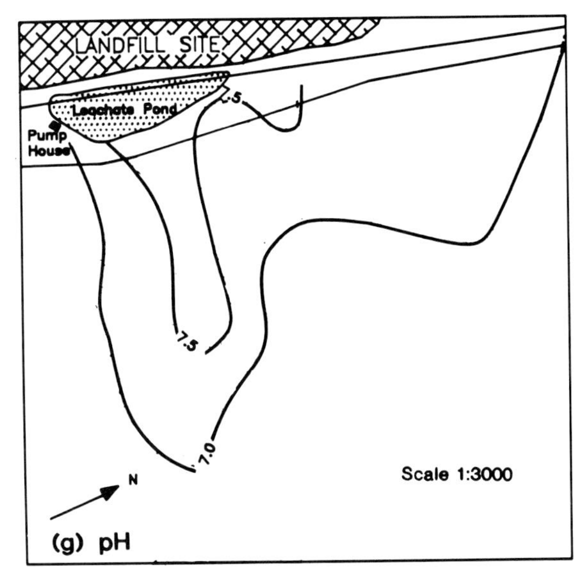

It was suggested that the subsoil drain should be constructed before the slurry wall to allow the level of groundwater in the valley to be lowered. This will also entail constructing the pump chamber and linking the associated piping with the local sewer system.

The subsoil drain is to be laid in a series of stages because of the variation in elevation and soil type. It will drain under gravity flow into the pump chamber. It will consist of a 150 mm diameter pipe surrounded by stone aggregate. The latter will be overlain by sand which will extend to the surface. The aggregate will be enclosed in geosynthetic fabric in an attempt to prevent clogging by fines, biological growths and the precipitation of inorganic substances. Blasting of rock will be required to the south of the pump chamber location in order to obtain the desired gradient. It is important that the subsoil drain and pump chamber are sited at a lower level than the base of the slurry

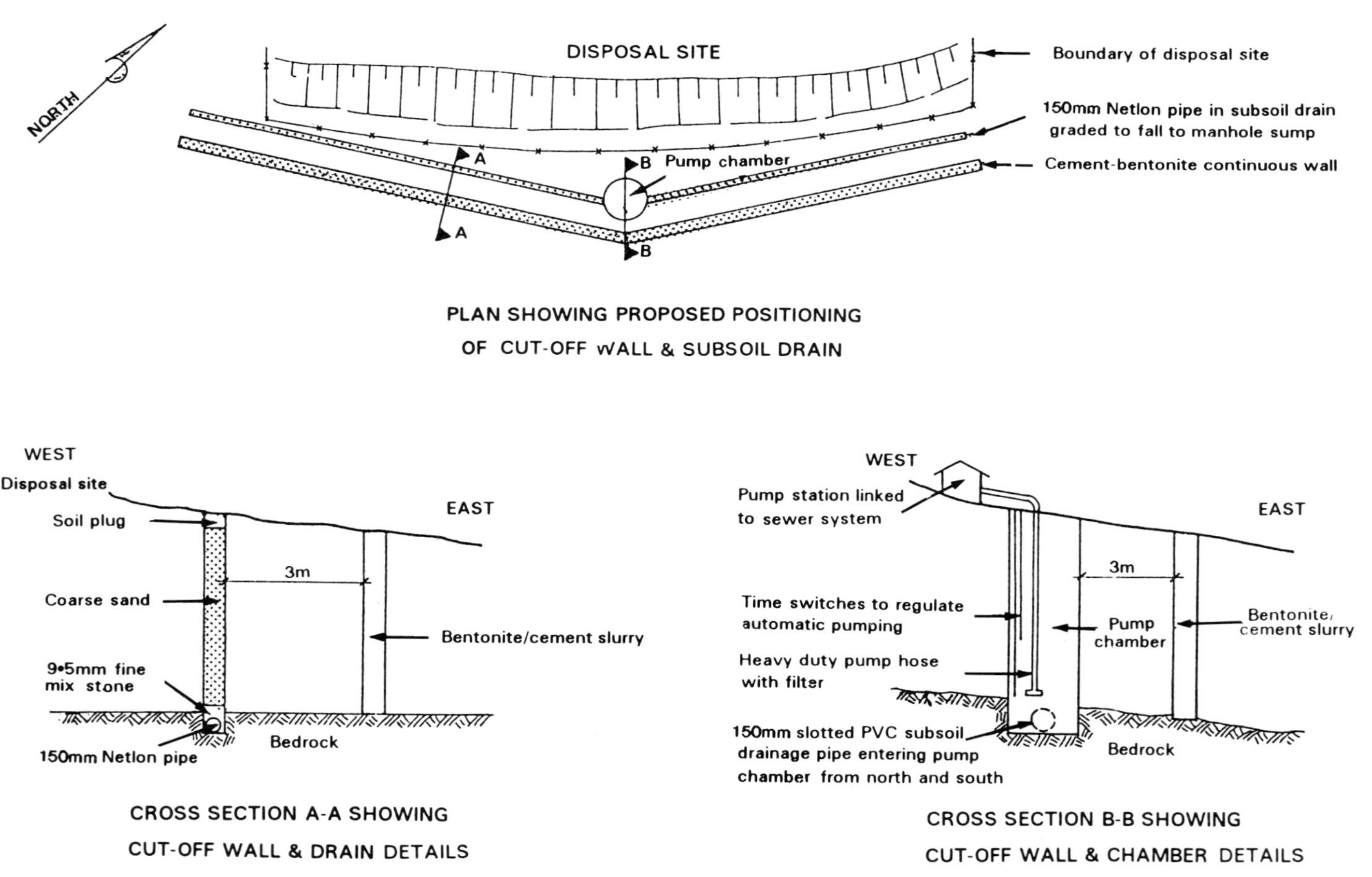

Fig. 14. Diagrammatic representation of the remedial works at Site 2.

wall in order to permit leachate to flow back from the wall into the drain.

The pump chamber is to be constructed at the centre of the valley where bedrock is at the lowest elevation to allow gravity flow of the leachate from the collection pipes. It will be constructed with the aid of a caisson since the soil in this area is saturated and has low strength. The chamber will be sealed on the inside. The joints between the leachate lines and the pump chamber will be flexible and valves will be installed on the leachate lines so that they can be shut off during periods of maintenance. Automatic submersible pumps will be used in the pump chamber.

Conclusions

The landfill at Site 1 has been operational since 1986. The valley in which the site occurs is occupied mainly by Berea Red Sand which is underlain by Dwyka tillite and sandstones belonging to the Natal Group. The permeability of the sand formation varies from 10^{-7} to $10^{-10}\,m\,s^{-1}$, as it contains an appreciable amount of clay. The permeabilities of the underlying formations are either low or virtually impermeable and so should provide effective containment of the disposed waste.

The quantity of leachate produced by a landfill is controlled by the amount of liquid entering in the form of percolating rainfall, groundwater or liquid disposed of on the one hand, and absorption by solids and evapotranspiration loss on the other. As evapotranspiration exceeds rainfall in the Durban area, a much more significant contribution to leachate production at Site 1 is made by liquid disposed of than by any percolating rainfall. Even though exact figures could not be obtained for water-balance determination, the contribution of liquid/sludge is very evident. At the same time, the nature of the liquid-sludge disposed of has a notable influence on the chemical character of the leachate produced. The appreciable quantities of hop waste placed in the landfill is largely responsible for the very high values of oxygen absorption, chemical oxygen demand, suspended solids, total dissolved solids and conductivity of the leachate analysed from the leachate sump. Such leachate could seriously pollute surface water or groundwater but is contained by the geological conditions and groundwater flow is directed towards the downstream neck of the valley by the bedrock topography. There the leachate is collected in a sump located at the toe of the landfill and from there conveyed to a sewerage treatment plant. There is no surface or groundwater usage within a radius of 1 km of the site.

Another landfill constructed in the greater Durban area was located on sandstones of the Natal Group and Dwyka tillite which are overlain by a veneer of residual, colluvial and alluvial soils. After its closure in 1989, seepage of leachate was noticed downstream of the landfill. This led to an investigation, which primarily consisted of excavating a number of inspection pits in the suspect area. Water samples were taken from these pits and analysed. The analyses revealed that a leachate plume had developed downstream of the landfill and was polluting the groundwater. The remedial measures proposed consist of a subsoil drain which will lead into a pump chamber. A cement–bentonite wall will be constructed immediately downstream of them to act as an impermeable barrier to any leachate seepage. The cutoff wall will be keyed into bedrock and will have a permeability of around $10^{-9}\,m\,s^{-1}$. In this way it is hoped to prevent further pollution of groundwater occurring.

References

ANON. 1993. *South African Water Quality Guidelines, Volume 1, Domestic Use*. Department of Water Affairs and Forestry, Pretoria.

BELL, F. G. 1993. *Engineering Treatment of Soils*. Spon, London.

CAMPBELL, D. 1983. Understanding water balance in landfill sites. *Waste Management*, **11**, 594–601.

Engineering properties of two types of pulverized fuel ash

M. Mahmoud & R. S. Morley

WS Atkins Consultants Limited, Woodcote Grove, Ashley Road, Epsom, Surrey KT18 5BW, UK

Abstract. The engineering characteristics of two types of pulverized fuel ash (PFA) are investigated in terms of their geotechnical properties. The two PFAs investigated are (1) a 'lagoon' PFA, subsequently stockpiled and then end-tipped for infilling an old gravel pit in Staffordshire, and (2) a 'conditioned' PFA used to create a mound adjacent to a coal power station in Yorkshire. Measurements of strength and compressibility characteristics are discussed with respect to predicting the behaviour of these two PFAs *vis-à-vis* slope stability and settlement. Charts are presented (a) for determining settlements of lagoon PFAs, and (b) for establishing strength parameters of conditioned PFAs.

There are two types of pulverised fuel ash (PFA): lagoon PFA and conditioned PFA. Lagoon PFA is formed as ash slurry which is pumped into a small reservoir (lagoon) via a pipeline from a power station. The ash subsequently settles out of suspension in the lagoon, the decantrate is discharged through an outfall system, and the ash is then stockpiled. Conditioned PFA is ash with a specified amount of water added to it as it leaves the power station. In engineering terms PFA may be classified in accordance with the Department of Transport Specification for Highway Works (Anon 1991a) as either Class 2E (lagoon PFA) or Class 7B (conditioned PFA). The design specification for PFA as an engineering material is documented in the Department of Transport Guidance Note HA44/91 (Anon 1991b). However, it is considered that HA44/91 is probably too restrictive (see e.g. Clarke 1992).

As an engineering material PFA has both advantages and disadvantages. In particular, PFA has the advantage that it is a pozzolanic material, thus benefiting from a self-hardening property. Since the strength of PFA compacted to engineering specification is known to increase with time, higher strength parameters than are measured in the laboratory may be adopted in slope stability analysis (Clarke 1992). A particular disadvantage of PFA is that it has the ability to absorb moisture through the rise of water by capillary tension and this may result in a loss in strength. Lagoon PFA has an advantage over conditioned PFA in that the concentration of insoluble chemicals and trace elements characteristically contained in PFA is known to decrease with time.

There is a considerable literature on the strength characteristics of PFA. However, most of the reported strengths relate to total stress parameters, with the c and ϕ values reported usually relating to specimens compacted to optimum moisture content using the 2.5 kg hammer (Anon 1991c). However, there is a dearth of published information on effective stress strength parameters.

Despite the extensive use of PFA by the UK construction industry since the 1950s, there is very little experience of building on PFA. Three case history records which relate to lagoon PFAs are worthy of note. Ballisager & Sorensen (1981) reported average settlements of about 175 mm and maximum values of 380 mm at the bottom of a large oil tank constructed directly on a 5 m depth of PFA subjected to an applied bearing pressure of 160 kPa. Charles *et al.* (1986) measured maximum settlements of 130 mm due to an estimated 80 kPa bearing pressure acting on an 11 m depth of PFA. Humpheson *et al.* (1991) assessed the field performance of ground-bearing foundations on a 10 m thick layer of PFA subjected to a bearing pressure of 25 kPa over a period of two months. Raft settlements averaged between 0.5 and 4.5 mm, with the maximum average settlement of 6.5 mm recorded over the following two years.

This paper is concerned with the engineering properties of a lagoon PFA and a conditioned PFA, concentrating on settlement considerations relating to the former and slope stability considerations relating to the latter.

Site history

Lagoon PFA site

An assessment of the history of the PFA-infilled old gravel pit in Staffordshire was made by means of a desk study which included aerial photographs dating back to 1963. Infilling of the gravel pit had commenced in 1981. The pit was backfilled with lagoon PFA from Rugeley power station, brought by truck to the site, end-tipped into the water-filled pit and then spread by bulldozers. The PFA had a moisture content of approximately 30%

From BENTLEY, S. P. (ed.) *Engineering Geology of Waste Disposal*,
Geological Society Engineering Geology Special Publication No. 11, pp 37–44

and was not compacted to an engineering specification. However, trafficking by dump trucks is believed to have increased compaction to around 90% maximum dry density to depths up to about 0.5 m below ground level (BGL). Filling of the site was complete by mid-1984 when the site is reported to have been infilled with 4–5 m of homogeneous light grey silt-sized PFA.

Conditioned PFA site

A contoured landscape mound composed of conditioned PFA has been gradually forming adjacent to a Yorkshire power station since the mid-1960s. It is up to 36 m in height with grassed slope angles at a maximum of 1(V) in 4(H). The PFA has been placed on a blanket of furnace-bottom ash. Placement of the PFA, at a moisture content of around 14%, is by boomstacker fed by conveyors. The material has been spread by bulldozer and lightly compacted by bulldozer with a towed vibratory roller.

Investigation

Lagoon PFA site

Ground investigations carried out at this site have included trial pits, cable tool boring and static cone penetration tests (CPT), together with associated standard penetration tests (SPT) and *in situ* permeability tests. Undisturbed U100 samples were taken for triaxial compression tests and oedometer tests.

Conditioned PFA site

An assessment of the engineering properties of the PFA formed part of a general study of the mound and its environs. The laboratory testing programme described below was performed on samples of PFA taken from the mound shortly after placement.

Ground conditions

Lagoon PFA site

Trial pit and borehole records describe the lagoon PFA as friable light grey and dark grey slightly clayey sandy silt with many fine and medium gravel size fragments of PFA. The PFA infill was generally a soft to very soft silt with a stiff to very stiff crust, with thicknesses in the range 2.3–4.8 m and 0.8–1.6 m respectively. The crust comprised cemented material. Cone penetration tests identified locally cemented layers within the main body of the PFA. The PFA was underlain by river terrace deposits varying in thickness between 0.6 and 3.8 m.

Standard penetration test N values were in the range 0 to 9, except for the top 1.5–2 m BGL where values of about 20 were recorded at 0.75 m BGL, these being attributed to layers where the PFA is highly cemented.

Typical graphs of cone resistance (q_c) and friction ratio (R_f) against depth are presented in Fig. 1. It can be seen that approximately 2 m of very stiff, becoming firm, cemented silt overlies 3.1 m of very soft to soft silt. Values of q_c (of up to about 40 MPa, with an average of about 16 MPa) and R_f (of an average 2.5%) were recorded in the top 1.5 m BGL, thus confirming the cemented nature of the near-surface deposits. At depths of 2–5 m BGL values of q_c and R_f generally averaged about 0.8 MPa and 1% respectively.

Also shown in Fig. 1 are typical CPT results obtained by Humpheson *et al.* (1991) from tests carried out in a lagoon PFA in Peterborough. It may be inferred that the Peterborough PFA was virtually free from cementation.

Piezometer monitoring records indicated that the groundwater table across the site was between about 1.5 and 3 m BGL, this corresponding to water strikes observed at similar depths.

Conditioned PFA site

The PFA mound rests on glacial clay underlain at depth by the Sherwood Sandstone.

Material properties

Physical properties

The index values and specific gravities are summarized in Table 1. All laboratory tests were carried out in accordance with B. S. 1377 (Anon 1991) unless stated otherwise.

Table 1. *Material properties*

Property	*Lagoon PFA (Staffordshire)*	*Conditioned PFA (Yorkshire)*
Plasticity index (%)	Non-plastic	3
In situ moisture content (%)	25–41 (range) 35 (average)	12–14
Specific gravity	2.27–2.33	2.21
In situ bulk density ($Mg\,m^{-3}$)	1.66–1.75	—
Dry density ($Mg\,m^{-3}$)	1.2–1.34	—
Maximum dry density ($Mg\,m^{-3}$)	1.29–1.43	1.52
Optimum moisture content (%)	21–29	14
c (UU)	0	—
ϕ (UU)		
First stage of test	31°	—
Lower and upper bounds	27–31°	
c' (CID)	—	0
ϕ' (CID)	—	25

UU: unconsolidated undrained multistage triaxial test; CID: consolidated drained triaxial test.

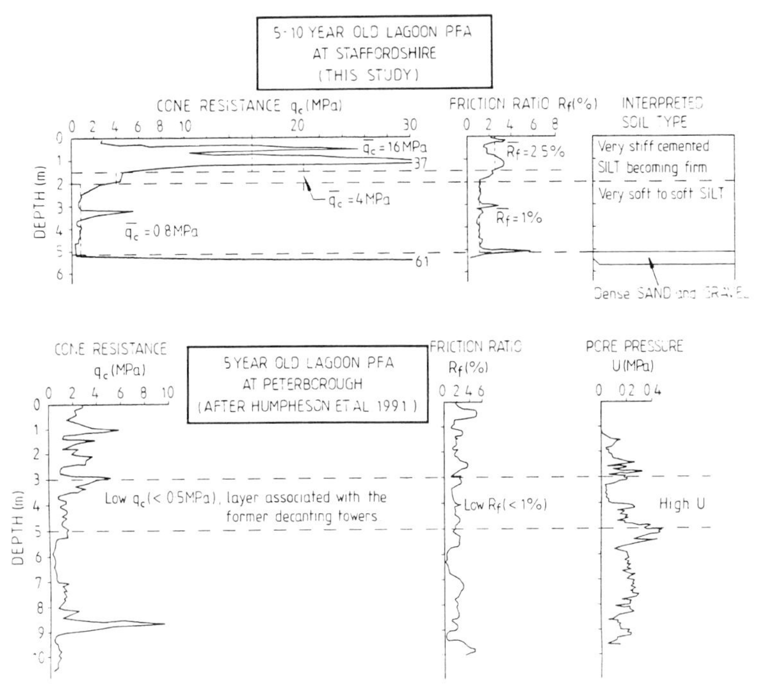

Fig. 1. Results of static cone penetration tests.

Lagoon PFA. The PFA samples were found to be non-plastic. Particle size distribution curves are presented in Fig. 2, which shows that the PFA primarily comprised very sandy silt, with no evidence of any variation in particle size distribution as a function of sample depth. Figure 2 also shows particle size distribution envelopes for the Peterborough PFA (Humpheson *et al.* 1991) where the PFA was described as either 'silt' or 'silty sand/sandy silt' PFA.

Conditioned PFA. Particle size distribution (Fig. 2) was obtained using a Malvern Particle Size Analyser, which determines gradation by measuring the intensity of a reflected laser source passing through a dispersed sample. The morphology was examined under a microscope. The PFA was fairly well graded with a uniformity coefficient (D_{60}/D_{10}) of about 10. It can be described as a fine to coarse silt with fine sand, and was virtually non-plastic. It was estimated that 98% of the particles were spherical.

Compaction

Maximum dry densities corresponding to values of optimum moisture content were determined using the BS Light (2.5 kg hammer) compaction test.

Lagoon PFA. Values of maximum dry density were in the range 1.29–1.43 Mg m^{-3}, with optimum moisture contents between 21% and 29% (Fig. 3). Humpheson *et al.* (1991) found that maximum dry density determinations made from lateral shaking tests and vibrating hammer were in the range 1.10–1.57 Mg m^{-3} whilst the BS Light test yielded values between 1.39 and 1.41 Mg m^{-3}, irrespective of grading. Therefore it may be inferred that the compactive effort from the 2.5 kg hammer is likely to have caused crushing of the PFA particles, thus resulting in possibly overestimated values of maximum dry density.

Conditioned PFA. A compaction test was carried out as a control for preparing suction, triaxial and Rowe cell specimens by weight as a percentage of the maximum dry density found in the test. The maximum dry density was 1.52 Mg m^{-3} with an optimum moisture content of 14% (Fig. 3). Figure 3 also shows the relationship derived by Clarke (1992) between maximum dry density and optimum moisture content for different conditioned PFAs.

Shear strength

Lagoon PFA. Shear strength parameters were determined from unconsolidated undrained multistage triaxial

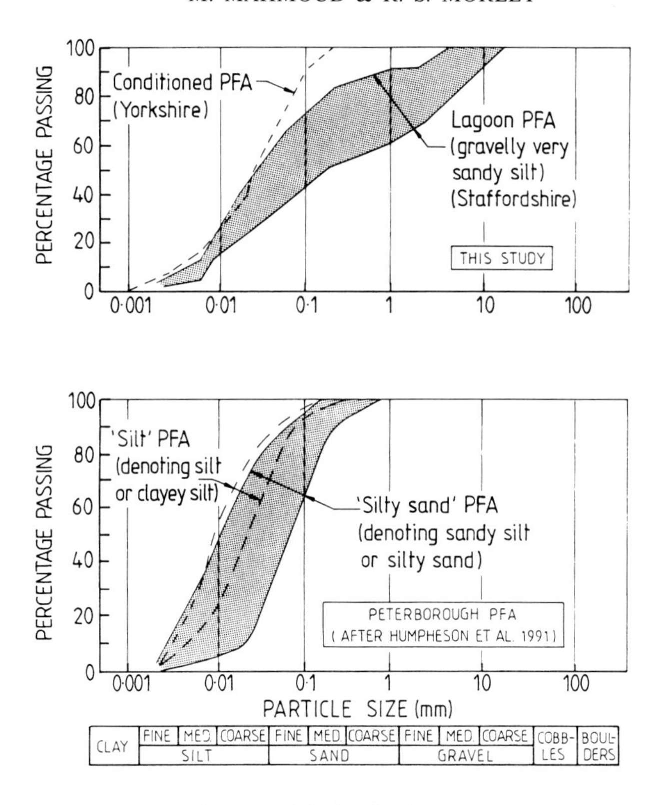

Fig. 2. Particle size distribution curves.

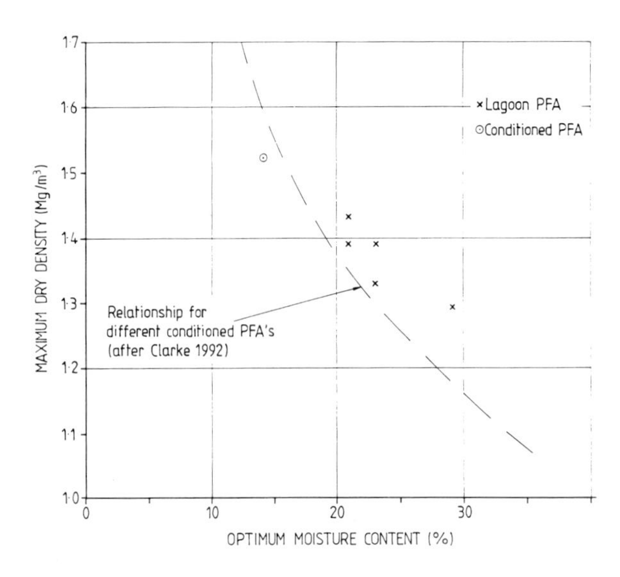

Fig. 3. Variation of maximum dry density with optimum moisture content.

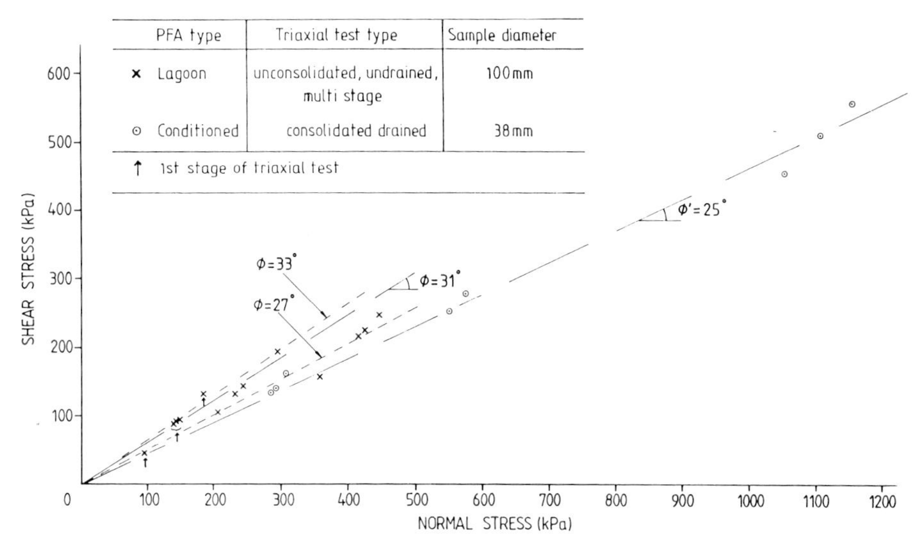

Fig. 4. Shear strength parameters.

tests on undisturbed 100 mm diameter samples. The failure envelope (Fig. 4) yields parameters $c = 0$, $\phi = 31^\circ$ corresponding to the first stage of the tests, with the failure envelope extremes of $\phi = 27^\circ$ and 33°. Consolidated undrained triaxial tests on 38 mm diameter samples for the Peterborough PFA (Humpheson *et al.* 1991) yield values of $c' = 0$, $\phi' = 33^\circ$.

Conditioned PFA. Consolidated drained triaxial tests (Bishop & Henkel 1962) were carried out on 38 mm diameter samples compacted at 85, 90 and 95% of maximum dry density. A back pressure of 600 kPa was required for saturating the samples. The failure envelope is presented on Fig. 4 which shows strength parameters $c' = 0$, $\phi' = 25^\circ$.

Strength due to suction (negative pore pressure) associated with partly saturated fine-grained soils was also considered. Soil suction arises mainly as a result of the surface tension of water absorbed by soil particles and decreases as the degree of saturation increases with increasing values of moisture content.

Suction tests were carried out using a Quick Draw Soil Moisture Probe (Sweeney 1982) with measurements taken at both maximum and 85% of maximum compactive effort for samples having moisture contents in the range 8–20%. A plot of suction versus moisture content is shown in Fig. 5(a).

The principle of unsaturated soil shear strength has been estabished through empirical testing. Results show that the basic Mohr–Coulomb relationship cannot be applied directly using negative pressures as this approach over-predicts the available strength of the material. Fredlund *et al.* (1978) proposed an extended Mohr–Coulumb failure criterion for unsaturated soils:

$$\tau = c' + (\sigma_n - u_a)\tan\phi' + (u_a - u_w)\phi^b \qquad (1)$$

where σ_n is the total normal stress; u_a is the pore air pressure (normally atmospheric); u_w is the pore water pressure; ϕ' is the internal friction angle; ϕ^b is the matrix suction friction angle; and c' is the effective cohesion.

The effect of ϕ^b can be visualized as either a friction angle or a cohesion component. Assuming the latter, the total cohesion of an unsaturated soil is then given by Equation (2):

$$c = c' + (u_a - u_w)\tan\phi^b. \qquad (2)$$

The value of ϕ^b can be determined through special triaxial testing in which the total stress conditions are held constant while the matrix suction is varied. ϕ^b is always less than ϕ' and normally lies in the range 13° to 20°, typically 15° (Fredlund 1987). In the absence of laboratory test results it was considered reasonable to take 15° as an approximation for purposes of analysis.

As described above, laboratory testing was undertaken to establish suction/moisture content relationships for the PFA at specific compactions. By assuming a ϕ^b value of 15° and taking various values of moisture content, it was possible to calculate the additional effective cohesion component $[(u_a - u_w)\tan\phi^b]$ as a result of unsaturated conditions. This additional effective cohesion can then be added to the material

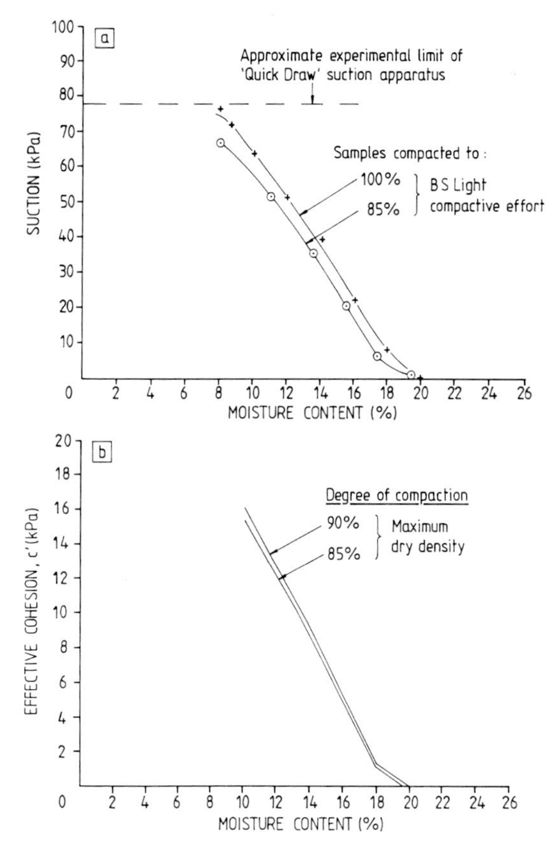

Fig. 5. Variation of suction/effective cohesion with moisture content-conditioned PFA.

effective cohesion (if any) given by the saturated effective stress tests.

The results of these calculations for PFA at 85 and 90% compaction, which most closely represent field conditions, are shown on Fig. 5(b) in terms of additional effective cohesion versus moisture content. At a moisture content of 14% (corresponding to that at the time of PFA mound construction) the effective cohesion of the material would appear to increase by 9 kPa.

Compressibility

Lagoon PFA. The results of one-dimensional oedometer tests on undisturbed samples are presented in Fig. 6 in terms of coefficient of volume compressibility, m_v, versus vertical pressure. Figure 6 also shows best-fit lines corresponding to 'first time load' and 'reload' data. m_v values for first time load data are in the range 0.12–0.04 $m^2 MN^{-1}$, with a trend of reducing m_v with increasing values of pressure up to about 150 kPa. 'Reload' data indicate m_v values in the range 0.06–0.01 $m^2 MN^{-1}$.

Also presented in Fig. 6 are the results reported by Humpheson *et al.* (1991) of one-dimensional oedometer and Rowe cell tests on samples of the Peterborough lagoon PFA. m_v values from the oedometer tests are in the range 0.4–0.05 $m^2 MN^{-1}$, whilst the best-fit line through the Rowe cell data compares well with the results presented herein, with m_v values varying between about 0.18 and 0.08 $m^2 MN^{-1}$ corresponding to stress levels up to 150 kPa.

Conditioned PFA. Conventional one-dimensional consolidation tests were carried out in a Rowe cell on two samples of PFA compacted at 85% and 90% of maximum dry density. Figure 6 shows m_v values in the range 4–0.04 $m^2 MN^{-1}$ corresponding to vertical pressures up to about 600 kPa.

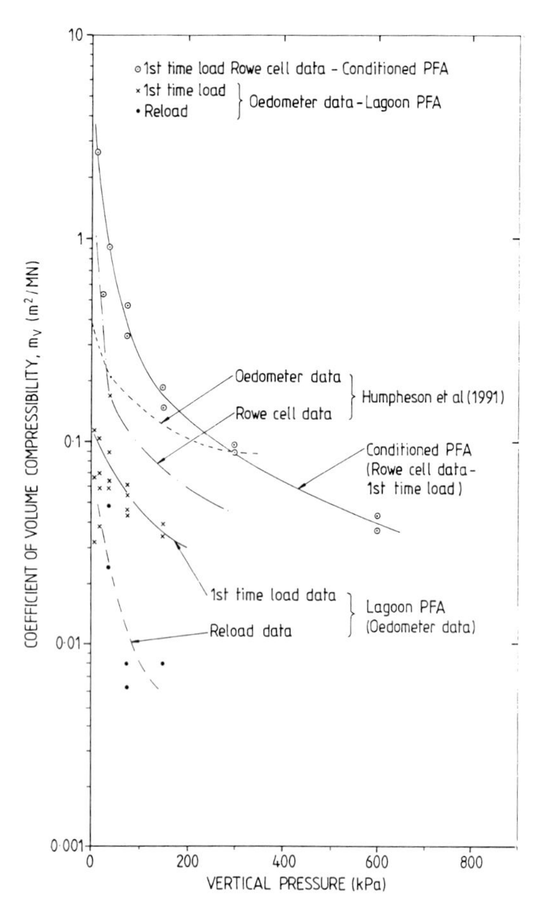

Fig. 6. Coefficient of volume compressibility versus vertical pressure.

Permeability

Lagoon PFA. An *in situ* falling head permeability test was carried out in a slotted standpipe piezometer installed at a depth of 3 m BGL. This gave a coefficient of permeability of $k = 1.4 \times 10^{-7}$ m/s for the PFA.

Conditioned PFA. Tests were carried out on compacted samples, nominally 38 mm diameter by 75 mm long, in a triaxial cell, as described in Head (1986). The constant-head method was used on fully saturated specimens at various confining pressures. Over the range of confining pressures 50–600 kPa values of k decreased from around $5 \times 10^{-7}\,\mathrm{m\,s^{-1}}$ to $10^{-8}\,\mathrm{m\,s^{-1}}$.

Engineering considerations

Lagoon PFA site

General. It is proposed to construct an overbridge which will cross an existing dual carriageway, linking roundabouts on both sides of the bridge. There will be a maximum embankment height of approximately 7 m over the edge of the PFA infilled gravel pit.

Embankment stability. Conventional slope stability calculations indicated that slope gradients of 1(V) in 2(H) would yield unacceptable factors of safety for slope stability. Hence geogrid reinforcement will be incorporated within the embankment PFA yielding adequate factors of safety of 1.3.

Embankment settlement. Maximum recorded settlements reported by Humpheson *et al.* (1991), Charles *et al.* (1986) and Ballisager & Sorensen (1981) are plotted against bearing pressure in Fig. 7. Maximum settlements of between 210 and 250 mm would thus be anticipated under bearing pressures from a 7 m high embankment constructed of suitable PFA fill. The majority of the anticipated 250 mm of total settlement is expected to occur 'immediately' i.e. during construction (see e.g. Humpheson *et al.* 1991).

Settlement calculations have also been carried out using the Buisman–DeBeer method (presented in Craig 1987). This method is based on correlations between average cone penetration resistance and the compressibility of granular materials; here it is assumed that PFA is a 'cohesionless' material. The Buisman–DeBeer method predicts that the 5.5 m thick layer of lagoon PFA will settle up to about 165 mm under a 7 m high PFA embankment.

There is also justification to presume that there will be some time-dependent settlement within the PFA; Humpheson *et al.* (1991) reported that there was an increase in the maximum average settlement of about 2–6 mm over a two-year period following construction. Based on assessments made from laboratory determinations of values of m_v (Fig. 6), Terzaghi's theory of one-dimensional consolidation calculates an anticipated settlement up to about 50 mm in the long term.

The total settlement calculated on the basis of the Buisman–DeBeer method (for immediate settlement) and Terzaghi's theory of one-dimensional consolidation (for long-term settlement) amounts to about 215 mm which falls within the band of the expected total settlement of 210–250mm based on previous observations of settlement of lagoon PFA deposits (Fig. 7).

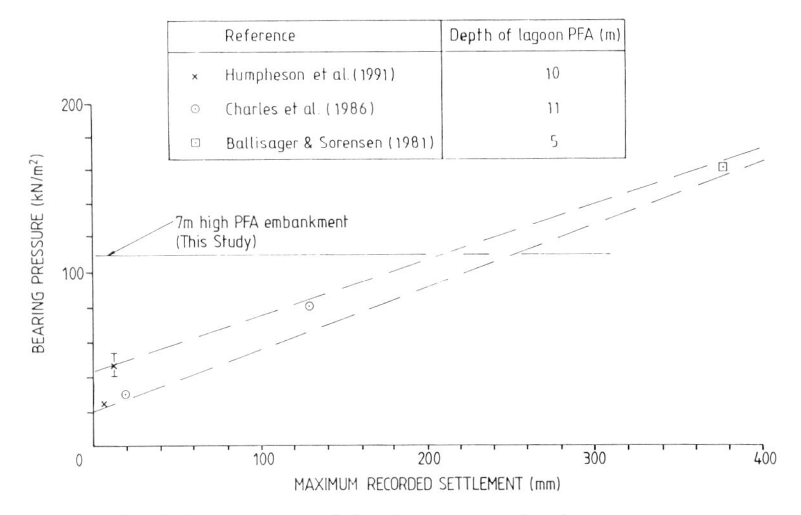

Fig. 7. Maximum recorded settlement versus bearing pressure.

Conditioned PFA site

General. The assessment of conditioned PFA properties formed part of a larger commission which ensured the landscaped mound was constructed according to best-known practice.

Stability considerations formed part of the investigation as discussed below.

Mound Stability. During mound construction the advancing face of PFA was formed at a slope of 1(V) in 3(H). Conventional circular stability analyses, using $c' = 0$ and $\phi' = 25°$, gave a factor of safety of 1.4. This value was considered adequate, hence it was not necessary to invoke suction or apparent cohesion to achieve a reasonable margin of safety. The overall factor of safety for the completed mound at slopes of 1(V) in 4(H) was 1.8, similarly satisfactory.

Conclusions

The static cone penetration test is considered to be a suitable technique for investigating sites of infilled lagoon PFAs.

Engineering properties of a five- to ten-year-old lagoon PFA and a one-day-old conditioned PFA have been investigated. Material properties of these two PFAs have not been compared as the variability in PFA generally from power station to station may be greater than between lagoon and conditioned PFAs.

Failure envelopes for both the lagoon and conditioned PFAs indicate no cohesion intercept, with values of $\phi = 31°$ for the lagoon PFA and $\phi' = 25°$ for the conditioned PFA. However, measurements of suction in compacted samples of partly saturated conditioned PFA suggest that the PFA from Yorkshire is capable of mobilizing higher strengths. This increase in strength due to suction may be represented in terms of an apparent effective cohesion c' which can be derived from Fredlund *et al.*'s (1978) relationship for partially saturated soils. It therefore seems reasonable to propound that the above-mentioned apparent c' is probably a function of both (a) suctions which some PFAs may sustain, and (b) the pozzolanic bonding which is known to develop in PFA in the presence of water.

Total settlement occurring within lagoon PFA deposits can be estimated by using the Buisman–DeBeer method for calculating immediate settlement and Terzaghi's theory of one-dimensional consolidation for calculating long-term settlement. Calculation of total settlement using the above approach yields results which are in good agreement with observations of maximum average settlements of lagoon PFA deposits reported in the literature.

References

ANON 1991a. *Manual of Contract Documents for Highway Works, Vol. 1: Specification for Highway Works.* Department of Transport, HMSO, London.

—— 1991b. *Guidance Note HA44/91.* Department of Transport, HMSO, London.

—— 1991c. *B. S. 1377. Methods of Testing Soils for Engineering Purposes.* British Standards Institution, HMSO, London.

BALLISAGER, C. C. & SORENSON, J. L. 1981. Flyash as fill material. *Proceedings of 10th International Conference on Soil Mechanics and Foundation Engineering, Stockholm*, **2**, 297–301.

BISHOP, A. W. & HENKEL, D. J. 1962. *The measurement of soil properties in the triaxial test*, 2nd edn. Edward Arnold, London.

CHARLES, J. A., BURFORD, D. & WATTS, K. S. 1986. Improving the load carrying characteristics of uncompacted fills by preloading. *Municipal Engineer*, **3**, 1–19.

CLARKE, B. G. 1992. Structural fill. *Proceedings of Conference on the Use of PFA in Construction*, Dundee, 21–32.

CRAIG, R. F. 1987. *Soil mechanics*, 4th edn. Van Nostrand Reinhold, Wokingham.

FREDLUND, D. G. 1987. Slope stability analysis incorporating the effect of soil suction. *In*: ANDERSON, M. G. & RICHARDS, K. S. (eds) *Slope Stability: Geotechnical Engineering and Geomorphology*. John Wiley, New York, 113–144.

——, MORGENSTERN, N. R. & WIDGER, A. 1978. Shear strength of unsaturated soils. *Canadian Geotechnical Journal*, **15**(3), 313–321.

HEAD, K. H. 1986. *Manual of Soil Laboratory Testing, Vol. 3: Effective Stress Tests.* Pentech Press, London.

HUMPHESON, C., SIMPSON, B. & CHARLES, J. A. 1991. Investigation of hydraulically placed PFA as foundation for buildings. *In*: GEDDES, J. (ed.) *Proceedings of 4th International Conference on Ground Movements and Structures*, Cardiff. Pentech Press, London, 68–88

SWEENEY, D. J. 1982. Some in situ soil suction measurements in Hong Kong's residual soil. *In*: MCFEAT-SMITH, I. & LUMB, P. (eds) *Proceedings of Southeast Asian Geotechnical Conference*, Hong Kong Institute of Engineers, 91–105.

Engineering properties and disposal of gypsum waste

A. R. Griffin

Brown & Root Environmental, Thorncroft Manor, Dorking Road, Leatherhead, Surrey, UK

Abstract. This paper describes the investigation and design of a site for the safe disposal of solid gypsum waste derived as a by-product from the manufacture of titanium dioxide pigment. The site is on the east coast of Malaysia located on an area underlain by soft marine clays. The final design of the containment structure and final disposal scheme was dictated by the nature of the ground and the consequences of the detailed environmental assessment. The waste material has some similarities with FGD gypsum and phosphogypsum but the presence of a high iron content provides the material with additional beneficial properties for landfilling.

Titanium dioxide (TiO_2) is a white non-toxic pigment. Its current main uses are in surface coatings (paint and powder coatings), plastics, paper, inks, pharmaceuticals and food additives. The pigment is produced in over 50 plants in 24 countries and the world production in 1988 was just over 2.9×10^6 tonnes. However, due to the recent worldwide recession, the industry has entered a period of extreme overcapacity and plant closures seem likely, the most vulnerable being those unable on economic grounds to comply with recent stricter environmental legislation. A major issue confronting producers using the 'sulphate route' is obtaining suitable safe solid waste landfill sites for disposal of the process residues. This paper describes the design concepts and studies carried out for the safe disposal of waste products from a new production plant owned by Tioxide (Malaysia) Sdn. Bhd, now constructed and in operation on the east coast of peninsular Malaysia. This plant was subject to a major environmental impact assessment in accordance with current Malaysian legislation and subsequently received approval from the Malaysian Department of the Environment in July 1989.

Waste products

The production of titanium dioxide pigment is carried out by two processes, namely the 'chloride' process and the 'sulphate' process. The sulphate process was adopted for the Malaysian plant and involves the digestion of the raw source material ilmenite ($FeTiO_3$) in 85–92% sulphuric acid. The source material used for this plant is obtained mainly from Malaysia and Australia. Apart from the main constituents of titanium and iron, ilmenite contains a number of trace heavy metals and radioactive compounds.

The titanium dioxide pigment is extracted from the ilmenite first by digestion with the sulphuric acid to produce soluble titanyl sulphate then followed by precipitation by hydrolysis as shown in the equations below:

$$FeTiO_3 + 2H_2SO_4 \rightarrow TiOSO_4 + FeSO_4 + 2H_2O$$

$$TiOSO_4 \xrightarrow{OH^-} TiO_2nH_2O + H_2SO_4$$

The hydrated titanium dioxide is leached, washed and calcined to produce the final TiO_2 pigment.

The main waste products from this processes are generated at a number of stages and include:

- Solid residues—unreacted ore residues from the digesters are removed by settlement and filtration;
- Liquid residues—two main liquid streams are produced: (i) the 'strong acid' stream from filtration after the hydrolysis stage (10–15%), and (ii) the 'weak acid' stream consisting of acid waters from hydrolysate washing and other process waters such as wash waters from flue gases;
- Gas residues—particulates, acid mists and sulphur oxides. The digestion and calciner gases are treated by limestone scrubbing processes.

The average liquid effluent received from the plant has a typical composition as shown in Table 1.

The effluent which is predominantly sulphuric acid and ferrous sulphate cannot be disposed of directly to the environment although this used to be done in Europe prior to the EEC Directives on the titanium dioxide industry. This effluent is now neutralized to produce an innocuous waste product—gypsum. Neutralization may be carried out in two stages depending on the requirement to separate pure gypsum ('white gypsum') for resale.

- First stage neutralization: the initial effluent is mixed with a slurry of finely ground limestone to a resulting pH of 2. White gypsum precipitates and is separated from the filtrate by filtration using drum or plate filters.

From BENTLEY, S. P. (ed.) *Engineering Geology of Waste Disposal*,
Geological Society Engineering Geology Special Publication No. 11, pp 45–55

Table 1. *Average liquid emission comosition*

Component	*Concentration* ($mg\,l^{-1}$)	
	Before treatment	*After treatment*
Acidity (as H_2SO_4)	125 000	(pH 6.5–7)
Fe	50 000	1.0
Solids	20 000	100
Cr (III and VI)	100	1.0
Cu	15	0.5
Mn	1500	1.0
Ni	10	0.5
Pb	25	0.5
V	200	1.0
Zn	100	0.2
Cd	0.1	0.01
Hg	0.05	0.01
As	0.5	0.05
Sn	0.5	0.1
Total SO_4^{2+}	225 000	5 000

- Second stage neutralization: the remaining filtrate is mixed with a calcium hydroxide slurry to produce a pH of 9–10. Additional gypsum precipitates along with iron oxides/oxyhydroxides which are separated by drum or plate filters; this is the 'red gypsum'.

A typical composition of treated filtrate is shown above in Table 1. Both this composition and the untreated composition does depend on the original ore feedstock used.

The neutralized 'red gypsum' cake from the second stage may contain some 18–20% iron (as Fe_2O_3). If a combined process is used and white gypsum is not removed the iron content would be in the order of 9–12% (as Fe_2O_3). The Malaysian plant at full production will produce about 50 000 t of pigment per annum with 400 000 t of gypsum requiring safe disposal.

Location of disposal site

The site for the pigment production plant had been located at Teluk Kalung, near Chukai in Terengganu State, Malaysia (Fig. 1). The factory site consisted of a flat coastal marshy area, part of which had been reclaimed by the pumping of marine sands from adjacent harbour development works. The actual production plant was to be built on this reclaimed portion. To complete the design of the plant and with the requirement to submit an environmental impact statement it was necessary to establish a suitable location and method for waste storage and disposal. Several potential sites were identified including valley infill areas and also the flat marshy unreclaimed area adjacent to the factory site. This area was bounded by a newly constructed highway built on wide embankments about 2–3 m above the original ground level.

The general framework for the approach to the first stage of the study is outlined in Fig. 2 (after Digioia & Gray 1979). This framework indicated the approach taken on technical and engineering grounds although it also formed the basis of the scoping exercise for the environmental assessment. On the basis of this approach, two possible sites for solid waste disposal were selected, one within the Bukit Pelandok valley to the north of the site and the other on the marshy land adjacent to the factory. While the former site had several advantages, the presence of overhead power lines restricted the height of fill, reducing the site's capacity to approximately $2 \times 10^6\,m^3$ (half of the target capacity). The second site was selected for further study and investigation.

Ground conditions

Full details of the ground conditions at the proposed solid waste disposal site have been given elsewhere (Griffin *et al.* 1991) and are summarized below.

The site is underlain by recent alluvial deposits. The flat swampy area is surrounded by hills of Lower Carboniferous metasediments (phyllites, slates, quartzites and schists) with intruded granites of Upper Carboniferous/Lower Permian age. These rocks also form the bedrock underlying the alluvial deposits.

The alluvial deposits are predominantly soft black to grey clays of marine origin, believed to have been deposited within the last 10 000 years, underlain by sands and stiffer clays representing older alluvial deposits. These soft clays are common throughout the Far East especially in the Mekong Delta, Chao Phraya Delta of Thailand and on the coastal plains of Malaysia (Cox 1968). At the selected waste disposal site the general soil profile, as established from an extensive drilling programme, consisted of 8–20 m of very soft to soft black to greenish grey silty clay with traces of sea shells and fine sand and thin horizons of peat with decayed vegetation, roots and tree fragments. The clays overlie a variable thickness (0–13 m) of fine to coarse sands. These in turn rest on the weathered bedrock profile. Towards the base of the soft clays there is a thin horizon of stiff light grey overconsolidated clay which may represent a period of exposure and dehydration during the recent depositional history. This horizon is usually at the base of the soft clays but in a few cases soft clays were found under this strata. A typical cross-section across the site is shown in Fig. 3.

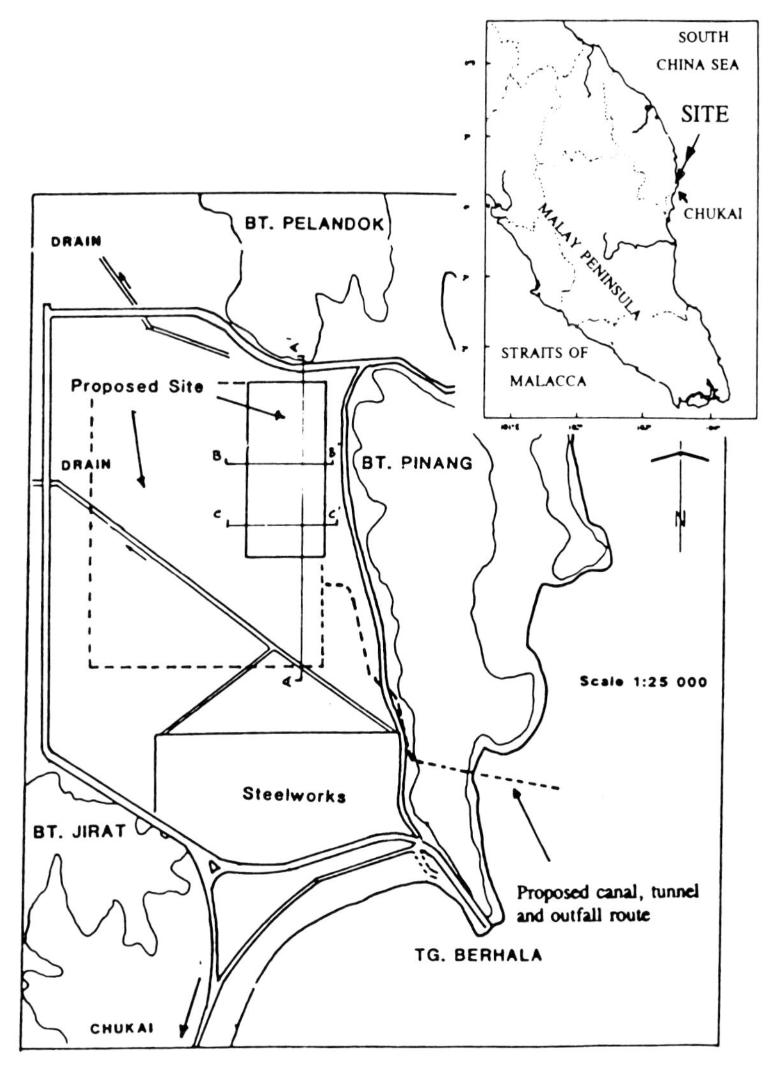

Fig. 1. Location of site.

Typical soil profiles and index properties are shown in Fig. 4 together with the results of corrected *in situ* vane tests. The clays are of high to extremely high plasticity with high liquidity indices. Particle size distributions indicate 20–50% clay size material and 5–15% fine sand. The corrected (Bjerrum 1973) vane shear strengths are generally consistent for the top 7 m at about 10 kPa, below which they increase linearly with depth with a Cu/Po ratio of about 0.35–0.40. The profile indicates a weathered crust with a depth of weathering of about 5–6 m, which is similar to the profiles for Bangkok clays (Cox 1968) but deeper than might be expected, for young clays close to the current coastline. As expected the clays are highly compressible, with the coefficients of compressibility ranging from 0.5×10^{-3} to $4.0 \times 10^{-3} \mathrm{m^2\,kN^{-1}}$ (Fig. 5).

Permeabilities of the clays were measured in the laboratory (calculated from the results of one-dimensional consolidation tests) and *in situ* from installed piezometers. The *in situ* values ranged from 5.3×10^{-8} to $9.7 \times 10^{-10} \mathrm{m\,s^{-1}}$. The permeabilities of the underlying sands ranged from 9×10^{-4} to $1.6 \times 10^{-5} \mathrm{m/s^{-1}}$.

The ground conditions present a number of problems to the design of any system. The soft marine clays have low shear strengths and high compressibilities which could cause excessive settlement to structures and foundation failures under even moderate bearing pressures. Most of the factory works are piled, as were

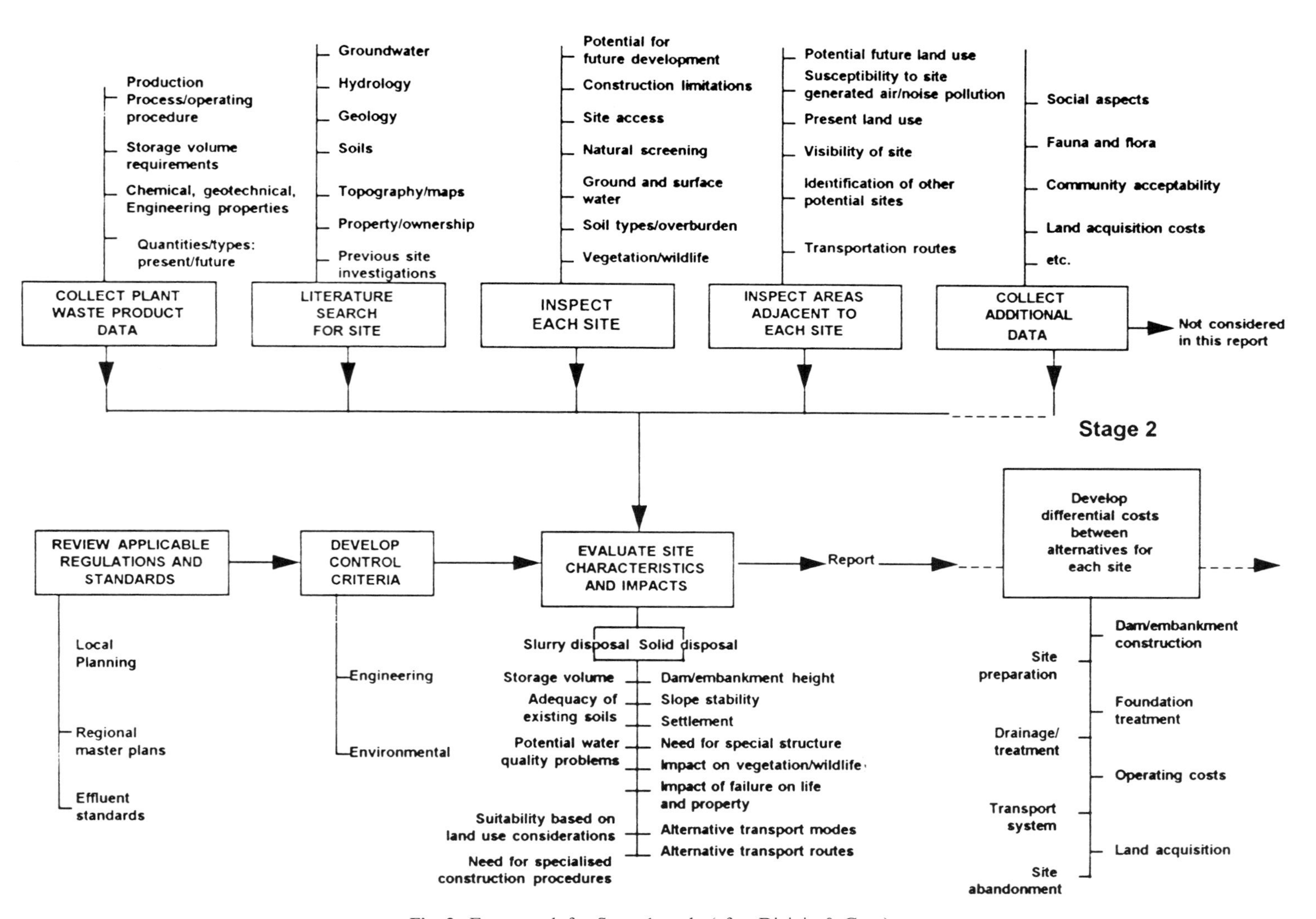

Fig. 2. Framework for Stage 1 study (after Digioia & Gray).

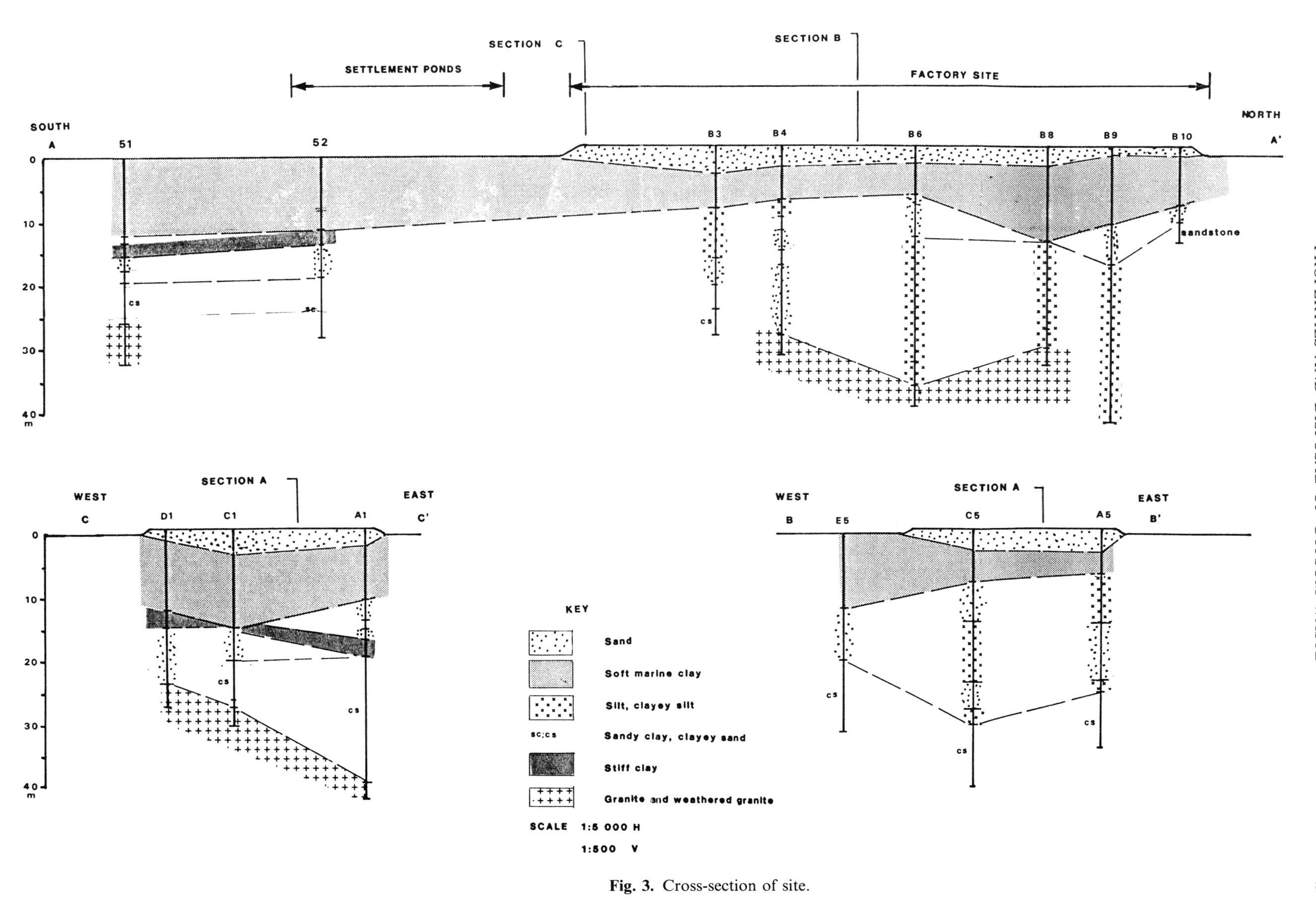

Fig. 3. Cross-section of site.

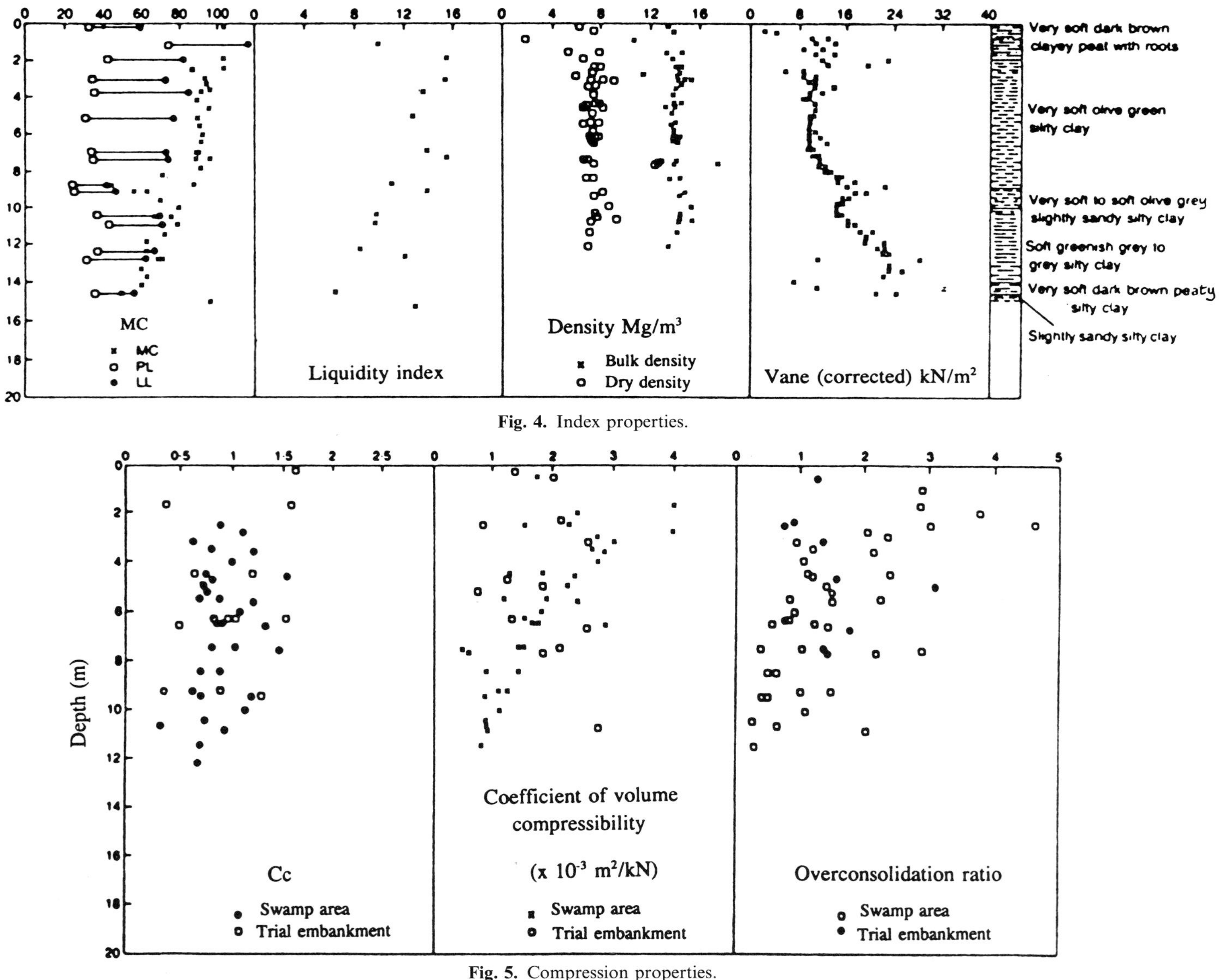

Fig. 4. Index properties.

Fig. 5. Compression properties.

some of the smaller structures for the waste containment system. However, the costs involved in piling of the main scheme works would have been prohibitive.

Design of disposal scheme

During full plant operation the neutralization plant is expected to produce about $63\,t\,h^{-1}$ ($41\,m^3\,h^{-1}$) of solid gypsum cake and about $155\,m^3\,h^{-1}$ of waste waters from the dewatering process. In addition, all runoff water from the factory area has to pass through a containment system allowing for treatment of any contamination. The waste treatment system was based on a principle of total containment. Two levels of containment were envisaged: an existing outer containment boundary represented by the highway embankment (shown in Fig. 1) and a proposed inner containment area of bunded lagoons developed within this boundary. The inner containment system consists of a 2.5 m high clay bunded area for the solid waste containment with a series of treatment and monitoring lagoons for liquid waste and runoff waters both from the factory site and from the solid waste landfill area. The lagoon system was designed for a residence time of about 11 days per lagoon which allows for shock loads and accidental or other spillages from the plant to be diluted or ponded for further treatment. It was originally proposed that the final effluent from the monitoring ponds, which would consist of clear water but with a slightly raised sulphate content, would be discharged in the swamp within the outer containment boundary and allowed to discharge through the natural existing drainage system to a nearby estuary. However, this final disposal route was not accepted at the public presentation of the EIA and an alternative disposal route was required. This involved the additional design of a concrete-lined canal system to carry the final waste water from the lagoons over a distance of 1200 m, through a 400 m long tunnel (excavated through Bukit Pinang) with final discharge eastwards to the sea via a marine outfall pipe. Because of the soft nature of underlying clays as described above, all structures had to be piled including the canal system, pump station, lagoon spillways, drawoff towers, etc.

The initial design of the lagoon embankments was checked for stability by conventional settlement and slope stability analyses (circular and non-circular). However, a stable design dimension for the structure was particularly sensitive to the shear strength values adopted for the stability calculations. The shear strength results obtained from *in situ* testing showed a wide scatter of values, particularly at shallow depths. In addition, the values for coefficient of compressibility for use in settlement calculations as obtained from one-dimensional oedometer tests appeared unreliable, particularly in the upper weathered (crustal) zone of the marine clay profile.

The design of the embankment containment system was therefore only finalized after the construction and monitoring of a trial embankment. This has been described in detail previously (Griffin *et al.* 1991) but in general consisted of a 2.0–2.5 m benched embankment fully monitored to assess foundation pore pressures, shallow and deep settlements and shallow displacements. The embankment was constructed in layers over a period of 40 days. The revised containment and lagoon system finally adopted is shown in Fig. 6. During monitoring of the 2.5 m high trial embankment, up to 600 mm of settlement was recorded. Similar settlements have now been recorded in the constructed embankments with a few sections reaching 1000 mm. Long-term settlement may continue as a consequence of secondary consolidation or creep, requiring continual maintenance to maintain the design freeboard. The trial embankment did not show any evidence of slope failure or shear failure in the upper foundation layers although some lateral displacement of the clays occurred close to the embankment toe. Pore pressures dissipated faster than expected allowing an increase in the effective strength of the foundation. The presence of peaty material, roots and fine sand laminae may have accounted for this phenomena. This therefore required a design modification to the bunds to include for a deeper central clay fill cut-off to prevent potential seepages of contaminants from the lagoons.

Properties of gypsum waste

To design an efficient solid waste handling and disposal system the properties of the waste needed to be understood. Key factors included the behaviour of the material during transport, compaction and its inherent long-term stability both with respect to a design slope and its behaviour when subject to excessive settlements.

Much background information is available on the properties of other types of gypsum wastes including flue gas desulphurization gypsum (FGD) (Hagerty *et al.* 1977; Morasky *et al.* 1980; Krizek *et al.* 1987; Rademacher *et al.* 1988) and phosphogypsum (Vick 1977; Gorle 1979), both of which are commonly disposed of in landfill. A summary of relevant engineering properties for sulphate-rich FGD and phosphogypsum are shown in Table 2.

The shear strength properties of these gypsums are dependent on the void ratio, i.e. the degree of compaction. The effective angle of friction ρ' for FGD gypsum ranged from 40° to 46° at void ratios of 0.9 to 0.5 respectively (Morasky *et al.* 1980). Other authors have reported lower effective stress values in the order of 33 to 37°. Effective strength tests on *in situ* material in an old phosphogypsum dump (Vick 1977) produced an envelope fitting the following equation:

$$\tau_f = 26 + \sigma \tan 32^\circ \text{ kPa}$$

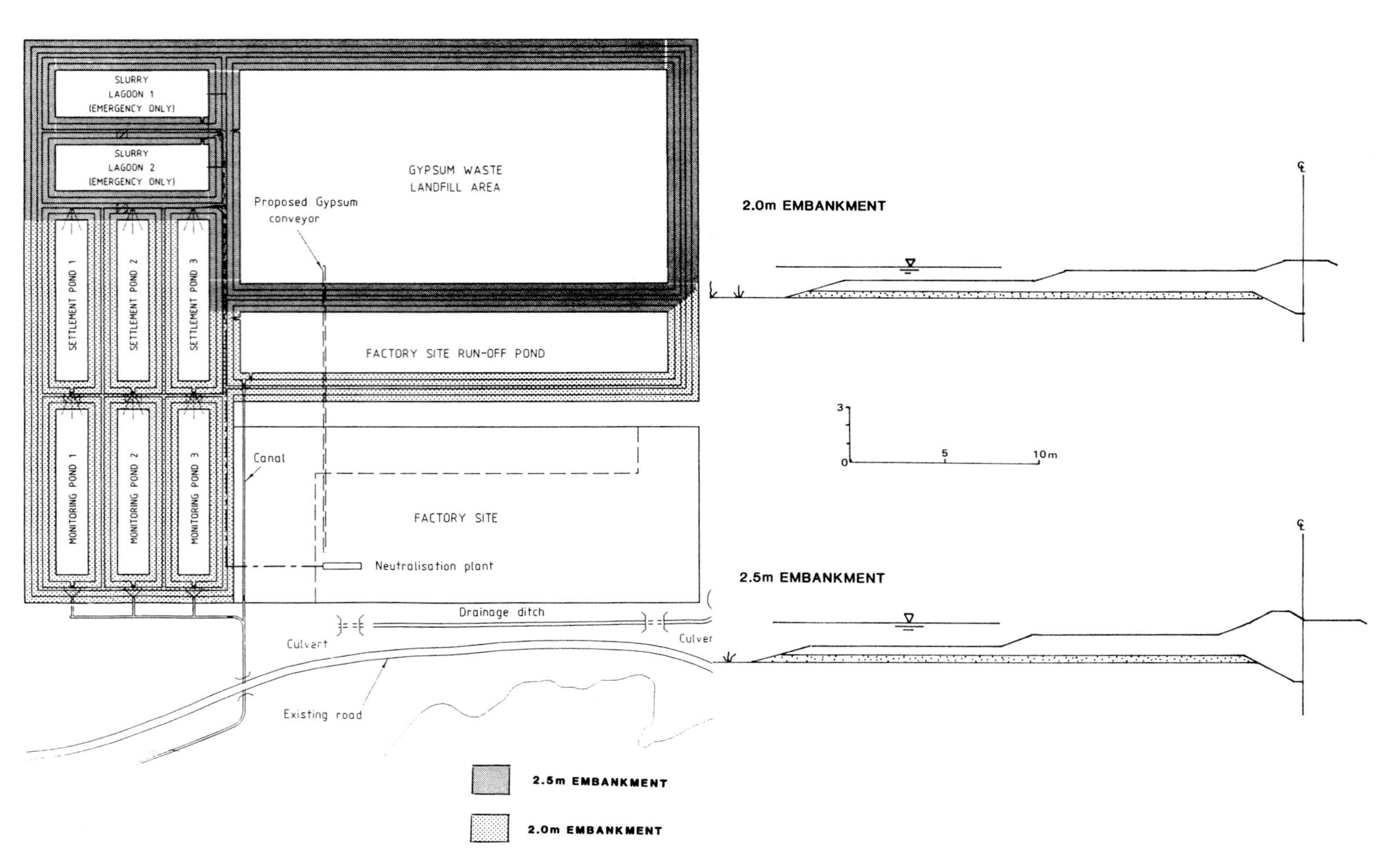

Fig. 6. Containment and lagoon system.

Table 2. *Properties of FGD gypsum and phosphogypsum*

Property	*FGD gypsum*	*Phosphogypsum*
Specific gravity	2.21–2.36	2.33–2.39
Bulk density ($Mg\,m^{-3}$)		
settlement	1.39–1.65	
vacuum filtration	1.48–1.78	
Particle size (%)		
clay	2–6	0
silt	66–76	80–95
sand	18–30	5–20
Compaction	(2.5 kg rammer)	(4.5 kg rammer)
max. dry dens ($Mg\,m^{-3}$)	1.43–1.55	1.42–1.57
optimum moisture content (%)	21–14	19–16

These tests indicated cementation of the *in situ* tailings, possibly as a consequence of solution/precipitation effects or fusion of the gypsum grains at high point stresses.

A fundamental difference between the above gypsums and the titanium waste gypsum is the presence in the latter of a high percentage of iron salts. In the initial waste the iron is present as ferrous sulphate which on neutralization with limestone and lime co-precipitates with the gypsum as a iron oxide or oxyhydroxide. However, the exact form of the precipitate will depend on a number of conditions including the rate of oxidation of the iron, neutralization pH and the concentration of intermediate sulphate complexes. The potential routes of mineral formation on neutralization is summarized in Fig. 7. The end products would not be the same if the original iron was in the ferric state prior to neutralization. Under these conditions the iron minerals would be either haematite (Fe_2O_3) or ferrihydrite ($Fe_{1.31}O_{0.94}(OH)_{2.06}$) depending on the pH and temperature of the mixture. Ferrihydrite ages to goethite (α-FeOOH).

The iron mineral produced at the Malaysian plant (at high neutralization pH) is initially a green rust complex which oxidizes to a reddish brown mineral. Both the green complex and brown mineral are magnetic and the latter is believed to be maghemite (γ-Fe_2O_3). Rapid oxidation of the green rusts produces another magnetic mineral (δ-FeOOH). The mineral has been confirmed from Mössbauer and magnetic susceptibility studies. The iron mineral forms a strong cementing agent on drying.

The waste gypsum is produced by pressure filtration on a plate press and is in the form of a soft to firm cake with a bulk density of about 1.45 $Mg\,m^{-3}$ and a moisture content of between 30 and 35%. The gypsum crystals formed are similar in dimension to the other gypsums described, consisting almost entirely of medium and coarse silt size particles. Maximum dry densities (from standard Proctor tests) ranged from 0.80 to 1.14 $Mg\,m^{-3}$ with optimum moisture contents of 88–54% and appeared to be related to the iron content of the samples. In the former case iron was present at about 25% as Fe compared to 8–12% in the latter. The high iron content appeared to induce thixotropic properties within the cake. With excessive handling they became soft and at high moisture contents behaved as a sludge, particularly after compaction in the Proctor mould. This property was less evident in the cakes with lower iron contents. However, the advantage of the iron content is the marked increase in shear strength of the material with ageing. The increase in strength (total stress conditions) of laboratory specimens is summarized below in Table 3.

On compaction the iron-rich gypsums have low permeabilities. Constant head (triaxial) permeability tests on laboratory samples compacted to bulk densities of 1.45 $Mg\,m^{-3}$ had permeabilities in the range of 1.6 to $3.0 \times 10^{-9}\,m\,s^{-1}$. The slightly less permeable material had a higher iron content but this may not be very significant. A pure gypsum sample without the iron and compacted to similar bulk density had a permeability of $3.6 \times 10^{-7}\,m\,s^{-1}$.

Similarly to FGD and phosphogypsum, compacted iron-rich gypsums have a high compressibility and consolidate rapidly under applied loads. Primary consolidation is believed to be extremely rapid, followed by larger and more sustained secondary consolidation. Hagerty *et al.* (1977) reported that in FGD sulphite sludge 90% of primary consolidation is completed in about 10 s. Primary consolidation of titanium gypsum was extremely rapid in oedometer samples with time to t_{90} in the order of 4–16 min. Long-term consolidation or creep effects would be expected in freshly placed red gypsum although with dehydration and cementation by the iron salts this creep effect may be markedly modified or reduced.

Impact of disposal scheme

The design of the scheme had to take into account both the difficult ground conditions for construction and the peculiar properties of the waste type.

The gypsum waste with its relatively high iron content was a difficult material to handle in the wet state and had poor compaction properties. With excessive handling it became sticky and difficult to manage. However, on drying this material increased in strength to produce a stable mass which could be stored in relatively steep piles. Because of its unstable nature when wet (and the potential for producing sulphate rich leachates), it needed to be isolated and contained. The natural ground conditions at the disposal site provided an impermeable base to the site which prevented leachate

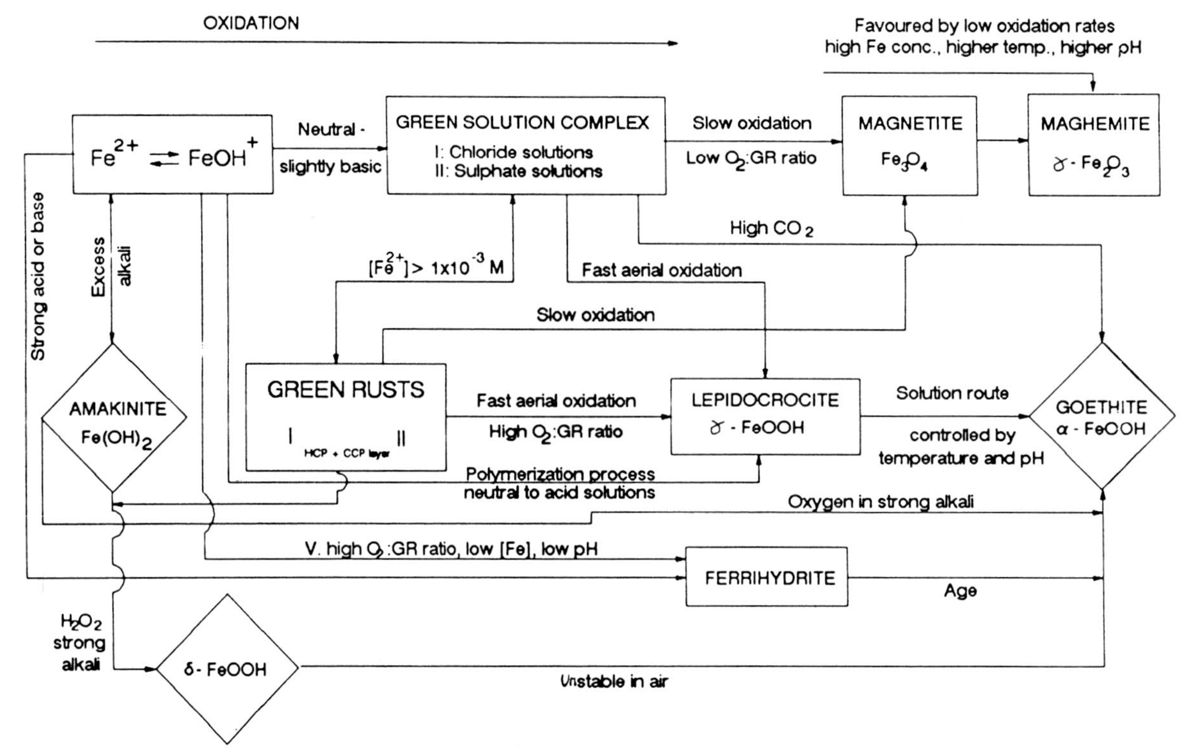

Fig. 7. Iron mineralogy.

migration. Containment bunds were successfully designed to retain the gypsum fill but due to the nature of the weak foundations only low structures could be employed and to maintain its stability restrictions had to be imposed on the height of the gypsum stack that could be placed behind the bunds. However, it is expected that, with time, long-term consolidation and strengthening of the foundation soils and the ageing effects of the gypsum will allow an extension of the stack height, increasing the capacity of the landfill area.

The installation of the full neutralization processes including the plant processes and the disposal system described above constituted about 20% of the total capital cost of the whole works. Production commenced in December 1992 and waste red gypsum is now being

Table 3. *Ageing effects on shear strength*

0% Fe		*6% Fe*		*15% Fe*		*25% Fe*	
MC %	*Shear strength (kPa)*	*MC %*	*Shear strength (kPa)*	*MC %*	*Shear strength (kPa)*	*MC %*	*Shear strength (kPa)*
34	146	15	381	58	139	96	5
16	309	1	626	28	696	62	146
8	379	0.3	909	3	1 037	28	948
		0.4	1 070	2	1 351	16	1 187
						3	1 533

Note: White gypsum (0% Fe) recompacted to same dry density at different moisture contents whereas others allowed to dry (age) from high moisture content.

produced. However, it is proposed that much of the gypsum will in the future be separated as the purer white gypsum for resale.

Acknowledgements. I would like to thank J. Graham, Project Manager of Tioxide (Malaysia) Sdn Bhd for permission to publish this paper and M. Richards, Environmental Manager of Tioxide for reviewing the draft. Thanks also go to P. Thompson, Tioxide Europe, for the supply of gypsum material for PhD research which the author has undertaken at Royal Holloway and Bedford New College under J. Mather. The bulk of the studies were carried out while the author was working for Binnie and Partners who are duly acknowledged.

References

BJERRUM, L. 1973. Problems of soil mechanics and construction on soft clays. *Proceedings of the 8th International Conference Soil Mechanics and Foundation Engineering*, Moscow, **1**, 111–159.

COX, J. B. 1968. *A review of the engineering characteristics of recent marine clays in South East Asia*. Research Report No. 6, Asian Institute of Technology, Bangkok.

DIGIOIA, A. M. & GRAY, R. E. 1979. Power plant solid waste—geotechnical aspects of disposal site selection and design. *Current Geotechnical Practice in Mine Waste Disposal, ASCE*, 113–180.

GORLE, D. 1979. Experimental embankment in phosphogypsum. *Proceeding of the Symposium on the Engineering Behaviour of Industrial and Urban Fill*, Birmingham, UK, D17–D28.

GRIFFIN, A. R., PENG, C. K. & GUDGEON, D. L. 1991. Waste retention embankments on soft clay. *Proceedings of the 6th Conference of the British Dam Society*, Nottingham, September 1990, 63–69.

HAGERTY, D. J., ULLRICH, C. R. & THACKER, B. K. 1977. Engineering properties of FGD sludges. *Proceedings of the Conference on Geotechnical Practice For Disposal* of Solid Waste Materials, ASCE, Ann Arbor, Michigan, 23–40.

KRIZEK, R. J., CHU, S. C. & ATMATZIDIS, D. K. 1987. Geotechnical properties and landfill disposal of FGD sludge. *Proceedings of Conference on Geotechnical Practice for Waste Disposal '87*, Ann Arbor, Michigan, ASCE Geotechnical Special Publication No. 13, 625–639.

MORASKY, T. M., INGRA, T. S., LARRIMORE, L. & GARLANGER, J. E. 1980. Evaluation of gypsum waste disposal by stacking. *Proceedings of the Symposium on Flue Gas Desulfurization*, Houston, Texas, 1031–1065.

VICK, S. G. 1977. Rehabilitation of a gypsum tailings embankment. *Proceedings of the Conference on Geotechnical Practice For Disposal of Solid Waste Materials*, ASCE, Ann Arbor, Michigan, 697–714.

Evaluation of geotechnical and hydrogeological properties of wastes

R. P. Beaven

Cleanaway Ltd, The Drive, Warley, Brentwood, Essex CM13 3BE, UK

Abstract. This paper uses evidence from three sources to evaluate the geotechnical and hydrogeological properties of waste. The effect of landfilling on leachate levels (through a reduction in refuse storativity) is illustrated using the results of leachate-level monitoring at two sites. The effect of landfilling on refuse permeability is shown by results from leachate pumping tests. The physical and hydrogeological properties of refuse were investigated using a large-scale compression cell. The design and operation of the compression cell are described and preliminary results from initial testing presented.

The design of landfill sites has, for a considerable time, been influenced directly by the physical and biochemical properties of wastes. For instance, the settlement of wastes has been attributed to both physical mechanisms and processes related to decomposition (Bjarngard & Edgers 1990; Morris & Woods 1990). Post-closure settlement of sites is a factor taken into account at a very early stage in the design of landfills. Planning applications for landfills within the UK routinely include 'pre-settlement' final landform contours, typically allowing for between 10% and 25% settlement. Post-closure settlement of landfills occurs principally through the long term degradation of wastes over many years or decades. Settlement or compression of waste also occurs within a landfill when layers of waste are loaded by, for example, a new lift of refuse (Edil *et al.* 1990). In this case settlement occurs by physical mechanisms over a relatively short time span of a few weeks.

Other important considerations in landfill design relating to the properties of wastes, concern the generation and control of leachate. Many wastes have the capacity to absorb an amount of water prior to the production of free-draining leachate. Water-balance calculations (summarized by Knox 1991) can be used to determine the size of a working cell within a site to control the amount of free leachate that will be produced from rainfall or other water inputs. This excess free draining leachate may then lead to the development of a zone of saturated refuse. The rate of build-up of leachate levels within the site and the amount of leachate held within the saturated zone is directly related to the effective porosity (n) or storativity (S) of the refuse. The ability to manage leachate within a site, through leachate drainage schemes and vertical pumping wells, is related to both the storativity and permeability of the wastes.

It has been demonstrated by Oweis & Khera (1990) that an increase in the dry density of waste materials occurs with depth in a landfill. This effect has been partially related to the primary mechanical compression of waste layers, as a response to increasing overburden pressure (Morris & Woods 1990, among others). Mechanical compression of wastes would result in a reduction in its void ratio and porosity. This could result in increased leachate levels and have a significant impact on the water-balance and leachate management of the site. Oweis (1986) reported evidence for a reduction in the permeability or hydraulic conductivity (K) of refuse with increasing density, and this relationship has been determined experimentally on milled urban refuse (Chen *et al.* 1977). This would imply that a decrease in K with refuse depth should be expected. Therefore, in terms of designing leachate-management schemes for landfill sites, not only is a knowledge of likely values for K and S of wastes required, but also an understanding of the manner in which they can be affected by the operations at the site over time.

Effect of landfilling on hydrogeological properties of refuse

Effect on leachate levels: empirical data

Leachate-level monitoring at an operational site indicated rises in leachate levels which could not be related to known water inputs to the site. The landfill at the time had a 6 m saturated zone and a total depth of refuse of up to 20 m. Leachate levels were monitored within perforated concrete ring 'chimneys' extended upwards from the base of the site as landfilling progressed. Leachate levels within one such monitoring point are depicted in Fig. 1. Stable leachate levels were recorded within this point over a period of 18 months from July 1984 to January 1986. Landfilling during this time was occurring in another area of the site.

Leachate levels then started to rise in early 1986, within two weeks of the resumption of landfilling in the

From BENTLEY, S. P. (ed.) *Engineering Geology of Waste Disposal*,
Geological Society Engineering Geology Special Publication No. 11, pp 57–65

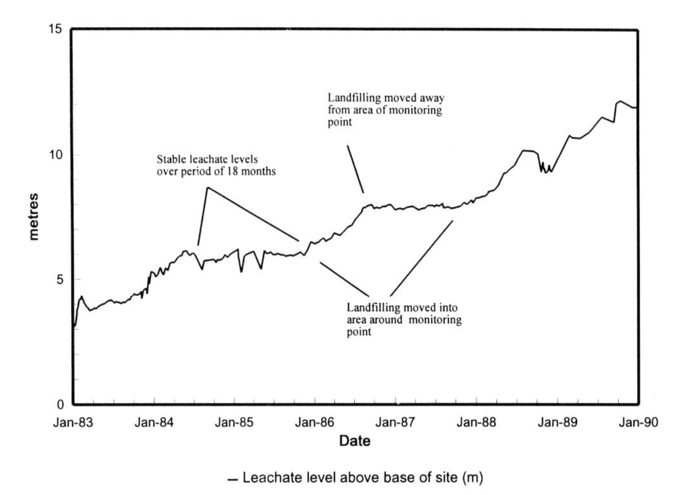

Fig. 1. Effect of landfilling on leachate levels.

area of the monitoring point. Over the following six-month period a further 4 m of refuse were emplaced and leachate levels increased by 2 m. When landfilling again moved away from the area being monitored, in June 1986, leachate levels stabilized. A further increase in leachate levels in January 1988 again correlated with the resumption of landfilling in the area.

On the basis of limited water inputs into the site, it is not thought that the rise in leachate levels was caused by an increase in the volume of leachate held within the landfill. The increases in leachate levels are considered to be a response to the surcharging of the landfill. There are two possible explanations to account for this. Firstly, the increase in levels may be caused by the consolidation of unconfined saturated refuse, with a reduction in the effective porosity and the upward displacement of leachate to saturate higher layers of wastes. Alternatively, the rise in leachate levels could reflect higher leachate heads without any increase in the depth of saturated refuse within the landfill. The presence of any horizontal confining low-permeability layers in the waste, which restricted the upward movement of leachate, could lead to an increase in recorded leachate levels through the build up of excess pore-water pressures or piezometric head. As the landfill site had received substantial quantities of clay material as intermediate cover and leachate pumping tests had indicated low sustainable yields of approximately 1 $m^3 day^{-1}$, the second option is considered more likely.

Effect on leachate levels: surcharge experiment

To investigate the above observations in more detail a controlled surcharging experiment was undertaken at another site. The landfill chosen consisted of a 24 m depth of refuse, with a 13 m saturated zone founded on low-permeability natural clay deposits. Two piezometers were installed into the refuse, one just into the leachate table and the other near the base of the site (Fig. 2). Over a period of two days a 3.5 m deep clay stockpile was emplaced symmetrically around the piezometers over an area of 6400 m^2 (80 m × 80 m). Leachate levels within the piezometers were monitored using pressure transducers and a data logger over a period of 10 months (Fig. 3). Settlement of both the stockpile and piezometers was also monitored intermittently during this time.

Leachate levels in the lower piezometer prior to the start of the experiment were approximately 0.5 m above the level in the upper piezometer, indicating the presence of vertical hydraulic gradients within the refuse.

Leachate levels within both piezometers reacted within minutes of the start of earthworks. Over the following two days levels in the lower piezometer increased by 1 m, whereas an increase of only 0.25 m occurred in the upper piezometer. The increase of 1 m in the lower piezometer may be indicative of a build-up of excess pore-water pressure within confined refuse at the base of the site. The increase of 0.25 m in the upper piezometer near the 'water' table may reflect an actual physical increase in the depth of saturated refuse, through a reduction in effective porosity. The rate of rise during these two days decreased when the earthworks stopped during the intervening night.

Leachate levels continued to rise within the lower piezometer for a period of 14 days to a peak of between 2 and 2.5 m above the original levels. This corresponded with the highest rate of settlement of 0.2 m in the first two weeks. Thereafter, barometric fluctuations of up to 0.3–0.4 m are superimposed on a slow 0.2–0.3 m decline in levels over the next 250 days.

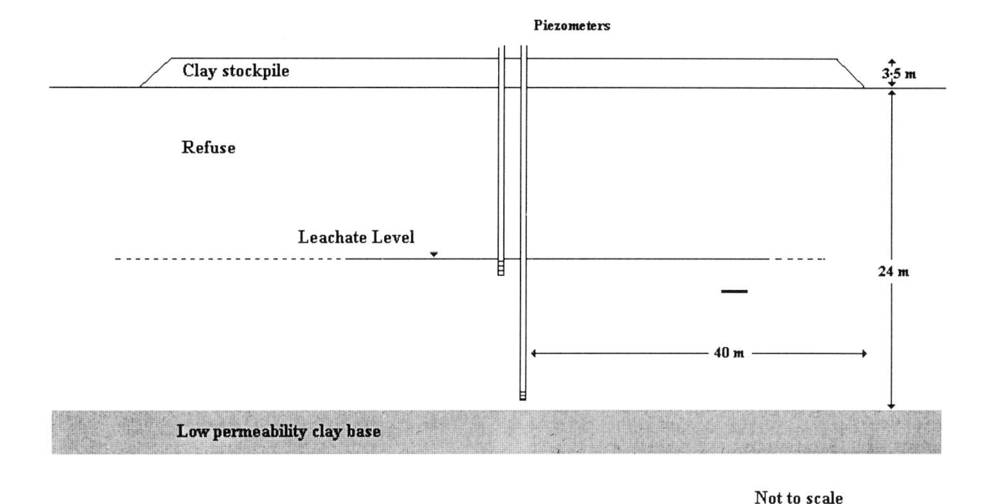

Fig. 2. Schematic cross-section through clay surcharge experiment.

Leachate levels in the upper piezometer continued to increase after the initial two-week period to a value of 1.3 m above the original levels after a period of 100 days. This increase is believed to have been caused by winter rainfall and was seen in other leachate monitoring points in the area.

The maintenance of a 0.75 m head difference between the upper and lower piezometers over a ten-month period indicates the presence of significant barriers to vertical flow within the refuse. The presence of clay soil materials used as intermediate cover materials is one likely cause.

Effect on permeability and storativity: leachate pumping tests

Leachate pumping tests were undertaken on another landfill to determine the hydrogeological properties of the refuse. The site consisted of a 9 m depth of refuse with a 5–6 m saturated zone. A well field was constructed with five piezometers situated at distances of between 5 and 75 m around a pumping well.

A pumping test was conducted for a period of five days at a pumping rate of $70\,m^3\,day^{-1}$ ($0.8\,l\,s^{-1}$) until drawdown in the pumping well reached the inlet to the

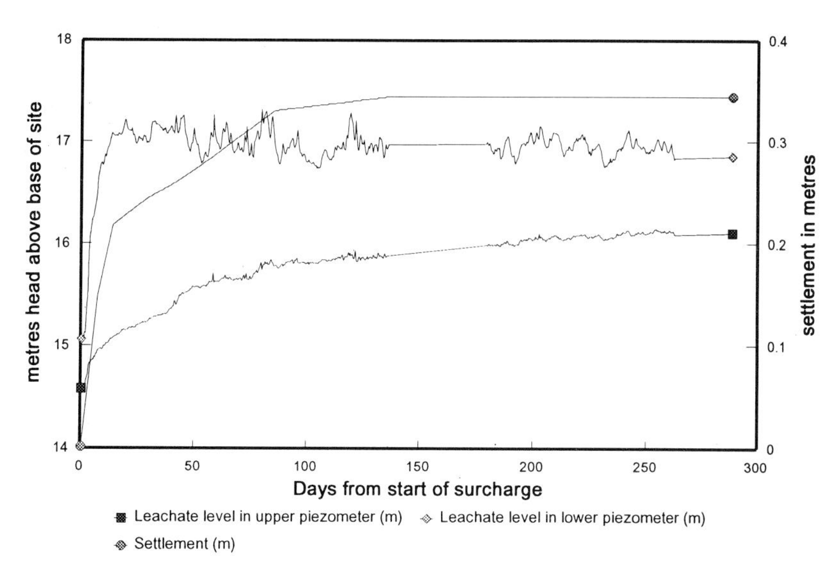

Fig. 3. Leachate level fluctuations and settlement during clay surcharge experiment.

pump. At this stage drawdowns in observation piezometers at 5, 25 and 50 m radii were approximately 1.0, 0.28 and 0.18 m respectively. Data from the piezometer at 5 m radius were analysed using a Boulton water-table type curve solution after applying Dupuits correction for large drawdowns relative to initial saturated thickness.

Both early and late data were matched satisfactorily to provide the following hydrogeological parameters:

Transmissivity	$T = 50\,m^2\,day^{-1}$
Hydraulic conductivity	$K = 9\,m\,day^{-1}$ $(1 \times 10^{-4}\,m\,s^{-1})$
Specific yield	$S_y = 0.04$
Storativity	$S_a = 6 \times 10^{-3}$

The pumping test was repeated nine years later when landfilling had increased the depth of refuse to 23 m and the saturated thickness to 6–7 m. Much of the original well field had to be redrilled, including the pumping well which was located within 5 m of the original. Step drawdown tests indicated that the pumping rate of $70\,m^3\,day^{-1}$ used in the original test far exceeded the sustainable yield of the well and a rate of $9.6\,m^3\,day^{-1}$ $(0.11\,l\,s^{-1})$ was used in the pumping test. Excessive drawdown in the pumping well caused the test to stop after 12 days, at which point drawdown in a piezometer at 25 m radius was 0.6 m. Analysis of the drawdown data from the various observation piezometers did not conform particularly well to water-table type curve solutions. Possibly reasons include the potential for high components of vertical flow, short circuiting and preferential flow paths in the vicinity of the pumping well. These factors were possibly reflected by high values of calculated storativity for observation piezometers close to the pumping well. However, comparison with a standard Theis type curve indicated the following average hydrogeological parameters:

Transmissivity	$T = 5\,m^2\,day^{-1}$
Hydraulic conductivity	$K = 0.7\,m\,day^{-1}$ $(8 \times 10^{-6}\,m\,s^{-1})$
Storativity	$S = 0.07$ (range 0.005–0.15)

The effect of the additional 14 m of waste on the hydrogeological properties of the refuse was to create an order of magnitude reduction in the hydraulic conductivity of the refuse. The effect on the storativity is not clear, as a reduction predicted from the results of the leachate-level data presented previously is not apparent.

Pitsea compression cell

Field evidence of the processes described above led Cleanaway Ltd to propose a research programme to investigate changes to the hydrogeological and geotechnical properties of refuse, when surcharged with loads up to an equivalent 50 m depth of landfill. The research is being jointly funded by the Waste Technical Division of the Department of the Environment and Cleanaway Ltd. The research has three main objectives:

- to investigate the effect of applied stress (depth of landfill) on refuse settlement and *in situ* refuse densities;
- to investigate the impact of applied stress and *in situ* refuse densities on absorptive capacity and effective porosity;
- to investigate the impact of applied stress and *in situ* refuse densities on hydraulic conductivity.

Experimental design

To achieve the above objectives it was concluded that a purpose built test cell would be required capable of exerting considerable vertical loads to simulate a 50 m depth of refuse. The following design criteria for the test cell were established:

- the cell should be of a size to accommodate the heterogeneous nature of the materials to be tested;
- the cell should allow the application of loads to the material undergoing testing to simulate a minimum 50 m depth of refuse, and be able to record compression and density variations;
- the cell should allow for the determination of the absorptive capacity and porosity of the material being tested at varying applied loads;
- the cell should allow for the determination of the permeability of the material being tested;
- the filling and discharging of the cell should be relatively straightforward.

Test cell size

The initial consideration, which controlled the nature of many of the other design features, was the appropriate size of the test cell. With the wide range of materials and material sizes that occur in refuse, a cylinder diameter of 2 m was chosen as a reasonable compromise between size and practicality of construction.

Standard oedometers used in soil mechanics testing laboratories have a height to diameter ratio of approximately 1 : 4 to prevent side-wall friction effects becoming significant within consolidation tests. To have maintained this ratio within the design would have led to a cell height of 0.5 m. With an emphasis in the experiments on determining the hydrogeological properties of the refuse, this height was considered inadequate, particularly for the successful operation of permeability constant head flow tests. Therefore, the height of the cylinder was increased to 3 m. Operational constraints meant that the maximum refuse depth within the cell would be 2.5 m, but even with the large amounts of compression anticipated, a minimum depth of refuse at the end of the tests of between 1 and 1.5 m should be maintained. It was recognized that these dimensions were a compromise. However, to satisfy every requirement would have required further increases in the diameter and height of the cylinder and would have been prohibitively expensive.

Test cylinder and supporting framework

The testing cell consists of a steel cylinder, 2 m in diameter and 3 m in height, suspended vertically within a steel support frame (Fig. 4). The base of the cylinder is sealed by a lower platen which is seated on an 'o' ring seal to create a watertight system. The lower platen is connected to a hydraulically operated telescopic piston cylinder, which allows the platen to be pushed upwards through the cylinder for the ejection of refuse. An upper platen, used to apply the simulate refuse loads, is again controlled by two hydraulic pistons and has a stroke which takes it from 0.5 m above the top of the cylinder to 2 m inside.

With the upper platen fully raised, the test cylinder can be rotated on a central pivot point within its supporting frame to a horizontal orientation. This is to facilitate the emplacement of refuse into the cylinder (45° to the vertical) and the discharge of refuse using the lower telescopic piston when horizontal. The whole of the test cell is supported on 'feet' at the four corners of the support frame. Load cells are situated under two of these feet, with pivot points under the other two, to allow the determination of the total weight of the structure at any time.

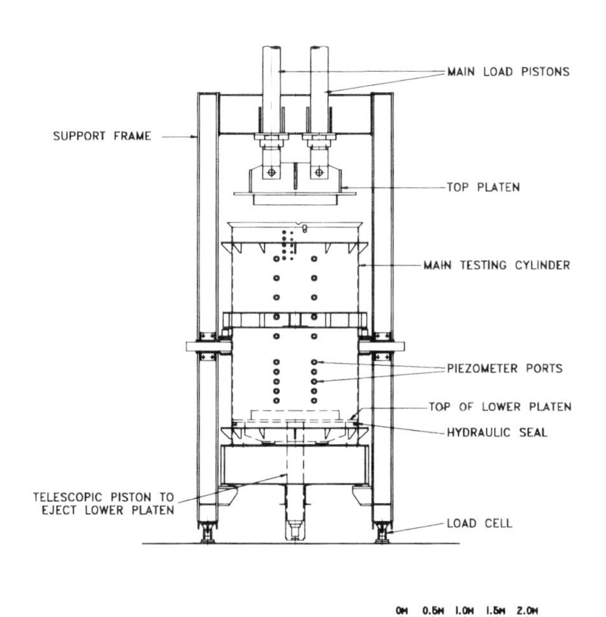

Fig. 4. Schematic section through large-scale compression cell.

Hydraulic system

A hydraulic system is used to control all movements of the cylinder and upper and lower platen. The hydraulic system consists of two hydraulic pumps, a control panel which operates a bank of solenoid valves that direct the oil to the appropriate pistons, and a 1000 litre reservoir tank for the oil. There are four sets of pistons on the test cell:

- two 200 mm diameter pistons to raise and lower the upper platen (used to apply the vertical stress to the refuse);
- one telescopic piston on the lower platen (to eject the refuse from the cell after testing);
- one piston for tilting the cylinder from vertical to horizontal;
- four small 'jack up' pistons which raise the cylinder prior to rotation.

Two hydraulic pumps are required for different operations. The 'main' pump is a high-flow pump used to operate the main movements of the platens and rotate pistons.

The 'load' pump is a low-flow pump used to simulate long-term loading of the refuse through the upper platen. A pressure-relief valve controls the pressure within the hydraulic circuit between 10 and 190 bar when the load pump is on. This equates to an applied load or stress on the refuse of between 9 and 600 $kN\,m^{-2}$.

Water flow and measurement system

Two 450-litre header tanks are situated on a scaffold tower adjacent to the compression cell to allow refuse to be saturated with water for porosity, absorptive capacity and permeability measurements. The tanks may be positioned up to 3 m above the top of the refuse within the cell.

Water is pumped into the header tanks from a 7000 litre reservoir tank. Water from the header tanks feeds upwards into the test column through two 'rings' of six 25-mm diameter ports in the lower platen. A similar arrangement of ports in the upper platen is used to carry the flow of leachate back to the reservoir tank during constant head tests. Electromagnetic flow recorders and totalizers are connected to both the inlet and outlet pipework.

There are two vertical lines of 18 piezometer ports up the side of the column with spacings between 150 and 400 mm. Piezometers are inserted horizontally to the desired depth into the refuse, by installing 6 mm nylon tubing through cable glands fitted in the piezometer ports. The piezometer ports are also used to measure differential compression at varying depths of refuse. Known lengths of cord inserted into the refuse through the ports allows differential compression to be measured at various heights within the refuse.

Table 1. *Size and category analysis of refuse*

Size (mm)	*Wt %*	*Pa/Cd %*	*PIF %*	*DP %*	*Tx %*	*Mc %*	*Mnc %*	*Gl %*	*Put %*	*Fe %*	*nFe %*	*<10 %*
+160	27.6	74.1	10.3	2.7	6.5	3.6	—	—	1.3	1.5	—	—
160–80	29.1	25.2	6.5	8.9	1.7	22.6	1.1	8.7	8.3	15.4	1.6	—
80–40	17.6	30.9	5.1	4.6	1.7	6.8	2.6	6.1	26.5	10.8	4.9	—
40–20	11.4	14.1	0.8	1.9	—	1.4	1.3	8.6	69.7	1.4	0.9	—
20–10	8.1	3.4	0.3	0.5	—	1.5	1.4	11.6	81.1	—	0.3	—
<10	6.2	—	—	—	—	—	—	—	—	—	—	100.0
Total Wt%	100.0	35.1	5.7	4.4	2.6	9.1	1.0	5.5	22.0	7.0	1.5	6.2

Moisture content of refuse MC = 34%.
Key: Pa/Cd, paper and card; PIF, plastic film; DP, dense plastics; Tx, textiles; Mc, miscellaneous combustibles; Mnc, misc' non-combustibles; Gl, glass; Put, putrescibles; Fe, ferrous metal; nFe, non-ferrous metal; <10, material <10 mm in size.

Compression tests on domestic refuse

Domestic refuse was tested within the cell as follows:

(1) A known weight of refuse, at its original moisture content, was loaded into the cell and then subjected to stages of increasing vertical stress. Compression of the refuse and changes in bulk density were monitored.

(2) The refuse was ejected from the cell and then reloaded to a similar starting density as above. The refuse was saturated and then drained to field capacity to determine its initial absorptive capacity and porosity. Rising head and constant head permeability tests were also undertaken. The refuse was then subjected to stages of increasing vertical stress, with further hydraulic tests being undertaken after each stage.

The refuse for the test was obtained directly from domestic refuse disposal vehicles as they discharged their loads at the site. Approximately 10 t of refuse was gathered by sampling a portion of the load from eight vehicles using a lorry with a hydraulic grab. The refuse

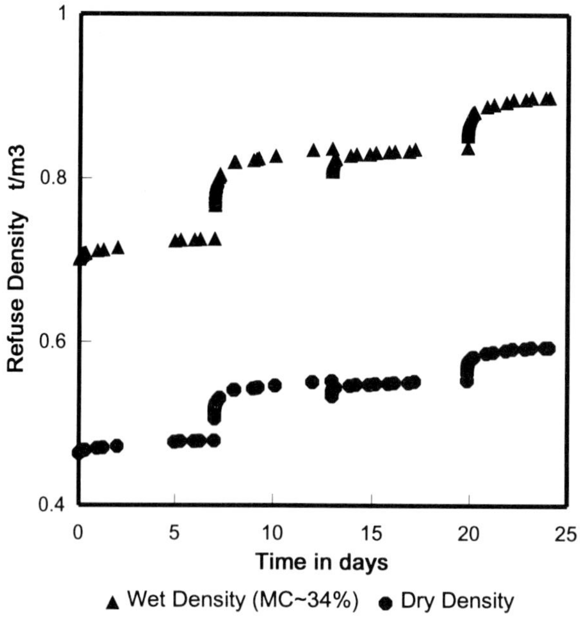

Fig. 5. Changes in refuse density in response to varying applied stress.

was placed on an area of hardstanding and samples taken from different positions to provide a more representative sample for testing. One subsample of approximately 4.5 t was taken for loading into the compression cell and another subsample of 3.5 t was sent to Warren Spring Laboratories for material classification. Table 1 provides a breakdown of the refuse constituents and size distribution. The moisture content (MC) of the refuse was determined at 34% (dried to constant weight at 105 °C), allowing the dry weight of the refuse to be calculated.

The refuse was loaded in stages into the compression cell to an approximate wet density (i.e. at initial moisture content) of 0.7 $t\,m^{-3}$.

The refuse in the cell was then subjected to a series of vertical loads applied from the top through the upper platen. Compression was measured against time for a given applied load, with measurements continuing until further compression was negligible (seven to ten days), whereupon the load would be increased. There were three increments of increasing load with one recovery phase. The magnitude of the load was controlled by a pressure-release valve on the hydraulic system.

The initial compression stage involved an applied load or stress of 40 $kN\,m^{-2}$ followed by stages of 165 $kN\,m^{-2}$, recovery (0 $kN\,m^{-2}$), 165 $kN\,m^{-2}$ and 322 $kN\,m^{-2}$. Based on an average bulk density of landfill of 10 $kN\,m^{-3}$, then the above stresses are equivalent to approximately 4, 16 and 32 m depths of landfill. The change in dry and wet weight densities during the various stages are illustrated in Fig. 5. At the completion of the test the *in situ* wet density of the refuse had reached 0.9 $t\,m^{-3}$. The corresponding *in situ* dry density was 0.59 $t\,m^{-3}$.

The refuse was ejected into a skip on completion of the tests described above, where it 'rebounded' to a loose state. Just over 4 t of refuse was then reloaded back into the test cell to a density of 0.54 $t\,m^{-3}$ after an initial 150 mm thick layer of gravel was placed on the bottom platen. Three oil-filled hydraulic pressure cells were also installed in sand pockets at various depths within the refuse to measure *in situ* vertical stress during the subsequent experiments. However, interpretation of the results derived from these cells has proved difficult. The recorded stress within the column drops off with depth from a value that is double the theoretically applied stress near the top of the refuse column, to 50% near the base of the refuse. Some reduction in vertical stress with depth would be expected due to the effects of side-wall friction. The over-reading of the cell near the top of the refuse is thought to relate to factors associated with the stiffness of the cell and sand pocket relative to the surrounding refuse. These results are being investigated further.

With an initial *in situ* density of 0.54 $t\,m^{-3}$, the refuse within the cell was saturated and then drained to field capacity to determine its absorptive capacity and effective porosity. The following results were obtained:

Full absorptive capacity of refuse	393 $l\,t^{-1}$ of refuse at original MC
Revised wet density of refuse	0.75 $t\,m^{-3}$
Effective porosity	28%
Void ratio	2.12
Air void ratio	0.39

The increase in wet density from 0.54 to 0.75 $t\,m^{-3}$ is solely due to the increase in the moisture content of the refuse due to uptake of its absorptive capacity.

Both rising head and constant head permeability tests were completed although problems with leakage from a seal on the lower platen meant that the calculated range of hydraulic conductivities of 10 to 50 $m\,day^{-1}$ is only approximate.

The refuse was fully drained before an applied stress of 40 $kN\,m^{-2}$ was exerted through the top platen. Compression of the refuse was monitored and then further saturation and permeability tests were undertaken while the applied load was maintained. Four further stages of increasing applied stress and saturation testing were undertaken, although progressive failure of

Table 2. *Summary of compression of refuse at field capacity*

Stage no.	*Applied stress ($kN\,m^{-2}$)*	*Dry denisty ($t\,m^{-3}$)*	*Wet density ($t\,m^{-3}$)*	*Volume of leachate drained in stage (litres)*	*Refuse wet wt (approx.) (kg)*	*Absorbtion capacity ($l\,t^{-1}$)*	*Absorbtion capacity w/w(dry) (%)*	*Effective porosity n (%)*
0		0.36	0.54					
Saturation		0.36	0.75		5645	397	37	28
1	40	0.43	0.89	131	5514	365	35	6–18
2	87	0.51	1.06	N/A	~5500	361	35	
3	165	0.56	1.15	N/A	~5500	361	35	
4	322	0.65	1.27	296	5204	288	30	
5	603	0.73	1.36	207	4997	237	26	

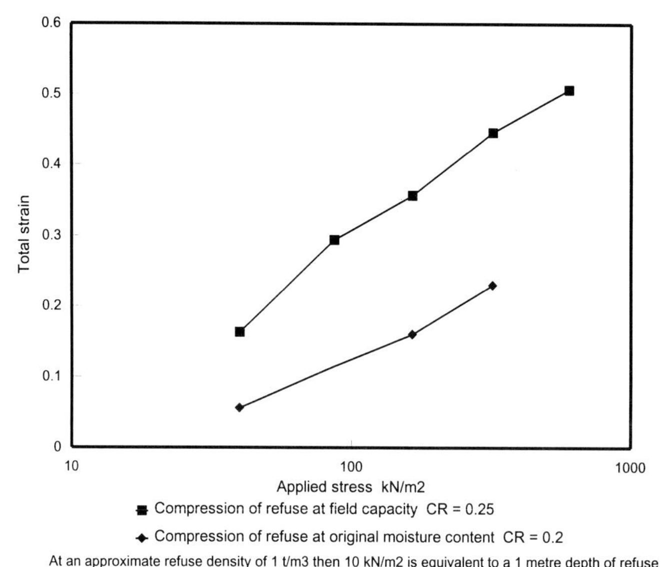

Fig. 6. The relationship between stress and strain for refuse at different moisture contents.

the seal on the lower platen meant that limited hydrogeological data were obtained. The results are summarized in Table 2.

During each compression stage, leachate was released from the column. However, whether this represents a continued reduction in the moisture content of the refuse is not clear as it was not possible to determine whether any further water was re-absorbed during the subsequent saturation tests.

The modulus of compression CR as applied to refuse is defined as the amount of strain (settlement) over 1 log cycle of applied stress. Figure 6 is a plot of total settlement or strain of the refuse from both sets of experiments against log stress. The calculated CR values are 0.2 for refuse at its initial moisture content and 0.3 for refuse at field capacity. These values are at the top of the range of values from 0.02 to 0.25 reported by Bjarngard & Edgers (1990). These values can be used to predict the amount of settlement or compression that will occur within a layer of refuse as a consequence to the emplacement of further lifts of refuse. An estimate of the initial stress conditions of the layer of waste is required, which would be related to compaction effort and *in situ* density. For instance, if a layer of refuse was emplaced at a low *in situ* density ($0.55\,t\,m^{-3}$, Table 2) equivalent to an applied stress of perhaps $10\,kN\,m^{-2}$, then that layer of waste would compress by approximately 25% in response to an additional 10 m of landfill (at $1\,t\,m^{-3}$), and by approximately 40% in response to a total additional 50 m depth of landfill. This corresponds to a CR of 0.25.

Conclusions

Assessment of the results from field tests and the preliminary results from the large-scale compression cell lead to the following conclusions:

- Considerable settlement or compression of wastes can occur as a result of surcharging existing wastes with new layers of refuse. A compression ratio of between 0.2 and 0.25 was determined within the compression cell for crude domestic refuse.
- The wet weight density of refuse can alter within a landfill, without any further compression or compaction, purely through a change in the moisture content or amount of water held as 'absorptive capacity'.
- Vertical hydraulic gradients can exist within landfill sites and when sustained for long periods of time are indicative of confining horizons (e.g. intermediate clay cover) within the waste.
- Increasing the overburden pressure or effective stress, by further landfilling, on saturated layers of refuse causes an increase in leachate pore pressure and leachate head by a reduction in storativity or void ratio. An increase in leachate head of 2 m was measured when a 3.5 m deep clay stockpile was emplaced on the surface of a 24 m depth of refuse with a 13 m saturated zone.
- A reduction in the hydraulic conductivity of refuse, by an order of magnitude, has been demonstrated as a result of increasing the depth of landfill from 9 to 23 m.

The opinions expressed herein are those of the author and do not necessarily reflect those of Cleanaway Ltd or the Department of the Environment.

References

BJARNGARD, A. & EDGERS, L. 1990. Settlement of municipal solid waste landfills. *In*: *Thirteenth Annual Madison Waste Conference*, 19–20 September 1990.

CHEN, W. H., ZIMMERMAN, R. E. & FRANKLIN, A. G. 1977. *Proceedings of the Conference on Geotechnical Practice for Disposal of Solid Waste Materials*. University of Michigan, ASCE.

EDIL, T. B., RANGUETTE, V. J. & WUELLNER, W. W. 1990 Settlement of Municipal Refuse *In*: LANDVA, A. & KNOWLES, G. D. (eds) *Geotechnics of Waste Fills*. ASTM STP 1070, 225–239.

KNOX, K. 1991. *Water balance methods and their application to landfill in the UK*. DOE Waste Technical Division Research Report No. CWM 031/91.

MORRIS, D. V. & WOODS, C. E. 1990. Settlement and engineering considerations in landfill and final cover design. *In*: LANDVA, A. & KNOWLES, G. D. (eds) *Geotechnics of Waste Fills*. ASTM STP 1070, 9–21.

OWEIS, I. S. 1986. Criteria for Geotechnical Construction of Sanitary Landfills. *In*: FANG, H. Y. (ed.) *International Symposium on Environmental Geotechnology*.

—— & KHERA, R. J. 1990. *Geotechnology of waste management*. Butterworth, Guildford.

Avian botulism associated with a waste disposal site

B. J. Lloyd

Centre for Environmental Health & Water Engineering, Department of Civil Engineering, University of Surrey, Guildford, Surrey GU2 5XH, UK

Abstract. An epidemic of botulism in gulls was reported in 1989 near a waste disposal site at Stewartby in Bedfordshire, England. Subsequently, it was demonstrated that all samples taken from the leachate treatment plain of the waste disposal site were contaminated with botulinum toxin, whereas control samples from the Stewartby lake edge, where gulls were seen dying, were negative. A causal relationship between botulism in the gulls and the waste disposal site was indicated. It is suggested that the organically enriched, anaerobic mud flats from which the toxins were isolated require ploughing in order to manage and reduce the risk of toxaemia arising from such sources.

Although rare in the human population, botulism in wild populations of animals, including birds, sometimes reaches epidemic levels. In 1970, five outbreaks of botulism in waterfowl occurred in the Netherlands (Haagsma 1974; Haagsma & Ter Laak 1979), and similar outbreaks were reported in the UK from 1969 onwards (Roberts 1977). Since then avian botulism has occurred every year in Holland, caused almost exclusively by type C toxin. Tens of thousands of birds have died, particularly during warmer summer months when temperatures exceed 20 °C.

Botulinum toxin is one of the most potent naturally occurring neurotoxins. As an example, 1 g is sufficient to kill 30×10^9 mice, which are 20 times more sensitive than pigeons to type C toxin (Prévot & Brygoo 1953). At least six immunologically distinct toxins (designated A–F) are produced by the bacterium *Clostridum botulinum*. This is a spore-forming Gram-positive staining bacillus which is only capable of growing under strictly anaerobic conditions. Whereas the bacterial endospore may survive for prolonged periods in the natural environment, it may be consumed, for example via fresh vegetable, and pass through the anaerobic environment of the gut without producing the toxin. Consequently, consumption of *Cl. botulinum* cells does not result in botulism. Human botulism is almost invariably associated with the consumption of contaminated and inadequately heated foodstuffs. It may also occur in canned or bottled foods in which bacteria have grown and released toxins after processing.

Botulism is not an infection, it is an intoxication. The preformed toxin must be ingested to produce toxaemia and the distribution and conditions under which sufficient toxin is produced in the natural environment are thus of considerable importance for wild and domestic animals. Soil, water and the intestinal tracts of animals are the natural habitats of *Clostridium botulinum*. Haagsma (1974), however, reported that no toxin was identified in mud, water or decomposing vegetation collected in places with a high mortality among waterfowl.

Smith *et al.* (1978) have made a detailed study of the distribution of *Cl. botulinum* in soils and report that in the UK, for example, type B toxin-producing strains may be isolated with a 5% frequency in soil samples and a 33% frequency in mud samples from the aquatic environment. Type C is much rarer and Smith (1987) reports that it occurred in only 3% of 554 mud samples in the UK.

Waste disposal sites which receive domestic refuse are characterized by populations of many thousands of gulls. These scavengers are notorious for their carriage of microbial pathogens such as species of the genus *Salmonella*. Whilst gulls are normally considered to be hardy birds and resistant to many infections, they sometimes die in epidemics associated with other agents such as the toxins of *Clostridium botulinum*.

The Stewartby L-field refuse disposal site

The roosting population of seagulls on the Stewartby & Brogborough lakes is estimated at 20 000–25 000 by the local park rangers. The Royal Society for the Protection of Birds (RSPB) first noticed seagulls dying in the spring of 1989, and this was estimated at a rate of about one death per day on the lakes, rising during the summer. Deaths thus continued through the summer and on 6 September 1989 the Assistant Pollution Officer for North Bedfordshire collected specimens, including lake water and soil specimens from the local L-field waste disposal site, where the gulls congregated for feeding on domestic refuse carted from London. The specimens were submitted to the PHLS Anaerobic (Bacteriology) Reference Unit at

From BENTLEY, S. P. (ed.) *Engineering Geology of Waste Disposal*,
Geological Society Engineering Geology Special Publication No. 11, pp 67–70

Luton, but the laboratory failed to isolate the pathogen using enrichment methods and only succeeded in isolating *Clostridium sporogenes*.

By the end of the long, hot summer several hundred water birds, principally seagulls, had died following continued observations of the classical avian botulism 'limber neck' symptoms. In November the author was invited to investigate the origins of the suspected botulism and visited the Stewartby lakes and waste disposal site on 15 December.

The L-field waste disposal site (Fig. 1) is skirted by a railway line whose embankment makes up the eastern boundary of the site and from which a spur railway line brings in refuse from London. The disposal site is characterized by two principal and distinct zones.

The upper refuse disposal zone may be subdivided into two areas: areas A and B (Fig.1). Area A is where dumping has ceased and the refuse has been covered with soil. In this soil-covered area, the organic refuse is undergoing anaerobic digestion and the methane produced is vented to the atmosphere and burnt off. Area B is where refuse is being continuously dumped and this area attracts the large actively feeding population of gulls.

Anaerobic liquor drains by gravity from under the upper zone to the lower zone which is composed of a mixture of ponds (area C) and mud-flats (area D) which are used as a leachate treatment plain. At the time of the site visit, the treatment plain had a shallow (5–20 cm) surface layer of liquid over about 20% of the area, partly as a result of the recent heavy rain but also due to the use of rain guns to spray the anaerobic liquor onto the leachate treatment plain. The gull populations were seen to move from the upper feeding zone to the lower treatment plain where they preened and also appeared to drink.

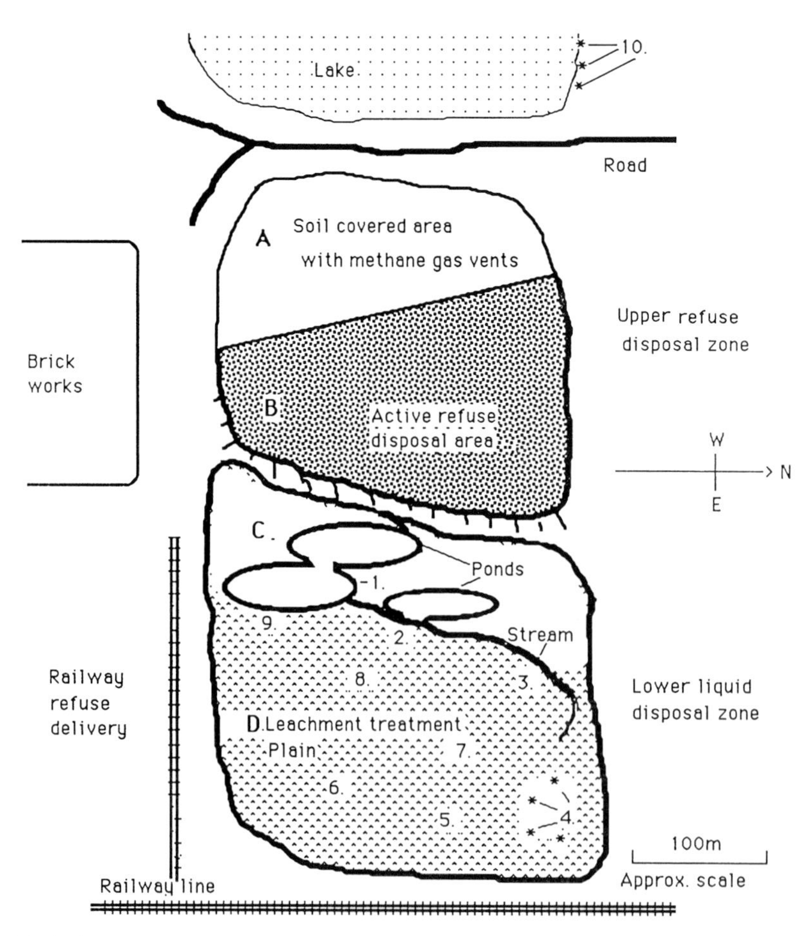

Fig. 1. Sketch map of L-field disposal site showing soil sample locations: A, B, C, D, Main areas of the waste disposal site; 1–9, disposal site locations in which soil was sampled for botulinum toxins; 10, lake site locations sampled for botulinum toxins.

Objective

An investigation of the L-field waste disposal site was undertaken in order to establish whether superficial anaerobic soil samples contained detectable botulinum toxins. The presence of toxins above the frequency normally encountered in soils might indicate that the source of botulism seen in gulls on the Stewartby & Brogborough lakes was the waste disposal site. If this proved to be the case, more intensive studies would be necessary and would indicate whether remedial action was required.

Sampling and identification of botulinum toxins

The leachate treatment plains were inevitably contaminated with substantial quantities of bird droppings. This organic enrichment of very shallow lagoons, together with the anaerobic liquor, seemed to create the most suitable environment for the continued survival and proliferation of anaerobic bacteria including the formation of *Clostridium botulinum* toxin. Soil samples were therefore collected from the leachate treatment plains where the soil showed signs of anaerobic conditions—typically saturated mud at the edge of areas of standing water.

The samples were taken by hand using inverted plastic bags which were plunged into up to 10 cm of blackened, sulphide-rich soil to remove samples of about 1 kg each, completely filling the bags. The sample bags were sealed to exclude air, double-bagged and labelled with a note of the location. The mud samples were transferred the same day to the laboratory of the Institute of Zoology of the Zoological Society of London, where samples were stored at −20 °C until they could be examined. The botulinum toxins were eluted from the soil samples in phosphate buffer at pH 7.0. The toxins were identified by neutralization and toxicity tests in mice by the methods described by Smith & Moryson (1975).

Results

The results are illustrated in Table 1.

Table 1. *Botulinum toxin types identified and sample location*

Sample no.	*Botulinum toxin type*	*Sample location*
1	D	At edge of ponds between upper deposition zone and lower treatment plain
2	B	Edge of treatment plain
3	D	Edge of stream running onto treatment plain away from the rain guns
4	C	Composite sample from treatment plain
5	C	Treatment plain (Area D of treatment)
6	D	Treatment plain (Area D)
7	C	Treatment plain (Area D)
8	D	Treatment plain (Area D)
9	B	Edge of pond adjacent to treatment plain
10	Negative	Composite sample from Stewart by lake edge

Area D: treatment plain area which receives anaerobic liquor from 'rain' guns.

Discussion

The anaerobic Reference Unit of the Luton PHLS failed to recover *Cl. botulinum* in six out of six (100%) samples of the Stewartby lake water, where gulls were seen dying from botulism. By contrast, subsequent analyses at the laboratories of the Institute of Zoology, London, showed that in nine out of nine (100%) soil samples collected during this investigation from the lower zone of the waste disposal site, botulinum toxins were present.

No toxins were isolated from a composite soil sample from the Stewartby lake shore, sampled on the same day as the waste disposal site.

Three types of botulinum toxin (B, C and D) were identified from the L-field waste site soil/mud samples. Type B, which is frequently implicated in human botulism, was the least frequent type, being identified in 22% of samples (2/9). Types A, B and E are the commonest causes of human botulism; types A and E were not isolated in this investigation.

Type C, which is the commonest cause of avian botulism, was recovered from 33% (3/9) of the waste site samples. Type C occurred at ten times the frequency encountered by Smith *et al.* (1978) in a survey of 554 mud samples from various parts of the country. However, he reported that Type C occurs plentifully at specific aquatic sites (e.g. the Mersey estuary), often in association with suspected or confirmed avian botulism. In a current study of 19 waste sites, Smith (pers. comm.) has encountered a 60% isolation rate of type C!

Type D toxin was the commonest type isolated (in 4/9 samples). This is uncommon in the Northern Hemisphere, where it normally occurs in a very low percentage of soil samples, and it was therefore surprising to find it in 44% of the waste site samples.

Conclusions

Botulinum toxin types C and D are notorious causes of botulism in animals and in this investigation there is strong circumstantial evidence to link the presence of type C, isolated at the waste disposal site, with the deaths of the gulls reported on the nearby lake.

Epidemic disease is characterized by the coincidence of a number of favourable environmental factors. The principal waste site factors on the leachate plain which could have created a suitable environment for the proliferation of *Cl. botulinum*, toxin release and subsequent death of wildlife are:

- oxygen-free conditions created in very superficial soil layers, provided by organically rich, sulpide-rich, high-BOD, anaerobic leachate liquors;
- enrichment and/or inoculation of clostridia by excreta from the high density of wildlife congregating on the site with additional nutrient inputs;
- persistent, hot (>20 °C) summer conditions, further lowering the amount of oxygen which can remain in solution in water, permitting growth of clostridia;
- shallow, standing water for drinking and hence ingesting toxin by waterfowl.

Consideration should be given to methods of creating an unfavourable environment for clostridia to grow and release toxin on waste disposal drainage areas. Three obvious improvements would include:

- efficient aeration of the anaerobic liquors before discharge to the site;
- periodically ploughing the mud-flats; and
- improving the drainage of the sedimentation plains.

It is considered that all of these recommended improvements would be constructive in preventing anaerobiosis.

Acknowledgements. The author is grateful to Dr Geoffrey Smith (Zoological Society of London) for carrying out the analysis of the soil samples and background information, and to Mr Steven Battersby for coordinating the site visit.

References

HAAGSMA, J. 1974. Ethiology and epidemiology of botulism in water-fowl in the Netherlands. *Tijddschrift voor Diegeneeskunde*, **99**, 434.

—— & TER LAAK, E. A. 1979. Avian botulism in the Netherlands. *Tijddschrift voor Diegeneeskunde*, **104**, 609.

PRÉVOT A. R. & BRYGOO, E. R. 1953. Nouvelles recherches sur le botulisme et ses cinq types toxiniques. *Annales de l'Institut Pasteur (Paris)*, **85**, 544–575.

ROBERTS, T. A. 1977. Unpublished report on *Clostridium botulinum* for the International Course in Food Microbiology, University of Surrey, July 1977.

SMITH, G. R. 1987. *In:* EKLUND, M. W. & DOWELL, V. R. (eds) *Avian Botulism. An international perspective.* Springfield Charles C Thomas, 73.

—— & MORYSON, C. J. 1975. *Clostridium botulinum* in the lakes and waterways of London. *J.Hyg. Camb.*, **75**, 371–379.

——, MILLIGAN R. A. & MORYSON, C. J. 1978. *Clostridium botulinum* in aquatic environments in Great Britain and Ireland. *J. Hyg. Camb.*, **80**, 431–438.

Remidial containment measures for a peat bog landfill site

K. McShane & B. J. Gregory

Kirk McClure Morton, Elmwood House, 74 Boucher Road, Belfast BT12 6RZ, UK

Abstract. Many landfill sites in Northern Ireland have been formed by infilling waste in low-lying bog areas, without any containment or control measures being incorporated. Typical of this is a working landfill site at Drumlough Moss in County Down, which has an areal extent of approximately 20 ha. It has been in operation for over 30 years and is situated on a peat bog, known locally as a 'moss', in an interdrumlin hollow.

Pollution of the nearby River Lagan, which is the major river flowing through the city of Belfast, was traced to the site. Consequently remedial measures had to be designed and constructed to provide effective containment so as to minimize future risks associated with the site. An understanding of the engineering geology and hydrogeological aspects of the site were critical and this paper describes those aspects and their influence on the design and construction of the works, not least of which was the selection of appropriate construction techniques.

Drumlough Moss is a 20 ha peat bog, known locally as a 'moss', situated in an interdrumlin hollow near Hillsborough in County Down (Fig. 1). For many years the bog land had been utilized for the unlicensed tipping of waste materials, arising from various sources including agricultural by-products and domestic and commercial refuse. In order to redress the situation and provide some form of effective control, the local authority elected to take over and develop part of the moss as a landfill site, and this has been in operation for over 30 years.

Due to the topography and nature of the site, the partial containment of surface water run-off and the build-up of leachates resulted in the formation of heavily polluted ponds of water both on the surface of the landfill mass and in adjacent areas. These had an adverse impact on the amenity of the area, particularly during the warm summer months, and resulted in a number of complaints by local residents. In addition, investigations by the Department of the Environment's Environmental Protection Division suggested that polluted water escaping from the site was responsible for a significant fish kill in the River Lagan.

A water-quality monitoring programme in the water-courses around the site rapidly confirmed that leachate was continuing to escape from the site at unacceptable levels. Consequently this established that the installation of containment and control measures would be necessary in order to minimize the risks associated with the site. A dual-purpose scheme was devised to keep clean surface run-off from the adjacent lands out of the landfill, and to keep leachate inside the landfill for treatment under a Phase II scheme. The local authority, recognizing these risks, acted accordingly, instructing the necessary works to be undertaken.

Geology

Solid geology

The underlying bedrock, which is exposed at the western edge of the moss, is Silurian greywacke, a turbidite comprising interbedded gritstones and mudstones. They have been subject to low-grade regional metamorphism, and are also intensely folded and steeply inclined along a NE–SW strike direction. The rock was formerly quarried from a small knoll, which is now buried within the landfill. A NNW–SSE trending basalt dyke, of Tertiary age, also occurs at the western end of the site.

Drift geology

The drift geology at Drumlough Moss comprises a sequence of peat and soft clays overlying glacial till. During the last (Midlandian) glaciation, ice of Irish origin flowed from a major ice axis which had formed over the Lough Neagh Basin and the Belfast Hills. This flowed northwards in north central Ulster and south-eastwards across County Down. One of the last major substages during deglaciation, know as the 'Drumlin Readvance' was responsible for the deposition of glacial till in the form of drumlins, the 'basket of eggs' topography typical of much of counties Down and Armagh.

Drumlough Moss itself occupies a glaciated hollow, bounded by drumlins to the north and south (Fig. 2). Glacial till or boulder clay cover, over the underlying greywacke in the immediate area of the site, is thin but increases in depth beneath the drumlins. The glacial till is a typical 'boulder clay', being predominantly a sandy silty clay but with particles ranging up to boulder size.

From BENTLEY, S. P. (ed.) *Engineering Geology of Waste Disposal*,
Geological Society Engineering Geology Special Publication No. 11, pp 71–75

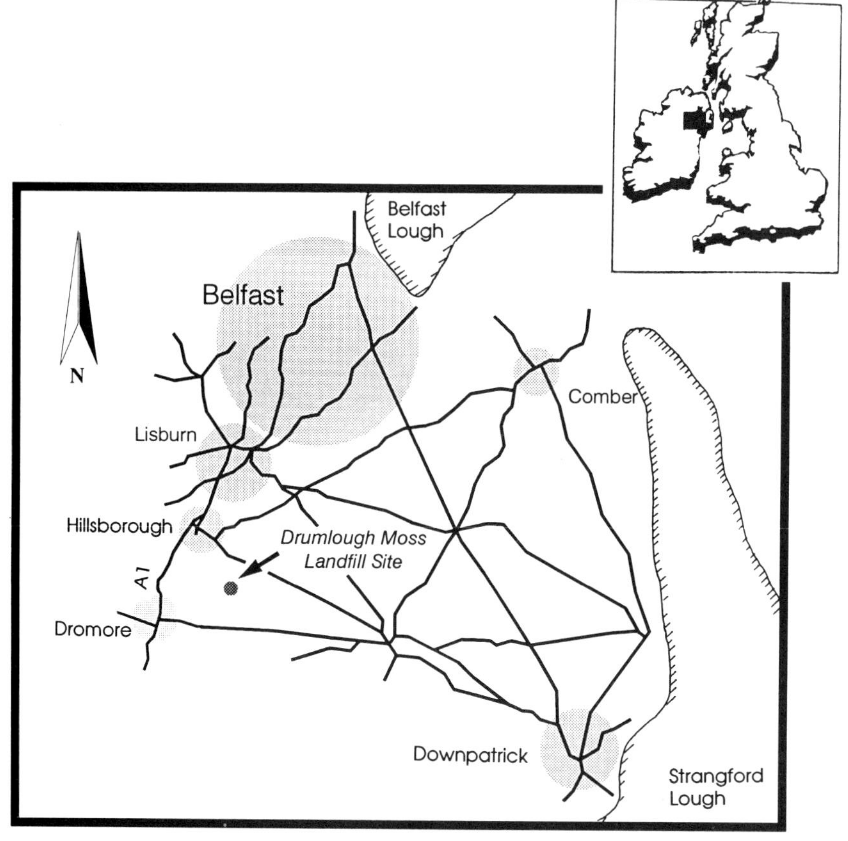

Fig. 1. Location plan.

In post-glacial times the interdrumlin hollow was occupied by a lake in which soft blue/grey clays were deposited before fen peat accumulation infilled the open water, leaving only a few isolated pools, resulting in peat deposits up to several metres in depth. Over an area of some 8 ha in the western part of the moss (Fig. 2) the peat has been covered by a mantle of landfilled wastes, up to some 5–7 m in depth, during the last 30 years.

Hydrogeology

The interbedded greywackes are typically of very low permeability and, given the intense folding and steep inclination, no significant groundwater movement is generally expected through them. In addition, the vertical basalt dyke at the western end of the site represents another barrier to movement of water through the rock. It was recognized also that typically the top of the greywacke has been subject to weathering and disturbance by glacial action resulting in a thin but more permeable horizon (by perhaps an order of magnitude or more) of 'weathered' greywacke.

The boulder clay overlying the greywacke also represents a very low permeability material capable, for all practical purposes, of providing effective containment for the landfill leachates. This view was reinforced by the presence of 'perched' ponds of water around the margins of the site, as described previously.

The peats and the weathered greywacke therefore represent the most permeable horizons within the profile, which could permit the movement of leachate.

The moss itself is actually situated on a minor watershed and drains both to the west and east via minor watercourses. Water levels are recharged by run-off from the surrounding higher ground. Notwithstanding the above it is therefore evident, given the site's location on the watershed, that it is surface water drainage from the site that poses the greatest risk of pollution, as opposed to leachate migration due to groundwater flow.

Design objectives

On the basis of an understanding of the geology and its engineering and hydrogeological implications, the design objectives could be stated simply as:

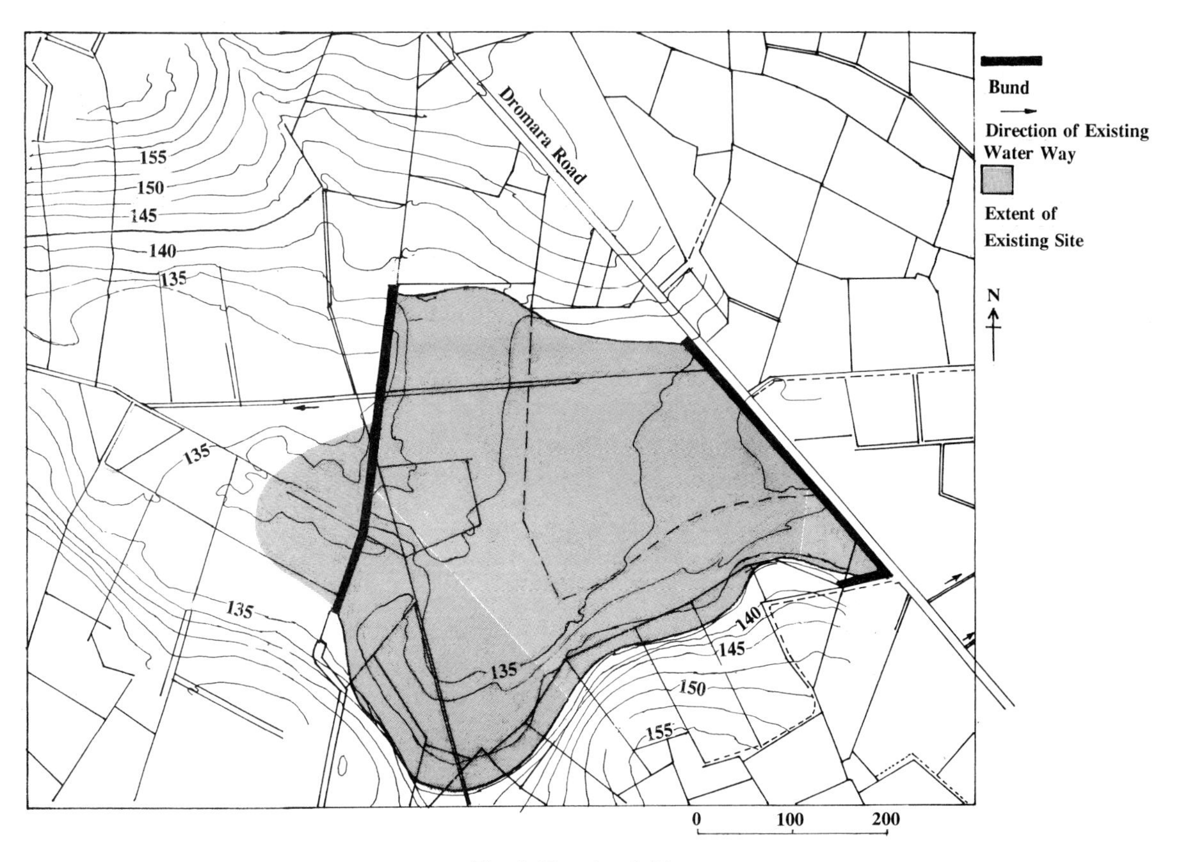

Fig. 2. Drumlough Moss.

- interception of surface water run-off from the surrounding higher ground;
- containment of the landfill mass and leachates;
- capping of the site to reduce the quantities of leachate generated;
- provision of leachate and landfill gas control systems.

It was decided at an early stage that these remedial works would be undertaken in two stages, with the progressive capping of the site and the provision of leachate and landfill gas control systems following the construction of the surface water drainage and containment measures. This later stage is not considered further in this paper.

Surface water

Interception of surface water run-off from the surrounding higher ground could be achieved by the construction of a peripheral drain around the site (Fig. 2). This drain will serve to intercept not only the run-off from the surrounding land but once the site is finally capped and sealed, it will also collect run-off from the restored landfill.

Containment mesures

As described above, the geological appraisal indicated that the boulder clay forming the drumlins to the north and south of the site would, for all practical purposes, provide effective containment for the retention of leachate. Therefore the indications were that 'total containment', could be achieved by the construction of impermeable cut-offs at the eastern and western ends of the site (Fig. 2).

In order to construct the cut-offs, several options were considered which were as follows:

- vertical HDPE (high-density polyethylene) cut-off wall;
- vertical bentonite cut-off wall;
- HDPE-lined bund.

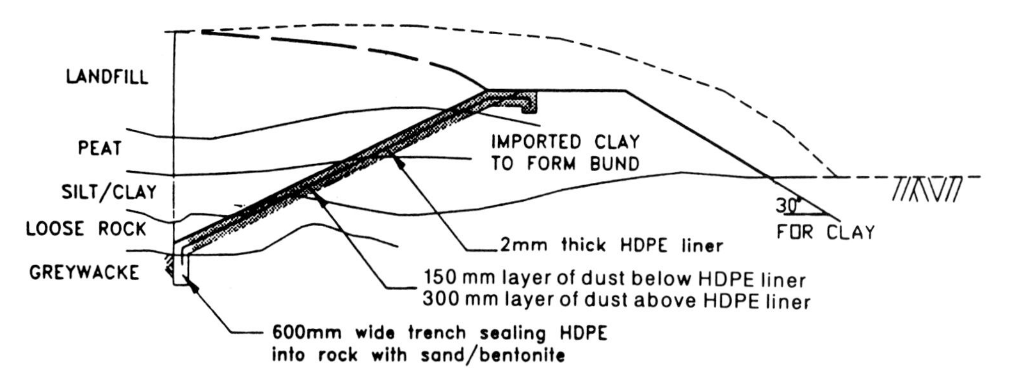

Fig. 3. Typical section through bund.

The first two options were assessed but rejected due to concerns about the ability to provide an effective key into the underlying greywacke. The landfill is underlain by peats and soft clays which are separated from the greywacke by a weathered permeable horizon at rockhead level. Difficulties were envisaged in ensuring that an adequate seal could be formed at the rockhead. In addition to the technical considerations, the costs associated with these two options were developed. In relative terms, they were also considered excessive and so these options were not pursued further.

The third option, the HDPE lined bund, illustrated in Fig. 3, proved to be the preferred solution. This form of construction permitted excavation of the overburden peats, and waste materials along the line of the proposed bunds, and the exposure of rockhead. This would allow the excavation of a trench, which would provide a seal for the permeable fissured greywacke while also providing an effective key at the toe of the slope for the HDPE membrane. The HDPE liner, 2 mm thick, would then be laid on the face of the bund as shown in Fig. 3. The embankment was originally conceived as rock fill, with a protective layer of quarry fines in order to prevent the possibility of any sharp edges piercing the liner which would then be sandwiched between two layers of geotextile.

The liner would then be sealed into the trench at the toe of the slope, as described above, and fixed at the crest of the slope by an anchor trench. A drainage layer on top of the geotextiles would then serve to allow both rapid drainage of perched leachate and venting of landfill gas. It should be noted that implicit in this design is an assumption that the control of groundwater and leachate flows into the areas of excavation could be maintained, relatively easily, by pumping from a number of sumps, as opposed to a sophisticated dewatering system. Costings of this form of construction also proved to be the most economically attractive, with the cost estimate being £730 000.

Construction techniques

Construction of the works has varied slightly from that originally specified. The major variation relates to the material used for construction of the bunds. Originally rockfill was specified but the main contractor, Gibson (Banbridge) Ltd, was able to negotiate a ready supply of boulder clay from an adjacent landowner.

By 'renting' several fields which lay on the side of a steep drumlin, the slopes could be 'eased' to give the farmer higher quality fields and in the process yield large quantities of boulder clay for use by the contractor. As a result of this proposal to construct the bunds from site won boulder clay, the slopes on the external face were eased. The impermeable nature of the clay had the added advantage in that should the HDPE liner fail for some reason, the clay would offer additional protection.

Bund construction was carried out by first removing any landfill material from beyond the line of the proposed embankment, and then excavating a 10 m wide trench to the top of the *in situ* boulder clay or rockhead level. The bund was constructed to a finished level by compacting the clay in 150 mm layers, thus effectively forming a clay barrier. This barrier also served to prevent water entering from outside the site. The remaining water within the site was then controlled by pumping from sumps to allow excavation of the rock trench, as described above.

The programme of excavating the trench at the toe of the slope has been described previously, but the site works served to confirm the design assumptions regarding the weathered permeable horizon in the greywacke. At first encounter the rock was loose, weathered and fissured with the characteristic red/brown staining of iron along joints indicative of groundwater movements. Using a Komatsu PC400LC excavator, variable depths of this loose rock were removed before clean, blue-grey greywacke was encountered. The intensity of jointing and fissuring reduced significantly at this depth also. This level was used to denote the top of the trench which was then hammered out to the required depth.

Additional research into the effective construction of the trench seal was also undertaken. The bentonite supplier, CBO International, investigated mix options and how they could be mixed on or transported to the site. The final solution comprised three stages

- sealing any cracks or fissures in the rock trench using cement/bentonite slurry
- placing the HDPE liner onto the clay bund and sealing with bentonite into the rock trench
- backfilling behind the bund and installation of a peripheral drain to prevent surface run-off entering the landfill site

The cement/bentomite slurry consisted of one part cement to four parts bentonite to 18 parts water, with the bentonite soaked for 2 h prior to the addition of 'slag' cement or Portland 'C' cement. This mixture was then poured into the trench and 'brushed' into cracks and fissures.

The actual filling of the rock trench was carried out using a sand/bentonite mixture. The research indicated that 7% bentonite added to a preselected sand, with an overall moisture content of 13%, would give a mixture with a permeability of 8×10^{-11} m s^{-1} provided that the maximum compaction of 1750 kg m^{-3} was attained. This was achieved by placing and compacting the mix in 150 mm layers. An angled fillet was then cut out to receive the HDPE liner, and backfilled again to provide an effective seal.

The 2 mm thick HDPE liner itself was installed on the landfill side of the bund, under standard quality assurance procedures, with material and weld testing undertaken to ensure compliance with the specifications. This provided the final variation from the original design, with the geotextiles being replaced by 150 mm layers of quarry fines, in order to provide effective protection to the liner. A 200 mm drainage layer was installed as the final phase and the excavation backfilled, on the landfill side, with selected wastes.

The external face of the western bund (Fig. 2) was topsoiled and grassed. For the eastern bund, however, construction was undertaken parallel to an existing road, bund construction being undertaken primarly below existing ground level, with final crest level at or about the road level. In this instance construction was restricted to short lengths, as opposed to the western bund which was carried out as one operation, with clay backfilling being undertaken immediately in order to minimize the risk of damage to the existing roadway.

Conclusions

The Drumlough Moss project has served to illustrate that relatively low-cost works, namely the provision of HDPE-lined bunds and peripheral cut-off drains, can be constructed to provide effective containment measures, provided that there is a clear understanding of the engineering geology considerations which impact upon the conceptual design principles. In this instance, surrounding drumlins formed from relatively low-permeability boulder clays provided effective containment to the site, and cut-offs (HDPE-lined bunds) were therefore required only at the western and eastern margins of the site. The design recognized the presence of a permeable horizon at the top of the greywackes which had to be sealed by a bentonite-filled trench, the final details of which were only resolved on site. This later aspect also emphasizes the importance of a 'team' approach, with the contractor and engineer, with the active support of the client, working together to develop suitable details. This was a key element in the success of the project.

Monitoring of water quality in the surrounding watercourse is currently being undertaken on a regular basis to ensure the efficacy of the remedial measures, as constructed. Initial results have proved promising with an improvement being indicated, although long-term trends are required to be certain. The final stage of site closure, namely capping and the provision of leachate and landfill gas control systems will then serve to ensure that the site is restored to useful agricultural use while presenting minimal environmental risks.

Acknowledgements. The authors wish to acknowledge the assistance of Lisburn Borough Council in the undertaking of this scheme and for permission to publish this paper.

Survey and containment of contaminated underwater sediments

B. T. A. J. Degen,[1] I. K. Deibel[2] & P. M. Maurenbrecher[3]

[1] GeoCom BV, PO Box 621, 2501 CP The Hague, The Netherlands
[2] Gemeentewerken Rotterdam, Galvanistraat 15, 3029 AD Rotterdam, The Netherlands
[3] Delft University of Technology, Faculty of Mining and Petroleum Engineering, PO Box 5028, 2600 GA Delft, The Netherlands

Abstract. The underwater sediments of the waterways, lakes and harbours of the Netherlands are sufficiently contaminated in many areas to necessitate their removal or isolation. The degree and nature of contamination and the general extent can be determined by sampling methods. To determine the thicknesses and distribution of the contaminated layers, shallow reflection geophysical surveys can be used in many situations, especially if previous erosion or dredging has created a slightly overconsolidated layer relative to subsequent deposition. The method is also used to aid investigation with respect to quantity control for sand placement to contain contaminated sediments *in situ* and to determine the depositional modes of hydraulically transported contaminated spoil in repository basins. Repository design for the Ketelmeer Lake allows for a certain amount of contaminant loss to the environment. The principal criterium is to significantly lessen contamination by 50%. Complete isolation would only improve the drop in contamination levels by a further small percentage.

The last ten years in the Netherlands have seen a rapid increase in site investigation techniques specially developed to detect and quantify polluted soils and groundwater. Techniques on land involve traditional methods with adaptations such as cone-penetration testing which measures the resistivity of the soil as well as taking small samples of the pore water. Geophysical techniques such as electromagnetic methods have been used to detect illegally dumped buried metal drums believed to contain toxic chemical wastes. More significant in the delta area of the Netherlands is the concentration of chemical waste derived from the industrial hinterland of Germany, France, Belgium and Switzerland as well as effluent added by riverside industries in the Netherlands. The northern estuary of the Rhine, the River IJssel, is situated at the head of the Ketelmeer Lake between the Noordoost Polder (North-east Polder) and Flevoland Polders and has become highly contaminated since its inception in 1965. Fish were discovered to have cancerous growths which were attributed to the high concentration of PCBs (poly-chlorobiphynylene) and heavy metals in the sediment. The locations of the most contaminated regions are shown in Fig. 1 (Bakker 1989).

More detailed distributions of contaminants in the significant estuaries of the Rhine–Maas delta (locations 1, 3 and 2 in Fig. 1) are shown in Figs 2, 3 and 4. The distributions are based on surveys using sampling techniques. The locations are the Ketelmeer Lake of the IJssel River estuary, the northern tributary of the Rhine (Fig. 2; from Winkels & van Diem 1991); the Haringvliet and Biesbosch the southern estuary of the Rhine/Maas (Fig. 3; from Lindijer 1987); and the Rotterdam Harbour complex, which can be regarded as the central estuary of the Rhine/Maas waters (Fig. 4; from Nieuwendijk and van Boxtel 1991). In all these three locations special depots, storage containment basins, have been constructed or planned so that the contaminated river bed silts can be removed by dredging and then transported to the basins for storage and, in some instances, for processing or partial processing.

Over-water geophysical reflection profiling surveys are used to obtain a more accurate estimate of the contaminant distribution and quantity as well as to determine the deposition of the dredged contaminants in the containment depots. Pilot surveys have been carried out in the Rotterdam area at the Geulhaven located at the confluence of the Oude-Maas River and the New-waterway to determine variations of the contaminated layers and at the Slufter contaminated dredged sediment repository. Background information about its design and construction is given by van Zetten (1987).

In addition to studying the source and terminal areas for the contaminants, a third application for the geophysical techniques is the use of monitoring sand placement over contaminated river beds to ensure their *in situ* containment. This has been done along the Lateraal Canal at Roermond where dredging for shipping is not a critical factor.

From BENTLEY, S. P. (ed.) *Engineering Geology of Waste Disposal*,
Geological Society Engineering Geology Special Publication No. 11, pp 77–86

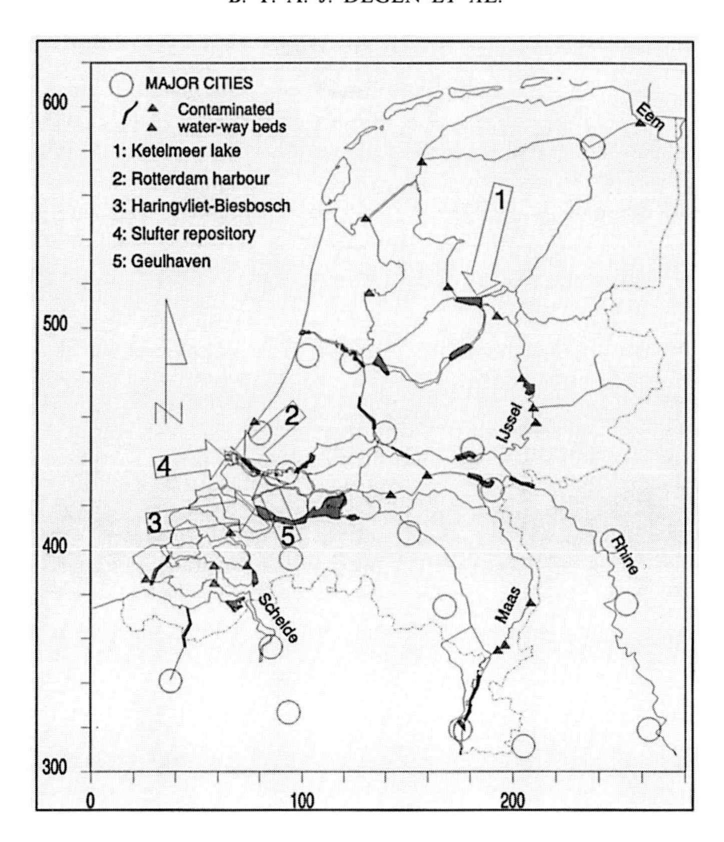

Fig. 1. Locations of most contaminated waterways in The Netherlands.

Equipment

Geophysical survey

Suitable equipment is required to investigate of up to 3–5 m depth. This can be achieved with the ORE Model 132P Pipeliner. Newer equipment with better on-line data-processing techniques is available but the costs are not easily offset due to the still-limited number of contracts for these types of surveys. The equipment components and specifications are listed in Table 1.

The selected seismic source frequency is dependent on the required depth of penetration. Usually a dual trace option is chosen as the 200 kHz frequency will reflect from the top of the suspended sediment whereas the lower 14 kHz frequency will penetrate the underlying loose/soft sediments to reflect at silt boundaries of the relatively young contaminated layers. These two traces can be superimposed to give a complete record from the suspended to the loose layers. Medium dense layers can be penetrated with the 3.5–5 kHz frequency range. Resolution decreases with a decrease in frequencies however. Consequently, in most surveys several runs are made to test which frequency range and combination is suited to the sediment conditions. The method has been found to be suited to conditions where previous dredging or river scouring has taken place so that slightly overconsolidated sediments are overlain by more recent normally or underconsolidated sediments.

The equipment weighs about 150 kg and is easily transportable by trailer. It can be mounted on small 5-m-long vessels though larger vessels make working conditions more acceptable.

Additional equipment consists of radio-controlled positioning equipment with an accuracy of 1 m over a range of 1 km. Less convenient methods of positioning can be done by optical methods. The radio equipment does not occupy much space as a result of considerable miniaturization in the electronics field. Accuracy of positioning is dependent on the distance of transmitter stations (10 km produces 0.3–0.4 m accuracy). Equipment is made for operation in all weather conditions.

Sampling

Various samplers for obtaining relatively stratigraphically undisturbed samples have been tried. Table 2 (van Diem 1988) gives an evaluation of samplers used for this purpose.

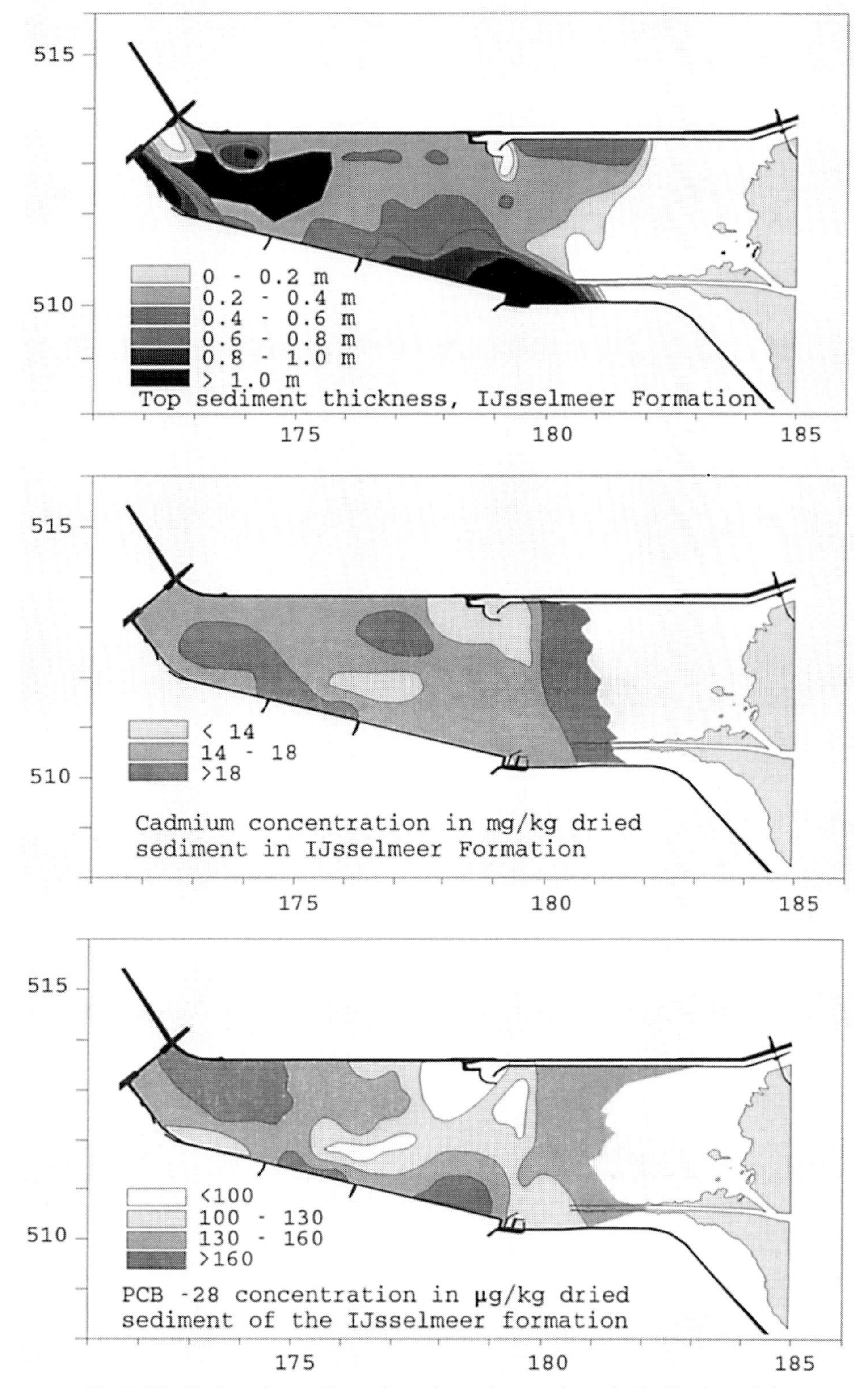

Fig. 2. Distribution of top sediment formation and contaminants in the IJsselmeer Lake.

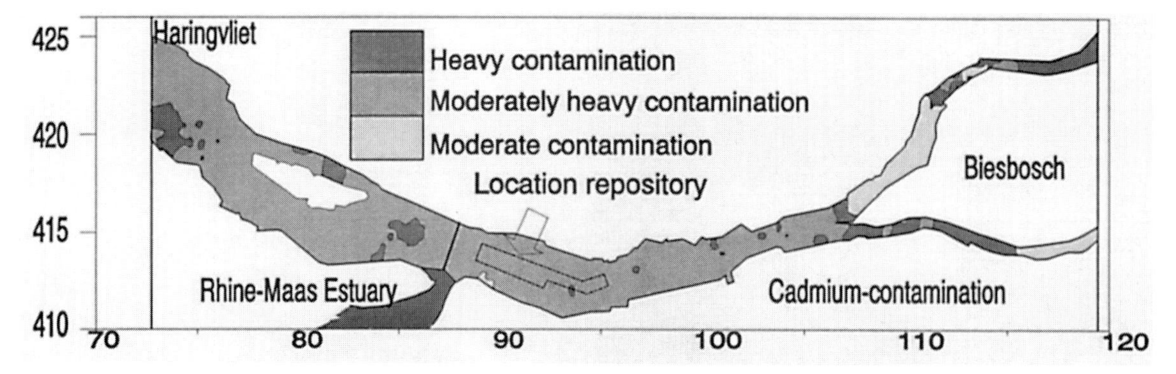

Fig. 3. Cadmium contamination distribution in the Rhine–Maas estuary.

Survey applications

In the wetlands of the Netherlands shallow reflection surveys have been carried out in connection with containment, disposal and storage of contaminated spoil or sediments. Three principal applications have been found for this type of survey:

(1) for quality and quantity control of cover material placed over underwater dumped contaminated dredged spoil;
(2) to determine the variation and amount of contaminated sediments;
(3) as an investigation aid for depositional processes of hydraulically transported dredged contaminated spoil in the Slufter repository basin.

An example of application (1) is a typical profile obtained from the Lateraal Kanaal shown in Fig. 5. Here both quality and quantity control is achieved through the use of shallow reflection profiling with the aid of samples for cover sand placed to contain contaminated dumped dredge spoil. From the analogue trace, interpretation lines are drawn to show the results of both the first sand cover layer and the second sand cover layer.

Further processing would normally be required as analogue trace scales are non-linear. For instance, to

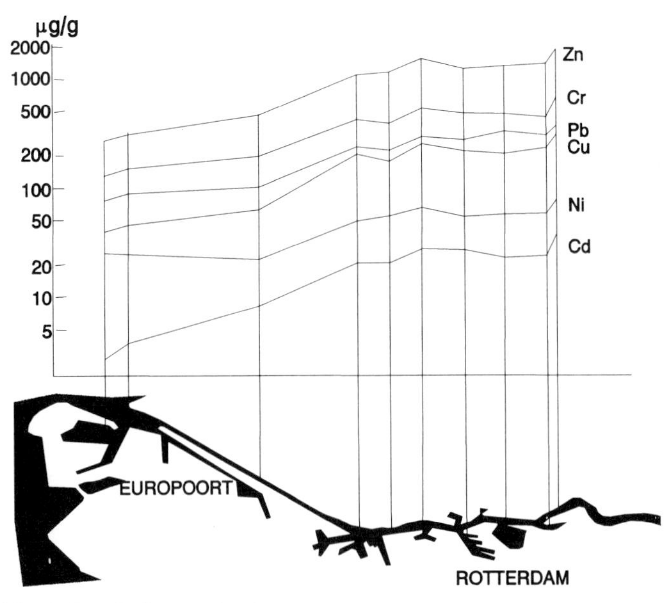

Fig. 4. Trend of contamination concentrations in the harbour of Rotterdam.

Table 1. *Pertinent specifications for shallow reflection seismic equipment*

O.R.E. Pipeliner Specification:				
Transducers (Model 132B) ***Over-the side Transducer Array***				
operating frequency kHz	Low Frequency	Medium	High Frequency	
	3.5	14	200	
	5			
Beam pattern	elliptical		+-3.5" conical to 3 db points	
athwartships	narrow	narrow		
fore/aft	wide (45"x90"@ 5 kHz)	wide (15"x 80")		
Input power rating	5000 watt	5000 watt	100 watt	
Transceiver (Model 310)				
Transmitter				
Operating frequencies	kHz			
single trace	3.5	5	14	200
dual trace	Channel 1		Channel 2	
	3.5 or 5 kHz	14 kHz	14 or 200 kHz	200 kHz
Power output	5 kW rms during kep pulse 1% duty cycle max. adjust 0-5 kW			
Receiver				
frequency kHz	3.5	14	200	
(selected by transmitter frequency)				
Band pass filter: kHz	5 @ 3.5, 5 & 14	10 @ 200		
sensitivity	1 mv receiver input produces 1 v rms output with 20 db signal/noise ratio			
TVG -time/variable/gain section *(amplification as signal looses strength)*, adjustable				
tracker	TVG ramp automatically starts when first bottom return is received			
Graphic recorder: various models				

Table 2. *Evaluation of soft sediment samplers*

Apparatus	*Criteria*				
	Ease of handling	*Sample disturbance**	*Accuracy*	*Depth sampling (≤cm)*	*Suitability*
Jenkins mudsampler	+	+ +	+ +	10	+ +
Box corer	+ −	−	+ −	40	+
Van Veen grab sampler	+ +	+ −	+	10	−
Eijkelkamp beaker sampler†	+ −	+ +	+ +	150	+
Vrij-Wit tube	+ +	+ +	+ +	150	+ +

* For a positive evaluation the sediment layering was not disturbed.
† This apparatus was at the time of the test a prototype so that in the meantime the practical handling has been improved.
+ + Very positive evaluation; + − moderate evaluation; − negative evaluation; + positve evaluation; − − very negative evaluation.

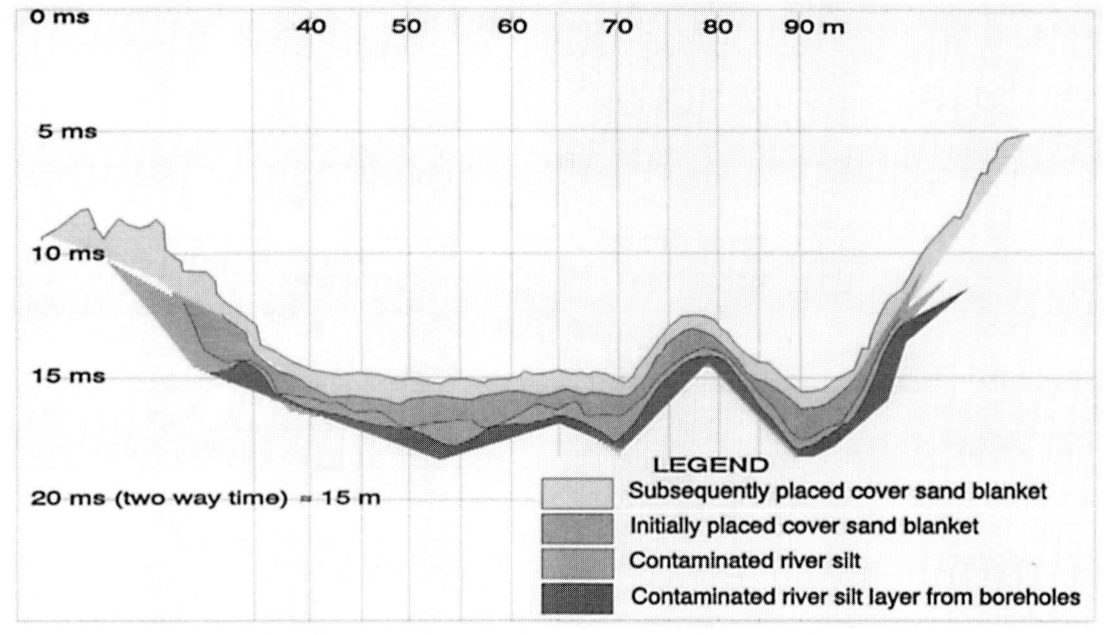

Fig. 5. Lateraal Kanaal quality/quantity control cover material.

estimate the amount of sand that the dredger had deposited for payment purposes, the cross-section has to be adjusted to a linear scale. The vertical scale represents the time for a particular sound signal to travel from the transducer to the receiver. By knowing the velocity of sound in water ($1500\,m\,s^{-1}$) and in soft sediments (clay/silt and sand) ($1500–1700\,m\,s^{-1}$), Fig. 5 can be adjusted to a linear scale (each horizontal bar represents a 10 ms timescale). The horizontal scale can be variable as it depends on the speed of the vessel and its route. The adjustment is usually performed by hand. Once the new profiles have been drawn the layer depths are plotted on a map and subsequent isopleth maps are made using a contouring programme which also determines volumes against a particular surface such as an overlay isopleth map of an underlying layer.

The principal application for surveys has been type (2), which is to determine the location and quantity of contaminated sediments. Applications cover a variety of situations: harbour entrances, harbours, canals, rivers and lakes. A pilot survey over an area of about 50 m by 150 m was carried out in the most contaminated section of the Geulhaven. Figure 6 shows the general contamination levels based on sampling (from Quaak & Reinking 1992) and the results from the survey. Contour maps shown in Fig. 6 were derived from eleven 150-m profile lines and five perpendicular 50-m profile lines. A typical geophysical reflection profile cross-section is shown in Fig. 7. The demonstration survey shows that considerable detail can be obtained with the aid of the geophysical survey. By selective dredging of the thicker sections, considerable volumes of material can be spared from transport to storage in the repository.

Additional information was obtained from 19 vibrocores to confirm the geophysical interpretations as well as to determine the degree of contamination of the sediments.

Application (3) was tried on an experimental basis at the main contaminated dredged spoil repository of the Slufter to determine the sedimentation distribution of the hydraulically transported dredged spoil. The results gave detailed cross-sections of the layering within the Slufter basin as shown in Fig. 8. The cross-sections show a highly irregular sediment bed surface and layering. The spreading process does not occur sufficiently evenly from the outflow units. The uneven spreading is attributed to segregation of the coarser fraction ($>63\,\mu m$) settling out within a 100 m radius and the finer fraction extending over a wider area of 400–500 m radius. The uneven spreading results in non-uniform consolidation of the dredged spoil so that the rate of intake capacity of the repository could be affected. With the more selective dredging of contaminated layers, through the aid of geophysical surveys together with the prognosis that the upstream sources of contamination will lessen, it is doubtful that the originally planned life-span of the Slufter repository would have to be adjusted.

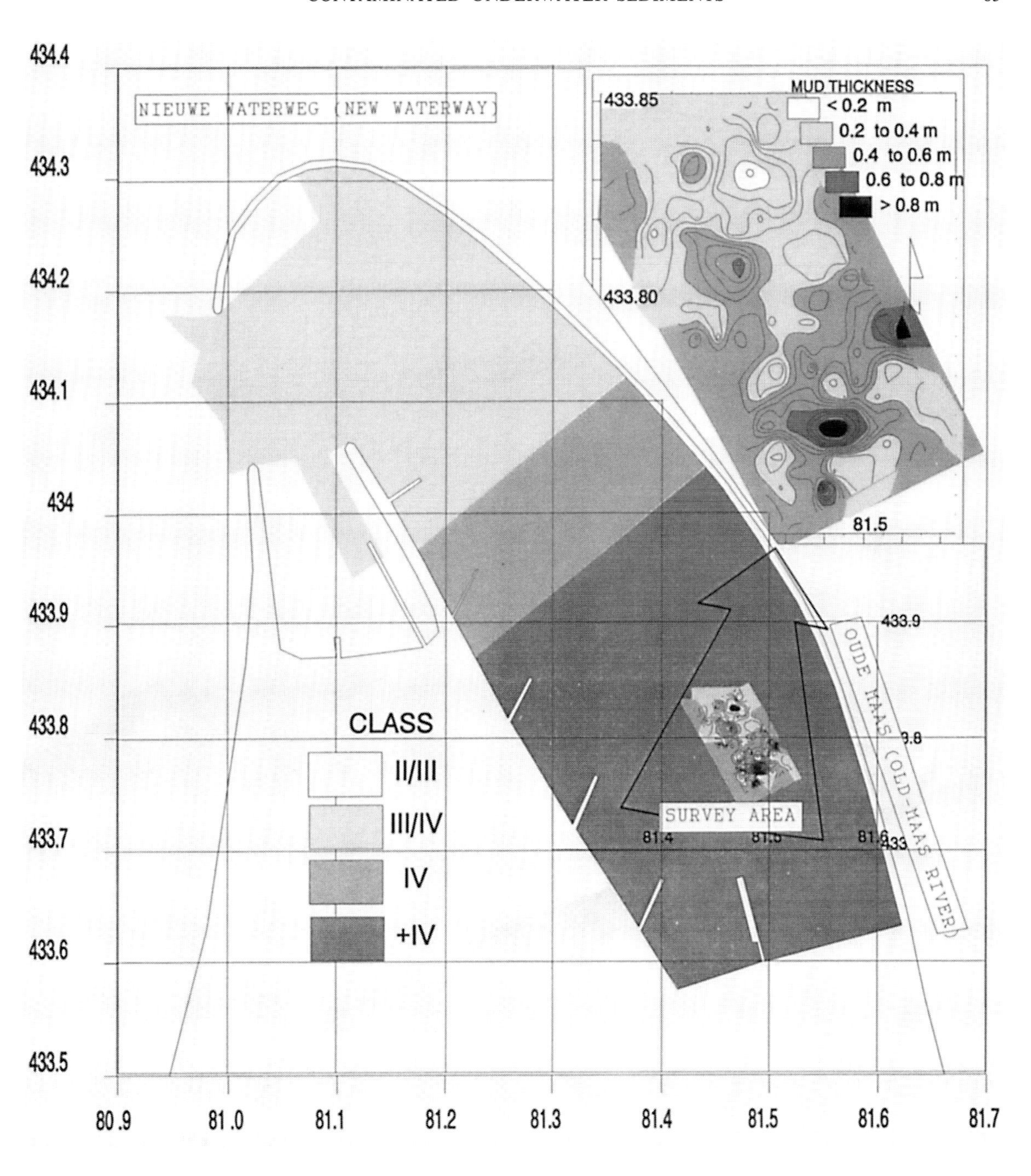

Fig. 6. Geulhaven contaminated mud isopach map.

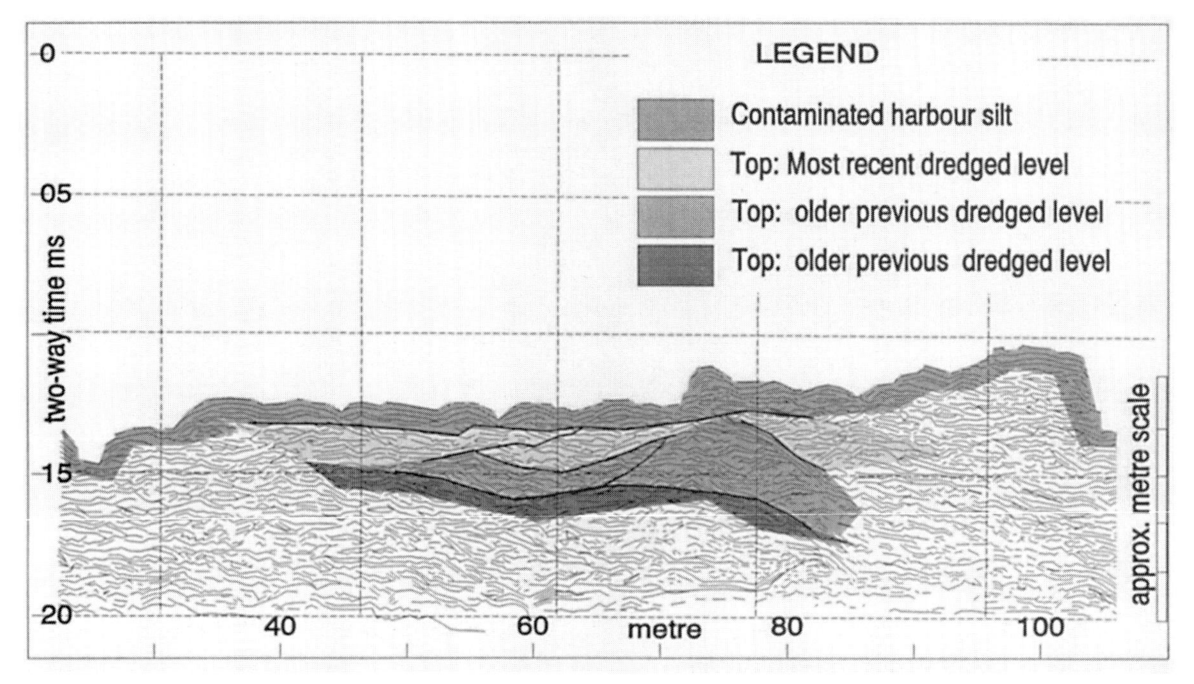

Fig. 7. Geophysical cross section at Geulhaven.

Further repositories of the Slufter type are planned for the Biesbosch–Haringvliet Estuary (Anon. 1989) on which the planned repository is shown approximately in Fig. 3 and the Ketelmeer Lake (van der Doef & Laboyrie 1992). The estimated consolidation with respect to contaminated spoil intake that would take place in the Ketelmeer Lake repository is shown in Fig. 9; the estimates would not be entirely dissimilar to those of the other river-bed sediment repositories such as the Slufter. The design aspects for the repository in the Ketelmeer Lake require several considerations to be taken into account, namely:

(1) The total volume increase ratio of the dredged sediment is 2.5 resulting from the net effect of the following factors:

- present estimates of *in situ* contaminated sediment volumes;
- dredging will result in an overbreak volume of less-contaminated layers (depending on variability of contaminated layer and method of dredging); overbreak is estimated at 40%;
- an extra 40% material will be added through the 20-year dredging programme as a result of sediment transport from the river IJssel;
- mixing of water and dredged material for hydraulic transport;
- consolidation will cause a volume decrease;
- decomposition of organic constituents will generate gases which will cause a volume increase, though ultimately through gas escape the volume should decrease.

(2) The geometry of the repository requires not only the above volume capacity considerations but also:

- in the proposed location (centre of the Ketelmeer Lake) the contaminated top layer must be removed and stored to allow for the deeper excavation (for storage volume as well as sand borrow for the repository embankments).

(3) The repository will need to consist of two compartments: the first compartment will supply sufficient sand for the ring dykes, and the contaminated dredged sediment will be temporarily stored elsewhere. As soon as the first compartment is complete, filling with contaminated sediment can start with the initial sediment coming from the location for the second repository. As soon as that sediment is removed the surplus sand (to be sold) will be removed in phases until the second repository capacity will be required.

The anticipated consolidation of the contaminated spoil is expected to both give it sufficient strength to contain itself (should the ring dykes fail) and seal

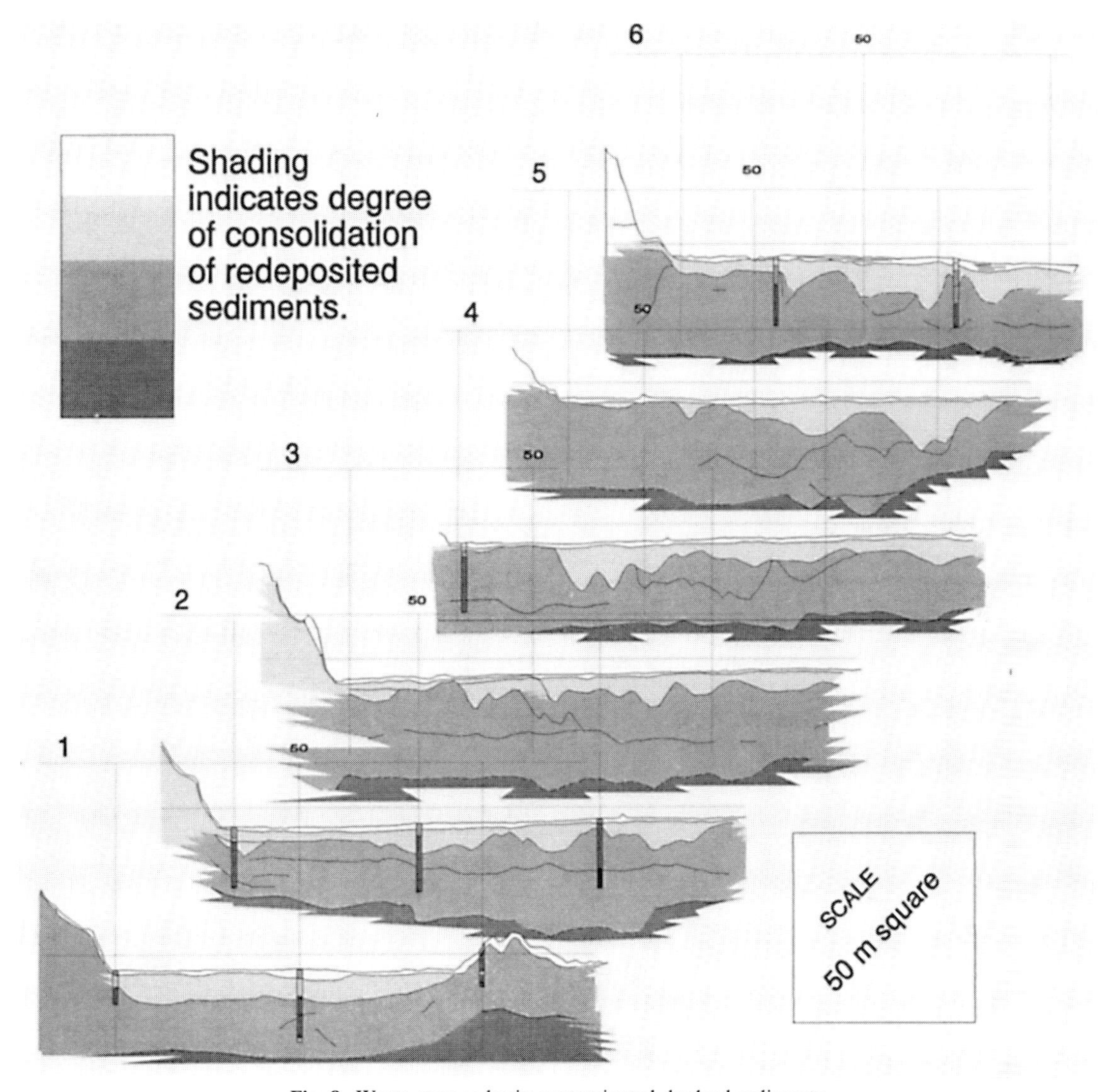

Fig. 8. Waste storage basin contaminated dredged sediments.

itself to cause minimal contamination through the groundwater. Leachate flow extruded by consolidation and from an excess vertical hydraulic gradient is calculated to be eight times less than the estimate for leachate flow from the existing Ketelmeer Lake situation. This does not mean that the contamination from leachate or overflow will meet the government's 'Algemene Milieukwaliteits Doelstelling 2000' (General Environmental Quality Objectives year 2000). The repository will cause a 50% reduction in environmental burden in 20 years. However, isolating the repository further would only add a further reduction of a few percent.

Conclusions

Increasing experience is being gained in surveying and containment of contaminated waterway sediments. Seismic reflection geophysical surveys can be used in suitable situations to obtain an accurate estimation of river bed sediment thicknesses and distribution, which could allow for more selective dredging and decrease the amount of sediment that requires transport and storage. The survey techniques are also used for *in situ* sand placement containment quality and quantity control and for investigating sedimentation processes within the repository basins. The method can be used in future

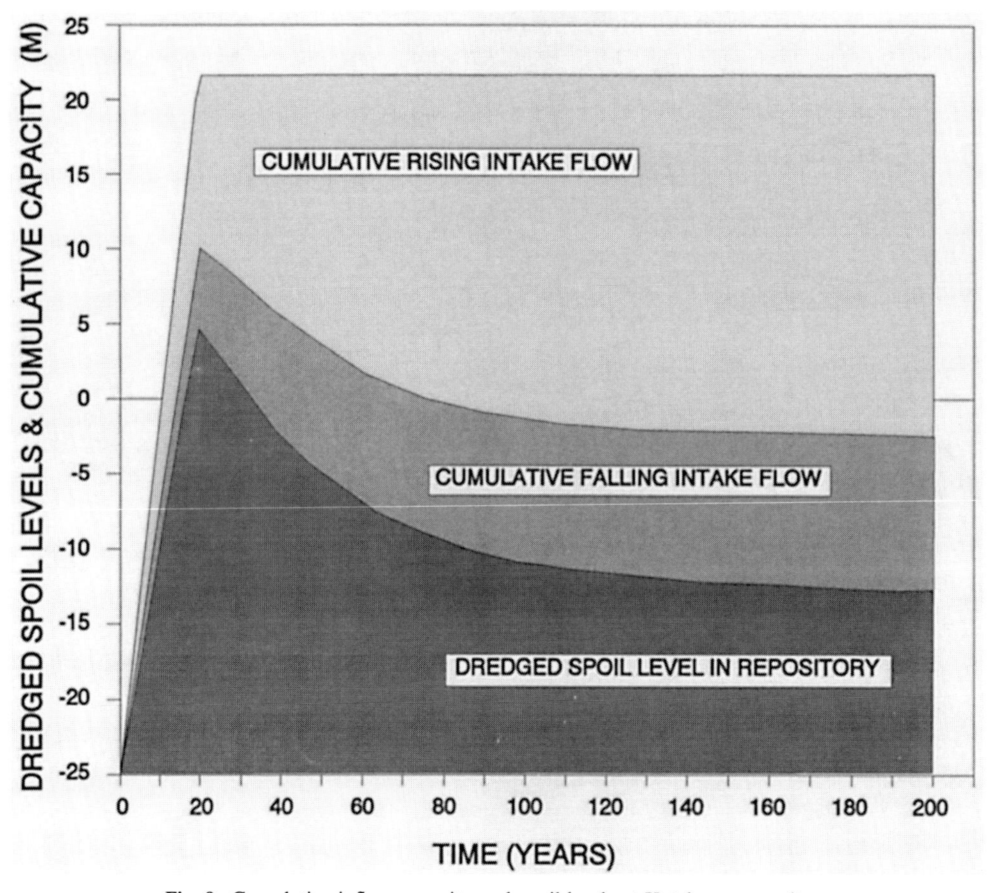

Fig. 9. Cumulative inflow capacity and spoil levels at Ketelmeer repository.

repositories being developed at Hollands Diep (Haringvliet–Biesbosch estuary) and in the Ketelmeer Lake (IJssel River estuary).

References

ANON 1989. Hollands Diep kan kostendekkend gereinigd (Hollands Diep can be cleaned cost-effectively. Ingenieurskrant KIvI-NIRIA, 6 April 1989.

BAKKER, T. 1989. *Inventarisatie van de vervuiling van waterodems van Rijkswateren* (Inventory contamination waterway beds of the state waterways). Waterbodems (waterway-beds), post technisch onderwijs (Post higher technical education series), Hogeschool, Heerlen.

DIEM, A. VAN 1988. De inventarisatie van waterbodem van het Ketelmeer (Inventory of the contaminated sediment of the Ketelmeer lake). Rijkswaterstaat-Directie Flevoland Report July 1988 and Memoir No. 73 Centre for Engineering Geology in The Netherlands, Faculty of Mining and Petroleum Engineering, TU Delft, MSc Thesis.

DOEF, M. R. VAN DER & LABOYRIE, H. P. 1992. Ketelmeer is trendsetter bij ontwerp speciedepots (Ketelmeer lake is the trendsetter with respect to dredged spoil repositories. *Land and Water*, **7** (July), 66–71

LINDIJER, G. 1987. Estuaria en zeën, een werkgebied in beweging (Estuaries and seas, a work area in movement). *PT Civiele Techniek*, **42**(3) (September), 68–73.

NIEUWENDIJK, K. & VAN BOXTEL, A. 1991. De problematiek rond baggerspecie in België, Duitsland en Nederland (The problems with respect to dredged sediments in Belgium, Germany and the Netherlands). *de Ingenieur*, **103**(6/7), 8–14.

QUAAK, M. P. & REINKING, M. W. 1992. Innovative dredging: techniques developed during the clean-up of Geulhaven in Rotterdam. *Terra et Aqua*, **48**, 3–11.

WINKELS, H. J. & VAN DIEM, A. 1991. *Opbouw en kwaliteit van de waterbodem van het Ketelmeer* (Build-up and quality of the sediments of the Ketelmeer Lake). Flevobericht No. 325, Ministerie van Verkeer en Waterstaat, Rijkswaterstaat Directie Flevoland.

ZETTEN, M. VAN 1987. Slufterdam disposal solution to Rotterdam's contaminated dredged material problem. *Dredging and Port Construction*, March, 31–35.

Stabilization of an urban refuse dump and its planned extension near Ancona, Marche, Italy

E. N. Bromhead,[1] L. Coppola[2] & H. M. Rendell[3]

[1] School of Civil Engineering, Kingston University, Penrhyn Rd, Kingston upon Thames, Surrey KT1 2EE, UK
[2] Dipartimento di Scienze della Terra, Università degli Studi di Siena, Sienna, Italy
[3] Geography Laboratory, University of Sussex, Falmer, Brighton, Sussex BN1 9QN, UK

Abstract. The paper describes the stabilization of a solid urban waste dump near Ancona, Marche Region, Italy. The 40 000 m^2 dump was established on the western slope of Monte Umbriano on a slope of 8°–10° over underlying materials comprising sandy–silty colluvial clays. Refuse was deposited to a maximum height of 13 m with a graded slope. Prior to placing the refuse, trench drains were constructed parallel to the direction of maximum slope. These drains were filled with 18–21 mm washed river gravel protected by geotextiles. The leachate from the tip was collected from the drains, and led to drainage shafts connected to a treatment plant.

In August 1988, active slip surfaces were encountered in the inspection shaft at a depth of about 9 m below ground surface. Following the installation of inclinometers, the slow mass movement of the whole of the slope was confirmed. The causes of the slip were studied via a series of field and laboratory investigations, and a series of stability analyses were undertaken.

The slope was stabilized by the construction of a system of deep drainage. The area of the existing tip was drained by installing subhorizontal drilled drains from pits near the toe of the waste dump, and the area of extension of the tip was drained by deep shafts from which an array of bored drains were installed. The extension of the tip over the drained area appears to have been successfully completed. Leachate from the tip has been collected and fed to the existing purification works for treatment.

The western slope of Monte Umbriano was selected as a site for an urban refuse dump in 1980, and it was developed in two phases. In March 1984, a study was commissioned in connection with a third-stage enlargement of the dump complex and it is with this particular part of the dump that this paper is concerned. In July 1988, it became apparent that the area of slope that included the third stage was sliding downhill, and a site investigation was carried out. In January 1989, Snamprogetti SpA were asked by the Ancona Council to prepare a scheme to:

- stabilize the area undergoing failure; and
- design a further enlargement of the waste dump with appropriate measures to ensure stability.

Their investigation included the verification of the depths of movement by inclinometer measurements. Three inclinometers located along the toe of the waste tip embankment showed downslope movements of 80, 95 and 25 mm respectively, occurring between 8 and 11 m below ground level in a one-month period. This paper reviews the original problem and both the remedial measures adopted to rectify it and new measures to permit a yet further enlargement of the tip site to be carried out in safety.

Geological and geomorphological setting

The general geology of the area around Ancona is shown in Fig. 1 (Cello & Coppola 1989). The third-stage enlargement of the Ancona urban refuse dump was placed on a slope of 8°–10° on the western side of Monte Umbriano, approximately 5 km southeast of the urban centre of Ancona. Before tipping, the slope and the surrounding area were typically hummocky terrain and tension cracks were common. The tip and the form of the pronounced toe heave of the mass movement are shown in Fig. 2. Positions of the site investigation boreholes are also marked on this figure. These boreholes revealed the following general succession:

- The colluvial horizon is 13–14 m thick and is a predominantly clay-rich horizon. Slip planes were

From Bentley, S. P. (ed.) *Engineering Geology of Waste Disposal*,
Geological Society Engineering Geology Special Publication No. 11, pp 87–92

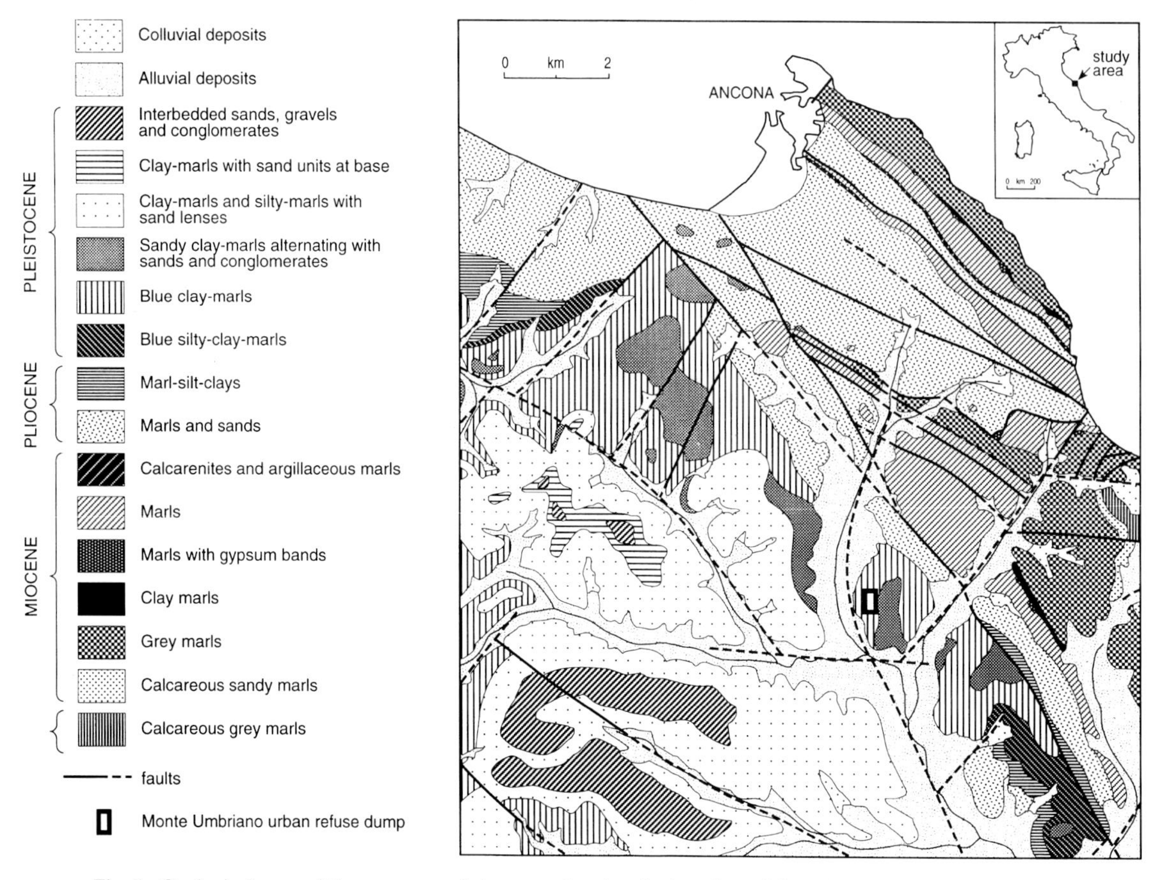

Fig. 1. Geological map of the area around Ancona, showing the location of the Monte Umbriano waste tip. Inset: location of Ancona within Italy.

encountered down to depths of 12 m below the ground surface. The current slip surface is at 9 m below ground level.

- The altered horizon is 4–5 m thick and comprises clays and marly clays. No stratification is visible but the material is fissured and some of the fissures are infilled with calcite.
- The unaltered stratified clay-marls are dry, intensely fractured and hard, and intercalated with rare thin bands of sand.

In the area immediately downslope of the dump, the colluvial cover is some 14 m thick. The colluvium represents earlier episodes of landsliding on the site which significantly pre-date its use as a waste dump. Given the climate of this area of Italy, groundwater conditions would be expected to be controlled by the mean annual soil-moisture deficit (resulting in the generally dry nature of the clay marls at depth) with ephemeral perched water-tables forming during the winter.

Material characteristics

Typical material properties were evaluated during the routine site investigation, and are listed below in Table 1. They demonstrate the exceptionally high undrained strengths of the very dry deeper materials. These are partly the result of the effective stress cohesion, which is commonly found in these materials and possibly results from calcite cementing. The high strengths are also the result of capilliary suctions arising from desiccation. Saturation processes occurring during effective stress testing raise the water content from its *in situ* values of typically 3–5% up to 21–24%.

Stabilization of the existing dump

Data for the period November 1988 to January 1989 from piezometers installed on site indicate that the groundwater levels in boreholes S2 and S3 were within

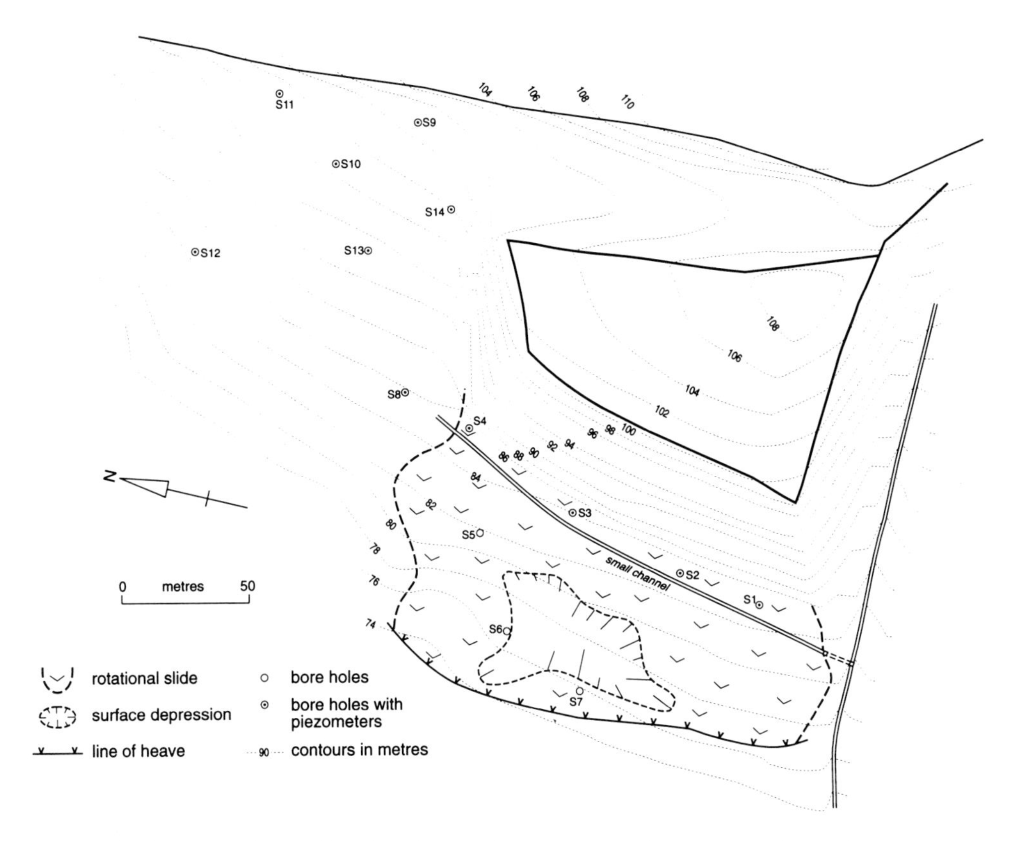

Fig. 2. Plan of Monte Umbriano tip complex, showing toe heave seen as slip developed.

about 1 m of the ground surface. This contrasted sharply with the levels in boreholes S1 and S4, which were, respectively, 8 m and 6 m below the surface, and also with the levels recorded in the area of the new extension, where water levels were between 7 and 12 m below ground level. In essence, the tip provided a source of water which was affecting the water-table elevation in its vicinity. Some of this water may well have been infiltration stored in the tip debris, but much of the leachate arose from decomposition of tip materials.

Stability conditions on the slope were assessed using routine stability analyses techniques. Assumed water levels at various elevations were examined, and these showed that at lower groundwater levels (i.e. below 8 m) the factor of safety rose above 1. It was therefore concluded that drainage of the site, to lower the level of the water-table, would be sufficient to ensure future stability.

Table 1. *A summary of the main geotechnical properties of the natural deposits*

	Colluvial clay-marls	*Altered clay-marls*	*Unaltered horizon*
Liquid limit	32–64%	60–70%	65–70%
Plastic limit	20–36%	20–35%	29–32%
Plasticity index	13–28%	35–45%	36–38%
c_u	35–100 kPa	280–325 kPa	325–650 kPa
ϕ' (drained)	23°–26°	25°–26°	
c'	20–36 kPa	50–80 kPa	300 kPa
ϕ' residual	6°–9°		

Installation of subhorizontal drains

Two sets of drains at angles of 8° and 5° to the horizontal were emplaced by drilling into the slope (Fig. 3). The upper drain is steeper (8°), and lies along the contact between the refuse material and the underlying slope; the lower drain is flatter (5°), and lies within the altered strata beneath the rubbish tip. Drilling was facilitated by

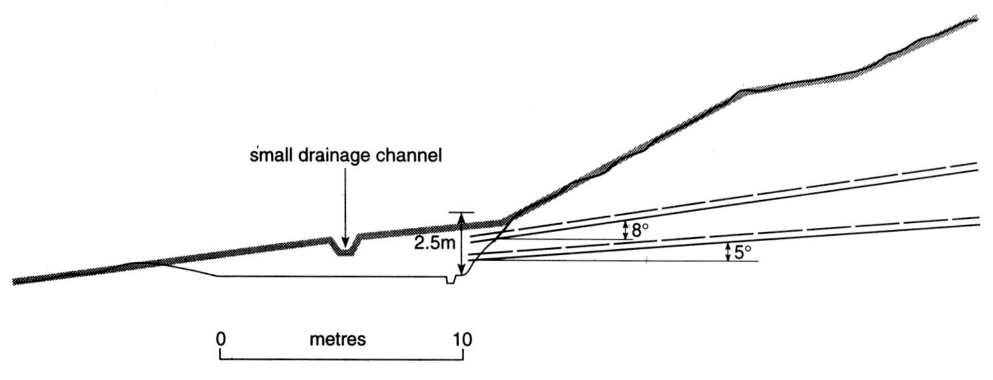

Fig. 3. Section through pit from which subhorizontal drains are drilled, showing their orientation.

clearing vertical faces 3 m wide and 2.5 m deep in pits on the downslope side of the waste tip. Drill diameters were 200 mm for the first 25–30 m and then 150 mm diameter up to 100 m. Drilling was undertaken using a self-propelled, track-mounted, hydraulic top-drive rotary rig. Casing was not required.

A series of 118 mm diameter PVC tubes were then introduced into the drill holes. These tubes are perforated with saw cuts for two-thirds of their circumference throughout the whole of their length. The tubes were covered by a geotextile to prevent the drain becoming blocked by fine material. The tubes join

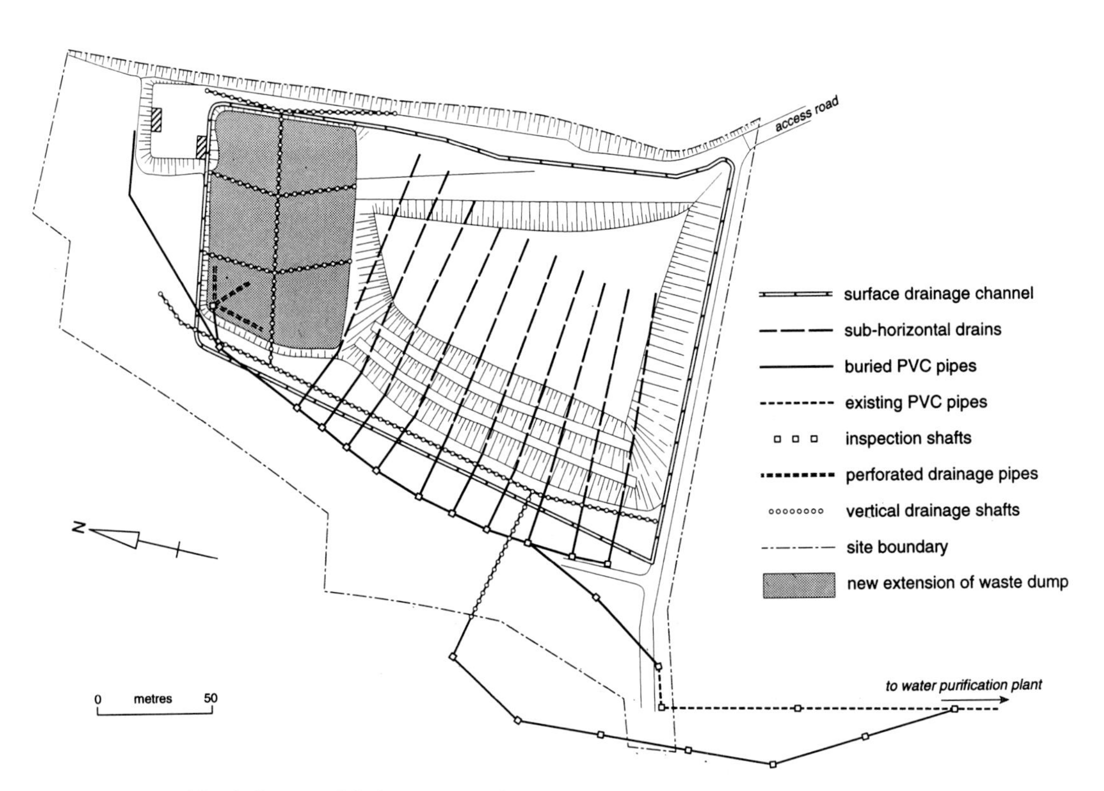

Fig. 4. Layout of drainage system for Monte Umbriano tip stabilization scheme.

immediately downslope of the tip and the effluent is directed to pre-existing drains or to new service tanks. Figure 4 shows a detail of the pits from which the bored drains were started, and the small collector channel at the reinstated toe of the tip.

Installation of other drains

The other measure used in order to reduce pore-water pressure was to install a series of drainage shafts at the foot of the slope allowing the collection of water from the colluvial horizons. The drainage shafts, with horizontal spacings of *c*. 4 m and internal diameters of 1.20–1.50 m, were augered to a depth of 15.5 m below ground level. This depth was in all cases greater than the depth of the slip surface, and the system of shafts acts, in effect, as a substantial and very deep counterfort drain. At 1.5 m from the bottom of each shaft a PVC tube of 0.85 m diameter connects each shaft with the adjacent one. The connecting tubes are inclined with falls of 1 in 50 (2%) to facilitate drainage. The drainage shafts are lined with a geotextile and filled with clean gravel, with the top of the shaft sealed with clay material. One in every four shafts is modified as an inspection shaft, with an inner corrugated steel liner with ladder allowing access to the base of the shaft and to the valves which regulate flow between the drainage shafts. Surface and subsurface waters are removed from the site through the network of pipes and shafts by gravitational flow. A small drainage channel around the perimeter of the refuse dump is designed to carry surficial flow away from the site and into a natural channel downslope of the main site. The layout of these shafts is shown on the drainage layout plan (Fig. 4).

Access for maintenance purposes was provided for both the subhorizontal drains, and for several of the drainage shafts.

Extension of the existing dump

The development of the urban refuse dump has to be in accordance with regional, national and EU norms in order to minimize pollution and any risk to public health. The extension uses compaction of the refuse to produce material with a density of between 0.8 and 1.0 t m^{-3}. The use of compacted waste has some major advantages, including an increase in the capacity and lifespan of the dump, control of superficial rill development and percolation, and elimination of potential infestation of the tip by animals and insects, and elimination of odour. It also enhances the possibility of eventual restoration of the site.

The design of the new extension includes the following drainage measures:

- a scarp 1.5 m high, 5 m long with an angle of 16° dipping into the slope and constructed along the downslope edge of the site, with a drainage tube running along slope at the base of the artificial backscarp;
- networks of drainage shafts down to depths of 12–16 m below the ground surface, below the depth of old and re-activated slip surfaces on the slope;
- linkage of drainage networks with those developed for the drainage programme for the adjacent area and therefore with the purification plant.

Biogas generation is not seen as a particular problem for this relatively small site as it is of limited dimensions, and is effectively isolated from urban areas.

Effluent chemistry

The chemistry of the effluent from the rubbish tip is monitored on a regular basis and results are compared with the norms for potable water. The data show considerable variations both in space and time. In particular, the limits for chlorides (200 mg l^{-1}), nitrates (50 mg l^{-1}) and mercury (0.001 mg l^{-1}) are exceeded in some of the inspection holes some of the time. Maximum values encountered during the monitoring period are 555 mg l^{-1}, 54 mg l^{-1} and 0.003 mg l^{-1} for chlorides, nitrates and mercury, respectively. The effluent pH is typically slightly alkaline, in the range 7.2–7.5.

Discussion

The original plans for the third-stage extension of the Ancona urban refuse tip clearly took into account the need to deal with any pollution from the site, and continued to function as intended throughout the remedial works.

The climatic conditions in the Ancona area are such that mass movements occur when there is some mechanism for moisture concentration on slopes. In the present case, the urban refuse, coupled with rainwater, appears to have allowed a perched water-table to develop within the colluvial cover on the slope. Site investigation also revealed that the colluvial cover contains many slip surfaces, and has a surface topography characteristic of relatively shallow mass movements. The solution to drain the site and to provide adequate drainage for a further extension of the site is unsurprising in a UK context, but not nearly so obvious in the context of a Mediterranean climate; for example, there was little, if any, sign of the emergence of leachate other than through the pre-existing tip drainage system.

Recently installed inclinometers, mostly in the area of the dump extension, show no consistent pattern of movements in the tip complex, and all apparent movements are readily accountable within the reading accuracy and repeatability of the inclinometer system. When considering the installation of an underdrainage system in a case such as this elsewhere in the world, it is essential to note that the discharge from the drains will be polluted and will require treatment. In the Monte Umbriano tip this did not cause additional expense.

References

CELLO, G. & COPPOLA, L. 1989. Modalità e stili deformativi nell'area anconetana. *Studi Geologici Camerti*, **XI**, 37–47.

Use of geophysical surveys during the planning, construction and remediation of landfills

John M. Reynolds[1] & David I. Taylor

Rust Environmental, Beech House, Park West, Sealand Road, Chester, Cheshire CH1 4QZ, UK
[1] Present address: Reynolds Geo-Sciences Ltd, 10 Bron y Nant, Mold, Clwyd CH7 1UX, UK.

Abstract. There are growing environmental pressures over the development, use and remediation of closed landfill sites throughout the UK, as well as overseas. There is also a reluctance to use direct investigative methods for fear of compromising the integrity of the landfill cap and liner systems. Consequently, there is an increasing role for geophysical surveys to be undertaken that are technically reliable, cost-effective and environmentally benign.

Geophysical surveys have been proven to be successful in the assessment of site suitability during the planning stage. For example, evidence of former mining, adverse geological structure, soft zones and areas with anomalous permeability may be detected. Geophysics can be used for mapping the thickness of mineral liners during the construction phase. Closed landfill sites can be investigated geophysically without invasion of the structure to determine the location of margins and the depth of fill, map possible leachate ponding and migration, and to assess the integrity of mineral and artificial liners, etc.

It is concluded that with current developments in data acquisition, processing, display and interpretation of geophysical data, the scope of environmental applications where multi-method geophysical surveys can be used is increasing rapidly. The range of geophysical methods currently available provides a powerful suite of investigative tools which complement direct observations.

Modern waste disposal facilities require elaborate and detailed investigations to ensure that appropriate design and safety precautions are undertaken. An increasing amount of legislation requires those responsible for waste disposal facilities to guarantee that their sites are suitably contained so as not to cause detrimental effects on the environment. The historical legacy of landfill sites throughout the UK is that many old tips have been long forgotten until they are rediscovered either by site investigation for redevelopment or by the emergence of ground problems (e.g. subsidence, gassing).

Consequently, there is a growing need for the ability to be able to investigate waste disposal sites cost-effectively and efficiently, and particularly using environmentally benign methods. Drilling boreholes through the base of a leachate-filled landfill is not acceptable without elaborate precautions to prevent leakage. Over the last 10–15 years, geophysical methods have been used increasingly. As both field data acquisition methods and interpretational procedures have improved dramatically over this time the potential for their future usage is greatly increased.

The objective of this paper is to provide an overview of the use of geophysical methods to investigate not only closed landfills, but sites being assessed to gauge their suitability for development as a landfill, or even during construction. Specific developments which have taken place since 1990 are described briefly to demonstrate the applicability of these geophysical tools.

Scope of geophysics

It is of vital importance when considering the use of geophysics, that the scientific as well as the commercial objectives of any survey are considered. Geophysical techniques should not be used prescriptively but instead incorporated within a broader site investigation strategy which may include direct methods (boreholes and trial pits). The role of geophysics should be examined during an initial desk study phase prior to any fieldwork. By so doing, the types of possible geophysical targets can be identified. On the basis of the likely contrasts in physico-chemical properties, the most appropriate geophysical method(s) can be selected. Furthermore, an equally important consideration is the nature of the existing site, i.e. how suitable it is for the deployment of geophysical methods. For instance, it is essential to ensure that there is sufficient space to undertake such a survey. Overhead power cables, buried utilities, metal fences and above-ground structures, vehicles, dense vegetation, etc., can

From BENTLEY, S. P. (ed.) *Engineering Geology of Waste Disposal*,
Geological Society Engineering Geology Special Publication No. 11, pp 93–98

Table 1. *Geophysical methods used in landfill studies*

- Electrical resistivity methods (vertical electrical sounding and sub-surface imaging)
- Self polarization
- Electromagnetic methods (ground conductivity, VLF, ground-penetrating radar)
- Magnetic methods (susceptibility, total field intensity, gradiometry)
- Gravity including microgravity
- Thermal mapping
- Seismic reflection and refraction
- Geophysical cones (seismic and conductivity)

all restrict the use of field methods in various ways and should be considered when designing any site investigation. This applies to all sites from greenfield to derelict urban wasteland.

It is important to consider the potential use of geophysics not just over closed landfills, but also over sites which are being investigated prior to and during the construction of new waste disposal facilities.

Geophysical methods available for the assessment of landfill and other waste disposal sites have been reviewed by Reynolds & McCann (1992) and are listed in Table 1. Of the methods cited, the most useful are those which respond to contrasts in conductivity (resistivity and electromagnetic (EM) conductivity) and in magnetic properties.

Self-polarization is seldom used as it requires very specific field conditions to be successful. It is also a very slow method to deploy and the interpretation is at best qualitative in most cases. All the other methods have specific applications but not all are appropriate for use over closed landfills. For example, ground-penetrating radar is seldom likely to yield useful results for two main reasons. First, if a clay cap is present over the site, the clay absorbs radar energy and little depth penetration can be achieved. Secondly, if no clay cap is present, but the landfill is wet, the ambient conductivities are likely to be high (typically in the range of 50–200 $mS\,m^{-1}$). Under such conditions, limited penetration into the landfill material can be achieved. At best, the top of the conductive zone may be detected. Thermal mapping is most suited to gassing landfills where methanogenesis, which is exothermic, can be detected as it will be apparent at the surface by localized 'hot-spots'. However, this method provides no information about subsurface thermal conditions. Borehole geophysical methods tend to be used in pre-construction surveys for geological correlation, and for ongoing monitoring of leachate migration around active landfills. A variety of case histories, largely from the USA, have been published in a series of publications edited by Ward (1990) to which reference should be made for further details. Geophysical cones, a development of the established Dutch cone technology, are being constructed to investigate the vertical variability in *P*- and *S*-wave seismic velocities and in low-frequency conductivity. The technology associated with conductivity cones is still being developed.

Particular descriptions of the various geophysical methods related to use on landfills and over contaminated sites have been given by Reynolds & McCann (1992) and by Reynolds & Taylor (1992) respectively.

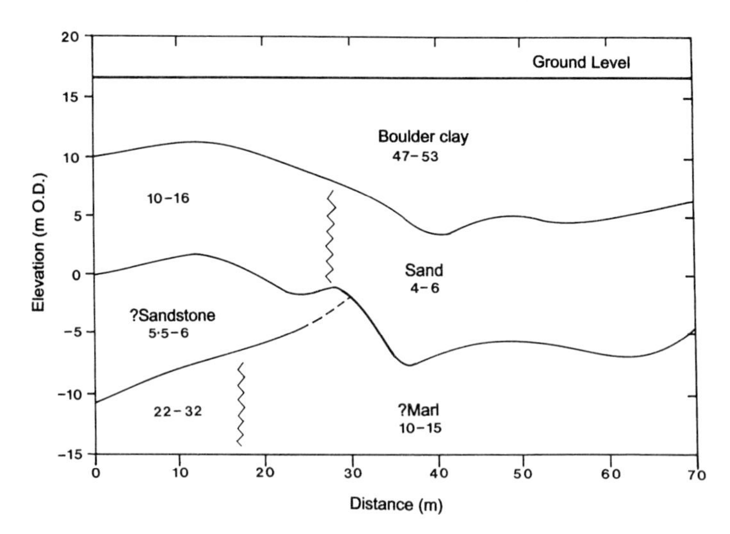

Fig. 1. Pseudo-geological cross-section derived from electromagnetic ground conductivity data. Values cited for geological materials are true conductivities in $mS\,m^{-1}$.

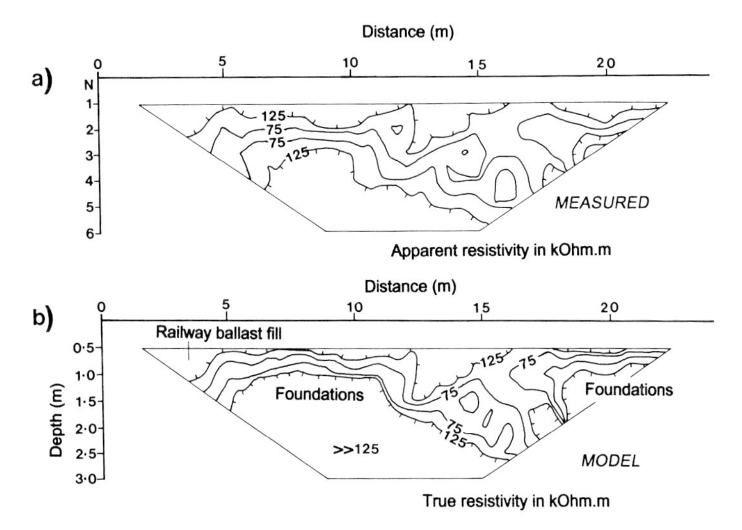

Fig. 2. Electrical resistivity sub-surface imaging pseudo-sections: (a) apparent resistivity profile, and (b) true resistivity–depth profile, over buried concrete slabs at 1 m depth.

More detailed explanations of many of the individual methods can be found in modern textbooks on geophysics, such as those by Kearey & Brooks (1991) and Reynolds (1995).

Electro-magnetic ground conductivity

Of particular importance to the investigation of landfill sites are electromagnetic ground conductivity mapping and electrical sub-surface imaging (SSI). While the former has been available for almost two decades, the methods by which the acquired data can be processed and interpreted have been developed substantially during recent years. Given adequate data quality and specific data coverage, it is possible to invert the apparent conductivity data which are measured directly in the field using equipment such as the EM31 and EM34 made by Geonics Ltd, Canada. The computer program which is used for the inversion process (EMIX34PLUS produced by Interpex Ltd, USA) creates a layered-earth model in which the thickness and true conductivity of each layer are estimated. The program computes a synthetic apparent conductivity value which is compared with the measured values. The software adjusts the layer parameters to reduce the error between the synthetic and observed data. The final display is a pseudo-geological cross-section with each layer represented being related to a known geological material by virtue of its true conductivity, which is a diagnostic physical property. With added borehole control the derived layered-earth model can be constrained quite tightly to produce a model which closely resembles the actual ground composition and structure. The software can be used where the layers are roughly horizontal and are laterally extensive. The modelling is not valid where there are three-dimensional structures smaller than the dipole size of the EM equipment being used. An example of the final interpretational display is given in Fig. 1. Other forms of data display as aids to interpretation include utilizing spatial display of apparent conductivity data and calculating the first and second horizontal derivatives of the apparent conductivity profiles.

Electrical resistivity sub-surface imaging

Barker (1992) has described a development of electrical array mapping in which a series of 25 electrodes connected by multicore cable can be addressed remotely using a laptop computer. The inter-electrode separation depends upon the required depth penetration and vertical resolution. The method utilizes the Wenner four-electrode array in which current is passed between two outer electrodes and the potential difference between the inner two is then measured. The ratio of the applied current to the measured potential difference gives a value of resistance. By multiplying this value by a geometric factor appropriate to the spacings of the electrodes, an apparent resistivity value is calculated. The SSI method addresses sets of four electrodes starting at one end and moving along the array, one electrode spacing at a time. The electrode spacing is then doubled and the process repeated. By increasing the spacing six times, a full pseudo-section of apparent resistivity values is collected and displayed (Fig. 2(a)).

An inversion software package is used to process the raw apparent resistivity data in order to produce a true resistivity–depth profile (Fig. 2(b)). Barker (1992) has presented a true resistivity–depth profile across a closed landfill from which it was evident that an anomalously low resistivity zone was present below the base of the landfill. This was interpreted to be associated with a leachate plume below the landfill.

The analytical procedures have been further modified (Barker, pers. comm.) using a new deconvolution method which reduces the effect of processing artifacts and which thus produces a more reliable resistivity–depth display. The full details of the deconvolution process have yet to be published.

Applications

The main role of geophysics is to provide areal coverage of a site, as well as depth penetration, by permitting interpolation between boreholes, and to provide potential targets for further direct investigation. For any waste disposal site, it is important to consider the potential unknowns for which the geophysics may be used. These unknowns are listed in Table 2.

There are three main times when geophysics can be used for the investigation of landfill sites. The first is during the initial planning stage when a particular site is being considered for its suitability for landfill construction. Geophysics can be used to assist with the determination of the ambient geological structure of the site and to detect the presence of any potential subsurface hazards, if appropriate. The second is at specific stages during the construction of a given landfill, in order to investigate particular aspects of a landfill. The third, and most common, is over closed landfills, for which there are many possible usages. These three aspects of the use of geophysics are discussed in more detail below.

Pre-construction surveys

A multi-method geophysical survey was used as part of a feasibility study for a proposed landfill in South Wales. At the time of the fieldwork, the site was still an operational farm. A complex geological structure below the site, complicated by the presence of shallow coal mineworkings, was evident from information from a limited number of boreholes. The geophysics was used to assist in the definition of the overall geological structure and to locate any possible artificial voids associated with the extraction of coal.

Where shallow coal workings were suspected, electromagnetic ground conductivity mapping was undertaken and the results displayed as a series of maps. From these it was possible to identify areas where the coal was still intact and where it had been worked at shallow depth (<10 m). A cavity was identified and confirmed by drilling. Basic display of the apparent conductivity data also revealed the dip direction of the local strata which was found to be consistent with the borehole results.

To aid the general structural interpretation, a series of electrical-resistivity vertical electrical soundings (VES) was completed. The processed data were interpreted to yield a vertical electrical stratigraphy which was used to help correlate the geological structures across the site in conjunction with existing borehole information.

In the middle of the site, no boreholes had been drilled and thus the complex structure was not adequately resolved. In order to assist with this, a digital ground-penetrating radar survey was undertaken using low-frequency antennae (300 MHz down to 35 MHz).

Although the local geology was not very suitable for radar work, due to the local Coal Measures strata, very weak reflections were identified which indicated the dip directions. One sub-surface feature was located which was interpreted as being due to coal workings at a depth of the order of 22 m. To confirm some of these findings, large-scale trenching was carried out and the structural details interpreted from the radargrams were corroborated.

The use of the multi-method geophysical survey in this project was viewed as contributing valuable information which would not have been achievable using direct techniques without considerable expense.

Table 2. *Potential unknowns for any landfill site (after Reynolds & McCann 1992)*

Type of initial void space	Former quarry, valley, engineered site, previous industrial site, etc.
Type of lining	None, mineral liner (compacted clay), artificial liner (e.g. HDPE membrane), combination liner (clay + membrane).
Type of capping (beneath topsoil)	None (natural venting), clay, membrane, combination, other, etc.
Site dimensions	Areal size, depth, shape (especially margins)
Site/tipping history	Style and degree of compaction and cover, age, types and likely mixtures of wastes, etc.
Geological factors	Types of substrates, local hydrogeology, sub-site faulting, seismicity (risk assessment), previous resource exploitation, sub-site cavities, site (slope) stability, etc.
Factors related to landfill materials	Degree of toxicity, saturation, gas generation, liquor/leachate generation, leachate mobility, compaction density and variability, material composition (inert builders' rubble, putrescible material, industrial refuse, liquids, sludges), radioactivity, biological hazards, etc.

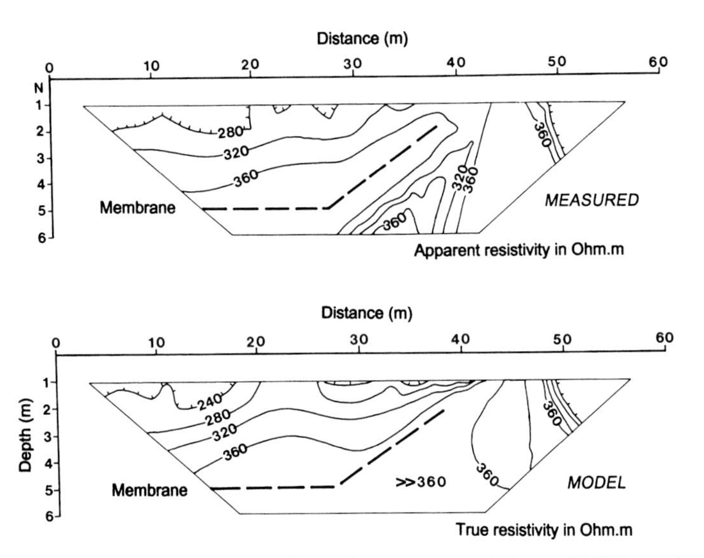

Fig. 3. Electrical resistivity pseudo-sections over an engineered structure containing an HDPE membrane (shown by dashed lines) indicating by the uniformity of the high resistivity values that its electrical (and thus hydraulic) integrity is intact.

Investigations during landfill construction

An electromagnetic survey was undertaken on a landfill site which was under construction in northwest England. The landfill was being engineered into a basal boulder-clay layer which was being used as a natural mineral liner. However, some concern was expressed about the variability in thickness of the clay layer which was known to overlie sand saturated with groundwater under artesian pressure. It was therefore entirely inappropriate to test the clay thickness by drilling. The geophysical survey was used to produce maps of the elevation relative to Ordnance Datum of the base of the boulder-clay/sand interface with an estimated accuracy of better than $\pm 10\%$ of the layer thickness. An example of the interpreted and processed conductivity data is presented in Fig. 1.

As a consequence of the geophysical survey, it was found that there were areas of thin boulder clay which affected the prospective construction of the individual cells within the overall landfill. This demonstrates that geophysics can also be used as part of the design procedure of a landfill. By incorporating the results of the geophysical survey, it was possible to redesign the location of the cell walls and define the lowest levels to which the boulder clay could be excavated so that the integrity of the mineral liner was not compromised.

At a site in North Wales, electrical resistivity Sub-Surface Imaging was used to investigate an engineered structure in which an artificial HDPE membrane was used to seal out local groundwater. The SSI was used in conjunction with ground-penetrating radar and electromagnetic ground conductivity mapping. The particular success of the SSI was that the hydraulic integrity of the membrane liner was checked successfully, as indicated in Fig. 3. As HDPE has a very high resistivity, of the order of 10^7 ohm m, any leak of groundwater through this liner would have resulted in an obvious low-resistivity zone. It is clear from Fig. 3 that the insulation afforded by the liner is intact as revealed by the high resistivity values present at the appropriate levels. The position of the HDPE membrane is indicated by a dashed line. It is important to make the distinction, however, that the SSI has been used to test the electrical integrity of the liner (and hence its hydraulic integrity), not to image the 4-mm-thick membrane which is not possible.

Table 3. *Examples of applications for which surface methods of applied geophysics can be used over closed landfills*

- Locating boundaries of old landfills
- Determining the depth to the base of the landfill
- Estimating the general composition of landfill
- Mapping areas with different tipping hsitories
- Identifying the presence of leachate
- Monitoring leachate migration
- Mapping groundwater contamination arising from leachate migration
- Locating buried metal (drums, pipes, scrap, etc.)
- Locating old slag deposits
- Investigation of old PFA lagoons
- Determining the depth of superficial spoil
- Determining the general structure of spoil tips

Investigations over closed landfills

Geophysical surveys over closed landfills are now commonplace. There are many uses to which geophysics can be put (e.g. Table 3). Perhaps the most common applications are to define the landfill edges, to estimate the thickness of the fill, and to investigate any particular matters related to the possible migration of leachate. While it is not yet possible to identify the presence of landfill gas build-up beneath a capping material within an active landfill, work is currently ongoing to test for this. Infill with high concentrations of landfill gases may have lower moisture contents relative to the surrounding material. This would lead potentially to a reduction in ambient conductivity (increase in resistivity) which may be detectable. Just how the signal due to the presence of gas would be differentiated unambiguously from the effects of variable fill is as yet not known.

Conclusions

It is clear that modern geophysical techniques can play a powerful role in the investigation of waste disposal sites. Geophysics can be used during the initial pre-planning application site-verification stage to ensure that the ground structure is appropriate for the development of a landfill. It has also been demonstrated that not only can geophysics be used to map the thickness of natural clay lining material, but by so doing, the information forthcoming can be used positively to modify the landfill design during the construction stage. Closed landfill sites are now being investigated routinely using environmental geophysics.

With the rapid development of modern data acquisition and interpretational methods, environmental geophysics promises to provide a suite of very powerful techniques for the investigation of waste disposal sites, during the planning, construction, operational and post-completion stages.

References

BARKER, R. D. 1992. A simple algorithm for electrical imaging of the sub-surface. *First Break* **10**, 53–62.

KEAREY, P. & BROOKS, M. 1991. *An introduction to geophysical exploration* (2nd edn). Blackwell, Oxford.

REYNOLDS, J. M. 1995. *An introduction to applied and environmental geophysics*. John Wiley, Chichester, in press.

—— & MCCANN, D. M. 1992. The geophysical methods for the assessment of landfill and waste disposal sites. *In*: FORDE, M. C. (ed.) *Proceedings of the Second International Conference on Construction on Polluted and Marginal Land*, 30 June–2 July 1992, Brunel University, 63–71.

—— & TAYLOR, D. I. 1992. The use of sub-surface imaging techniques in the investigation of contaminated sites. *In*: FORDE, M. C. (ed.) *Proceedings of the Second International Conference on Construction on Polluted and Marginal Land*, 30 June–2 July 1992, Brunel University, 121–131.

WARD, S. H. (ed.). 1990. *Geotechnical and environmental geophysics*, Vols I–III. Investigations in Geophysics No. 5, Society of Exploration Geophysics, Tulsa, Oklahoma.

Application of the STRATA3 visualization software to the investigation of landfill sites

A. Al Masmoum [1] & S. P. Bentley[2]

[1] Department of Civil Engineering, University of Makkah, Saudi Arabia
[2] School of Engineering, University of Wales, Cardiff, South Glamorgan CF2 1YF, UK

Abstract. STRATA3 is a PC-based computer program which provides a powerful means of storing and interrogating quantitative information on subsurface ground conditions. The input data may be any items of geological or geotechnical information that can be defined spatially; the program is at its most powerful when it is used to create three-dimensional models of subsurface environments. Two categories of data can be extracted from the models: qualitative data in the form of visual images and graphics and quantitative data in the form of precise measurements displayed in numeric form. The practical applications of the STRATA3 program to civil engineering practice are described and two case studies are presented. The first concerns the collation of site investigation data in an environmental geology project. The second case study illustrates the use of the program for managing chemical contamination data at an old gasworks adjacent to Cardiff Bay.

STRATA3 is a PC-based computer program that provides a powerful means of storing and interrogating quantitative information on the subsurface environment. The expression 'subsurface environment' is used purposefully because although the system was primarily designed to manage data on superficial geology together with water-table position, it can also be used to manage data on engineering parameters within the ground and chemical concentrations within the ground.

The program is at its most powerful when it is used to create three-dimensional models of the subsurface environment. Once created these models can be interrogated to provide graphical images in the form of perspective views, sections, contours and isopachytes; additionally the program provides a comprehensive range of functions for obtaining numerical information. For example, it is possible to make detailed measurements between any points on or inside the three-dimensional models.

The STRATA3 data management software has been developed as a collaborative effort between Wallace Evans Limited (Engineering and Environmental Management Consultants) and the School of Engineering at the University of Wales Cardiff. The intent was to develop software for internal use by Wallace Evans Limited. Development started in 1988 and subsequently several researchers from the University of Wales Cardiff (Greenshaw 1989) and programmers and engineers from Wallace Evans Limited and its parent company, Welsh Water plc, were involved. By early 1992 the basics of the software had been established but the user interface and certain other functions required further refinement. Prior to this final stage of work, market reseach was conducted so that the technical refinements would accord with industry needs. In early 1994, STRATA3 was purchased from Welsh Water plc by Cheadle-based MBA Geosoft and considerable program development has ensued.

Data points and interpolation

The procedure for creating subsurface models can best be illustrated with a geological example. Inputs for producing a geological model are generally twofold: geological information and geological experience or knowledge. Any item of geological information that can be defined spatially by x, y and z coordinates can be input to the program. Boreholes and trial pits provide rigorous items of information but the models must accord with all available geological map and cross-section information so this too can be input.

The layers within the model are produced by interpolation between data points. Three different interpolation methods are used; these are weighted average, triangulation and projected slope. There is a fine-tuning command which allows the user to specify the number of nearest neighbours used in the weighted average and projected slope calculations (the default is six), the weighting power applied to the distance weighted average calculation, and to enable or disable an octant search function for weighted average calculations. The octant search function divides the area around each point into eight equal segments and the nearest neighbour in each segment is chosen in preference to the nearest neighbour in absolute terms. This has the effect of smoothing out anomalies caused by borehole concentrations.

From Bentley, S. P. (ed.) *Engineering Geology of Waste Disposal*,
Geological Society Engineering Geology Special Publication No. 11, pp 99–102

The aim of the interpolation is to fit the data points precisely using shaped surfaces which are geologically acceptable. In certain cases, a second stage of data input is valuable, and this is the input of geological experience or knowledge. The procedure is as follows. A geologist interrogates the model by calling up cross-sections in a number of different orientations. Modifications to the shape of the layers are made by introducing 'artificial' data points. It tends to be an iterative procedure: artificial data points are introduced, the data set is reinterpolated and the effects are examined. In practice, each iteration is performed relatively quickly. We are not altogether happy with the term 'artificial' data point because they are not strictly artificial; they are judged to be essential and hence real by the geologist who is building and verifying the model. A better descriptive term might be that these are 'knowledge-based' data points.

Some construction professionals are wary about the introduction of artificial points and prefer to work with models which, although not strictly geologically correct, contain only proven, factual information. Clearly, the type and 'accuracy' of the models will depend on their end-use and the end-users.

A feature of the STRATA3 system is that a precis of any data point can be requested and displayed in the corner of the screen. For a borehole, this would show soil types and values of reduced levels.

Output facilities and model interrogation

Once a three-dimensional model has been finalized and approved it can be used to provide a wide range of data. The data that can be extracted from the models fall into two categories: qualitative data in the form of visual images and graphics, and quantitative data in the form of precise measurements displayed in numeric form. For all methods of extracting data the operations are requested from a menu using the mouse; the information requested is displayed virtually immediately.

Output facilities include:

- perspective views with the surface geology defined;
- perspective views with any of the upper layers stripped away; for example, if details were required of a soil type which occurred in the depressions on the rockhead then all the upper soil layers could be stripped away to reveal the distribution of this soil type on the bedrock. This interrogation tool is valuable during construction of the models when geological shapes and features are being verified;
- cross-sections, which can be obtained very quickly using the active mouse to define the direction change points and end-points. To define and draw a cross-section takes about 10 s.
- contour maps showing the elevation of any surface (e.g. rockhead, top of gravels, water-table), and isopachytes showing thicknesses of chosen soil layers;
- geological maps; the smoothness and accuracy of the geological map can be controlled by the user increasing the tile resolution;
- horizontal slice maps showing the geology at any subsurface level selected by the user; this function is useful for assessing foundation needs and for monitoring pollution with depth;
- volume calculations: the program calculates volumes within any defined polygon and between any specified levels within the ground model;
- independent layers which can be viewed on the cross-sections and as individual surfaces in perspective. These layers can be used to portray water-tables or contaminant levels; any number can be incorporated into models;
- the ability to obtain precise x, y and z coordinates by moving the cursor inside the model, over surfaces or on cross-sections. This is perhaps the most powerful of all the interrogation functions. A simple application is that it can be used to predict the geology in a proposed borehole within the modelled environment.

Many of the functions of STRATA3 can be implemented by inputting commands at the keyboard using 'hot keys'. This has the advantage of increasing speed for the competent user because it bypasses the need to access the menu system and in some instances provides for greater accuracy.

Any section, plan, perspective or combination thereof can be plotted and the user has control over scale, hatching, annotation size, legend style, paper size, colours, etc. The various parameters can be changed in the plotter submenu of the settings main menu. STRATA3 can download data to MOSS, the highways design package, and can create DXF files for export to CAD software, e.g. Autocad, or word-processing packages such as Wordperfect. When the original forms of input data are as databases or spreadsheets it is generally straightforward to directly link these storage systems to the modelling program.

Practical applications

One of the most important aspects of the program is that it enables engineers and geologists to construct models which accord with all available geoscience information about an area. The models can be built for urban areas where there are sometimes thousands of historical borehole and trial pit records, or the models can be built for new construction sites where large numbers of boreholes are to be drilled.

The vast numbers of historical borehole records that exist for cities and urban areas are largely under-utilized by engineers and geologists during the desk study stage of a site investigation. Long-term cost benefits would be made if historical records were systematically collected and collated and used to produce three-dimensional models.

As a geoscience data management system, the STRATA3 program has a wide range of applications both in civil engineering and urban planning practice. From a planning viewpoint, an important advantage of the program is the provision of three-dimensional graphics which permit enhanced appreciation by developers, planners and other non-specialists of complex geological and geotechnical conditions. In the field of civil engineering the main applications are in:

- Site appraisal and investigation, where the rapid provision of geological and geotechnical data from historical records can provide substantial cost benefits for the feasibility study and the desk study stage of investigations. The program can also be used to rapidly interpret new borehole data within the context of existing records.
- Preliminary foundation design of buildings and highway structures. The program provides an indication of the depth to suitable bearing strata and permits, at an early stage in any project, estimates to be made of the likely foundation requirements.
- Preliminary costing of groundworks and earthworks.

The program also has utility in urban planning practice, most notably in land-use planning and development control in hazardous environments. In land-use planning, the influence of ground conditions on development costs can be taken into account so as to reduce the impact of unforeseen and inappropriate ground conditions. Local ground conditions can be optimized when planning the layout of structures within individual sites and, if appropriate, a similar logic can be applied to the selection of sites for particular types of development. For example, three-dimensional ground models can assist in the selection of optimum sites for car parks, green belt/recreation land, subsurface structures, multistorey structures, etc.

With regard to development control in hazardous environments, the STRATA3 program can be used to gauge the degree of hazard from natural phenomenon such as ground instability or flooding. For example, such hazards may relate to the distribution of topography together with particular geological sequences and/or to the occurrence of a high watertable. The modelling program highlights the individual factors that might be detrimental to the long-term security of developments and permits an interpretation of their effect, or combined effects, in the absence of speciality hazard maps.

Construction of geological and geotechnical models of this type uses historical information that has already been paid for; the modelling is a means of obtaining additional value for money from this existing information.

Environmental geology

As part of a rolling programme of environmental geology mapping, the Department of the Environment funded a study concerning 550 km^2 of coastal wetlands around the Severn Estuary. The aims of the study were:

- to compile and interpret information on engineering and construction problems, remedial measures and their costs in a form which could be applied generally to comparable settings;
- to present geological, geotechnical and related information in a form in which it could be readily taken into account in land-use planning, development and development control.

Output from the study (Hornby *et al.* 1993) included a set of 43 thematic maps, covering themes such as:

- data coverage
- geology
- geomorphological features
- mineral and water resources
- surface hydrology
- infrastructure
- statutory and other planning constraints

Five of the maps were compiled to illustate how earth science information could and should be used by different professionals involved in the development process. These maps concerned:

- earth science and safety constraints for planning;
- earth science information for engineering/construction;
- earth science information for development;
- earth science and infrastructure information for mineral development;
- earth science and other control information for conservation.

For these maps, guidelines were written to assist each user group in obtaining maximum benefit from the information.

Additionally, STRATA3 was used to construct a computer model of the three-dimensional geology of a portion of the study area which has been earmarked for major investment and redevelopment. This provided an illustrative example of how the mapped earth science information relevant to the user groups could be 'brought to life' as a three-dimensional model. It was also possible to demonstrate how the program could be used to interrogate the model to provide quantitative decision and design data.

Contamination study

This case study illustrates the use of the program for managing chemical contamination data at an old gasworks of some 15 ha adjacent to Cardiff Bay, UK. The construction of a barrage across the mouth of Cardiff Bay which will impound the waters of the rivers Taff and Ely has commenced. The creation of this 300 ha freshwater lake will cause a permanent rise in the surrounding groundwater levels (Thomas 1996). The gasworks study aimed to investigate the distribution and concentration of contaminants under the site and to assess the implications of water-table rise.

In total, 91 soil samples were tested for the following parameters: pH, total cyanide, acid-soluble sulphate, sulphide, toluene-extractable matter, phenols, lead, arsenic, selenium, copper, nickel, zinc, chromium, cadmium, barium, beryllium, molybdenum, strontium, vanadium and manganese.

Some 36 water samples, 24 from trial pits and 12 from boreholes were tested for the following paramenters: pH, electrical conductivity, cyanide, sulphate, sulphide, total organic carbon and polcyclic aromatic hydrocarbons.

In environmental audits of this kind such large testing programmes are common and clearly the data are difficult to appreciate and assess when presented conventionally.

A number of simple enhancements to the functionality of STRATA3 were necessary to enable the processing and presentation of chemical data. First, a function was needed to provide a continuous definition of concentration values down the sampling profiles. A simple function of connecting known values with straight lines within the limits of determination was used for this purpose. This function permitted interpolated chemical concentration data to be obtained for any depth for all data locations. The data obtained were then used by existing program routines to provide contour maps of horizontal slices through the site. The contour intervals used in each chemical map can be made to coincide with threshold and trigger levels corresponding to each contaminant (EEC 1979; ICRCL 1986).

The information obtained from the contaminant mapping exercise was used to assess the ground and groundwater within the zone of potential inundation associated with the barrage impoundment. This, together with consideration of possible pollutant transfer mechanisms, allowed identification of the implied risk resulting from the predicted increase in groundwater levels.

Conclusions

As a geotechnical data management system, the STRATA3 program has a wide range of uses in civil engineering. Practical applications include: site appraisal and investigation, preliminary foundation design and preliminary costing of groundworks and earthworks. The program can incorporate new items of ground information easily and quickly, thereby creating up-to-date working models to assist decision-makers. By its ease of use, the program also encourages the good practice of interpreting new ground information in the context of historical data. In terms of communication between construction professionals and with non-specialists, an important advantage of this program is the provision of three-dimensional graphics which permit enhanced appreciation of complex geological and geotechnical conditions.

STRATA3 has all the qualities necessary for it to become a routinely used software package in civil engineering practice.

Acknowledgements. The authors wish to thank members of the Geotechnical Department of Acer Wallace Evans, Penarth.

References

EEC 1979. *The protection of groundwater against pollution caused by certain dangerous substances.* EEC Directive 80/68, Brussels.

GREENSHAW, L. M. 1989. *A multi-surface graphics package for application in geotechnical data management and land use planning.* MPhil thesis, University of Wales, Cardiff.

HORNBY, R. P., BENTLEY, S. P., EDWARDS, R. J. G. & RICE, S. M. M. 1993. *The presentation of earth science information for planning, development and conservation.* Acer Wallace Evans.

INTERDEPARTMENTAL COMMITTEE ON THE REDEVELOPMENT OF CONTAMINATED LAND 1986. *Notes on the redevelopment of gaswork sites.* ICRCL 18/79, London.

THOMAS, B. R. 1996. Possible effects of rising groundwater levels on a gasworks site: a case study from Cardiff Bay, UK. *Quarterly Journal of Engineering Geology*, in press.

Coal carbonization in northeast England: investigating and cleaning-up an historical legacy

R. A. Forth[1] **& D. Beaumont**[2]

[1] Department of Civil Engineering, University of Newcastle, Drummond Building, Newcastle upon Tyne, NE1 7RU, UK
[2] Civil and Geotechnical Laboratory, Durham County Council, County Hall, Durham DH1 5UL, UK

Abstract. The combustion of coal in an oxygen-deficient atmosphere produces a complex mixture of gaseous, liquid and solid components depending on the coal composition and the temperature of combustion. Large quantities of coke, attaining a peak production of some 30×10^6 t per annum in 1956, were produced, generally on large sites. Hundreds of small gas works existed throughout the British coalfields in the nineteenth century and the first half of the twentieth century. Liquids such as tar, amoniacal liquors and crude benzol were produced. Spent oxides, containing cyanides ('blue billy'), produced during the final stages of gas production were frequently dumped near the gas works on available land. Local authorities have, in some cases, been given the task of identifying the location of these long-abandoned and heavily contaminated sites to bring them into productive use either as residential, industrial or recreational sites. This paper discusses the investigation and treatment or disposal of the hazardous waste products. It recommends that engineers and planners discuss redevelopment plans and remedial measures for sites at an early stage and maintain close and continuing liaison throughout the site investigation and construction phases of redevelopment programmes.

Coke has been produced in Britain since the early 1900s following the rapid growth in the iron and steel industries, particularly in northern England. Hundreds of derelict sites where coke was produced are scattered throughout the northeast coalfield, ranging from sites of tens of hectares, such as at Hawthorn Colliery, County Durham, to sites of less than an acre where coke was a by-product of gas works which provided lighting for towns and villages.

Coke achieved its maximum production level of about 30 million tonnes in 1956, declining to about 8 million tonnes in 1984. Hardly any coke at all is commercially produced at the present time.

Production

The combustion of coal in an oxygen-deficient atmosphere produces a complex mixture of gaseous, liquid and solid products, proportions of which depend on the coal composition and the temperature of combustion (Peebles 1980).

Gas, ammoniacal liquors and tar is produced by condensation of the non-solid products when leaving the combustion chamber and entering a hydraulic main. Gas is partially purified in the main and then further purified to remove ammonia, cyanides and sulphur compounds.

Broadly speaking the higher the temperature of combustion, the more gas (but with a lower calorific value) is produced. However, less tar and ammoniacal liquors are produced.

The final stage of gas purification involved the use of ferric oxide which reacted with the gas to form iron sulphides and iron cyanides—the spent oxides or 'blue billy', so called because of its colour.

Tar and benzol were refined in distillation and chemical treatment processes involving alkali and acid washes and distillation. Sulphur-free benzol was distilled to produce a variety of chemicals including toluene and heavy naptha.

Waste products

Any of the chemicals produced (ICRCL 1986, 1987) may be found on coal carbonization sites due to dumping of wastes, leakage or spillage during production, and during the demolition process when, in an effort to provide a graded site, contaminants may have been spread around.

Some of the contaminants (such as tar and 'blue-billy') are generally obvious to the naked eye either at the surface or when encountered during site investigation. However, the liquid wastes, such as the ammoniacal liquors, cannot usually be detected in this way. Other

From BENTLEY, S. P. (ed.) *Engineering Geology of Waste Disposal*,
Geological Society Engineering Geology Special Publication No. 11, pp 103–110

Table 1. *Hazardous by-products from coal carbonization sites (from DOE, 1987)*

Hazard	*Form*	*Contact*	*Hazardous concentration*
Tar	Semi-solid	Skin	500 mg kg^{-1}
Phenols	Liquid	Skin	50 g l^{-1}
	Soil		20 000 mg kg^{-1}
Spent oxides	Solid	Skin	40 000 mgkg^{-1} (ingestion)
			50 000 mg kg^{-1} (skin irritation)
Thiocyanites	Gas and liquid	Inhalation and skin	1–4 ppm*
Free cyanide	Soil	Ingestion	2500 mg kg^{-1}
	Gas	Inhalation	300 mg kg^{-1}
Sulphide	Soil	Skin	50 000 mg kg^{-1}
	Gas	Inhalation	only at very high concentrations
Pitch	Dust	Inhalation	?

* Cyanogen chloride.

solids such as ash, flue dust and high-silica bricks can be identified visually although high-silica bricks cannot be distinguished from normal bricks by sight.

Hazards

The following by-products are considered hazardous: tar, phenols, spent oxides/thiocyanites, free cyanide, sulphides, sulphur, acid liquids, methane and pitch dust (see Table 1). In addition, acidic soils with pH values as low as 2 can produce skin irritation and lead to the production of toxic gases such as hydrogen cyanide. The main deleterious effect, however, is on concrete in buildings and services.

Sites in the Durham coalfield

The range of sites investigated in the Durham coalfield extends from relatively small sites of less than an acre which produced gas for lighting for villages and towns, to sites of tens of hectares where coal and coke were produced in large quantities. The local authority has been delegated the responsibility of identifying the hazards within some of these sites and carrying out appropriate remedial action to bring the sites back into economic production either for residential or commercial development. Large sites clearly have the greatest potential for development; the cost of remediation on some of the smallest sites would preclude redevelopment in many instances.

Site investigation

The historical development of the sites is in the first instance investigated by a study of the old maps and plans, where available, and photographs. For example, Ordnance Survey maps dated 1854, 1894, 1912, 1920, 1940 and 1961 were used to trace the development of the site of the proposed Spennymoor Police Station (Fig. 1). This enables the former industrial use to be ascertained and is of considerable benefit in identifying the actual position of buildings and hence the probable areas of contamination. Nearby subways and old voids/sewers could also be filled with contaminants. The geological strata can normally be identified by reference to the British Geological Survey maps and records which provide information on boreholes in the site vicinity as well as a geological interpretation. Aerial photographs are available covering the Durham coalfield from 1945 and 1961. Other photographic records are usually available in the local council offices or museum, and elderly local people are often a useful source of additional information. A walk-over survey is also essential.

After obtaining all the relevant background information the most appropriate and cost-effective site investigation can be planned. This often does not follow strictly the Guidelines of DD175 (BSI 1988) as to do so would involve excessive expenditure. DD175 recommends that a minimum of three samples per sampling point are taken. The number of sampling points varies from 15 for a site of 0.5 ha to 85 for a site of 5.0 ha. Instead the investigation is carried out in phases and initially is concentrated on areas of likely contamination identified from the desk study and walk-over survey. Considerable benefit is gained from the use of trial pits which are relatively cheap to carry out and provide the investigator with an excellent visual appraisal of the site. Pits are logged from the surface if greater than 1.2 m in depth and unsupported by shoring (Health and Safety Regulations). Normally, bulk samples from the trial pits would be taken at regular intervals. However, chemical testing would be limited to those samples which clearly show contamination

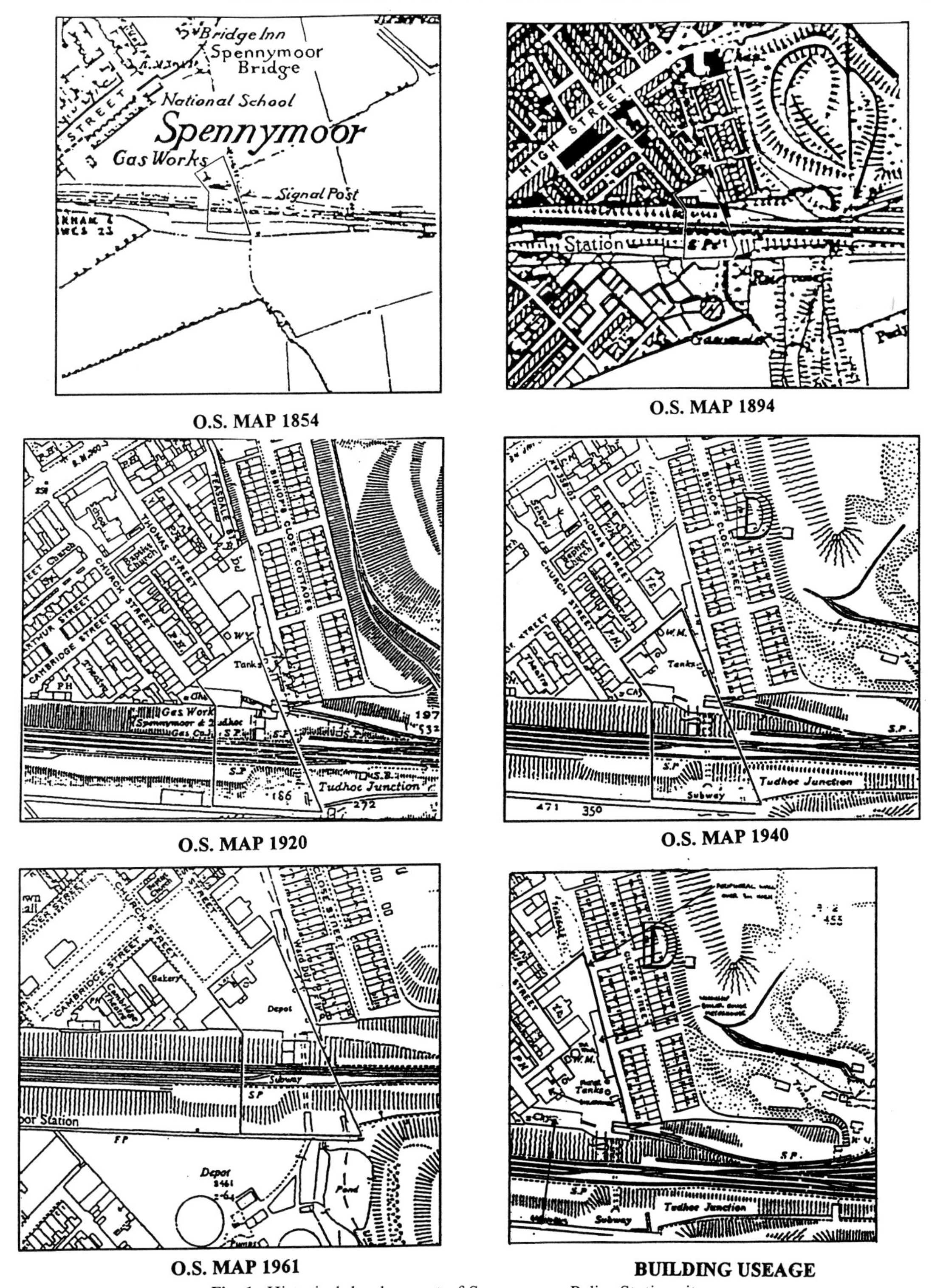

Fig. 1. Historical development of Spennymoor Police Station site.

visually or by smell, and on or adjacent to those parts of the site where, from the desk study, contamination is expected. This is a useful screening exercise prior to assessing the information and carrying out more detailed further phases of site investigation.

A large number of trial pits, down to about 4.5 m, can be carried out for the cost of one borehole. Boreholes, however, are required to investigate the ground below the range of the mechanical excavator. Cable percussion boring is often extended until rockhead or an obstruction is encountered. At the sites of former coal carbonization work such obstructions are common as a result of demolition and the presence of underground storage facilities, notably for tar, and old foundations.

At least one borehole would be extended through into rock to confirm the geological strata below the overlying sequence (normally made ground and glacial deposits). In the Durham coalfield several coal seams may be present beneath the site and a mineral valuer's report is always obtained before site investigation proceeds. Depending on the proposed ultimate use of the site, rotary cored holes, using air flush, are used to confirm or otherwise the presence of exploited coal seams. Investigations are normally confined to the top 30 m of the bedrock. Air flush is the best method of drilling to locate voids as the response (loss of air pressure at the drill head) is instantaneous.

The site investigation is therefore not only directed towards determining the degree of contamination of the site but also the geotechnical design parameters. It is also of critical importance to determine groundwater levels, the degree of contamination and the direction of flow of groundwater, particularly if there are water courses adjacent to the site. This may well involve siting boreholes for monitoring purposes off-site, and maintaining a monitoring programme over an extended period.

In some cases where gases such as methane or carbon dioxide are suspected from made ground it may be appropriate to insert gas monitoring wells and monitor these at regular intervals.

Sampling of liquids and soils is carried out according to a strict sampling protocol, which includes the use of fixing agents and various sampling containers for storage.

Safety aspects

Investigations of former coal carbonization sites are potentially hazardous and it is essential to take precautions to protect site personnel. Skin contact can be avoided by the use of overalls with elastic cuffs and sealable pockets. Boots should be fastened at the top to prevent ingress of contaminants. Goggles may be necessary in some instances. Washing facilities, and perhaps in some cases, showers, should be provided on site. Before leaving site, overalls, boots and gloves should be removed. No smoking or eating is permitted in the works area. To prevent inhalation of contaminated dust or vapour, face masks may be necessary.

On the largest sites of several hectares where site control may be difficult it may be necessary to establish restricted areas where only properly equipped personnel may enter. Guidelines for investigations on contaminated land have recently been published (Site Investigation Steering Group 1993).

Chemical analyses

The following chemical analyses are typically carried out on former coal carbonization sites: pH, phenols, arsenic, boron, cadmium, chromium, lead, mercury, nickel, zinc, copper, selenium, sulphides, thiocynate, total cyanide, total sulphate and toluene extractable matter (TEM). Screening for polyaromatic hydrocarbons (PAH) is normally carried out first. Total PAH is detected to a limit of $10\,mg\,kg^{-1}$, the determination being by ultraviolet light. A more accurate and expensive method of analysing for 16 priority pollutant PAHs to a limit of $1\,mg\,kg^{-1}$ is carried out by gas chromotography. Water samples are analysed where appropriate and tested for the same suite of contaminants as the soils (excluding TEM). Where samples of tar are encountered these may also be analysed. The cost of suite of chemical tests is of the order of £100 per sample. The results of chemical testing is interpreted according to the ICRCL (1987) trigger values and, for groundwater, the Dutch standards (NVPG 1990). Interpretation is carried out with the end-use of the site borne in mind.

Remediation of the sites

The contaminated land, once identified, can either by treated *in situ*, removed or encapsulated, or a combination of methods may be used. Commercial pressures often preclude *in situ* treatment. At the present time, the option of treatment *in situ* by biological or chemical methods is unproven although it is claimed that the toxic wastes can be neutralized by calcium oxide or quicklime. Developments in this area can be expected over the next few years.

The choice, therefore, usually lies between encapsulation or removal. The preferred solution depends on a number of factors such as volume and toxicity of contaminants, proximity and availability of 'special waste disposal' sites, and end-use of the decontaminated site.

Encapsulation requires the construction of an engineered containment structure into which the

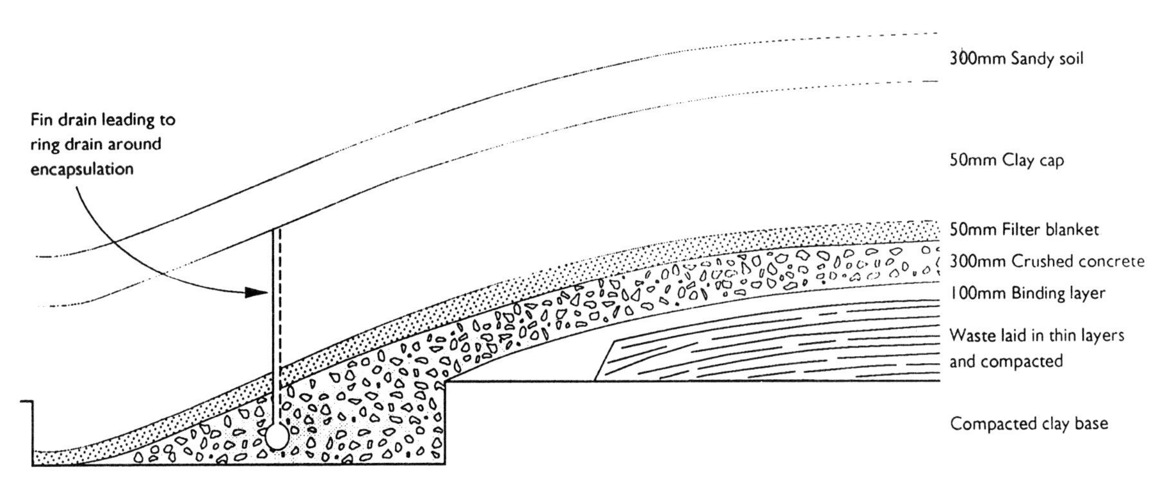

Fig. 2. Typical encapsulation design (after Cairney 1992).

contaminated material can be placed and sealed with an impermeable cap. The long-term integrity of the structure must be assured. A multi-layer structure such as described by Cairney (1992) using clay bunds and/or geomembranes (HDPE) is the commonest solution (Fig. 2). A layer is incorporated into the structure to prevent the upward and downward movement of capillary water. This layer could consist of a 300 mm thick layer of crushed concrete, rubble or gravel.

The encapsulation system may be wholly or partially below ground level, depending on site conditions and proposed land use. An impermeable cap consisting of about a metre of compacted clay fill is invariably employed, with suitable drainage around the structure.

Case histories

Hawthorn Coke Works, County Durham

A desk study was carried out on the site of the former coke works at Hawthorn, which is at present being considered for industrial development. The site is very large, extending over some 46 ha. Examination of old Ordnance Survey plans show that prior to 1954 the site was a greenfield area. Hawthorn shaft was sunk between 1954 and 1957 and the coke works opened in 1957. The coke works closed in 1984 and was demolished in 1987. The colliery continued to produce coal until 1991 and demolition included the removal of all foundations to a depth of 600 mm.

The geological profile of the site shows glacial deposits about 15–20 m thick overlying fine-grained calcitic dolomite of the Permian Middle Magnesian Limestone. Coal was mined from six seams at depths in excess of 200 m and ground movement from these operations should by now have ceased.

The majority of the site is covered with up to a maximum of 6 m of colliery spoil. During demolition, the by-products area was covered with a thin layer of clay, approximately 0.5 m thick. Contamination consists of coal tar, much of which is contained in nine tar wells which are 3–4 m diameter and over 2 m deep. There are, however, other areas of tar-saturated soil as a result of indiscriminate dumping. Spent oxides have been deposited in two areas. Tanks marked containing ammonia are present on the site and hence there is a strong probability of ammoniacal liquors. Near the tar wells there is an area of distinctive red colouring of unknown origin and content.

The site has been subdivided into areas of high, moderate and low contamination. Site investigation will be targeted on the following areas of concern:

- coal and coke storage area: 15–20 boreholes to 20 m depth
- mine area: 5–7 boreholes to 20 m depth
- by-products area: 5–7 boreholes to 8 m depth.

Standpipes will be installed to monitor gas and water quality. In addition, an extensive programme of trial pitting will be carried out as this is easily the most

Table 2. *Sampling Density (after Ferguson, 1992)*

Sample no./hectare	*Circle radius (m)/area (m^2)*
128	5/78
30	10/314
14	15/707
10	17/1000
8	20/1257
5	25/1963
4	30/2827
3	35/3848

cost-effective way of exploring areas of shallow contamination.

The sampling pattern and density is determined from the results of the desk study. The by-products area, extending over some 6 ha, is the main area of high contamination. A herringbone sampling distribution (Ferguson 1992) with 14 sample points per hectare will detect a circular target with a 95% probability (Table 2). Thus, with 84 trial pits in this area, the total number of samples would be 252 (i.e. three samples per pit).

A second phase of sampling may be appropriate depending on the results of this investigation.

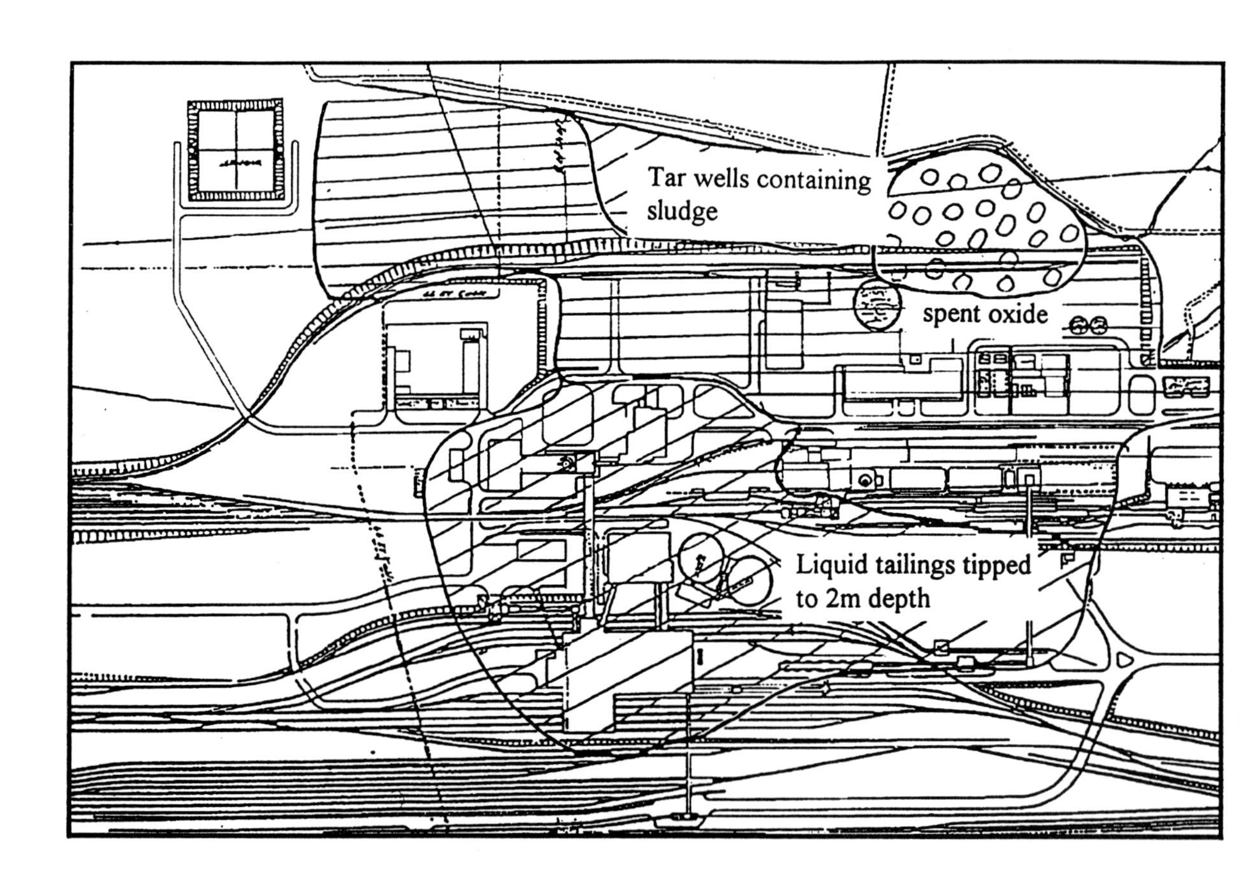

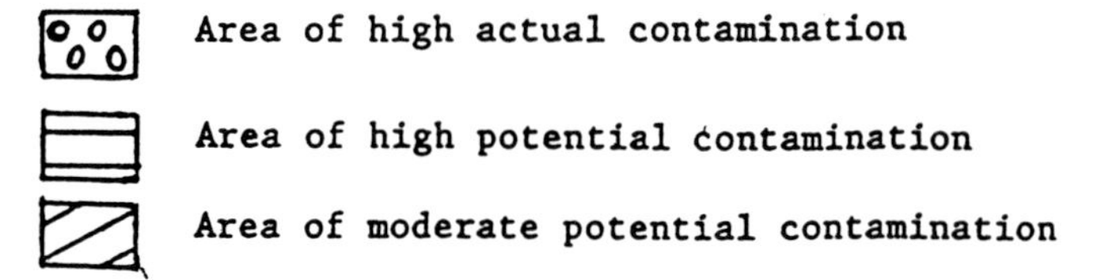

Fig. 3. (a) Plan and (b) photograph of Hawthorn coke works.

Fig. 3. Continued.

The moderately contaminated area of about 5 ha can be sampled at a density of five samples per hectare, which would detect a circular target of 25 m radius with a 95% probability. The total number of trial pits would be 25, and, with three samples per pit, a total of 75 samples would be taken. In the event of any areas of high contamination being located, closer density sampling can be instigated.

The rest of the site, some 25 ha, is considered to be of low contamination potential and a structured sampling programme may not be necessary.

The total cost of the pitting and sampling operation on this basis would be around £35 000, of which, with the exception of about £1000 (hire of JCB), would be accounted for by testing.

Spennymoor Police Station

The history of the Spennymoor Police Station site, located adjacent to the Spennymoor Leisure Centre, can be traced from a perusal of the Ordnance Survey maps from 1854 to the present day (Fig. 1). The 1854 map shows an east–west railway crossing the southern part of the site whilst in the northwestern corner of the site a gasworks is shown. By 1894 the site had changed dramatically with the development of the Tudhoe Ironworks to the east, a gasometer to the south and residential properties to the north and east. A circular structure (possibly a tar tank) is present at the southern end of Thomas Street, although this structure does not appear on the 1912 map. The gasworks is clearly shown on the 1920 plan together with what appears to be a conveyor belt system from the adjacent railway siding. By 1940 this conveyor belt system had been demolished and it is possible that by this time the gasworks had been resited to the south of the railway. By 1961 the gasworks building had been demolished and the site was being used as a council depot. A site inspection revealed the presence of an old subway beneath the former railway embankment containing a 16 in. (40 cm) gas main, a culvert passing beneath the site in a north–south direction and the presence of a buried tank with two manhole covers.

A study of the geological maps indicates that glacial till overlies rocks of the Middle Coal Measure Series. The nearest colliery, Whitworth Park No. 2 Pit, was found to have 33 m of drift overlying the Coal Measures. The mineral valuer reported that three coal seams in excess of 90 m depth had been worked in the past, but that surface movements associated with the mining should have ceased.

Cable percussion boring and trial pitting were carried out and an underground tar tank was discovered. A borehole adjacent to the tank encountered tar in the soil suggesting that the brick sides of the tank were leaking.

Eighteen samples were tested for phenols, arsenic, cadmium, chromium, lead, nickel, zinc, sulphide, thiocyanate, cyanide (total), copper, toluene-extractable matter and pH. Only one phenol sample, six arsenic samples and ten copper samples exceeded the threshold values for buildings laid down in the ICRCL document (1987). The toluene-extractable matter (TEM) was found to vary between 0.05% and 2.46% from the soil samples tested. On the basis that the polyaromatic hydrocarbons (PAH) content can normally be taken as approximately half of the TEM content, then all samples were found to exceed the threshold value of 50 mg kg^{-1} for domestic gardens, allotments and playing areas. PAH screening confirmed these results.

Site investigations revealed that building rubble comprised a large proportion of the made ground around the site of the former gas works building. In 1977 the Leisure Centre was constructed, and land forming the disused railway embankment removed. Part of the site was landscaped.

Results of the investigation revealed contamination in the made ground on-site and it is suspected that the area of contamination extends off-site also to the west and north of the site at shallow depth.

The buried tar tank was found to be about 12 m in diameter and within the tank contaminated water to a depth of about 2 m was found to overlie about 3 m of tar. A buried brick arch structure, some 2–3 m wide, was also found passing beneath the wall of the adjacent site. This structure was found to be loosely filled with bricks and rubble. A buried filter tank (12 m diameter, 5 m deep) full of coke, ammoniacal liquors, phenols and 'blue billy' was found during site excavation.

Most of the contamination was found to be concentrated on the northern end of the site where the former gas works was located. In view of the relatively small volume of tar, and the construction proposals, the recommendation was for its removal off-site to a 'special waste tip'.

Summary

This paper illustrates problems associated with the investigation and remediation of contaminated land on the site of former coal carbonization works in northeast England. Examples are given of the intensity of investigation on two sites of vastly different dimensions, which illustrate the scale of the problems faced by local authorities. In order to provide the most cost-effective investigation it is essential that planners and developers liaise at the earliest opportunity with the geotechnical engineer, and continue this association throughout the investigation. In this way the investigation can be tailored with the future use of the site firmly in mind.

References

British Standards Institution 1988. *Draft Code of Practice for the identification of potentially contaminated land and its investigation.* DD175, British Standards Institution, Milton Keynes.

Cairney, T. 1992. *Environmentally secure encapsulations of contaminated land.* Proceedings of 2nd International Conference on Construction on Polluted and Marginal Land, London.

DOE 1987. *Problems arising from the redevelopment of gas works and similar sites.* Department of the Environment, London.

Ferguson, C. C. 1992. The statistical basis for spatial sampling of contaminated land. *Ground Engineering*, June.

ICRCL 1986. *Notes on the redevelopment of gas works sites*, 5th edn. Interdepartmental Committee on the Redevelopment of Contaminated Land 18/79.

ICRCL 1987. *Guidance on the assessment and redevelopment of contaminated land*, 2nd edn. Interdepartmental Committee on the Redevelopment of Contaminated Land 59/83.

NVPG 1990. *Soil purification.* Dutch Association of Soil Treatment Companies.

Peebles, M. W. H. 1980. *Evolution of the Gas Industry*, The Macmillan Press, Hong Kong, 1–235.

Site Investigation Steering Group 1993. *Site investigation in construction—4. Guidelines for the safe investigation by drilling of landfills and contaminated land.* Thomas Telford, London.

Settlement of Beddingham Landfill

A. B. Di Stefano

Contest Melbourne Weeks, The Oasts, Newnham Court, Bearsted Road, Maidstone, Kent ME14 5LH, UK

Abstract. Traditionally, the settlement of landfills has been empirically determined by observation and measurement conducted on landfills after final restoration. Since measurements undertaken in this manner do not normally commence until several years after the commencement of deposition, settlement characteristics immediately after placement cannot be determined. Further, the range of typical settlement values currently adopted, 10–35% of deposited waste thickness, is too great to enable accurate planning and design of restoration profiles.

A long-term research programme has been instigated, utilizing pressure cells buried within the refuse mass as landfilling progresses, to enable the determination of settlement characteristics from immediately after deposition has taken place through to final waste stabilization several decades into the future.

The programme has been running for approximately two years to date, and the data collected in the first year are presented, together with a preliminary analysis.

An essential part of the initial stages of any proposed landfill is an assessment of planning, legal, environmental and economic factors to determine viability and impact on surrounds. This must involve, *inter alia*, a preliminary design of the facility to ensure that all facets of the proposed development have been considered.

Such a preliminary design requires assumptions, as not all site-specific data would be available; some data would be too costly to obtain, or would involve an inordinate timescale to obtain. One such assumption is the expected settlement of the waste within the completed landfill facility.

Settlement assumptions, generally made at the conceptual stage of a project and rarely updated throughout its life, must be regarded as a critical input factor in facility design as site economics are directly affected by the quantity of waste input.

The scarcity of available published settlement data relating to UK waste and landfills that could be utilized for site-specific design prompted Blue Circle to commence a long-term monitoring programme encompassing four sites at differing stages of their life cycles. Generally, Blue Circle landfill sites accept only household, industrial and commercial wastes. Whilst the sites have differing site-specific plant, the waste-compaction procedures are similar, in line with current accepted good landfill practice, to ensure good compaction and thus maximize void space. Generally speaking, compacted waste densities immediately after deposition are of the order of $1\,t\,m^{-3}$ at all sites.

This paper presents site-specific results of the Beddingham Landfill for the first year's settlement monitoring data, and attempts a preliminary categorization, understanding and overview analysis. As it is based on relatively little short-term data, no attempt has been made to develop a prediction formula. This will be undertaken once several years' data are available from all four sites.

Installation details and methodology

Traditional settlement measurement has relied on the installation of survey monitoring points on a completed landfill surface, and thence the use of surveying equipment. This technique suffers from two disadvantages.

- Settlement measurement does not commence until landfilling is complete. In a deep site (say 40 m) which has a number of phases and cells resulting in alternate active and passive filling areas, a significant time period may elapse before settlement monitoring is commenced. Hence, the settlement of the waste during the initial period immediately after deposition is not monitored. It could be argued that it is total landfill settlement that is pertinent, not progressive. This issue will not be addressed here other than to say a complete understanding of the settlement process is required in order to develop accurate prediction models.
- Any fixed monitoring points (whether on the completed restoration surface or on an intermediate temporary waste surface) are prone to damage by aftercare and/or operational activities (if not by direct vandalism).

From Bentley, S. P. (ed.) *Engineering Geology of Waste Disposal*,
Geological Society Engineering Geology Special Publication No. 11, pp 111–119

Table 1. *Settlement data from Beddingham Landfill*

Date	*Monitor Level (mAOD)*						*Actual settlement (m)*					
	A		*B*		*C*		*A*		*B*		*C*	
	1	*2*	*1*	*2*	*1*	*2*	*1*	*2*	*1*	*2*	*1*	*2*
05-Jul-91	42.94		42.55		44.91		0.00		0.00		0.00	
12-Jul-91	42.82		42.31		44.79		0.12		0.24		0.12	
23-Jul-91	42.58		41.71		44.53		0.36		0.84		0.38	
26-Jul-91	42.67		41.66		44.46		0.27		0.89		0.45	
30-Jul-91	42.58		41.66		44.46		0.36		0.89		0.45	
02-Aug-91	42.62		41.64		44.47		0.32		0.91		0.44	
06-Aug-91	42.22		41.48		44.26		0.72		1.07		0.65	
09-Aug-91	42.55		41.44		44.22		0.39		1.11		0.69	
16-Aug-91	42.65	48.58	41.37	47.64	44.13	47.46	0.29	0.00	1.18	0.00	0.78	0.00
20-Aug-91	42.16	48.38	41.35	47.57	44.10	47.40	0.78	0.20	1.20	0.07	0.81	0.06
28-Aug-91	42.18	47.80	41.28	47.43	43.97	47.22	0.76	0.78	1.27	0.21	0.94	0.24
04-Sep-91	42.06	47.88	41.07	47.12	44.10	46.92	0.88	0.70	1.48	0.52	0.81	0.54
06-Sep-91	42.03	48.06	41.01	47.07	43.68	46.86	0.91	0.52	1.54	0.57	1.23	0.60
10-Sep-91	42.04	48.14	40.99	47.04	43.63	46.82	0.91	0.44	1.56	0.60	1.28	0.64
13-Sep-91	42.12	47.63	40.94	46.98	43.59	46.72	0.82	0.95	1.61	0.66	1.32	0.74
17-Sep-91	41.97	48.04	40.83	46.83	43.42	46.52	0.97	0.54	1.72	0.81	1.49	0.94
20-Sep-91	41.97	47.99	40.79	46.79	43.37	46.46	0.97	0.59	1.76	0.85	1.54	1.00
24-Sep-91	41.94	48.06	40.73	46.73	43.29	46.38	1.00	0.52	1.82	0.91	1.62	1.08
27-Sep-91	41.93	47.99	40.71	46.71	43.26	46.34	1.01	0.59	1.84	0.93	1.65	1.12
01-Oct-91	41.88	47.83	40.63	46.63	43.19	46.27	1.06	0.75	1.92	1.01	1.72	1.19
04-Oct-91	41.86	47.85	40.59	46.59	43.15	46.22	1.08	0.73	1.96	1.05	1.76	1.24
08-Oct-91	41.85	47.78	40.52	46.57	43.12	46.18	1.09	0.80	2.03	1.07	1.79	1.28
11-Oct-91	41.85	47.73	40.55	46.56	43.10	46.14	1.09	0.85	2.00	1.08	1.81	1.32
15-Oct-91	41.83	47.71	40.51	46.50	43.04	46.08	1.11	0.87	2.04	1.14	1.87	1.38
18-Oct-91	41.82	47.66	40.48	46.47	43.00	46.04	1.12	0.92	2.07	1.17	1.91	1.42
05-Nov-91	41.72	47.16	40.19	46.18	42.69	45.66	1.22	1.42	2.36	1.46	2.22	1.80
13-Nov-91	41.64	47.27	40.03	46.07	42.58	45.49	1.30	1.31	2.52	1.57	2.33	1.97
26-Nov-91	41.61	47.07	39.92	45.87	42.43	45.24	1.33	1.51	2.63	1.77	2.48	2.22
04-Dec-91	41.55	46.99	39.81	45.75	42.31	45.16	1.39	1.59	2.74	1.89	2.60	2.30
10-Dec-91	41.51	46.90	39.75	45.67	42.22	45.34	1.43	1.68	2.80	1.97	2.69	2.12
16-Dec-91	41.47	46.71	39.66	45.62	42.15		1.47	1.87	2.89	2.02	2.76	
23-Dec-91	41.44	46.62	39.57	45.44	42.12		1.50	1.96	2.98	2.20	2.79	
03-Jan-92	41.43	46.50	39.46	45.34	42.03		1.51	2.08	3.09	2.30	2.88	
09-Jan-92	41.39	46.42	39.43	45.29	41.97		1.55	2.16	3.12	2.35	2.94	
16-Jan-92	41.36	46.09	39.32	45.21	41.90		1.58	2.49	3.23	2.43	3.01	
28-Jan-92	41.41	46.53	39.43	45.40	41.90		1.53	2.05	3.12	2.24	3.01	
06-Feb-92	41.36	46.49	39.23	45.33	41.81		1.58	2.09	3.32	2.31	3.10	
27-Feb-92	41.53	45.80	39.38	45.06	41.85		1.41	2.78	3.17	2.58	3.06	
05-Mar-91	41.25	45.79	39.98		41.48		1.69	2.79	2.57		3.43	
20-Mar-92	41.22	45.94	38.97		41.40		1.72	2.64	3.58		3.51	
08-Apr-92	41.15	45.84	38.82		41.25		1.79	2.74	3.73		3.66	
15-Apr-92	41.11	45.54	38.77		41.21		1.83	3.04	3.78		3.70	
01-May-92	41.12	45.83	38.78		41.15		1.82	2.75	3.77		3.76	
22-May-91	41.13	45.80	38.76		41.11		1.81	2.78	3.79		3.80	
17-Jun-92	41.10	45.73	38.66		41.02		1.84	2.85	3.89		3.89	
13-Jul-92	40.96	45.73	38.42		40.81		1.98	2.85	4.13		4.10	
03-Aug-92	41.00	45.52	38.44		41.03		1.94	3.06	4.11		3.88	
03-Sep-92	40.93	45.47	38.33		40.91		2.01	3.11	4.22		4.00	
23-Oct-92	40.83	44.98	38.11		40.64		2.11	3.60	4.44		4.27	

Original relative depths must be entered in appropriate row 9.
A base, 32.30; B base, 11.80; C base, 11.30.

Relative settlement at A						*Relative settlement at B*						*Relative settlement at C*					
1-Base waste depth		*2-Base waste depth*		*2 − 1 Waste depth*		*1-Base waste depth*		*2-Base waste depth*		*2 − 1 Waste depth*		*1-Base waste depth*		*2-Base waste depth*		*2 − 1 Waste depth*	
m	%	*m*	%	*m*	%	*m*	%	*m*	%	*m*	%	*m*	%	*m*	%	*m*	%
10.64	0.00					30.75	0.00					33.61	0.00				
10.52	1.13					30.51	0.78					33.49	0.36				
10.28	3.38					29.91	2.73					33.23	1.13				
10.37	2.54					29.86	2.89					33.16	1.34				
10.28	3.38					29.86	2.89					33.16	1.34				
10.32	3.01					29.84	2.96					33.17	1.31				
9.92	6.77					29.68	3.48					32.96	1.93				
10.25	3.67					29.64	3.61					32.92	2.05				
10.35	2.73	16.28	0.00	5.93	0.00	29.57	3.84	35.84	0.00	6.27	0.00	32.83	2.32	36.16	0.00	3.33	0.00
9.86	7.33	16.08	1.25	6.22	−4.82	29.55	3.90	35.77	0.20	6.22	0.80	32.80	2.41	36.10	0.17	3.30	0.90
9.88	7.17	15.50	4.78	5.63	5.14	29.48	4.14	35.63	0.58	6.16	1.83	32.67	2.81	35.92	0.66	3.26	2.25
9.76	8.30	15.58	4.30	5.82	1.80	29.27	4.82	35.32	1.45	6.05	3.46	32.80	2.42	35.62	1.49	2.82	15.26
9.73	8.58	15.76	3.20	6.03	−1.72	29.21	5.02	35.27	1.59	6.06	3.32	32.38	3.67	35.56	1.66	3.18	4.44
9.74	8.51	15.84	2.71	6.10	−2.93	29.19	5.09	35.24	1.68	6.05	3.44	32.33	3.82	35.52	1.77	3.19	4.08
9.82	7.71	15.33	5.82	5.51	7.05	29.14	5.24	35.18	1.84	6.04	3.64	32.29	3.93	35.42	2.04	3.13	5.95
9.67	9.12	15.74	3.30	6.07	−2.39	29.03	5.59	35.03	2.25	6.00	4.27	32.12	4.43	35.22	2.59	3.10	6.85
9.67	9.12	15.69	3.61	6.02	−1.55	28.99	5.72	34.99	2.37	6.00	4.27	32.07	4.58	35.16	2.76	3.09	7.15
9.64	9.40	15.76	3.18	6.12	−3.24	28.93	5.92	34.93	2.53	6.00	4.27	31.99	4.82	35.08	2.98	3.09	7.15
9.63	9.49	15.69	3.61	6.06	−2.23	28.91	5.98	34.91	2.59	6.00	4.27	31.96	4.91	35.04	3.09	3.08	7.45
9.58	9.96	15.53	4.59	5.95	−0.37	28.83	6.24	34.83	2.81	6.00	4.27	31.89	5.12	34.97	3.29	3.08	7.45
9.56	10.15	15.55	4.47	5.99	−1.05	28.79	6.37	34.79	2.92	6.00	4.27	31.85	5.24	34.92	3.42	3.07	7.75
9.55	10.24	15.48	4.90	5.93	−0.03	28.72	6.60	34.77	2.98	6.05	3.48	31.82	5.33	34.88	3.53	3.06	8.05
9.55	10.24	15.43	5.21	5.88	0.81	28.75	6.50	34.76	3.01	6.01	4.11	31.80	5.39	34.84	3.64	3.04	8.65
9.53	10.43	15.41	5.33	5.88	0.81	28.71	6.63	34.70	3.18	5.99	4.43	31.74	5.56	34.78	3.81	3.04	8.65
9.52	10.53	15.36	5.64	5.84	1.48	28.68	6.73	34.67	3.26	5.99	4.43	31.70	5.68	34.74	3.92	3.04	8.65
9.42	11.47	14.86	8.71	5.44	8.23	28.39	7.67	34.38	4.07	5.99	4.43	31.39	6.61	34.36	4.97	2.97	10.75
9.34	12.22	14.97	8.03	5.63	5.03	28.23	8.20	34.27	4.37	6.04	3.64	31.28	6.93	34.19	5.44	2.91	12.55
9.31	12.50	14.77	9.26	5.46	7.89	28.12	8.55	34.07	4.93	5.95	5.07	31.13	7.38	33.94	6.13	2.81	15.56
9.25	13.06	14.69	9.75	5.44	8.23	28.01	8.91	33.95	5.27	5.94	5.23	31.01	7.74	33.86	6.36	2.85	14.35
9.21	13.44	14.60	10.31	5.39	9.07	27.95	9.11	33.87	5.49	5.92	5.55	30.92	8.00	34.04	5.86	3.12	6.25
9.17	13.82	14.41	11.47	5.24	11.60	27.86	9.40	33.82	5.63	5.96	4.91	30.85	8.21				
9.14	14.10	14.32	12.04	5.18	12.65	27.77	9.68	33.64	6.14	5.87	6.41	30.82	8.30				
9.13	14.19	14.20	12.78	5.07	14.50	27.66	10.05	33.54	6.41	5.88	6.19	30.73	8.57				
9.09	14.57	14.12	13.26	5.03	15.14	27.63	10.15	33.49	6.55	5.86	6.51	30.67	8.75				
9.06	14.85	13.79	15.29	4.73	20.24	27.52	10.50	33.41	6.78	5.89	6.06	30.60	8.96				
9.11	14.40	14.23	12.60	5.12	13.66	27.63	10.15	33.60	6.26	5.97	4.78	30.60	8.96				
9.06	14.88	14.19	12.87	5.13	13.52	27.43	10.81	33.53	6.46	6.10	2.74	30.51	9.23				
9.23	13.25	13.50	17.08	4.27	27.99	27.58	10.31	33.26	7.20	5.68	9.41	30.55	9.10				
8.95	15.88	13.49	17.14	4.54	23.44	28.18	8.36					30.18	10.21				
8.92	16.17	13.64	16.22	4.72	20.40	27.17	11.64					30.10	10.44				
8.85	16.82	13.54	16.83	4.69	20.91	27.02	12.13					29.95	10.89				
8.81	17.20	13.24	18.67	4.43	25.30	26.97	12.29					29.91	11.01				
8.82	17.11	13.53	16.89	4.71	20.57	26.98	12.26					29.85	11.19				
8.83	17.01	13.50	17.08	4.67	21.25	26.96	12.33					29.81	11.31				
8.80	17.29	13.43	17.51	4.63	21.92	26.86	12.65					29.72	11.57				
8.66	18.61	13.43	17.51	4.77	19.56	26.62	13.43					29.51	12.20				
8.70	18.23	13.22	18.80	4.52	23.78	26.64	13.37					29.73	11.54				
8.63	18.89	13.17	19.10	4.54	23.44	26.53	13.72					29.61	11.90				
8.53	19.83	12.68	22.11	4.15	30.02	26.31	14.44					29.34	12.70				

In order to overcome these two deficiencies, it was deemed necessary to install settlement monitoring stations within the landfill mass. The author is unaware of any similar approach being undertaken in the UK, apart from the studies of Watts & Charles (1990).

A market survey was undertaken to determine the most appropriate monitoring equipment for this application. A hydraulic/pneumatic pressure cell was deemed the most appropriate, offering the advantages of simplicity of design and installation, and a readily accessible readout facility.

The equipment consists of a pressure cell buried within the refuse mass, connected via two pairs of nylon tubes to a readout station located at a higher elevation than the cell, and on stable non-settling ground. One pair of tubes is water-filled, applying a constant hydraulic pressure to the cell via a header tank situated in the readout station. The other pair of tubes is utilized only during readings, when it is pressurized using nitrogen gas from a portable readout unit. When the nitrogen pressure equalizes the hydraulic cell pressure a direct reading of metres water is obtained, being the height difference between the readout station header tank and the pressure cell.

The main advantages of this system are as follows

- it is simple in design;
- no rigid rods or tubes are required; the nylon tubing is flexible which results in ease of installation and accommodates post-installation landfill mass movements; operational accidents are also hopefully eliminated;
- system accuracy is ±20 mm, far in excess of landfill requirements (±250 mm?).

The system's disadvantage is cell inaccessibility once installed, and the unknown long-term effects of the aggressive landfill environment on system components.

Six cells have been installed at Beddingham. Three cells were installed at 5 m, 20 m and 30 m (approximately) from the quarry face on an existing refuse surface, below which some 25 m of waste had been deposited over a number of years. These cells were designated Level 1, A, B and C respectively (Fig. 1).

These cells were then buried under 6 m of refuse and a further three cells were installed vertically above the original three, designated Level 2, A, B and C. Further refuse was placed above these cells immediately after commissioning.

The Level 1 cells were installed on waste that had been deposited over a period of years. That is, it had already undergone some settlement and biodegradation. The quarry base beneath the cells is not precisely known, but has been interpolated from historic records.

The Level 2 cells were installed on freshly deposited waste. An accurate depth of waste between the Level 1 and Level 2 cells is known.

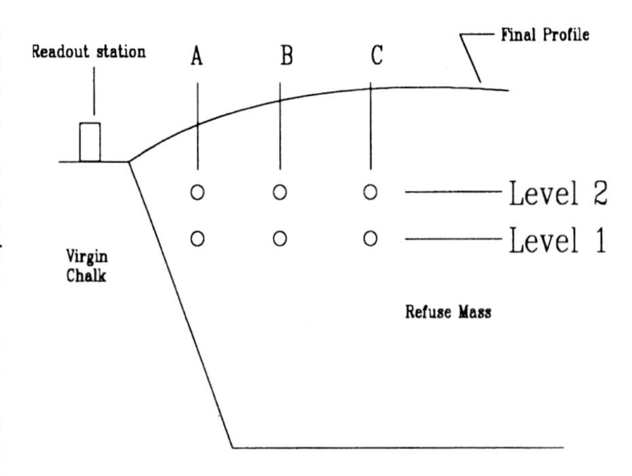

Fig. 1. Schematic installation layout.

Once final restoration levels are achieved, the installation of surface survey stations immediately above the A, B and C stations is planned. Thus two layers of freshly deposited waste and one layer of 'old' waste can be monitored.

Monitoring results

The collected monitoring data from July 1991 to October 1992 inclusive are presented in Table 1. The first six columns are the actual reduced readings obtained from the monitoring installation. Units are in metres AOD. The second set of six columns records the actual vertical movement of each cell, with respect to its original installed level. The remaining columns tabulate waste thickness and percentage settlements. Percentages are calculated with respect to the original thickness of waste at the date of monitor installation. The first six columns of Table 1 are presented graphically in Figs 2–4.

Preliminary analysis

Intuitively, the total settlement of a waste mass can be categorized into three distinct stages, although two or more of these stages may be occurring simultaneously. These stages, as described by Charles (1993), are:

- immediate physical compression, caused by the weight of successive fill layers;
- long-term biodegradation causing volume reduction;
- long-term physical creep compression.

In the case of the monitored Beddingham results, settlement of the new waste as recorded by the difference

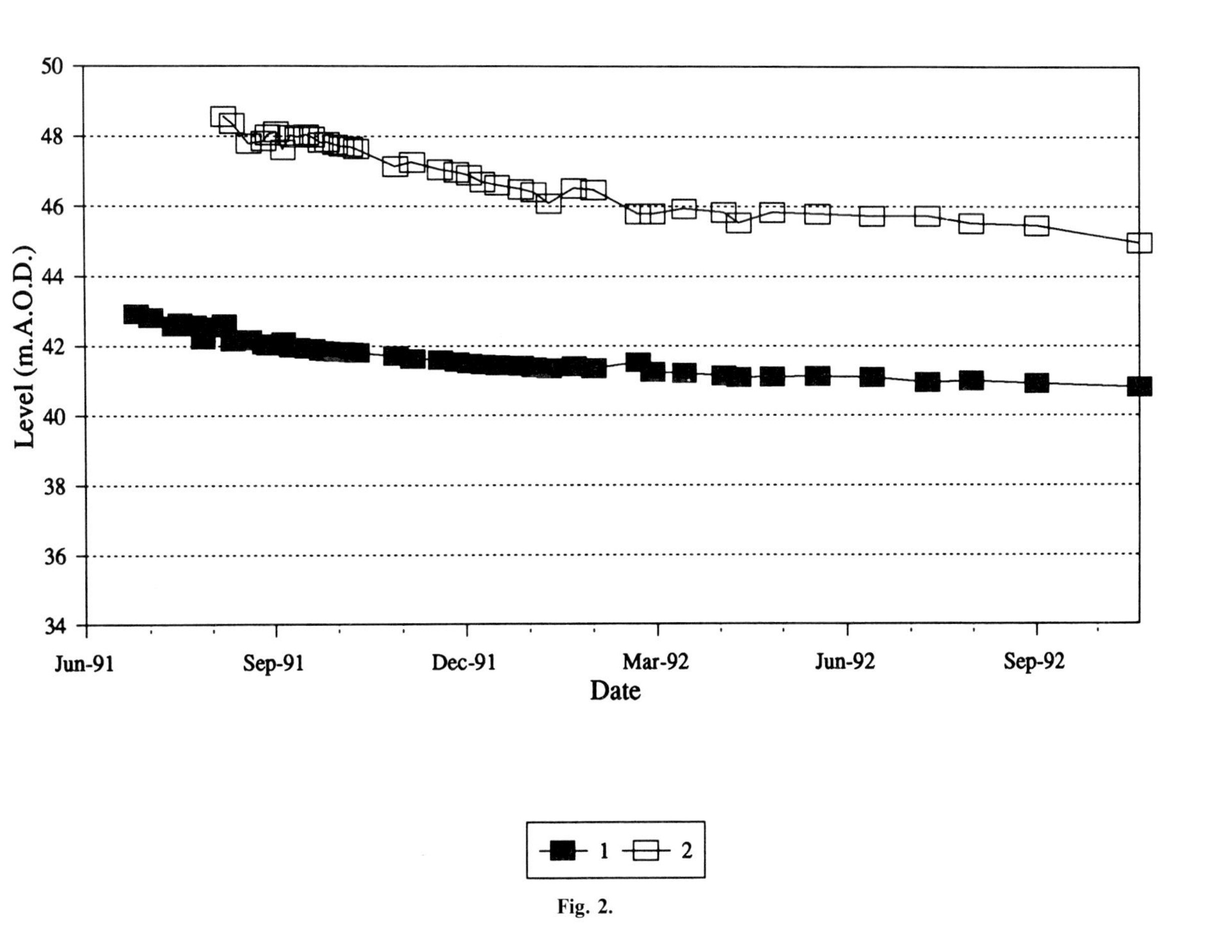

Fig. 2.

between the Level 1 and Level 2 cells without doubt fits into the first of the above categories, immediate physical compression, for the initial monitoring period, up to *c.* early 1992, when a change in settlement rate is discernible. After this point settlement can probably be attributed to a combination of both biodegradation and physical creep compression.

The 'old' waste settlement rates appear to follow a similar trend, but ostensibly for different reasons. This waste, having been deposited for several years, has undergone some biodegradation, but without surcharging. The author believes that the void ratio increased rather than settlement occurring. Thus when the 'new' waste was deposited the immediate settlement effects noted on the 'old' waste were due to a reduction in this void ratio—a 'catching up' effect on the long term biodegradation volume reduction. After early 1992, biodegradation was probably augmented with long-term physical creep compression.

If this model is correct, and the author's interpretation of the Beddingham results is valid, then these results, particularly the A monitors, represent all three stages.

Reviewing the total settlements recorded, the A cells have settled 20–30%, whilst both the B and C monitors have recorded percentage settlements of 5–14%, approximately half. This phenomenon has been noticed by previous researchers (e.g. Morris & Woods 1989) and to date has only been described as an 'edge effect'.

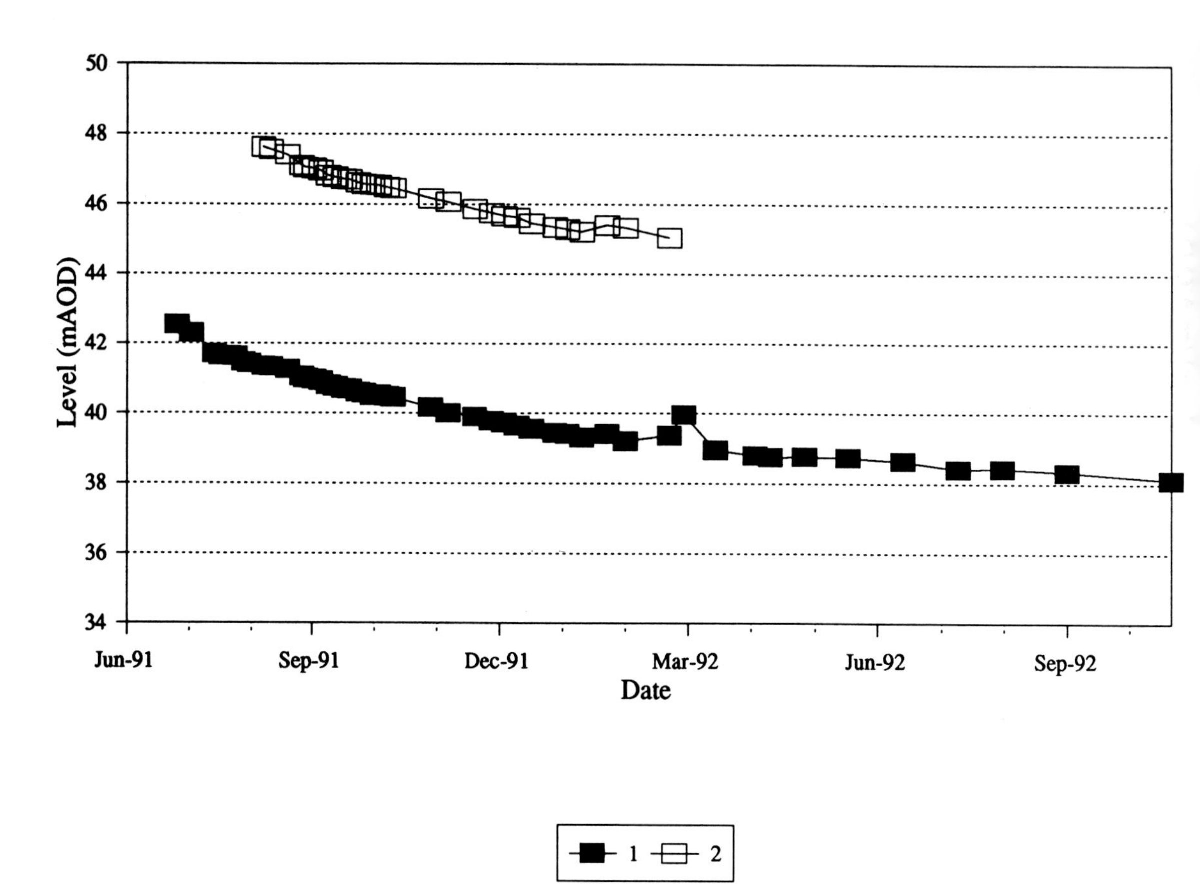

Fig. 3.

Irrespective of the cause, the phenomenon needs to be taken on board when setting surcharge levels at the planning application stage. It is generally not.

If the same percentage settlement is applied over the entire site, the 'edge effect' will result in depressions around the periphery of the site. Ideally, what needs to be incorporated into the design is an increase in percentage surcharge levels around the site periphery. At present this is generally achieved within the aftercare programme by surcharging with cover soils, over a number of years, sometimes as much as 25 m from the landfill edge. This is not only costly, but also disrupts the long-term restoration regime. Figure 5 schematically shows the present practice and consequences, and the possible solution suggested.

A review of the overall results raises doubts on conventional wisdom as to the total settlement that may be expected several decades into the future.

Considering the 'new' waste between Levels 1 and 2, the immediate physical compression has resulted in settlements at A, B and C of 9.1%, 5.6% and 6.3% respectively over the common time period August 1991 to December 1991. Over a longer time base, A has undergone a settlement of 30% in the first 15 months since deposition. The rate of settlement currently shows no sign of abating. The long-term physical creep and biodegradation ('old' waste below Level 1) has undergone 13.4%, 9.1% and 8.0% over the period August 1991 to December 1991, with the A1 cell settling 19.8% in the first 15 months.

Beddingham Landfill

Monitor C

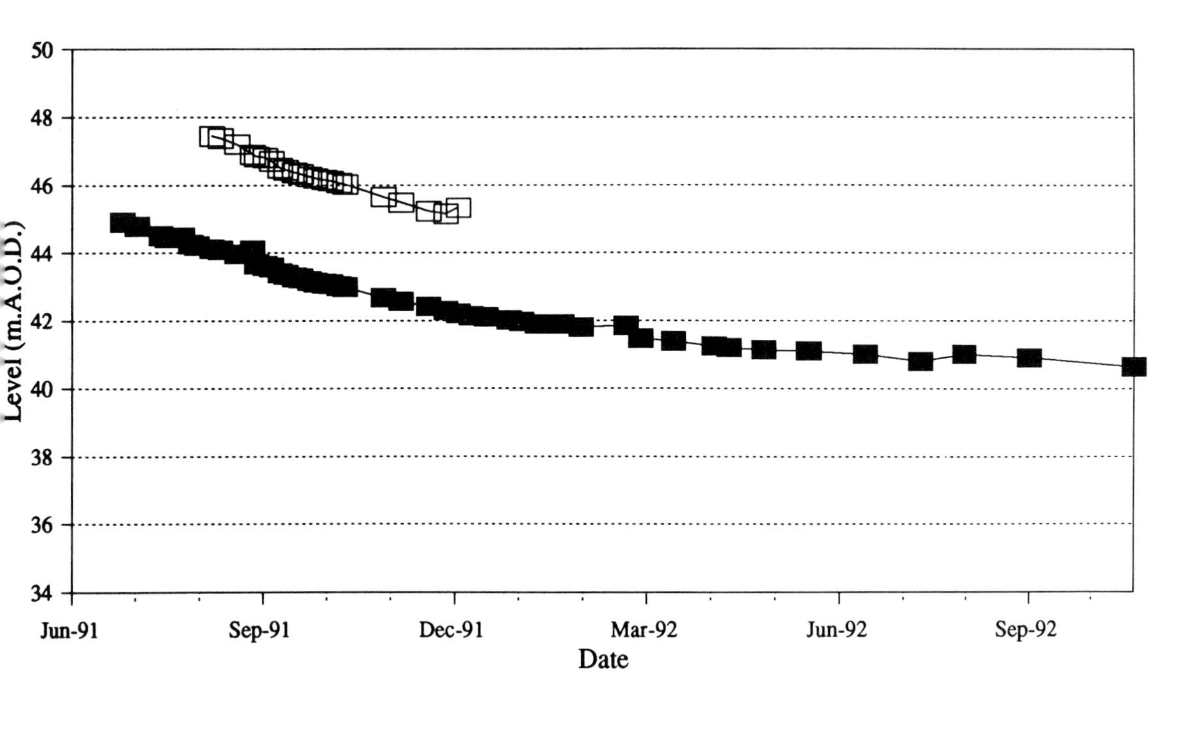

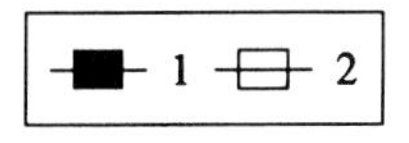

Fig. 4.

Whilst a direct addition of these settlements may not be strictly valid, it certainly gives an indication of the possible settlements that may occur. Over the shorter time period, this results in settlements of 22.5%, 14.7% and 14.3%, and over a longer period at position A, 49.8%.

Typical settlement guidance values currently adopted lie in the range 10–35% of deposited thickness. Waste Management Paper No. 26 (Anon. 1986) quotes 10–25%. Based on the Beddingham results these current guidance values underestimate the settlement potential of landfills, as in a four-month period settlements away from the edge of the landfill were recorded at 14%, those adjacent to the edge 25% (which is already at the upper bound of current accepted practice). The recorded settlement at A of 50% in 15 months also must give some cause for concern.

Conclusions

The settlement of landfills takes several decades to complete, if not longer. This current study is a long-term monitoring programme which will hopefully also span several decades to enable a detailed picture of landfill settlement to be developed. This paper presents the results of the first year's work on one site. As such, only a superficial assessment of landfill characteristics can hope to be presented. Indeed, the study findings to date

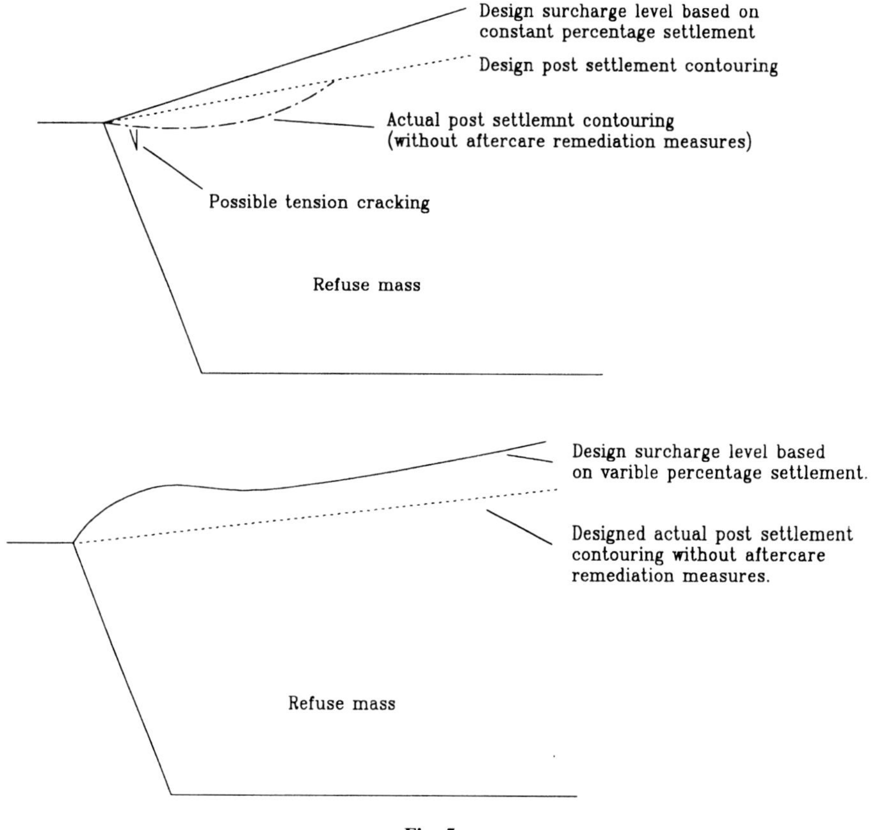

Fig. 5.

raise several issues requiring further analysis, rather than presenting solutions to commonly known problems. Complete results are presented, with a minimal of analysis, to encourage other researchers to undertake their own analysis of these presented data.

The principal findings of this study to date are as follows. Long-term biodegradation of the refuse mass may occur without significant settlement. If the waste is not surcharged (by other waste or by inerts) then an increase in voids may occur, resulting in sudden settlements when surcharging is applied at a later date. This could pose particular problems to shallow sites. Consideration should be given to surcharging completed landfills with inerts (to be removed later) to promote rapid stabilization of the waste mass, thus enabling final restoration to be undertaken sooner than would normally be expected. Such a course of action would obviously create short-term conflict between operators and statutory authorities, and local residents, though the long-term benefits of a quickly restored site should not be underestimated.

'Edge effects' can present restoration profiling and aftercare difficulties. The Beddingham study indicates that settlements at the edge of landfills (adjacent to quarry faces) may be double the settlement experienced within the main body of the landfill. Consideration should be given to incorporating variable percentage settlements at landfill edges at the planning application stage, to enable realistic post-settlement contouring compatible with the proposed afteruse to be incorporated into the initial landfill design.

Current accepted landfill practice allows for 10–35% settlement of total waste depth. The Beddingham study results indicate settlements of the order of 14% for a four-month period, with peak settlements of 50% for a 15 month period. It is therefore suggested that accepted current practice may underestimate the settlement potential of waste masses.

Acknowledgement. The author gratefully acknowledges permission to publish from Blue Circle Wastes Management, Landfill Division, Snodland, Kent ME6 5PH, UK.

References

ANON. 1986. *Landfilling wastes*. Waste Management Paper No. 26, Department of the Environment, London.

CHARLES, J. A. 1993. *Building on fill: geotechnical aspects*. Building Research Establishment Report.

MORRIS, D. V. & WOODS, E. W. 1989. *Settlement and engineering considerations in landfill and final cover design*. Proceedings of Conference, Geotechnics of Waste Fills—Theory and Practice, Pittsburgh, 9–21.

WATTS, K. S. & CHARLES, J. A. 1990. *Settlement of recently placed domestic refuse landfill*. Proceedings of the Institution of Civil Engineers, 971–973.

SECTION 2

CONSIDERATION FOR THE DESIGN OF NEW LANDFILLS

Pre-design evaluation of disused quarries as landfill sites

P. Lyle,[1] P. J. Gibson,[2] A. R. Woodside[1] & W. D. H. Woodward[1]

[1] School of the Built Environment, University of Ulster at Jordanstown, Newtownabbey, Antrim BT38 OQB, UK
[2] Department of Geography, St Patrick's College, Maynooth, Kildare, Republic of Ireland

Abstract. This paper outlines how faults and fracture zones may be detected with use of a total field precession magnetometer during the initial and preliminary stages of the assessment of disused quarries as possible landfill sites. The technique has been applied across a range of lithologies and can help in the more accurate assessment of potential pollutant conduits.

The use of standard permeability tests for natural liners and tensile tests for synthetic liners in the pre-design evaluation phase are discussed. The approximate costs of using such tests and geomagnetic profiling in pre-design evaluation are discussed.

Disused quarry sites are commonly considered as potential waste disposal sites. With stricter environmental controls being applied to the design and siting of landfill facilities, there is a need for more reliable assessment methods in the pre-design evaluation of potential sites. The use of geophysical methods for environmental investigations has increased in recent years and a total field proton precession magnetometer has been used to recognize geological features such as faults, fracture zones and igneous intrusions which are relevant to the geological and hydrogeological assessment of a potential landfill site. These features have been identified across a range of common rock types, often in the absence of visible outcrop and through vegetation and overburden cover (Gibson & Lyle, 1993).

Magnetometer use in pre-design hydrogeological assessment

To obtain planning consent for landfill operations it is normally a requirement to carry out a hydrogeological survey of the proposed site. This survey will assess the nature, disposition and permeability of the bedrock and will ascertain the patterns of groundwater movement. This is normally done by conventional methods of groundwater monitoring and the use of pump testing. The movement of groundwater through rock can be affected by the presence of zones of fractured or broken rock which may be associated with faulting on various scales. Gibson & Lyle (1993) have shown that these fracture zones in parts of northeast Ireland can be more than 1 km wide and may have a lateral extent of several tens of kilometres. Work carried out on a relatively impermeable sandstone in this area (Lyle & Gibson 1994) has shown that transmissivity values can be increased by up to a factor of ten if the rock has been fractured by adjacent fault movements.

While such zones are often clearly visible and can be detected by standard geological mapping techniques, their existence or full extent can often be obscured by vegetation or loose soil or rock. As such zones can be an important factor in the hydrogeological regime of the area under investigation and are clearly potential pollutant conduits in the event of any escape of leachate during the life of the site, it is important to assess as far as possible their full extent.

Details of the technique

The Earth's magnetic field varies from around 30 000 nanoteslas (nT) at the Equator to 60 000 nT at the Poles, with a value in the British Isles of around 49 000 nT. Departure from this average value is caused by a number of factors including variation in the concentrations of magnetic minerals such as the iron oxide magnetite in the bedrock. Such changes in mineral concentrations are likely to be concentrated along faults or associated fracture zones and consequently the magnetic signature in such zones is often markedly different from the normal background reading for the area. The magnetism is measured using a portable total field proton precession magnetometer consisting of a container of hydrocarbon fluid (decane) through which an electric current is passed. This causes the protons within the fluid to spin and act as small magnetic dipoles and to be aligned by the generated magnetic field. When the current is removed the spin of the protons causes them to precess about the direction of the Earth's magnetic field and this precession generates a signal whose frequency is proportional to the magnetic field intensity. The magnetometer measures the geomagnetic

From Bentley, S. P. (ed.) *Engineering Geology of Waste Disposal*,
Geological Society Engineering Geology Special Publication No. 11, pp 123–126

field to an accuracy of 0.1 nT. In general, the repeatability of readings at any one station is of the order of ±0.5 nT and traverses are made with a station spacing of 10–20 m, although this can be reduced to smaller intervals for better definition in the vicinity of magnetic anomalies.

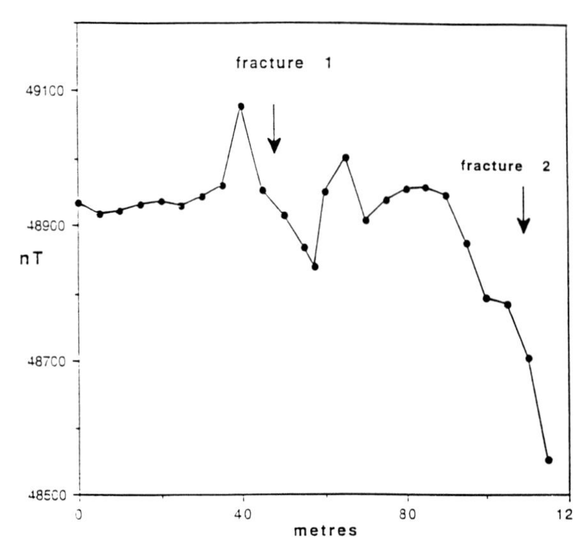

Fig. 2. Magnetic profile across a disused sandstone quarry showing the occurrence of two fracture zones associated with higher transmissivity values (after Lyle & Gibson 1994).

Typical magnetic anomalies

Fracture zone associated with large-scale faulting (Fig. 1)

This represents a major fault zone in northeast Ireland and is marked by a sharply defined 1.2 km wide zone of fractured basalts where the magnetic signature shows high-frequency anomalies of more than 2000 nT. This pattern is very different from the unfractured basalt and the sedimentary rocks which occur to either side of the fault (Gibson & Lyle 1993). Such a region is likely to be unsuitable for landfill sites but fracturing over such a wide area could not necessarily be anticipated from conventional geological mapping techniques in poorly exposed areas.

Fracture zones associated with small-scale fracturing

The magnetic profile shown in Fig. 2 is across a disused quarry in Silurian sandstones in County Down (Lyle & Gibson 1994). This quarry was a potential landfill site and was investigated using conventional geological mapping techniques, pump tests and magnetic profiling. Mapping showed the presence of a 3 m wide zone of fractured rock in the quarry face, Fracture Zone 1. A series of pump tests carried out on the floor of the quarry showed that this zone had higher transmissivity values than the relatively impermeable sandstone adjacent. The pump tests indicated a further zone of higher transmissivity, suggesting a second fracture zone (Fracture Zone 2), parallel to the first (Table 1).

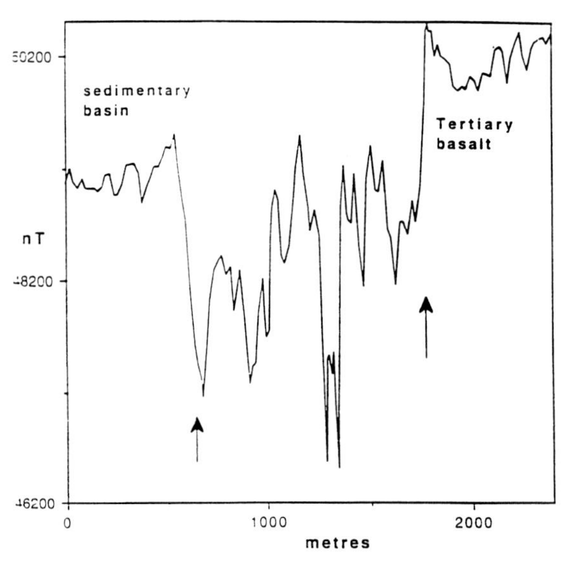

Fig. 1. Magnetic profile associated with large-scale faulting between basalts and sedimentary rocks in northeast Ireland.

Magnetic traverses made across the quarry clearly showed the presence of both fracture zones. For example, in Fig. 2, Fracture Zone 1 is marked by an anomaly of approximately 220 nT, while Fracture Zone 2 has values more than 350 nT below the normal value for the area. Geomagnetic profiling also showed that the fracture zones traversed the quarry floor where exposure was obscured by rubble and vegetation and continued beyond the limits of the site under investigation.

Small-scale igneous intrusions (Fig. 3).

Small-scale igneous intrusions such as dykes are capable of influencing groundwater movement. The magnetic profile shown in Fig. 3 is across a dolerite dyke intruded into Devonian sandstones which act as an aquifer. The

Table 1. *Pump test results*

	Transmissivity (m^3)
Fracture Zone 1	245
Fracture Zone 2	260
Unfractured Sandstone	36–56

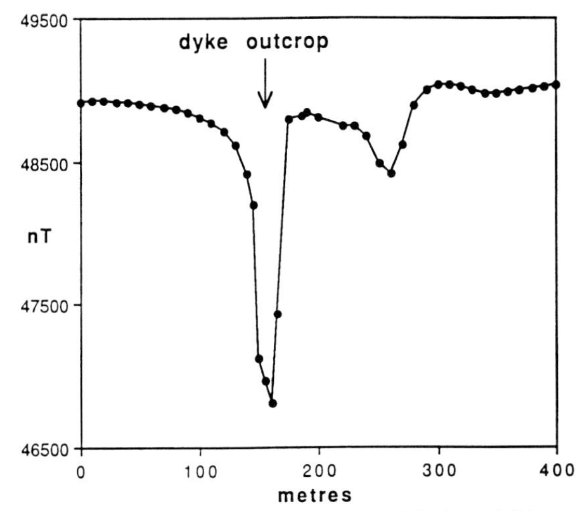

Fig. 3. Magnetic profile across a dolerite dyke intruded into sandstone.

dyke in this case is approximately 70 m wide and since it is impermeable relative to the intruded sandstone, it will act as a near-vertical barrier to the free movement of water. The dyke produces a negative anomaly of 2000 nT.

The use of magnetic profiling in the pre-design evaluation of a potential waste disposal site can, therefore delineate faults or fracture zones in the vicinity of the site; can identify minor intrusions which may influence groundwater movement; can produce a more accurate assessment of potential pollutant conduits when used in conjunction with other conventional techniques, and can therefore allow for more efficient placement of groundwater monitoring wells.

The technique has already been used to investigate disused basalt and sandstone quarries and current work is assessing the potential of the method for other lithologies such as limestone and granite.

Pre-design evaluation of liners

Clay liners

The lining of a landfill site may be a natural material which ideally has a permeability less than $1 \times 10^{-9}\ \mathrm{m\,s^{-1}}$. This is normally clay which may be present on site or imported, or in some cases alternatives have included materials such as pulverized fuel ash (PFA). When clay is to be used as a lining material care must be taken to ensure that it is laid at a moisture content which gives the required permeability. This is usually a few per cent greater than the optimum, i.e. for maximum dry density. Should the moisture content of the material be less than the optimum, compaction difficulties will be experienced.

It is essential to carry out the British Standard Test for permeability on extracted samples if the appropriate authorities are to be satisfied.

Synthetic liners

Over the last ten years many landfill sites have used synthetic liners in the form of butyl rubber, chlorosulphanated polyethelene, ethelene copolymer bitumen, ethylene–propylene rubber, neoprene, polyethylene or polyvinyl chloride (PVC). Each of these materials has advantages and disadvantages and as part of the pre-design evaluation process it is advisable to consider the suitability of such materials. The most popular form of synthetic lining in use today appears to be high density polyethelene (HDPE) sheeting with a thickness up to 3.2 mm. This material is not only capable of providing good chemical resistance and withstanding tensile stresses caused by moderate settlement of refuse or even the subgrade material; it is also commercially available throughout the UK. Another characteristic of HDPE is its visco-elastic properties and once a design has been carried out the designer must ensure that the material properties of the geosynthetic liner to be used on site are representative of the design criteria. It is recommended that preliminary sample testing and confirmatory tests during the contract are carried out.

It has been the authors' experience that consultants having specified the use of HDPE have insisted that the materials and the seams be tested in accordance with ASTM D638 and ASTM D413 Type B Method which provides a value for tensile strength and film tearing bond (FTB) or peel (PEEL). Certain UK contractors have considered it necessary in the past to have this work carried out in the USA. However, most material research laboratories in the UK have the facilities for testing such materials and the use of the Instron tensile testing frame is recommended for such purposes. As results are often required within a few days at most, to enable site work to progress, some laboratories may need to provide a faster, more efficient service to meet the needs of industry.

Cost of pre-design evaluation

The approximate cost of the pre-design assessments outlined above and excluding the cost of planning procedures is minimal when compared to the overall costs of developing a modern landfill site. The cost of a magnetometer survey as detailed above would be approximately £2000, permeability tests are approximately £200 per sample and preliminary tensile load frame tests are £50 per sample. Consequently a pre-design evaluation could cost less than £3000 and could save the developer a considerable sum in the recognition, for example, of an unsuitable quarry site.

References

ANON 1975. B. S. 1377: *Methods of test for soils for civil engineering purposes. Part 5. Compressibility, Permeability and Durability tests.* British Standards Institution, HMSO, London.

——1982. ASTM D413-82: *Standard test methods for rubber property—adhesion to flexible substrate.* American Society for Testing Materials.

——1986. ASTM D638-86: *Standard test methods for tensile properties of plastics.* American Society for Testing Materials.

——1989. ASTM D3083-89: *Standard specification for flexible PVC plastic sheeting for pond, canal and reservoir lining.* American Society for Testing Materials.

GIBSON, P. J. & LYLE, P. 1993. Analysis and interpretation of major magnetic anomalies within the Tertiary basalts of northeast Ireland. *Irish Journal of Earth Sciences*, **12**, 149–154.

LYLE, P. & GIBSON, P. J. 1994. Magnetometer profiling in the evaluation of potential waste disposal sites. *Proceedings of the Institution of Civil Engineers Geotechnical Engineering*, **107**, 217–221.

Waste disposal in steep-sided quarries: geomembrane-based barrier systems

D. R. V. Jones

The Nottingham Trent University/Golder Associates (UK) Limited, Landmere Lane, Edwalton, Nottingham NG12 4DG, UK

Abstract. The use of geomembrane liners on landfill sites in the UK has been widespread since the mid-1980s. Various types and thicknesses exist although high-density polyethylene (HDPE) liners are the most widely used for basal and side lining. Planning authorities still regard the import of waste materials as one of the best means of restoring worked mineral voids. The waste disposal industry must therefore continue to develop old quarry sites in order to provide a reasonable land bank of refuse disposal sites for the community. The engineering of these sites has become increasingly more difficult as the requirements for lining have tightened due to the need to contain leachate and landfill gas. This paper discusses the use of geomembrane liners for landfill applications, with particular reference to steep sided-quarries.

The waste disposal industry in the UK has relied for many years on the infilling of worked mineral voids as the primary means of disposal. These sites now have to be designed on a fully contained basis to prevent the migration of leachate and landfill gas. The barriers used are typically either low-permeability clay, bentonite-enriched soil, geomembrane or various combinations.

The lining of shallow-sided quarries such as clay, sand and gravel extraction pits for landfill, can provide the designer with problems, notably in side slope and leachate drainage design. Such problems, however, are magnified when considering steep-sided quarries. The use of hard rock quarries for waste disposal is becoming increasingly popular due to the huge potential void space. The engineering design of suitable lining systems for steep-sided quarries needs to be carefully considered.

This paper sets out current design practice for shallow-sided landfills with particular reference to the importance of geomembrane properties. Engineering considerations for the design of barrier systems incorporating geomembranes for steep-sided quarries will be discussed, together with a brief discussion of the importance of construction quality assurance.

The principle of containment

Containment, in terms of landfill engineering, means the prevention of migration of leachate and uncontrolled escape of landfill gas. Details of a typical composite clay/geomembrane barrier for a contained landfill is shown schematically in Fig. 1. The base liner is of paramount importance for the protection of groundwater from leachate. A typical section for a base liner is shown in Fig. 1(a) which comprises the following.

- The low permeability ($<1 \times 10^{-9}\,\mathrm{m\,s^{-1}}$) clay liner, typically 1 m thick. In areas where suitable clay is not available, a bentonite/sand mixture is becoming increasingly popular, and since its permeability can be made lower than the clay, a thinner layer may be suitable.
- The geomembrane, usually 1.5–2.5 mm thick high-density polyethylene (HDPE). The contact between the clay layer and the geomembrane has a significant effect on the performance of a composite liner. A good contact can only be achieved by supervised subgrade preparation and rigorous geomembrane construction quality assurance.
- The leachate collection and removal system is designed to prevent the build-up of leachate greater than a specified head (typically 1.0 m). Recent work by Brune *et al.* (1991) has raised questions regarding the permeability of sand drainage layers in the long term due to encrustation. In order to maintain the permeability of the system, fine to medium gravel may be needed. Collection pipes are then used to remove the leachate to a sump for treatment, either on or off site, or for recirculation.
- The first waste lift should comprise selected material to ensure that the liner is not damaged during this phase.

The side-slope liner prevents the escape of leachate and the migration of landfill gas from the site. HDPE geomembranes are again typically used, with textured HDPE geomembranes becoming increasingly popular due to the higher frictional properties. Drainage and protection measures are also required on side slopes; however, placement on sloping surfaces is more difficult than on the base. The drainage layer needs to be connected with the leachate drainage and collection

From Bentley, S. P. (ed.) *Engineering Geology of Waste Disposal*,
Geological Society Engineering Geology Special Publication No. 11, pp 127–131

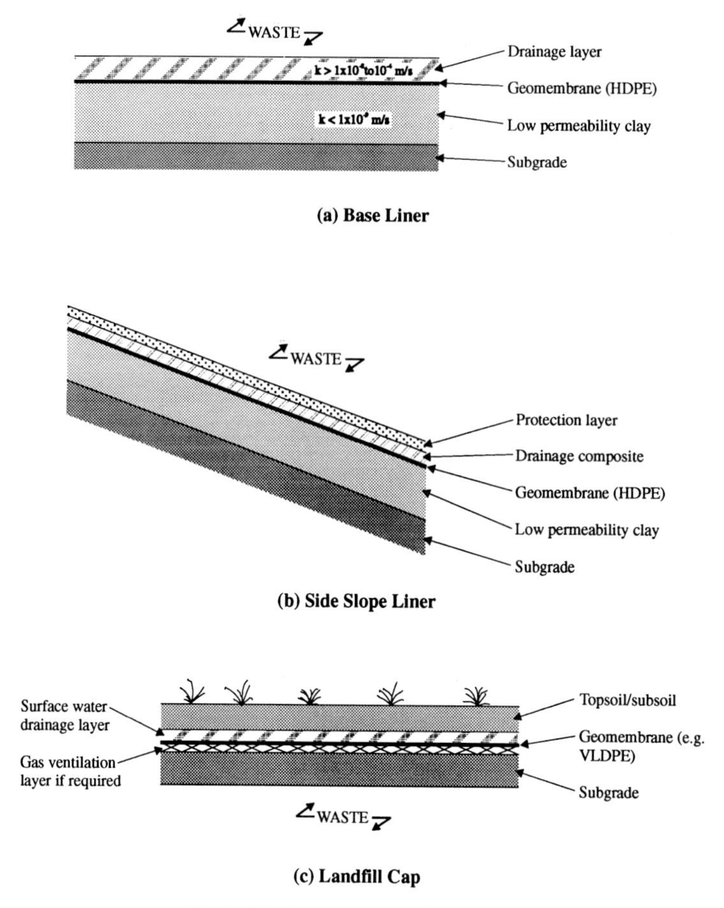

Fig. 1. Typical composite lining system.

system downslope. A typical system is shown in Fig. 1(b).

The landfill cap is constructed as soon as is feasible on completion of filling in a certain area. A typical cap is shown in Fig. 1(c) and comprises the following layers:

- a subgrade layer placed over the refuse to provide an intermediate cover and a smooth surface for final capping;
- a gas ventilation layer of sand/gravel to allow the collection and removal of landfill gas: if an active gas ventilation system such as pumping from gas wells is employed, this layer is omitted;
- A barrier to minimize precipitation percolation into the refuse: this can be a mineral layer or a very flexible geomembrane such as very low-density polyethylene (VLDPE);
- a drainage layer to remove infiltrated water;
- a topsoil layer and vegetative cover to increase evapotranspiration and reduce infiltration.

The above are conceptual designs for shallow-sided sites. Extending these to steep-sided quarries requires particular attention to the design of the side-slope liners.

Geomembranes

Geomembranes are relatively impermeable synthetic materials usually made from continuous polymeric sheets. Most geomembranes are thermoplastic polymers, thermoset polymers or a combination of both. High-density polyethylene (HDPE) is the most common geomembrane used for landfill barriers. HDPE's properties include good chemical resistance, strength and permeability and it can be produced at a reasonable cost when compared with other polymers. Manufacturers are now overcoming the two main disadvantages of HDPE, i.e. low frictional properties and high thermal expansion/contraction, by producing various types of textured and heat-reflective sheets.

The chemical resistance of geomembranes has been shown by many manufacturers to be very good. Chemical resistance charts have been produced which list generic chemicals against many common geomembranes, and Koerner (1990) lists HDPE as one of the most effective geomembranes. However, the composition of leachate is variable and only waste-specific tests truly give an indication of chemical resistance performance. Recent work at the Geosynthetic Research Institute in the US indicate that the life of an HDPE geomembrane may be much longer than initially thought, possibly up to several hundred years.

Geomembrane strength is a broad term that encompasses tensile behaviour, seam behaviour, tear resistance and puncture resistance. Three tests are commonly used to establish the geomembrane's tensile strength: the dumb-bell shape, the uniform width and the three-dimensional axisymmetric tests. The dumb-bell shape test is an index test while the remaining two are performance tests.

Geomembrane seams, where two or more panels are joined together, can control the performance of the geomembrane on site. The strength of a geomembrane seam is usually measured in shear and in peel, by testing a sample in a tensile testing machine. The tear resistance of a geomembrane is not usually a problem but some thin geomembranes may need to be tested to establish their tear strength.

The puncture resistance of a geomembrane can be a critical factor in the design of the barrier if the subgrade contains stones or any debris. Falling objects can also penetrate the geomembrane. The index tests currently used for puncture resistance evaluation are not representative and do not model the actual behaviour of geomembranes. Hullings & Koerner (1991) propose a test method which simulates the puncture mechanism using hydrostatic pressure to compress the geomembrane over a representative subgrade.

Permeabilities, measured by water-vapour transmission tests, of around 1×10^{-15} m s^{-1} are typical for HDPE geomembranes, but it must be noted, however, that in addition to leakage due to permeation through geomembranes, there is also leakage through defects. Leakage can be modelled by the method proposed by Giroud & Bonaparte (1989), and is dramatically reduced when good contact between geomembrane and clay can be assured.

Side-slope lining for steep quarries

There are several design and construction considerations that need to be addressed for lining the side slopes of quarries.

Clay/geomembrane composite barriers

The conventional design for shallow slopes, typically at angles of around 18°, i.e. one vertical to three horizontal, relies on the friction between the clay layer and the subgrade, and the geomembrane and clay layer, for overall stability. Furthermore, the geomembrane is anchored into a trench normally located a few metres behind the crest of the slope. If such a conventional design was used on a deep steep-sided quarry, both the material requirements in forming the 1 on 3 slope and the void space lost would be significant. Clay barriers could be placed on steeper slopes, up to 1 on 2 say, for relatively small heights; greater heights and steeper slopes would lead to slope failure within the clay. If constructed in short lifts ahead of the main refuse placement, overall stability could still pose problems since the low density and high compressibility of the refuse would result in large deformations occurring before the passive resistance could be mobilized fully.

Quarry Face

The quality of the quarry face will significantly affect the barrier design. Depending on the geology of the quarry, the rock face may be jagged and have overhangs that could damage the geomembrane, or may be riddled with discontinuities which provide an easy flow path for leachate and landfill gas. The face is likely to be irregular and, particularly in older quarries, may be dangerously unstable with considerable amounts of loose material.

To overcome these problems, a combination of rock face stabilization and geomembrane protection is required. The rock face should initially be cleaned of loose debris or even pre-split, and it may then be possible to stabilize the face using rock bolts, wire mesh and shotcrete. If the profile of the face is particularly difficult then thought must be given to creating an artificial surface. This may be achieved using metal or textile gabion baskets or reinforced earth to create a steep-sided wall with a relatively smooth surface; however, the internal stability of such systems would have to be ensured. Although potentially more expensive than the 1 on 3 clay-lined option, the increased void space available may make it a viable solution.

Whatever form of rock face preparation is carried out, the geomembrane must be protected from damage. For a shotcreted face a heavy non-woven geotextile, possibly with a geonet for strength, may form a suitable layer between the face and the geomembrane. The use of gabions may result in construction and long-term maintenance problems.

Geomembrane design

The stress/strain behaviour of HDPE geomembranes follows the curve shown in Fig. 2. There is a well-defined

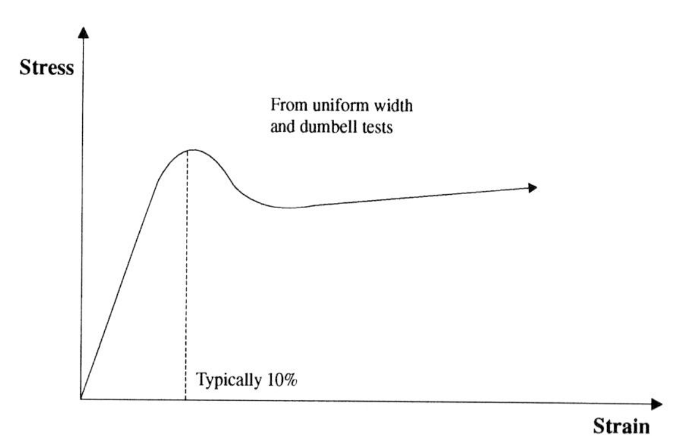

Fig. 2. Typical stress/strain behaviour for HDPE.

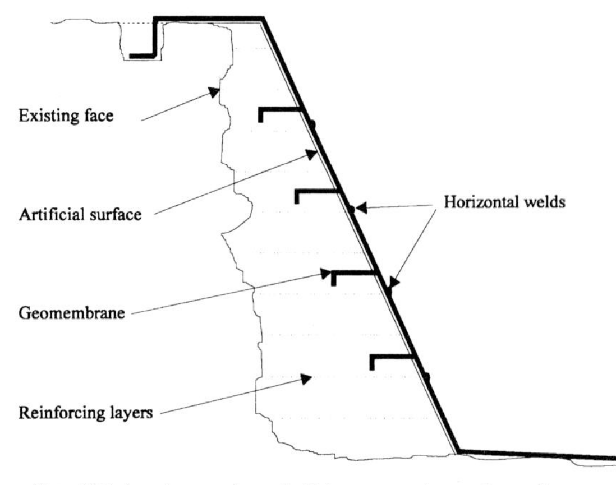

Fig. 3. Schematic steep slope lining system.

yield point to which a factor of safety can be applied and which becomes a target value for design purposes. For conventional side-slope design, the following stresses are induced in the geomembrane:

- tensile stress due to the difference between the shear stress induced by the weight of refuse, and the shear stress transmitted through friction to the geomembrane below;
- shear stress due to differential settlement of the refuse.

Settlement of refuse is characteristically irregular, and difficult to predict. Initially there is a large settlement within a month or two of completion of filling, followed by a substantial amount of secondary compression over an extended period of time. Edil *et al.* (1990) indicate that under its own weight, refuse settlement typically ranges from 5 to 30% of the original thickness. Large shear stresses can thus be induced in the geomembrane and, since the settlement is irregular, differential stresses can be induced along the quarry face, possibly damaging the geomembrane panels and also the panel seams.

For artificially created surfaces, such as gabion or reinforced soil walls, it may be possible to adapt the conventional anchorage design. This would involve anchoring the sheets in lifts during construction of the walls, producing a series of overlaps (see Fig. 3). The overlapping sheets could then be welded together to provide the leachate and gas seal. A major disadvantage, however, is that this would require horizontal seams along the slope, which are notably difficult to construct and would be subjected to the large refuse settlement induced stresses.

An alternative method that is being considered is fixing the geomembrane to a prepared steep slope. If the quarry face had been prepared with shotcrete as suggested earlier, it may be possible to fix the geomembrane directly to the face using rock bolts for example. To ensure the integrity of the barrier, the outside face of the fixing would need to be sealed, possibly with a welded geomembrane patch. Such a method requires detailed analysis of the stresses induced in the geomembrane. Stress concentrations at the fixing locations would result from expansion/contraction of the geomembrane during the operational phase of the landfill, and from the refuse settlement which may last many years after landfill completion.

Drainage and protection

The geomembrane placed on the slope must be protected from possible damage during refuse placement. The leachate drainage layer can often provide this protection. For shallow side slopes, this can be a sand layer laid directly on the geomembrane, a gravel layer on a geotextile, or a geosynthetic drainage composite. Containing sand or gravel on a steep slope would prove difficult and could only be achieved either by placing a heavy barrier such as a gabion wall in front of the layer, or by placement of the sand or gravel immediately ahead of the waste lift.

In some instances, there may have been drawdown of the water-table during the quarrying operations which has recharged or is in the process of recharging. Also, perched water-tables may exist in the surrounding rocks. In order to prevent the build-up of water pressure behind the geomembrane, a further drainage system may be required.

As discussed above, large and irregular settlements can be expected to take place within the refuse. There is an opportunity to reduce the shear stresses induced in the geomembrane by building in a low friction surface. However, attention must be paid to the stability of the waste during construction when it is often placed against the side slopes with a temporary steep face.

Construction quality assurance

Landfill design must ensure that adequate construction quality assurance (CQA) is possible. Hall & Marshall (1991) discuss the role of CQA in the installation of geomembrane liners, and identify geomembrane seaming as being the most critical area to control.

There are two principal methods of forming seams: the 'hot wedge' fusion method and the extrusion method. The fusion weld consists of an electrically heated 'wedge' passed between two sheets and as the surface melts, two rollers are applied. This forms two parallel seams with a space between which can be subjected to an air test. In extrusion welding, a ribbon of molten polymer is extruded between the two sheets to be welded. Some of the sheet material is liquefied and the entire mass is then welded together. This can then be tested by a vacuum box, or if a strand of copper wire is placed between the sheets, a spark test can be carried out.

Non-destructive tests such as the air, vacuum and spark tests can only verify seam integrity. Destructive tests are needed to verify seam strength, both qualitatively on site and quantitatively in the laboratory. On-site testing is carried out on narrow 'tab' seam samples which are tested in both peel and shear but can only verify that the seam is as strong as the sheet. Laboratory tests, carried out on larger seam samples, give peel and shear strengths for the seams.

Good CQA is essential in ensuring that the landfill is constructed as designed. For quarry sites, the work of the contractor and CQA engineer is more difficult due the geometry of the slope. The CQA programme should include monitoring and verifying all works during landfill construction, from the preparatory works on stabilizing the rock face, through to the first waste lift.

Conclusions

This paper has reviewed the current design practice for shallow-sided landfills. The extension to the design of steep-sided quarry landfills is not simple and the main design issues have been discussed. The use of steep-sided quarries for landfilling will increase in the foreseeable future, and such sites will have to be designed on a fully contained basis. This poses interesting challenges for the design engineer to provide technically sound and economically affordable designs. Research work currently in progress at the Nottingham Trent University, in collaboration with Golder Associates (UK) Ltd, is investigating the issues raised in this paper.

Acknowledgements. The author is currently employed as a Teaching Company Associate on a joint research project between Golder Associates (UK) Ltd and The Nottingham Trent University. The guidance and help in preparing this paper from Miss Morag Aiken and Dr Z Al-Dhahir at Golders and Dr Neil Dixon at the University are gratefully acknowledged.

The views expressed in this paper are those of the author and are not necessarily the views of either Golder Associates or the Nottingham Trent University.

References

BRUNE, M., RAMKE, H. G., COLLINS, H. J. & HANERT, H. H. 1991. *Encrustation processes in drainage systems of sanitary landfills. Proceedings Sardinia '91.* Third International Landfill Symposium, Cagliari, Italy, 999–1035.

EDIL, T. B., RANGUETTE, V. J. & WUELLNER, W. W. 1990. *Settlement of municipal waste In*: LANDVA, A. & KNOWLES, G. D. (eds) *Geotechnics of Waste Fills—Theory and Practice ASTM STP 1070.* Philadelphia, 225–239.

GIROUD, J. P. & BONAPARTE, R. 1989. Leakage through liners constructed with geomembranes—part 1. Geomembrane liners. *Geotextiles and Geomembranes*, **8**(2), 27–67.

HALL, D. H. & MARSHALL, P. 1991. The role of construction quality assurance in the installation of geomembrane liners. *Proceedings of the Conference on Planning and Engineering of Landfills.* Midlands Geotechnical Society, 187–191.

HULLINGS, D. & KOERNER, R. 1991. *Puncture resistance of geomembranes using a truncated cone test.* Geosynthetics '91, Atlanta, USA, 273–285.

KOERNER, R. M. 1990. *Designing with geosynthetics*, 2nd edn. Prentice Hall.

Geological and other influences on the design of containment systems in hard rock quarries

D. C. Mann

Applied Geology Ltd, Cranford, Kenilworth Road, Blackdown, Leamington Spa CV32 6RG, UK

Abstract. It is common practice in the UK to restore abandoned quarries by landfilling. Modern landfill facilities are designed on the principle of containment of wastes, leachate and gases, but, in hard rock quarries, there are particular problems which must be overcome. This paper describes the engineering design of the containment systems proposed for use in three hard rock quarries located in different parts of the UK. At each of the three sites the design has been influenced by the local geology and hydrogeology, and the use of indigenous materials has been a priority. The regulatory authorities have a statutory responsibility in the licensing of new landfill facilities, but are becoming increasingly involved in the early planning of these sites. Their influence on the evolution of the three designs is assessed by comparing the three containment systems.

In the UK, abandoned quarries are often restored by landfilling. Today, landfill facilities are designed to ensure the containment of wastes, leachate and landfill gases. However, abandoned hard rock quarries can often present particular difficulties to their development as contained landfill sites due to the presence of high vertical rock faces, fractured, highly permeable rock around the quarry edges and complex hydrogeology. In addition, often there are no local materials available for the construction of low permeability natural liners.

In the design of landfill facilities it is necessary to take into account a number of criteria including the geology, hydrogeology, hydrology and topography of the site, together with the regulations and guidelines produced by statutory bodies such as the National Rivers Authority (NRA) and local waste regulation authorities (WRA). However, it is considered that the use of indigenous materials should be a priority if economic design solutions are to be achieved.

The geology within the UK is diverse and many different minerals have been excavated from quarries over many years. An examination of three hard rock quarries, located in different parts of the UK and in different geological settings, demonstrates the influence the local geology and hydrogeology can have on the design of containment systems. However, the local preferences of NRA/WRA are also highlighted when the three systems are compared.

Factors influencing the design of containment systems

There are numerous factors which influence the design of containment systems for landfill sites, but perhaps the most important are those listed below.

Geology

Ideally, there will be a source of low-permeability lining materials either on the site or nearby, but often it is necessary to study geological maps to identify potential sources away from the site and often at some distance (perhaps up to 80 km). However, it is important to consider the workability of any material as well as its ability to achieve low permeabilities when compacted and tested in a laboratory. If the map study or testing fails to identify a source of natural low-permeability material for use in a containment system, then artificial liners must be considered.

The *in situ* permeability of the underlying strata at any site can also influence the design of containment systems. The presence of fissures/joints, and/or faults, will increase the permeability of strata underlying the landfill and may encourage the migration of leachate and gas away from the site. Alternatively, where mass permeabilities are low, the strata may be considered to be contributing to a containment system.

The integrity of any containment system can be affected by ground movements and it is necessary to consider the influence of any compressible material, mineworkings or natural cavities below a site.

Hydrogeology

The location of the groundwater table and the affect the construction of the landfill facility may have on the groundwater regime must be assessed during the design of the containment system. Construction below the groundwater table may pose some problems during the early stages of development of a site but on completion of filling, the positive head provided by a high external groundwater table may be considered to contribute to the containment of leachate within the site.

From Bentley, S. P. (ed.) *Engineering Geology of Waste Disposal*,
Geological Society Engineering Geology Special Publication No. 11, pp 133–139

NRA groundwater protection policy

The NRA (1992) has published a national (England and Wales) policy for groundwater protection. It is intended that this policy will provide a framework for decision-making, but it is also stressed that it is not prescriptive and any assessment must take into account site-specific considerations.

The policy has been developed on the basis of identifying source protection zones. In terms of landfill acceptability, the factors to be considered are summarized in Table 1, and should be considered in the planning of any new landfill facility.

Planning and other requirements

In many instances, the granting of planning permission for the development of landfill facilities is accompanied by a series of conditions, some relating to the import/export of materials to the site, and these can affect the design of containment systems, particularly if there are no natural lining materials available on site. In addition, the operation of crushing and screening plant will require the approval of Her Majesty's Inspectorate of Pollution.

Quarry one

The quarry is located in Scotland and was previously worked to produce roadstone. Operations ceased in the early 1980s and the quarry has remained derelict since then. The site covers an area of approximately 10 ha and the depth of the quarry varies between 10 and 30 m

Table 1. *NRA landfill acceptability matrix*

Site type	*Source protection*		
	I *Inner zone*	*II* *Outer zone*	*III* *Catchment zone*
1. High pollution potential (landfills accepting domestic, commercial and industrial waste either individually or on a co-disposal basis)	Not acceptable	Not acceptable	Only acceptable with engineered containment and operational safeguards
2. Medium pollution potential (landfills accepting construction, demolition industry wastes and similar)	Not acceptable	Acceptable subject to evaluation on a case by case basis and adequate operational safeguards	Acceptable subject to evaluation on a case by case basis and adequate operational safeguards
3. Low pollution potential (landfills accepting inert, uncontaminated waste)	Not normally acceptable	Acceptable only with adequate operational safeguards	Acceptable only with adequate operational safeguards
Site type	*Resource protection*		
	Major aquifer	*Minor aquifer*	*Non-aquifer*
1. High pollution potential (landfills accepting domestic, commercial and industrial waste either individually or on a co-disposal basis)	Only acceptable with engineered containment and operational safeguards	Only acceptable with engineered containment and operational safeguards	Acceptable only with adequate operational safeguards. Engineering measures may be necessary in order to protect surface waters
2. Medium pollution potential (landfills accepting construction, demolition industry wastes and similar)	Acceptable only with adequate operational safeguards	Acceptable only with adequate operational safeguards	Acceptable only with adequate operational safeguards
3. Low pollution potential (landfills accepting inert, uncontaminated waste)	Acceptable	Acceptable	Acceptable

below the surrounding land. The side slopes are generally vertical, although there are berms located along two of the sides of the quarry. It is estimated that $2 \times 10^6 m^3$ of waste will be required to complete the restoration and that this would be placed within a ten-year period.

The geological maps of the area indicate the site to be underlain by basalt lavas of the Calciferous Sandstone Age. Within the site, localized areas of tipped waste materials were present. The natural strata comprise basalts and these are highly weathered over the top 2–3 m, the remainder being very strong rock material, with variable blocky, columnar and 'pillow lava' jointing.

During the working life of the quarry, pumping was required to maintain the water level near to the base. In addition to groundwater entering the quarry, a small stream discharges down one of the slopes and since abandonment, the quarry has become flooded. It is expected that, without pumping, the groundwater will rise to a level 10 m above the base, near to the level of the lowest rim of the quarry.

As the site is located in Scotland, it is not covered by the NRA's Groundwater Protection Policy. Although there are several reservoirs within the area, the nearest, less than 1 km from the site, lies above the top of the quarry. Others are between 3 and 4 km from the site but are thought to be fed by surface water. The site location may be considered to lie within an area that would correspond to the NRA's definition of a non-aquifer.

As part of the planning permission granted for this site, it was a condition that, during development, all earthworks materials must be found on site and the export of surplus materials was not permitted. There are no indigenous low-permeability materials on site and therefore a double synthetic flexible membrane liner was proposed for the base and lower slopes up to 10 m above the base, as shown in Figs 1 and 2. As the quarry base is below the groundwater table, it is necessary to provide an underliner drainage system which takes water to a sump for pumping to the surface. The primary and secondary liners are made from 2.5 mm thick high-density polyethylene (HDPE) sheet, separated by a leakage detection layer. A leachate-collection/liner-protection layer is provided above the primary liner. All of the drainage materials will be produced on site by crushing and screening the basalt and any waste will be stockpiled for use as daily cover during landfill operations.

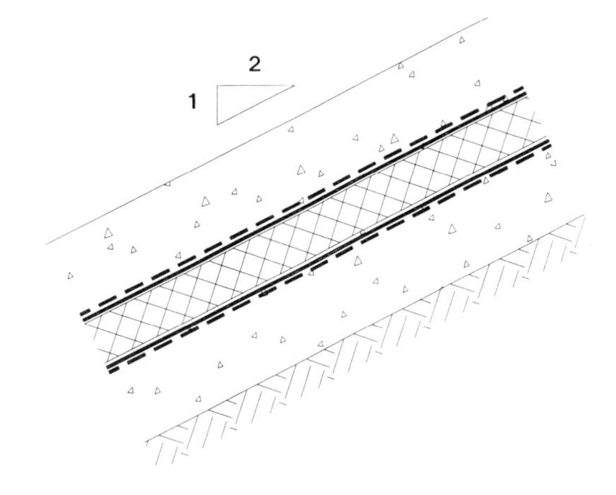

Fig. 2. Quarry One, side liner (lower slopes).

Over the lower 10 m the quarry will be profiled with side slopes of 1 in 2, as shown in Fig. 2. However, above this level the vertical slopes of the quarry will be retained and a single HDPE flexible liner will be used in a 'Christmas Tree' configuration, as shown in Fig. 3.

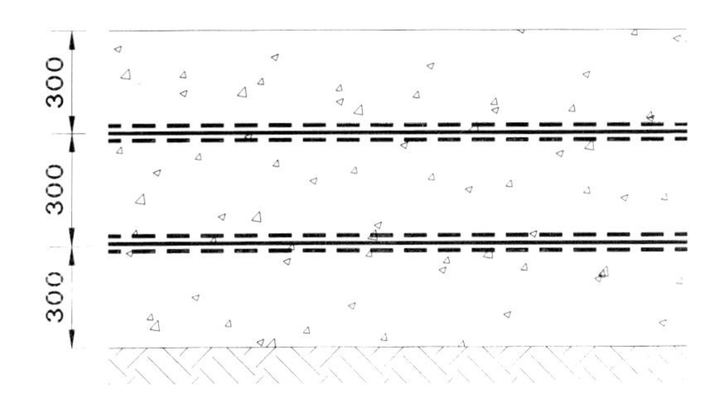

Fig. 1. Quarry One, base liner.

Explanation of symbols used in figures

PREPARED SUBGRADE
CLAY LINER
DRAINAGE LAYER
GEOCOMPOSITE DRAIN
SAND PROTECTION LAYER
SAND/BENTONITE LINER
HDPE MEMBRANE
NEEDLEPUNCH GEOTEXTILE
GEOTEXTILE SEPARATOR

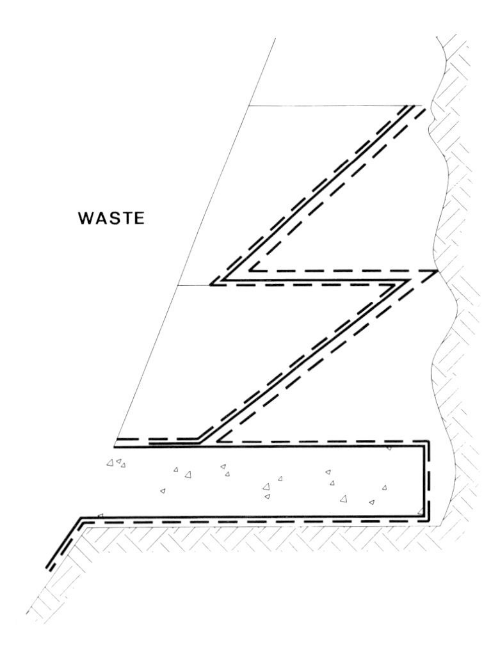

Fig. 3. Quary One, side liner (upper slopes).

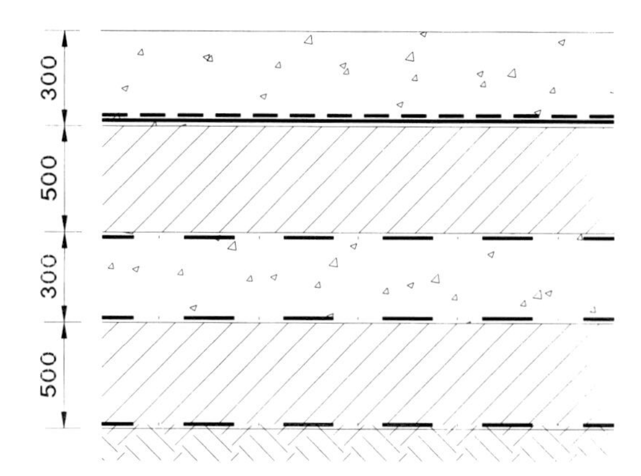

Fig. 4. Quarry Two, base liner.

Quarry two

The quarry is located in the north of England and during its lifetime two different materials were worked. The upper strata provided clays and shales for brick-making, and building stone was excavated from sandstone strata below the shales. The site covers an area of approximately 13 ha and extends to depths of between 10 and 30 m. The side slopes are generally vertical, down to the quarry base. At the time of abandonment, there were substantial volumes of reject materials on the quarry floor. The quarry is estimated to have up to $3 \times 10^6 m^3$ of void space and restoration should be complete within 10–15 years.

The geological maps of the area indicate the site to be underlain by strata belonging to the Middle Coal Measures of Carboniferous Age. The rocks are principally sandstones, siltstones, mudstones and coal. Within the site boundaries, the top 8 m consist of highly to moderately weathered, weak to moderately weak, laminated siltstones and mudstones. Below 8 m there are massive, moderately weathered, moderately strong to strong sandstones. Workable coal seams are known to exist at depth.

During the operation of the quarry it is believed that there were no groundwater problems. Monitoring boreholes in and around the site have indicated the groundwater table to lie several metres below the lowest floor level in the quarry.

According to the NRA policy, Coal Measures strata are considered to form a minor aquifer. There are no reservoirs or borehole abstraction points within the immediate vicinity of the site.

In the planning permission granted for the site there were no restrictions placed on movement of earthworks materials to and from the site.

The only materials on site which could possibly be used to form a low-permeability liner are the weathered siltstones and mudstones rejected for brick-making. Laboratory tests indicate that these materials can only achieve a minimum permeability of $1 \times 10^{-8} m s^{-1}$ but their use was proposed in a composite liner together with a 2.5 mm membrane on the base and lower slopes, as shown in Figs 4 and 5. The leachate-leakage detection layer permits the collection of liquid and directs it to a sump, for pumping to a treatment works. Stone for the leachate-drainage layer and leachate-leakage-detection layer will be provided from the sandstones on site following crushing and screening.

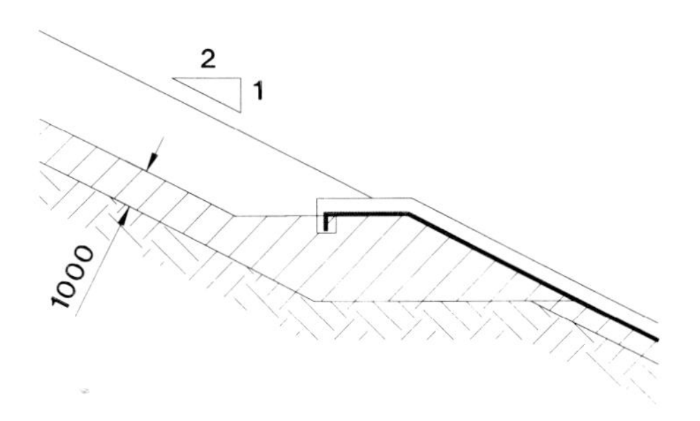

Fig. 5. Quarry Two, side liner.

Reprofiling of the slopes is required to permit placing of the 1 m thick mineral liner above the composite base, and waste material from within the quarry will be utilized for this purpose. There is no material on site capable of attaining permeabilities of less than $1 \times 10^{-9}\,m\,s^{-1}$ and therefore clay soils will need to be imported to the site. Material required to support the 1 m-thick clay liner on the slope during construction will be found from within the waste material in the quarry.

Quarry three

The quarry is located in the Midlands and was previously worked from the 1920s to produce sand for a variety of uses. Operations ceased in the early 1990s. The site covers an area of approximately 20 ha and the depth of the quarry varies between 25 and 30 m, with the side slopes generally near vertical. Restoration of the quarry will involve the placing of $1.9 \times 10^6\,m^3$ of waste over a period of about 12 years.

The geological maps for the area indicate the site to be underlain by strata belonging to the Triassic Sherwood Sandstones Group. The rocks generally are weak, fine-grained sandstones with occasional thin mudstone. Around the fringes of the quarry there are superficial deposits of glacial till comprising coarse silty gravels.

The quarry remained dry during normal operations and monitoring boreholes have identified the groundwater table to lie approximately 6 m below the base of the existing quarry.

The site lies within Zone III of a major aquifer as defined by the NRA. There is a licensed abstraction borehole approximately 1 km from the site.

Planning permission has not yet been granted for this site but it is believed that no formal objection has been made by the NRA.

There are no indigenous materials on site to form a natural, low-permeability liner. Desk studies failed to locate any suitable, and readily available, materials within a 50 km radius of the site. Details of the proposed lining system are shown in Fig 6 and 7. Over the base and the lower slopes up to 10 m above the base, a

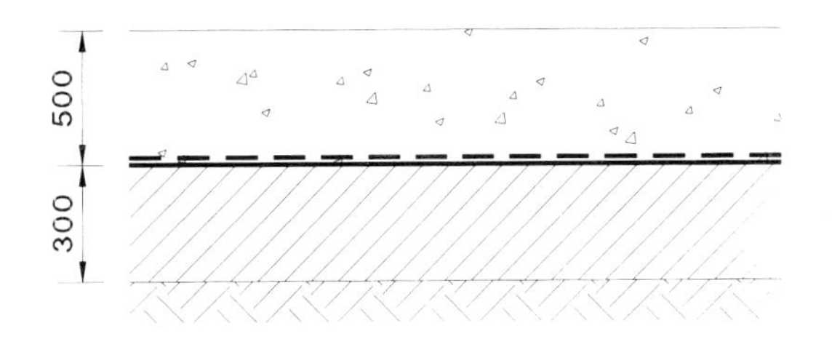

Fig. 6. Quarry Three, base liner.

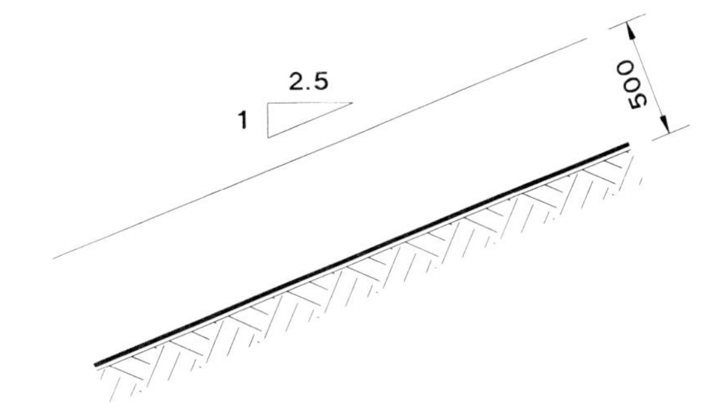

Fig. 7. Quarry Three, side liner (upper slopes).

composite liner will be used. This will comprise a 300-mm-thick layer of bentonite-enhanced sand (BES) using material from within the quarry and a 2.5-m-thick HDPE, flexible membrane.

The lower slopes will be constructed to a gradient of 1 to 3 to assist with the placement of the BES, but above 10 m the slopes will be 1 in 2.5 (Fig. 7). Berms are required, principally to assist in placing the membrane liner on the slopes. A single 2.5-mm-thick HDPE membrane, placed directly onto reprofiled slopes, will form the barrier in the upper slopes.

Comparison of containment systems

Table 2 summarizes those aspects of the design of the containment systems which are considered to be the most important. It can be seen tnat, for the basal liner in each of the quarries, either a composite or double synthetic liner is proposed, a single liner being unacceptable given the nature of the strata below each site.

Quarries One and Three are similar in that a source of low-permeability natural liner is not available in the immediate vicinity, but the containment systems proposed are quite different. The reasons for the differences are primarily due to the different hydrogeological setting. Due to the location of Quarry Three on a major aquifer, it was considered that a composite lining system was most appropriate for this site. In Quarry One the double synthetic liner, together with the positive head provided by external groundwater, will ensure that leachate is contained within the quarry. However, it should be noted that the presence of the interliner drainage layer permits the extraction of any leakage which may occur through the liners.

In Quarry Two, use has been made of indigenous clay to form the natural liner below the HDPE membrane, although the material can only achieve a permeability of $1 \times 10^{-8}\,m\,s^{-1}$. However, the provision

Table 2. *Details of containment systems*

	Quarry One	*Quarry Two*	*Quarry Three*
Base liner			
Type	Double	Composite	Composite
Materials	FML 2.5 mm HDPE membrane	2 × 500 mm thick clay ($k < 1 \times 10^{-8}$ m s^{-1}) plus 2.5 mm HDPE membrane	300 mm thick BES ($k < 1 \times 10^{-10}$ m s^{-1}) plus 2.5 mm HDPE membrane
Liner to lower slopes			
Type	As base	As base	As base
Extent	10 m above base	5 m above base	10 m above base
Gradients	1:2	1:2	1:3
Liner to upper slopes			
Type	Single FML	Mineral	Single FML
Materials	2.5 mm HDPE membane	1000 mm thick clay ($k < 1 \times 10^{-9}$ m s^{-1})	2.5 mm HDPE membrane
Gradients	Vertical	1:2	1:2.5

of a leachate-leakage-detection layer means that an adequate containment system is provided, and this, taken together with the location of the site over the Coal Measures (minor aquifer), has resulted in its acceptance by the NRA and local WRA.

It is interesting to note the requirements in each of the quarries for the provision of double/composite lining systems over the lower slopes. In each of the quarries, the proposed operating licence requires that the head of leachate on the liner shall not exceed 1 m, and the leachate management system proposed for each site is similar. In Quarry One the double lining system extends to 10 m above the base which coincides with the expected external groundwater level at the site, and may be considered reasonable. However, in Quarry Three the composite liner must also extend to 10 m above the base to satisfy the NRA, and in Quarry Two, to 5 m above the base to satisfy the local WRA. It appears that if the maximum permitted head is 1 m above the liner, and there are no particular external hydrogeological factors to be considered then there is no need to extend the double/composite liner more than 2 m above the base at the most.

Although consideration must be given to the provision of stable slopes on which to place any lining system, it is often the case that the slopes are chosen to facilitate placing of the lining system rather than on the basis of slope stability criteria. Thus Quarry Three lower slopes are provided at a gradient of 1 to 3 to assist in placing the BES. Steeper slopes are acceptable at Quarries One and Two due to the provision of different lining systems.

Above the lower double synthetic/composite-lined slopes, there are also differences in the three quarries. In Quarry One and Quarry Three a single 2.5-mm-thick HDPE membrane is to be utilized but in Quarry Two a nominal 1 m thick clay liner is required, although this has to be imported to the site. There do not appear to be any geological or hydrogeological reasons for the difference, the system in Quarry Two being used as a result of preferences by the local WRA. The placement of mineral liners (as shown in Fig. 5) is very difficult in deep quarries and it is considered that the use of a single HDPE membrane offers considerable construction advantages. The risk of leachate leakage through the single liner is extremely low given the relatively steep slopes and the general requirement to limit leachate to a maximum depth of 1 m above the base. In addition, it is normal to install landfill gas extraction systems, thus creating a negative pressure within the contained landfill and reducing risks of landfill gas migration through the single liner.

Summary

In developing designs for containment systems it is important to utilize indigenous materials whenever possible, and to take into account the local geology and hydrogeology. An examination of the systems proposed for three hard rock quarries has demonstrated that it is possible to use local materials (either as a liner or in drainage layers) and that the selection of individual components can be influenced by the geology/hydrogeology (Quarries One and Two). However, it is also clear that, in some instances, the design can be influenced by the NRA or local WRA and early consultation with these bodies is recommended. Whilst the authorities can provide positive assistance in the development of the design of containment systems, it is important that the fundamentals of the design are based on aquifer protection requirements and the need to control landfill gas migration, together with the local geology and hydrogeology.

Acknowledgements. The author would like to thank all of the clients who permitted the use of design details in this paper, and who provided many helpful comments and suggestions during its preparation.

References

DEPARTMENT OF ENVIRONMENT 1992. *Landfilling wastes.* Waste Management Paper 26, DOE, London.

NATIONAL RIVERS AUTHORITY 1992. *Policy and practice for the protection of groundwater.*

Permeability considerations about rock foundation on waste disposal

A. Foyo & C. Tomillo

University of Cantabria, E.T.S. Ing. de Caminos, 39005 Santander, Spain

Abstract. The use of the traditional permeability test (the water pressure test) to analyse the permeability of rock foundations for waste disposal, is not applicable in all materials because the high pressures used in the test may produce hydraulic fracturing. This paper describes a new permeability test, called the low pressure test or LPT, which is used to obtain the true permeability of a waste disposal foundation on weak and impervious rock. Finally, the results obtained with the use of the LPT to analyse the permeability of a waste disposal foundation are described.

Usually, Packer tests or water pressure tests are the most common methods to analyse the hydraulic conductivity or permeability of the fractured rock masses.

In parts of Spain there are weak and impervious geological formations which have a cleavage as the main structural characteristic. In this type of rock mass, experience shows that hydraulic fracturing may occur when the traditional water pressure test is carried out to determine the permeability of a banished zone. In fact, in a rock mass foundation of, for example, Cambrian and Silurian metamorphic schist, Carboniferous and Permian slates, or clay-marls of Cretaceous and pre-orogenic Tertiary, hydraulic fracturing may occur between pressures of 2 and 3 bar during the WPT test.

Knowledge of the permeability of a waste disposal rock foundation, must be achieved in terms of true permeability, so the phenomena of hydraulic fracturing must be avoided. Other conditions such as the depth of the test zone, the water-table depth and the loss pressure head in the tube must also be taken into account.

The low pressure test

At the 6th International Congress of the International Association of Engineering Geology (Foyo & Cerda 1990) and the XVII International Congress on Large Dams (Foyo *et al.* 1991), the low pressure test or LPT was defined as a method to determine the permeability of large-dam foundations when cleavage is the main structural discontinuity.

Following the definition of the 'critical permeability' as the 'water absorption under hydraulic fracturing conditions', the LPT was specified as follows:

- a test section length of 3 m;
- a pressure-level sequence of between 1 and 5 bar;
- 5-min intervals as the load time for each pressure level.

The true or total critical pressure P_T (Fig. 1) can be obtained from the expression:

$$P_T = P_M + P_H - P_W - P_R$$

or, when the test section is below the water-table,

$$P_T = P_M + P_H - P_R$$

where: P_T is the total critical pressure in the test zone, P_M is the critical pressure measured in the manometer,

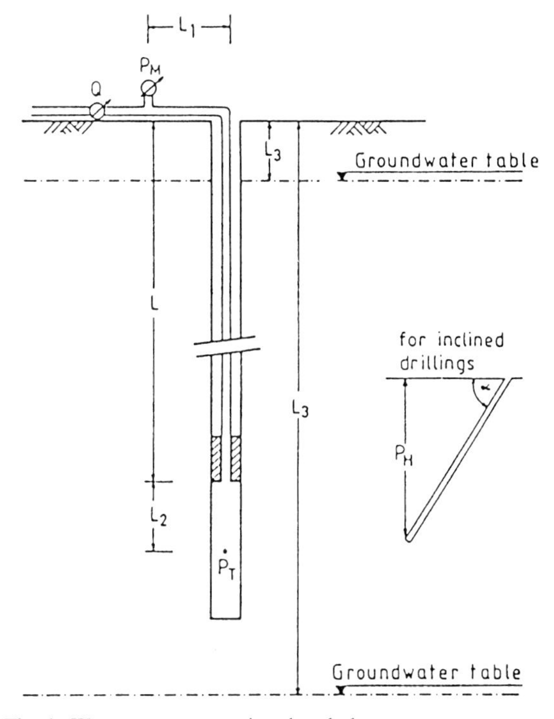

Fig. 1. Water pressure test in a borehole.

From Bentley, S. P. (ed.) *Engineering Geology of Waste Disposal*,
Geological Society Engineering Geology Special Publication No. 11, pp 141–143

P_H is the hydrostatic pressure head, P_W is the pressure due to water-table depth, and P_R is the head loss along the pipes.

The equation for the head loss due to hydraulic friction of a straight tube is, according to Ewert (1985):

$$P_R = \lambda \frac{v^2(L + L_1)}{2gd}$$

$$\frac{1}{\sqrt{\lambda}} = -2\lg\left(\frac{2.51}{Re\sqrt{\lambda}} + \frac{k}{d}\frac{1}{3.71}\right)$$

where v is the velocity of the flow, $L + L_1$ is the length of the tube (see Fig. 1), d is the diameter of the tube, k is the coefficient of roughness, g is the gravity, λ is the coefficient of friction influenced by d, k and Re, and Re is the Reynolds number.

With the test as specified above, when hydraulic fracturing conditions occur, the critical pressure will be defined as well as the corresponding flow in the test zone, which is also called the critical permeability unit or ULC (Fig. 2).

Taking into account that in weak and impervious fractured rock masses, hydraulic fracturing appears between pressures of 2 and 3 bar when executing a LPT, the hydraulic conductivity (in terms of permeability coefficient) could be expressed as follows:

$$1.00\,\text{ULC} = 0.25 \times 10^{-6}\,\text{cm s}^{-1}$$

Analysis of the results

The waste disposal rock foundation where the test has been studied was a sequence of clay-marls and sandy-marls of the Paleoce epoch, of the Lower Tertiary

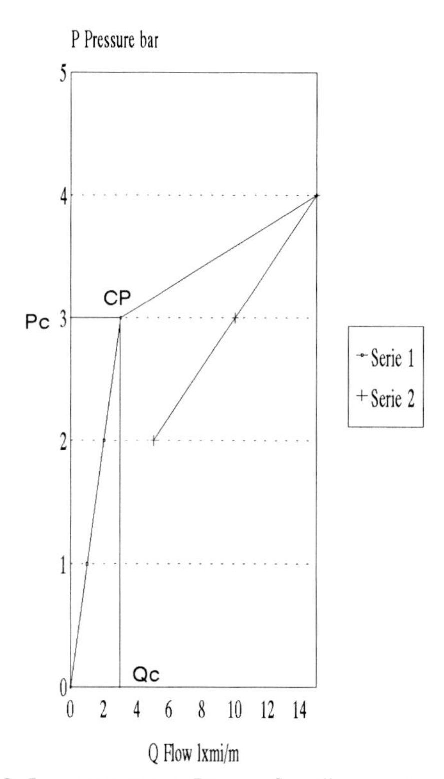

Fig. 2. Low-pressure test. Pressure-flow diagram. P_c, critical pressure; Q_c, critical permeability; CP, critical point.

(a)

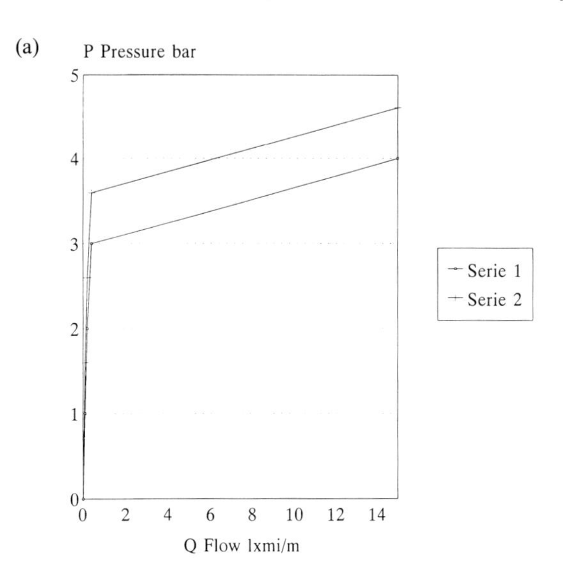

(b)

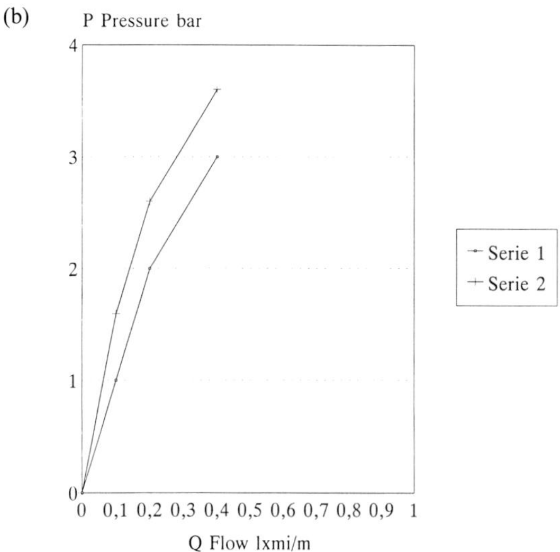

Fig. 3. (a) Low-pressure test. Borehole S-3, 8.50 m. Pressure-flow diagram. Serie 1, manometric pressure; serie 2, total pressure. (b) Details of Fig. 3(a) pressure-flow diagram. Serie 1, manometric pressure; serie 2, total pressure.

period. The main structural characteristic of the rock mass is defined by the narrow spacing of bedding. Consequently, the secondary permeability of the rock mass could be considered very low.

All the LPTs, with a diameter about 86 mm, were carried out below the water-table level.

In borehole S-3, the results of the LPT carried out at 8.50 m depth show the facility of this kind of water test to determine the critical pressure and, as consequence, the critical permeability in the test zone (see Figs 3(a) and (b)). In Fig. 3(a), the low permeability or water absorption capacity is evident under the first three pressure levels, 1, 2 and 3 bar. Furthermore, as shown in Fig. 3(b), at pressure levels above 3.62 bar, the total water absorption levels reveal that the hydraulic fracturing occurred during the first stages of the test.

The critical permeability about 0.4 l min m^{-1} (0.1×10^{-6} cm s^{-1}) could be interpreted as a natural consequence of the geological and structural characteristics of the material. The modified horizontal scale of Fig. 3(b), shows that the hydraulic fracturing phenomena appear from the first pressure levels during the test.

Conclusion

The results obtained in the analysis of the permeability of a waste-disposal rock foundation, by means of the low pressure test or LPT, have indicated the suitability of the test to the determination of the critical pressures and the critical permeability at which hydraulic fracturing appears during the test. Furthermore, the critical permeability must be considered as the true water absorption capacity of the rock mass, defined as critical permeability unit or ULC.

Taking into account the criteria as described by Bruce & Millmore (1983), to express the critical permeability unit, ULC, as equivalent Lugeon units, ULE, the following expression may be used:

$$1\ \mathrm{ULE} = \frac{Q}{P} 10$$

where ULE is the equivalent lugeon unit, l min m^{-1}, Q is the water flow corresponding to the critical pressure, ULC, l min m^{-1}, and P is the critical pressure (bar).

References

BRUCE, D. & MILLMORE, J. 1983. Rock grouting and water testing at Kielder Dam, Northumberland. *Quarterly Journal of Engineering Geology, London*, **16**(1), 13–29.

EWERT, F. K. 1985. *Rock Grouting with emphasis on dam sites*. Springer, Berlin.

FOYO, A. 1993. *Utilización del Ensayo L.P.T. en la determinación de la permeabilidad de formaciones esquistosas*. IV Jornadas Españolas de Grandes Presas, Murcia, 63–75.

—— & CERDA, J. 1990. *Critic permeability. New criteria for the determination of permeability on large dam foundations*. 6th International Congress of the International Association of Engineering Geology, Amsterdam, vol. 2. Balkema, Rotterdam, 1177–1184.

——, TOMILLO, C. & CERDA, J. 1991. *The low pressure test. Determination of the permeability and groutability of slate rocks in large dam foundations*. XVIII International Congress on Large Dams, Vienna, Q.66, R.5.

WARWICK, D. W. & GELDENHUIS, S. J. 1990. *Engineering geology and its application to the selection of sanitary wastes sites on the Wiwastersrand, South Africa*. 6th International Congress of the International Association of Engineering Geology, Amsterdam, Symposia. Balkema, Rotterdam, 175–181.

A groundwater trace study using a fluorescent dye

M. Townend[1] & R. Aldridge[2]

[1] Pell Frischmann Consultants, George House, George Street, Wakefield, West Yorkshire WF1 1LY, UK
[2] Pro Soil Surveys Ltd, Elton Lodge, Newton Road, Leeds, West Yorkshire LS7 4HE, UK

Abstract. The test procedure for a groundwater trace study in a sandstone aquifer is described. The field test was simple, quickly implemented and carried out within a limited budget. A result was obtained and this formed part of the hydrogeological assessment of the site.

A groundwater trace test was carried out at Manywells Quarry, Cullingworth, near Bradford in West Yorkshire. The objective was to establish whether a proposed landfill would pose a significant risk to Manywells Spring, a licensed abstraction. The study of groundwater flow by means of a tracer was requested by the National Rivers Authority. The test was designed to determine the direction and rate of flow in the area adjacent to the quarry.

Geology and hydrogeology

Manywells Quarry is located about 10 km west of Bradford. The site is underlain by the Rough Rock and Rough Rock Flags of Namurian age. At this location the Rough Rock is a fine-grained, thickly bedded, well jointed sandstone. The underlying flags are similar but, as the name suggests, they exhibit thick laminations.

The southern boundary of the quarry is formed by a near-vertical face up to 26 m high. Structural data obtained near the base of the quarry indicate that the rock generally dips to the southeast at less than 5°. The sandstone is generally thickly to very thickly bedded with some units up to 4 m thick. These are generally cross-bedded but some remain massive throughout. The sandstone is well jointed, showing two orthogonal joint sets orientated approximately northeast to southwest and northwest to southeast. Major joints, i.e. those which penetrate a number of bedding planes, are very widely spaced, up to 3 m or greater. Many minor joints exist within beds and these are generally medium spaced, at approximately 0.5 m. Most joints are near vertical with planar and rough surfaces. Some larger curved surfaces are also present. Measurements near the base of the quarry show bedding planes to be tight or with 1–2 mm aperture sizes. Joint apertures are of a similar size where stable surfaces can be identified. Wider joint apertures are also present but it is considered that these are as a result of instability of the quarry face.

Long-term monitoring of groundwater levels indicates that the water-table is in excess of 14 m below the floor of the quarry. Flow lines were constructed, based on water levels in boreholes, and from these it was considered that groundwater would flow from the quarry in a northeast direction, possibly avoiding Manywells Spring.

The success of the field test depended on the tracer being detected at one or more of the monitoring points. The ability of the aquifer to retain a tracer by absorption and/or adsorption was a concern during planning. It was decided that sodium fluorescein, which is a water-soluble dye, would be used. The dye is safe to use, easy to handle and suitable for quantitative analysis. Further description of the properties of this dye are given by Siddle *et al.* (1986) and B. S. 5857 (Anon 1980).

The use of a radioactive isotope tracer was rejected due to the environmental sensitivity of the site. Sodium chloride was also rejected, as calculations indicated that a large amount would be required and because the salt could have been retained by clay minerals.

Monitoring points

The location of the monitoring points is shown in Fig. 1. These were a combination of boreholes, natural springs and issues from drains.

Manywells Spring was a licensed abstraction used by Yorkshire Water, but is now only used as a compensation supply for Hewenden Beck. The rate of abstraction was approximately 2.3×10^5 l day^{-1}. The flow of water from the spring is concealed in an adit and culvert before entering Hewenden Reservoir. Records indicate that the spring may also be connected to drains constructed from old shafts located to the southwest of this area (W. B. Woodhead & Sons 1893).

Test procedure

Borehole B was chosen to inject a single slug of sodium fluorescein dye into the groundwater. From long-term

From Bentley, S. P. (ed.) *Engineering Geology of Waste Disposal*,
Geological Society Engineering Geology Special Publication No. 11, pp 145–147

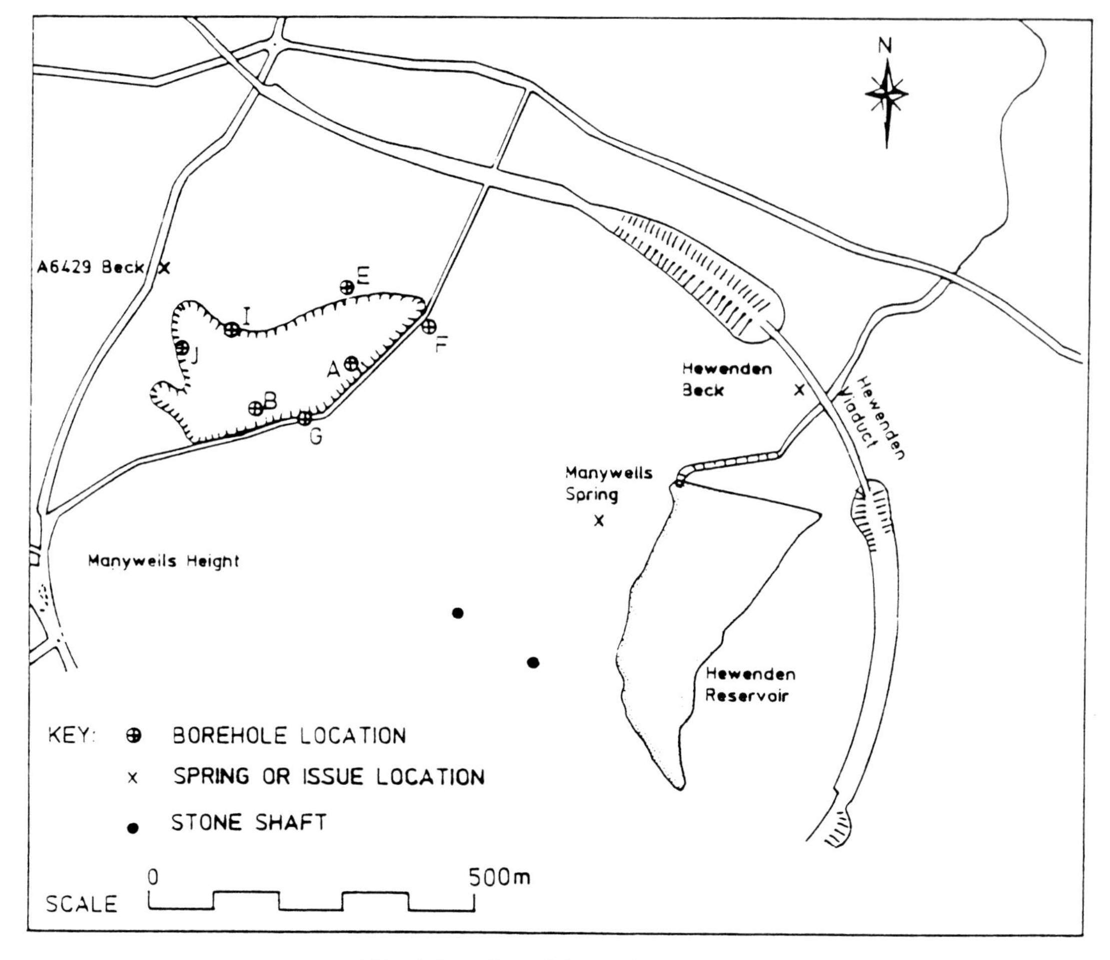

Fig. 1. Locations of the monitoring points.

monitoring of groundwater levels, this borehole was found to have a relatively high water level and was responsive to changes in the aquifer. This borehole was also located in a suitable position in relation to Manywells Spring and other monitoring points.

Prior to commencement of the test, background levels of fluorescein were checked over a period of one month with two sets of water samples from each monitoring point. A schedule was prepared with the intention of sampling over an eight-week period. Carbon bag samplers were suspended down boreholes and immersed in springs to enable a simple method of continuous monitoring. These consisted of norite pellets contained in a nylon bag.

Injection of the dye was carried out on 30th November 1992. Approximately 4 kg of sodium fluorescein powder was mixed with 1000 l of clean water. The dye was introduced by gravity feed with samples taken from the supply during the injection period. One person performed the mixing operation and then left site to avoid the risk of contamination of subsequent samples. Samples were taken from Borehole B at hourly intervals to measure the dilution at the source. Samples from the monitoring points were taken after 48 h and then at weekly intervals.

The limits of detection of sodium fluorescein, as determined by the laboratory, were $0.02\,mg\,l^{-1}$ in water and $0.5\,mg\,kg^{-1}$ on charcoal.

Results

Background samples taken in the month prior to the field test indicated fluorescein levels below the limits of detection. The concentration of fluorescein injected into the source borehole was measured as $3.97\,g\,l^{-1}$.

Table 1. *Level of fluorescein from groundwater samples* ($mg\,l^{-1}$)

	Sample date					
Borehole or location	*30.10.92*	*23.11.92*	*02.12.92*	*07.12.92*	*14.12.92*	*24.12.92*
A	—	—	0.14	0.03	—	—
B	—	—	—	—	—	—
E	—	—	—	—	—	—
F	—	—	—	—	—	—
G	—	—	0.17	0.04	—	—
I	—	—	—	—	—	—
J	—	—	—	—	—	—
B6429 Beck	—	—	—	—	—	—
Manywells Spring	—	—	0.09	0.03	—	—
Hewenden Beck	—	—	—	—	—	—

No reading indicates below limits of detection.

Evidence of fluorescence was detected at Manywells Spring using a portable ultraviolet light on the 2nd, 7th and 12th of December 1992.

The results from analysis of water samples, given in Table 1, show that fluorescein was present in small amounts in borehole A, borehole G and Manywells Spring, a short time after the test commenced. The levels of dye in the boreholes indicate dilution of between 23 000 and 28 000 times the original source concentration. The level in Manywells Spring indicates a dilution of 44 000 times the original source concentration. Detectable levels of fluorescein remained in the groundwater for a period of no more than 14 days.

Conclusion

The trace study established that groundwater below the site has a traveltime to Manywells Spring of less than two days. The direction of flow was found to be different from that predicted by monitoring of groundwater levels.

The field test indicates that groundwater flow is primarily by 'fissure flow', which is the generally accepted mechanism for this aquifer. It is probable that a major joint set, trending northeast to southwest, was responsible for the rapid transit of the dye towards the spring. It is also possible that drainage from old stone shafts located to the south of the quarry has had an influence on groundwater flow.

The sodium fluorescein dye was found to be a suitable trace element for this site and the sandstone aquifer. Its fluorescent properites allowed an instant visual assessment of progress, and quantitative data were obtained to confirm the result.

References

ANON. 1980. B. S. 5857: *Methods for measurement of fluid flow in closed conduits, using tracers. Section 1.2 Constant rate injection method using non-radioactive tracers.* British Standards Institution, HMSO, London.

SIDDLE, H. J., JONES D. B. & WARREN, C. D. 1986. The use of tracer techniques to assess groundwater flows in site investigations. *In*: HAWKINS, A. B. (ed.) *Site Investigation Practice: Assessing B. S. 5930.* Geological Society, London, Engineering Geology Special Publication, **2**, 375–384.

W. B. WOODHEAD, & SONS 1893. Case of Bradford Corporation verses E. Pickles. Defendant's Plan. Unpublished plan held by West Yorkshire Archive Centre, 15 Canal Road, Bradford, West Yorkshire.

Ground investigation and design for a landfill at Seater, Caithness

A. J. Croxford

Sir William Halcrow and Partners Ltd, Burderop Park, Swindon, Wiltshire SN4 0QD, UK

Abstract. This paper describes the site investigations and design considerations for a landfill development near Wick in Caithness, Scotland. The site is underlain by the Caithness Flagstone Series of the Middle Devonian with a thin cover of boulder clay and peaty topsoils. The location is on raised ground sloping gently to the north with poor natural drainage. The near-surface geology was investigated with a combination of ground-probing radar and trial pitting. Radar profiles from traverses across the site were correlated with the lithological information from the trial pits to provide an estimate of the thickness and nature of the superficial deposits. Groundwater and surface water conditions were investigated and a water balance calculated for design considerations. The final design allows for a 25-year landfilling facility operated on the containment principle. Perimeter, sub-level and internal drainage systems were incorporated in the design.

Sir William Halcrow and Partners, Scotland Ltd were commissioned in 1989 by Caithness District Council to carry out a preliminary study of an area at Seater, located 15 km northwest of Wick in Caithness, to determine the suitability of the area for development as a domestic and commercial waste landfill. The area consisted of a central depression surrounded by higher ground on three sides. The ground elevation varied between approximately 40 and 80 m AOD.

Further investigation of part of the area at Seater and the submission of an application for planning consent were recommended in the initial study. In the summer of 1989 Halcrow undertook an in-depth second-stage study of the selected area in support of a detailed planning application. The study included the following components:

- a detailed topographic survey of the site to produce a plan and sections;
- a site investigation to determine the ground conditions, and the volume of materials on site;
- hydrological and hydrogeological studies;
- the production of a preliminary landfill design and tipping plan.

The site investigations are described in the following sections.

Site Investigations

Stage 1

Stage 1 investigations started with an appraisal of existing data sources. There were no records of any previous ground investigations in the area. All existing data were obtained from published geological maps and information was adequate for an initial desk study.

In order to determine the more precise nature of the deposits, their thickness and the hydrogeological relationships between units, seven trial pits were excavated across the site. The solid geology was as expected from published data. The bedrock is the Caithness Flagstone of the Middle Devonian, a carbonate-rich, laminated, fine, silty sandstone, heavily weathered at outcrop.

The nature of the drift deposits and the interface between the base of drift and the bedrock was found to be variable across the site. The flagstones were overlain by boulder clay of variable composition with a blue clay forming the upper part of the unit on low ground. The boulder clay was overlain by peat on lower slopes and a thin soil on higher ground. The top of the flagstones was found to be fractured and in places mixed with boulder clay.

The stage 1 geotechnical investigations comprised the excavation of the trial pits, the detailed physical description of the lithologies encountered and several Mackintosh probes on the lower part of the site.

Information on the site hydrogeology was also obtained from the trial pits. Groundwater occurred in all three lithological units. In the flagstones, groundwater storage was concentrated in the fractured and weathered portions of the unit, with the fresh rock having a negligible remaining primary porosity. Groundwater occurred in the boulder clay although only in noticeable quantities where sand, gravel lenses and stringers were intercepted during trial pitting. The overlying peat, where thickest on the lower ground, was usually saturated. Two groundwater units were identified: a confined unit below the boulder clay and a water-table above. Confined groundwater in the top of the flagstones frequently overflowed above ground level when the boulder clay was removed during trial pitting.

The data obtained from the trial pitting were sufficient to construct a conceptual model of hydrogeological conditions although were inadequate to determine the

From BENTLEY, S. P. (ed.) *Engineering Geology of Waste Disposal*,
Geological Society Engineering Geology Special Publication No. 11, pp 149–152

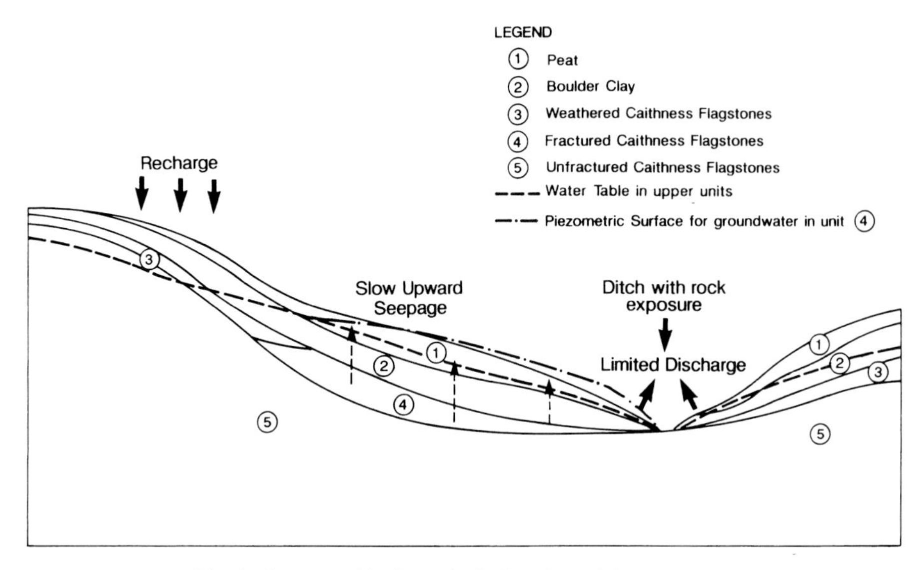

Fig. 1. Conceptual hydrogeological section of the Seater area.

spatial distribution of lithological units. A conceptual section is given in Fig. 1.

The preliminary hydrological studies identified rainfall and wind conditions, the site catchment, channels flowing at the time of the study and approximate mean annual flood volumes.

Stage 2

Hydrogeological and hydrological investigations. The second part of the hydrogeological investigation was designed to obtain an estimate of the spatial distribution of lithological units and a better approximation of groundwater heads.

An additional 50 trial pits were excavated on a grid layout across the site. Groundwater was encountered in 36 pits and piezometers were installed in 30.

The distribution of lithologies was noted from the trial pit data and approximate groundwater heads were determined for each hydrogeological unit. The heads measured below the boulder clay are likely to be approximate due to the difficulty in retaining the original geological structure during backfilling after installing the piezometers.

In the extended hydrological studies mean monthly rainfalls were derived for the area and predictions made of a range of storm durations and return periods for the site. Evapotranspiration was also estimated and a water balance calculated. Flow estimates were derived for runoff from the site during storm events using calculations from the *Flood Studies Report* (NERC 1975) procedures for ungauged catchments. A plan was constructed showing details of the existing site drainage.

Geotechnical investigations. Most geotechnical data were derived from the trial pitting and from a ground probing radar survey.

Representative samples of the drift deposits on the site were recovered from the trial pits and a programme of laboratory testing was carried out to aid the classification of the materials and to determine their relevant engineering properties. Compaction, particle size distribution, Atterburg limits and moisture content tests were carried out. Detailed logs were kept of all trial pits and full descriptions produced of each of the lithologies encountered on the site.

Ground-probing radar (GPR) was used successfully to provide longitudinal and transverse sections to the bedrock surface along the trial pit grid. The radar survey was carried out on the grid along seven parallel lines, spaced at 60 m intervals across the site and along three intersecting survey lines. The total length of profiling was approximately 6.4 line km. A three-dimensional model of the site was successfully constructed from a combination of the radar and trial pitting results.

The GPR survey was carried out by Oceanfix International Ltd with an SIR 8 system, an Oceanfix colour digital processor unit and an 80-MHz centre frequency antenna. The data were recorded on video tape, analogue tape and graphic recorder. During surveying the radar signal was displayed on a colour video monitor. The equipment was mounted in an all-terrain vehicle and an antenna towed behind on a purpose-built wooden sledge.

The survey lines were set out prior to the start of the survey with pegs in the ground at 10 m intervals and markers every 100 m. The survey data were marked on the recorder as the antenna passed over each peg. A total of five days was needed for the survey, including setting out.

All data were proofed prior to demobilization from the site. The data were viewed to check for complete coverage and satisfactory quality. Checks were made, using trial pit data, to ensure that the resolution of survey output was sufficient to identify geological boundaries. The field data were then post-processed off site and reduced to linear vertical profiles for each survey line. All the data were 'ground truthed' using trial pit data where the radar profiles coincided with the trial pits and a linear regression was used for the reduction of the data between pits. Data were also cross-correlated where cross profiles intersected the main longitudinal profiles. The post-processing included filtering to remove background noise and enhancement was used on areas of weaker data. A software package for an HP9920 computer was used to interrogate the data and identify interfaces and targets.

The GPR was a successful method of investigation. By correlation with trial pit logs, the radar results allowed the construction of a three-dimensional model of drift deposits on the site. Layering in the peat and till, interfaces between peat, till and bedrock, exfoliation, weathering and structures in the bedrock were identified using this method. Shallow hollows, interpreted as glacial striations, and trending southwest–northeast, were identified in the top of the bedrock. The bedrock surface was the most well-defined boundary on the radar profiles. The use of a low-frequency during the survey (80 MHz) gave better resolution at lower levels in the profiles, and less detail of the more superficial strata. The use of two frequencies in the survey, high and low, would have given improved overall data recovery.

Design considerations

The design of the Seater landfill included consideration of the following factors:

- the optimum use of the site materials for landfill operation;
- the control of surface and groundwaters;
- the control of groundwater pollution.

Optimum use of materials

With the three-dimensional data available from the site investigation the volume of each type of deposit could be estimated. From this information, consideration was given to the optimum use of materials. Provision could be made for:

- a total resource for restoration of the site after the planned 25-year life of the facility;
- total volumes of rock/boulder clays for bund construction;
- a three-year supply of cover and capping materials;
- rock for crushing, processing and use as permanent and temporary road-making materials.

These materials have been stockpiled on site.

Ground and surface water control

Surface water entered the site area from the west and the high ground on the northern and southern sides. The confined part of the flagstones was also recharged from these areas. A peripheral ditch was proposed for installation around the site to divert surface and groundwater inflows. The drain was designed to cut into both the superficial deposits and the bedrock to drain the waters potentially entering the site.

A network of drains was installed below the lined part of the landfill. Four dendritic networks were installed with 225 mm diameter porous concrete pipes laid parallel to the sub-grade base contours to intercept groundwater. Pipe trenches were backfilled with broken rock and capped with a geotextile filter sheet to reduce the migration of sand into the drains.

Control of groundwater pollution

The local groundwater system would be severely affected by pollution by leachates. Several wells are used for groundwater abstraction in the local area. There are also two sites with SSSI status in the area, Lock Watten and the River Wick Marshes.

The site was therefore designated to be operated on a full leachate containment principle using an artificial membrane to line the site.

Conclusions

The site investigations carried out at Seater provided sufficient information on the geotechnical, hydrogeological and hydrological conditions to design the landfill and estimate material resources.

The GPR survey was a valuable extension to the site investigation. The thin drift deposits on the site allowed adequate resolution to define the bedrock surface. Thicker overlying clay units may preclude the use of this method in other environments. The use of two frequencies in the operation of the GPR would have given improved resolution at shallow depths.

Cooper *et al.* (1991) give additional detail on the design and operational aspects of the Seater landfill.

References

NERC 1975. *Flood studies report.* Natural Environment Research Council.

COOPER, E., BOYD, R. D. & FULLER, A. N. 1991. *The planning, design and construction of a landfill site at Seater, Caithness.* Proceedings of the Midland Geotechnical Society conference on The Planning and Engineering of Landfills, 137–142.

Stability considerations in landfill lining design

J. H. Dixon[1] & S. P. Bentley[2]

[1] Netlon Limited, New Wellington Street, Blackburn, Lancashire BB2 4PJ, UK
[2] School of Engineering, University of Wales, Cardiff, South Glamorgan CF2 1YF, UK

Abstract. Modern landfills are generally designed with a composite base and side lining system consisting of a low-permeability mineral layer overlain by a geomembrane. Methods of installation, quality assurance procedures and chemical compatibility are well addressed in most contract specifications. However, fundamental stability considerations including the control of tensile strain are often overlooked in the design. These factors may be critical to the performance of the lining system. This paper examines designs using structural geogrids to avoid stability problems associated with the lining system.

The long-term performance of a containment landfill is highly dependent upon its lining system. Relatively low levels of tensile strain in the lining may cause cracking within the mineral sealing layers and/or accelerated degradation of the geomembranes. This can result in a significant increase in the permeability to leachates and landfill gases. Information on the permissible levels of long-term tensile strain in the lining systems is limited but is a subject of growing interest. Current information indicated that safe design strain levels are below 3% and for critical structures an order of magnitude less (Genske *et al.* 1993).

Required characteristics of structural geogrids

Geogrids are planar polymer structures formed by a regular network of tensile elements having apertures which permit interlock with adjacent soil. The function of the geogrid is to carry long-term tensile load.

There are four essential geogrid design properties:

- long-term stress–strain characteristics which are used to identify the safe design strength;

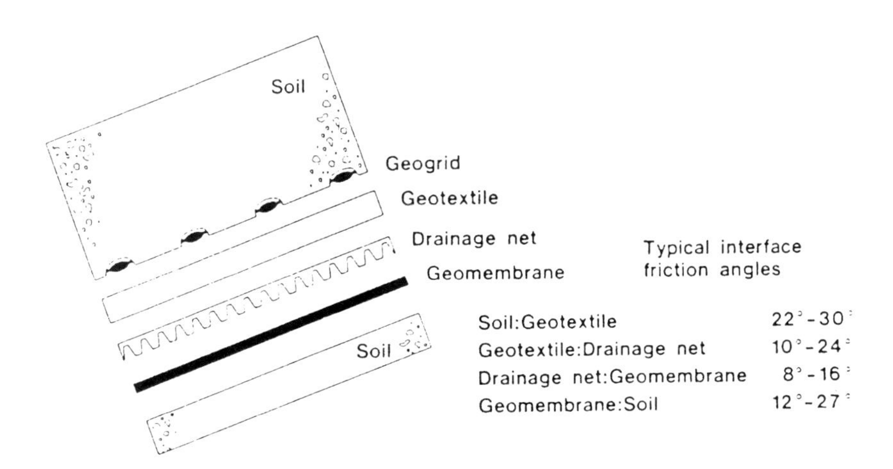

Fig. 1. Typical interface friction angles of landfill lining system.

From Bentley, S. P. (ed.) *Engineering Geology of Waste Disposal*,
Geological Society Engineering Geology Special Publication No. 11, pp 153–157

- the junction strength through which the soil load is transferred via the transverse bars of geogrid to the load-carrying longitudinal ribs;
- robustness during installation;
- polymer durability throughout the design life.

As with all products incorporated within the landfill, the geogrids should be manufactured in accordance with independently certified quality assurance procedures.

Sloping lining systems: reinforced soil veneers

Particular attention to stability is necessary when contemplating lined slopes steeper than 1 : 7 (vertical : horizontal). Linear slippage between the individual components of the lining system (which often include geosynthetic protection and drainage layers) or the soil layers may be a problem. The interface shear strength between these materials can be very low, with interface friction angles of 8° or less (Fig. 1).

Such low values of interface friction have led to a number of landfill failures including the Kettleman Hills Landfill, USA (Mitchell *et al.* 1990).

The sliding forces may be controlled by altering the geometry. If this is not feasible then structural geogrids may be introduced to support the component weight of overlying soil and temporary construction surcharging with the necessary factor of safety. Non-structural components within the lining system should not be designed to carry tensile load.

It is recommended that product-specific testing is carried out to establish the interface friction angles between the individual lining components. If the minimum friction angle occurs above the geomembrane then the geogrid design strength is based on the performance limit strain of the geogrid, typically 10%. If however, the minimum friction angle occurs below the geomembrane then the value of geogrid design strength

Fig. 2. Geogrid installation in a reinforced soil veneer.

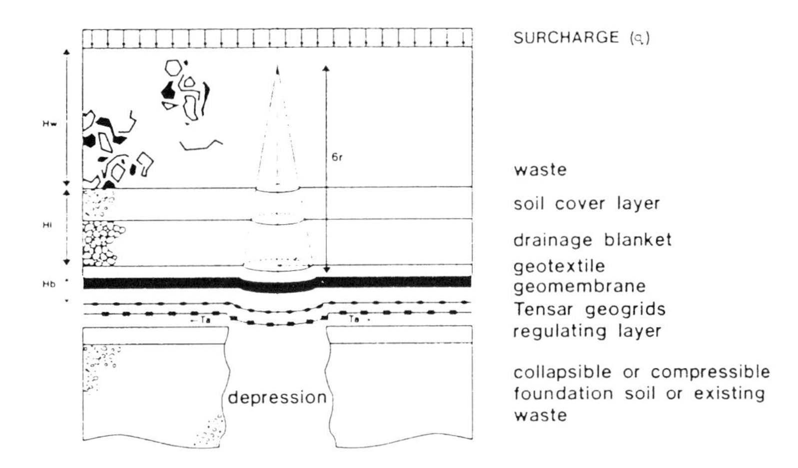

Fig. 3. Schematic of lining support system.

Fig. 4. Tensar geogrid lining support, Suffolk.

selected must be compatible with the performance limit strain of the geomembrane, typically 3%.

A reinforced soil veneer was employed at Brackletter Landfill, Scotland (Hall & Gilchrist 1993). It was necessary to place a 1.0-m-thick soil cover layer over a geomembrane to cap the landfill which slopes at 1:2 (vertical: horizontal). Tensar SR110 geogrid was placed directly on top of the geomembrane and anchored at the crest (Fig. 2) prior to placing the soil on the slope.

Vertical extensions to existing landfills ('piggy-backing')

The shortage of suitable land for landfill sites is increasing the demand to place waste over existing landfills. This practice began in the USA in the late 1980s and is sometimes termed 'piggy-backing'.

In general the existing landfill would have been operated on a dilute and disperse principle, and would be unlined. The new extension would therefore need a base containment lining system placed over the cap of the existing landfill. Structural support must be provided to this lining to control its localized deformation caused by the formation of voids as the underlying waste degrades (Fig. 3). Allowance should be made for the collapse of large objects within the existing waste, such as refrigerators which can implode.

A similar structural support application is presented in a paper by Stephens & Bodner (1991). This describes a Pennsylvania municipal landfill sited over former coal workings and subject to subsidence. The first known UK geogrid support layer to a landfill 'piggy-back' was in Suffolk in 1993 (Fig. 4).

Support over weak and variable soil foundations

Structural geogrids may be used to support linings over weak and variable formations. In addition to controlling differential settlement, they may also provide the only

Fig. 5. Tensar Geocell foundation support, Hausham.

practical means of construction access. In extreme cases, Tensar geogrids can be fabricated on site into a Geocell (a continuous open-top cellular mattress) which is filled with granular material. This solution was adopted at Hausham, Germany, where a landfill was formed over extremely weak silty tailings (Fig. 5). The Tensar Geocell provides a stable and stiff foundation for the composite lining system.

Lining support to the sides of steep quarries

Many UK landfills are sited in disused steep quarries. Operators wish to maximize void space and to construct the lined landfill sides as steep as possible.

'Conventional' reinforced soil methods can be used to stabilize steep mineral bunds constructed against the quarry face, with due allowance for drainage. These bunds may be formed in 3–4 m high lifts with intermediate horizontal berms. A geomembrane can then be placed against the stable, smooth bund face and anchored on the berm. Design consideration should also be given to controlling in-service geomembrane deformation from 'drag-down' forces as the waste decomposes and settles.

Summary

Structural considerations are fundamental to the design of landfill lining systems. Designs should be based on serviceable strain limits. Structural geogrids with carefully selected characteristics may be used to satisfy the long-term stability and serviceability criteria.

References

GENSKE D. D., KLAPPERICH, N. P. & THAMM, B. 1993. *Historical landfills and old sites—remediation and design.* Sardinia '93.

HALL, C. D. & GILCHRIST, A. J. T. 1993. *Steeply sloping lining systems—Stability considerations using reinforced soil veneers.* Green '93, Bolton.

MITCHELL, J. K., SEED, R. B. & SEED, H. B. 1990. Stability considerations in the design and construction of lined waste repositories. *In: Geotechnics of Waste fills—Theory and Practice.* ASTM (PCN 04-010700-38), 209–224.

STEPHENS, A. & BODNER, R. 1991. *Geogrid liner support at Empire sanitary landfill.* Geosynthetics '91, Atlanta. Industrial Fabrics Association International, 15–22.

Laboratory investigations into designated high-attenuation landfill liners

M. I. Bright, S. F. Thornton, D. N. Lerner & J. H. Tellam

Hydrogeology Research Group, School of Earth Sciences, University of Birmingham, Edgbaston, Birmingham B15 2TT, UK

Abstract. The present UK landfilling policy of total containment of waste is likely to present long-term problems in terms of liner integrity, so that additional leachate control measures may become a necessity. The design of economic liners from natural materials with high-attenuation characteristics that could serve as a back-up for containment sites is discussed. Laboratory column experiments are currently underway to test the performance of proposed liner recipes using a range of widely available materials. The results obtained should indicate the extent to which it is possible to design a liner material on the basis of simple measurements.

In recent years the historical reliance in the UK on attenuation of potential leachate pollutants by *in situ* materials has been gradually replaced by the philosophy of total isolation of waste from the environment, in response to changes in groundwater protection policy and EU directives. However, while an engineered containment site provides a short-term solution to groundwater pollution, the integrity of the liner and leachate collection system in the long- term cannot be guaranteed, as is evident from recorded failures (Johnson *et al* 1981; Gray 1989; Mott & Webber 1991; Seymour 1992), and it therefore seems important to consider back-up measures. Such measures could, in principle, take the form of an engineered liner composed of a mixture of natural materials known to possess properties that would maximize attenuation.

This paper briefly summarizes attenuation mechanisms in natural materials and discusses the chemical design of low-cost, high-attenuation liners as a second line of defence for containment sites. Laboratory column studies currently being carried out to test the feasibility of proposed liner recipes (different combinations of selected natural materials in varying proportions and configurations) under anaerobic conditions are discussed. Preliminary results are presented.

Attenuation mechanisms

There are a number of processes capable of attenuating leachate during its migration through natural materials. These processes, principally comprising sorption and ion exchange, precipitation, degradation and dilution, are summarized in Table 1.

Not all processes, however, are beneficial. For example, if ferric oxyhydroxide coatings are present, an anaerobic leachate will tend to dissolve them, resulting in the mobilization of adsorbed heavy metals (e.g. Korte *et al.* 1976). Ferric oxide is also a more favourable electron acceptor than CO_2, and its presence in large quantities in liner material could inhibit the normal methanogenic reduction of CO_2. Occluded salts present in earth materials are also capable of dissolving in leachate, thereby increasing competition for available adsorption sites.

Chemical design philosophy

The design of an attenuating liner might be tackled by a detailed hydrogeochemical investigation of individual leachate species. However, the complex interactions between species as well as between each species and the liner renders such an approach impracticable. A more pragmatic approach is the recognition, based on the processes summarized in Table 1, that attenuation is likely to be maximized by a high clay content (to increase inorganic species sorption), a high organic carbon content (to increase organic and heavy metal species sorption) and a high carbonate content (to buffer pH at levels where metal mobilization is limited, cation exchange can occur and methanogenic bacterial growth conditions are optimized). Maximization of these components might be achieved by selecting a natural material rich in clay, organic carbon and carbonate. However, few materials possessing these prerequisites are likely to be available, so that the liner would have to be constructed from a locally available base material or 'substrate' and modified by the addition of readily available clay and/or carbonate and/or organic-rich materials.

Although extremely simplistic, the approach may well work, especially if designs are supported by the results of simple batch-style experimentation on the available

From BENTLEY, S. P. (ed.) *Engineering Geology of Waste Disposal*,
Geological Society Engineering Geology Special Publication No. 11, pp 159–164

Table 1. *A summary of major attenuation processes that occur in the vicinity of a landfill*

Mechanism	*Attenuation properties/characteristics*	*Solutes attenuated*	*Favourable conditions*
Sorption/ion exchange by clays	High particle surface areas, moderate to high cation capacities (CECs)	Major ions and trace heavy metals	Alkaline pH
Sorption by particulate organic matter	Degree of partitioning directly proportional to fraction of organic carbon (fo_c) and inversely related to solubility of individual organic compounds[1]	Hydrophobic organic pollutants	
	Very high particle surface areas	Heavy metals[2,3]	Alkaline pH
Precipitation	Precipitation of metals as hydroxides and carbonates onto soil particle surfaces and into pore water	Heavy metals	Alkaline pH[4–7]
Organic transformation	Abiotic transformations at slow flow rates, although they can be activated by biologically derived enzymes[8,9]	Aliphatic and aromatic compounds	
	Biogradation—typically a much faster process	Organic compounds	A near natural environment (pH 7–8) a prerequisite for establishment of methanogenic bacteria.[10–12] Sufficient substrate and nutrients necessary for microbe survival
	Methanogenic bacteria responsible for the degradation of volatile fatty acids (VFAs) and other organic compounds. But where VFA concentrations are high and CO_2 dissolution in the unsaturated zone occurs, acid conditions, unsuitable for methane bacteria development, prevalent		
	Adequate residence time necessary for microbial colonization		Slow flow rates
Dilution	A direct result of hydrodynamic dispersion of solutes in subsurface environment	All contaminants	

[1] Karikhoff *et al.* (1979). [2] Reimers *et al.* (1978). [3] Fuller (1980). [4] Griffin *et al.* (1976). [5] Korte *et al.* (1976). [6] Artiola & Fuller (1980). [7] Labauve *et al.* (1988). [8] Bouwer (1984). [9] Vogel *et al.* (1987). [10] Farquhar & Sykes (1982). [11] Hoeks & Borst (1982). [12] Mather (1989).

materials and by simplified theoretical calculations. To test the approach, an extensive suite of laboratory column experiments is currently being carried out as detailed below.

Experimental design

The general approach

A set of columns have been packed with example substrates and additives in various proportions and configurations. Leachate is fed through the columns under anaerobic conditions, and the effluent is analysed for a wide range of determinands. The effects of temperature and, to a limited extent, residence time are examined. Details of the choice of recipes (substrate + additives) and of the column design are given below.

Choice of substrates

Three natural materials widely available in the UK were selected to form the basis for the recipes. At one end of the spectrum Triassic Sandstone, characterized by low CEC, very low organic carbon content and fairly

Table 2. *Typical properties of materials selected for column experiments*

	Triassic Sandstone	*Coal Measures Clay*	*Oxford Clay*
Organic carbon (% wt)	0.023	2.6	3.9
$CaCO_3$ (% wt)	<1	<1	14.8
CEC (meq/100 g)	2.75	14.5	35.6
pH	4.8	2.45	7.7
Leachable salts	low	moderate	low
Fe oxides	high	low	low

low buffering capacity, and by a high Fe oxide content, typifies materials which would make very unsuitable substrates without modification. In contrast, Oxford Clay appears to possess optimal properties as an attenuating liner, requiring no modification. Coal Measures clay represents substrates of intermediate properties (Table 2).

Recipes tested in column experiments

Based on the considerations outlined above, recipes were designed as indicated in Table 3 and designated 'bad', 'improved' or 'good' according to the extent to which each is predicted to approach the 'optimum attenuation' requirement.

Table 3. *Substrates tested in column experiments*

Substrate	*Materials added*	*Recipe type*
Inert sand (>98.5% SiO_2)	—	Blank
Triassic Sandstone	—	Bad
Oxford Clay (5, 10, 15, 15%)	Inert sand*	Good
Triassic Sandstone	Oxford Clay (10, 15, 15, 20%)	Improved
Coal Measures Clay (10%)	Inert sand* + $CaCO_3$ (admixed and layered)	Improved

* Added simply to increase flow rates for laboratory experimentation.

Column experiments

Twenty UPVC columns were packed with the various test recipes and fitted with collection systems designed to maintain anaerobic conditions in the effluent, prevent volatile loss and serve as an outflow regulator (Fig. 1). Of these columns, 18 are of standard 1 m length × 10 cm diameter dimensions providing a residence time of 25 days at a flow rate of 140 ml day^{-1} To evaluate the influence of residence times on attenuation, one of the recipes has been repeated in a 1 m × 6.5 cm as well as a 1 m × 15 cm column, providing residence times of 8 days and 45 days respectively. Three of the recipes have also been repeated with the column lagged with fibre-glass insulation to maintain internal column temperature conditions. Columns were initially saturated with tap water (80 μs cm^{-1} conductivity).

Methanogenic leachate (1200 l) was spiked with stock solutions of selected compounds normally present in leachates but not in significant quantities in the leachate utilized. The liquid was pumped into six CO_2/N_2

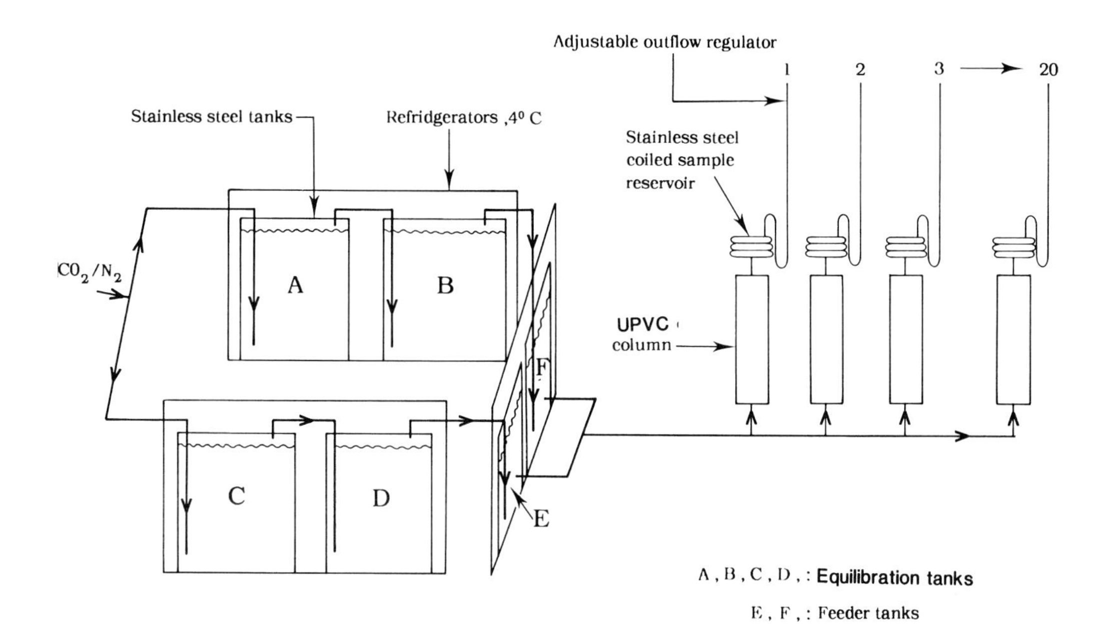

Fig. 1. Schematic diagram of column apparatus.

preflushed stainless-steel tanks contained in chest freezers maintained at 4 °C to minimize chemical and microbial changes in the leachate (Korte *et al.* 1976; Kjeldsen 1986). Four of the six tanks serve the sole purpose of equilibrating the supply tanks to prevent partitioning of volatiles into the headspace (see Fig. 1). The columns were fed from the supply tanks via upward flow through stainless steel pipes under constant CO_2/N_2 pressure. Concentrations of unspiked and spiked species in the leachate are listed in Table 4.

Column effluents are sampled twice a week via sampling valves located at the top of the columns and analysed employing the techniques listed in Table 5.

Preliminary results and discussion

The results of the analysis of effluent samples for selected columns and parameters spanning an average of two pore volumes each are shown as breakthrough curves in Figs 2(a)–(e). (It is proposed to pass a minimum of 10 pore volumes through each column.)

In columns 1 (100% inert sand) and 5 (inert sand + 15% Oxford Clay), the conservative nature of chloride is reflected in its early breakthrough (Figs 2(b) and (d)). The early breakthrough of ammonium in column 1 (Fig. 2(b)) as compared with its retardation in column 5 is consistent with the expected attenuation behaviour of the two recipes. The high CEC of Oxford Clay can account for NH_4^+ removal in column 5. However, the higher evolutionary trend for alkalinity in column 1 compared with column 5, suggests a complex interplay of factors and species which can only be unravelled as investigations progress.

Figure 2(c) provides evidence that organics are being removed from solution in column 4 (inert sand + 5% Oxford Clay), which can be attributed, at least in part, to the high organic carbon content of Oxford Clay (3.9% wt). Reduced immobilization in column 6 (Fig. 2(e): 100% Triassic Sandstone) is consistent with expected behaviour although attenuation is still probably greater than would be expected given the very low fo_c of the sandstone (0.023% wt). In column 1 evidence of organics attenuation inspite of the supposed inertness of sand (Fig. 2(a)) suggests the operation of mechanisms other than sorption on to soil organic carbon. While biodegradation is a possible explanation, the considerably long period normally found to be necessary for the establishment of microbial communities (Mather 1989) points to other possibilities, one of which could be the formation of insoluble metallo-organic complexes at alkaline pH. Another possibility is adsorption onto the mineral component of the soil, a phenomenon which has been observed to occur by several authors (e.g. Schwarzenbach & Westhall 1981) at very low fo_c. However, no definite interpretations can be made at the present time.

Table 4. *Solute concentrations before and after spiking of leachate*

	Unspiked concentrations	*Equilibrium concentrations after spiking*
Organics ($\mu g\, l^{-1}$)		
Benzene	27.2	270.7
Toluene	17.4	715.3
1,1,1-Trichloroethane	11.9	1037.7
Trichloroethene	46.5	703.6
1,1,2,2-Tetrachloethane	96.7	126.7
Tetrachloroethene	12.7	392.4
1,2-Dichichlorobenzene	25.3	520.8
1,2,3-Trichlorobenzene	<0.1	151.0
Napthalene	0.29	119.8
Lindane	0.31	86.8
Dieldrin	<0.1	15.9
Hexachlorobenzene	<0.1	47.4
Hexachlorobutadiene	<0.1	277.2
Inorganics ($mg\, l^{-1}$)		
Iron (Fe^{2+})	10.35	13.9
Manganese (Mn^{2+})	0.17	0.17
Cadmium (Cd^{2+})	nd	2.65
Chromium (Cr^{2+})	nd	0.39
Nickel (Ni^{2+})	nd	4.34
Zinc (Zn^{2+})	nd	2.25
Calcium (Ca^{2+})	127	124.5
Magnesium (Mg^{2+})	139.5	134.5
Potassium (K^+)	510	489.5
Sodium (Na^+)	1450	1340
Ammonium (NH_4^+)	1048	1029
Chloride (Cl^-)	1955	1965
Sulphate (SO_4^{2-})	5215	52.08

nd = Not detected.

Table 5. *Analytical techniques employed*

Analytes	*Technique*
Organics	
Volatiles	Static head space/GC-MS
Semi-non-volatiles	Liquid liquid extraction/GC-MS
COD	Dichromate reflux method
Inorganics	
Major cations and heavy metals	ICP
Anions	Ion chromatography, autoanalyser colorimetry
Alkalinity	Titration
pH	pH meter
Conductivity	Conductivity meter

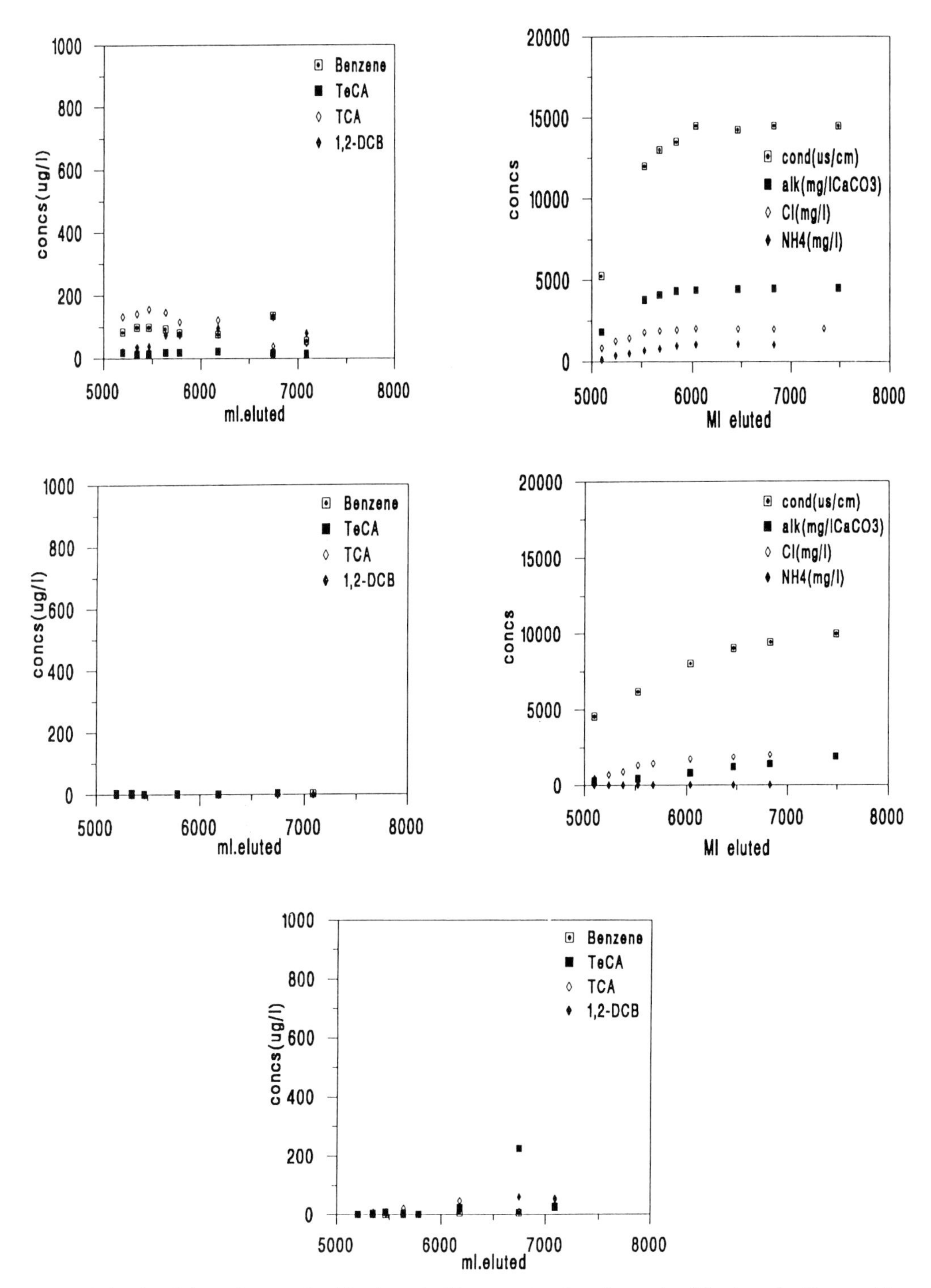

Fig. 2. Solute breakthrough curves for (a) column 1 with respect to selected organics; (b) column 1 with respect to selected parameters; (c) column 4 with respect to selected organics; (d) column 5 with respect to selected parameters; (e) column 6 with respect to selected organics.

Conclusion

Preliminary evidence suggests that attenuation is occurring within the columns. However, a complex interplay of factors and species appear to be in operation, and the overall success of the methodology employed for liner design (measurements of CEC, foc, carbonate content, etc.) is yet to be assessed. Whether or not undesirable solutes are leached from the liners should become clearer after a longer period of experimentation. A measure of the extent to which attenuation is permanent (particularly with respect to sorption reactions) will be studied at the end of the leachate loading period.

The approach, if successful, should permit the engineering of high-attenuation, low-cost back-up liners for containment sites from widely available materials.

References

ARTIOLA, J. & FULLER, W. H. 1980. Limestone liner for landfill leachates containing beryllium, cadmium, iron, nickel and zinc. *Soil Science*, **129**, 167–179.

BOUWER, E. J. 1984. Biotransformation of organic micropollutants in the subsurface. *In*: *Proceedings of the NWWA/API conference on petroleum hydrocarbons and organic chemicals in groundwater—prevention, detection and restoration.* Texas.

FARQUHAR, G. J. & SYKES, J. F. 1982. Control of leachate organics in soils. *Conservation and Recycling*, **5**, 55–68.

FULLER, W. H. 1980. Soil modification to minimize movement of pollutants from solid waste operations. *CRC Critical Reviews in Environmental Control*, **9**, 213–270.

GRAY, D. H. 1989. Geotechnical engineering of land disposal systems. *In*: BACCINI, P. (ed.) *The Landfill: reactor and final storage.* Springer-Verlag, 145–173.

GRIFFIN, R. A., SHRIMP, N. F., STEEL, J. D., RODNEY, R. R., WHITE, W. A. & HUGHES, G. M. 1976. Attenuation of pollutants in municipal leachate by passage through clay. *Environmental Science and Technology*, **10**, 1262–1268.

HOEKS, J. & BORST, R. J. 1982. Anaerobic digestion of free volatile fatty acids in soils below waste tips. *Water, Air and Soil Pollution*, **17**, 165–173.

JOHNSON, T. M., CARTWRIGHT, K. & SCHULLER, R. 1981. Monitoring of leachate migration in the unsaturated zone in the vicinity of sanitary landfills. *Groundwater Monitoring Review*, **1**, 55–63.

KARIKHOFF, S. W., BROWN, D. S. & SCOTT, T. A. 1979. Sorption of hydrophobic pollutants on natural sediments. *Water Research*, **13**, 241–248.

KJELDSEN, P. 1986. *Attenuation of landfill leachate in soil and aquifer material.* PhD thesis, Technical University of Denmark.

KORTE, N. E., SKOPP, J., FULLER, W. H., NIEBLA, E. E. & ALESII, B. A. 1976. Trace element movement in soils: influence of soil physical and chemical properties. *Soil Science*, **122**, 350–359.

LABAUVE, J. M., KOTUBY-AMACHER, J. & GAMBRELL, R. P. 1988. The effect of soil properties and synthetic municipal landfill leachate on the retention of Cd, Ni, Pb and Zn in soil and sediment materials. *Journal of the Water Pollution Control Federation*, **60**, 379–385.

MATHER, J. D. 1989. The attenuation of the organic component of landfill leachate in the unsaturated zone: a review. *Quarterly Journal of Engineering Geology*, **22**, 241–246.

MOTT, H. V. & WEBER, W. J. 1991. Diffusion of organic contaminants through soil–bentonite cut-off barriers. *Journal of the Water Pollution and Control Federation*, **63**, 166–176.

REIMERS, R. S., ENGLANDE, A. J., KRENTEL, P. A., DANFORTH, R. A., LEFTWICH, D. B. & LO, C. P. 1978. *The effectiveness of land treatment for removal of heavy metals.* Proceedings of the Industrial Waste Conference, Purdue University, **32**, 1013–1034.

SCHWARZENBACH, R. P. & WESTHALL, J. 1981. Transport of nonpolar organic compounds from surface water to groundwater: laboratory sorption studies. *Environmental Science and Technology*, **15**, 1350–1367.

SEYMOUR, K. 1992. Landfill lining for containment. *Journal of the Institute of Water and Environmental Management*, **6**, 389–396.

VOGEL, T. M., CRIDDLE, C. S. & MCCARTY, P. L. 1987. Transformations of halogenated aliphatic compounds. *Environmental Science and Technology*, **21**, 22–36.

Geomembrane landfill liners in the real world

S. J. Mollard,[1] C. E. Jefford,[2] M. G. Staff[2] and G. R. J. Browning[3]

[1] Wardell Armstrong, 22 Windsor Place, Cardiff CF1 3BY, UK
[2] Wardell Armstrong, Lancaster Building, High Street, Newcastle-under-Lyme, Staffordshire ST5 1PQ, UK
[3] Geology Division, Staffordshire University, ST4 2DE, UK

Abstract. This paper provides an introduction to high-density polyethylene (HDPE) geomembranes, emphasizing some practical aspects essential for the successful installation of such liners at landfill sites. Installation procedures are discussed, including quality of the sub-grade, methods of laying panels, seaming techniques and destructive and non-destructive methods of testing. The need for construction quality assurance (CQA) is highlighted as a means to minimize leakage and to ensure that geomembrane liners function as specified throughout the life of the landfill facility.

Geosynthetics comprise products that consist almost exclusively of synthetic materials designed to perform one or more of the following functions:

- separation
- reinforcement
- filtration
- drainage
- liquid/gas barriers

Liquid/gas barriers form a range of products known as geomembranes, also termed flexible membrane liners (FML), which are manufactured predominantly from thermoplastic polymers and are virtually impermeable.

Geomembranes are widely used in tunnel lining, heap-leach treatment cells, canals and ditches, water storage, lining/capping landfills and as vertical or horizontal cut-off barriers isolating contamination. This paper considers the application of high-density polyethylene (HDPE) geomembranes to the UK landfill industry.

Landfill liners

Landfilling is the main method of waste disposal in the UK. Increased environmental concern and resulting legislation have driven the technology of landfilling away from the philosophy of dilute and disperse to the fully engineered containment system.

Early containment landfill facilities comprised a low-permeability mineral liner, usually clay. There has been concern over the effectiveness and durability of such liners when exposed to some of the more aggressive leachates and liquids associated with waste disposal.

In the early 1970s the US Environmental Protection Agency (EPA), because of these concerns, became interested in the use of various synthetic materials as possible alternatives to the mineral liner. Materials such as PVC, polyethylene and butyl-rubber had been successfully used as essentially impermeable membranes in water storage and water conveyance projects. However, the durability of these materials in the landfill environment for the life-time of such a facility was questioned. Research carried out under the auspices of the EPA, as discussed by Tissinger *et al.* (1991), demonstrated that not all the geomembrane products on the market were chemically compatible with certain types of liquid waste products and solvents. For both household-waste-derived leachate and hazardous waste liquids it was found that polyethylene material was the least affected. The more concentrated and aggressive the leachate, the higher the density of polyethylene required; this has led to the wide acceptance and common use of HDPE as a basal liner to landfill and waste storage facilities.

HDPE geomembranes

In the UK, HDPE geomembranes, when used as landfill basal liners, are 2 mm thick or greater. The major manufacturers are located in Europe and North America, although UK manufacturers now exist. This has led in the past to excessive packing, handling and transportation of the material destined for use in the UK.

The effectiveness of the HDPE geomembrane to act as a barrier to gas/leachate migration and groundwater ingress and to perform its containment function depends upon the quality and integrity of the sheeting, including all seaming, being maintained from production through to site installation and subsequently to the life of the facility. While HDPE sheeting has a laboratory permeability of about $1 \times 10^{-15}\,m\,s^{-1}$, it is generally accepted that the overall permeability of an installed

From BENTLEY, S. P. (ed.) *Engineering Geology of Waste Disposal*,
Geological Society Engineering Geology Special Publication No. 11, pp 165–170

HDPE geomembrane is greatly increased by leakage through holes or other defects, such as imperfect seaming. Where such defects occur, the flow of leachate through the membrane can be instantaneous and there is no natural self-sealing mechanism. Consequently, it is becoming common practice for a HDPE geomembrane (primary liner) to be used in conjunction with an underlying mineral liner (secondary liner). This arrangement is known as a composite lining system.

Installation of landfill liners

Site preparation for lining with HDPE

The installation of a landfill geomembrane commences with the design of the panel layout for the site. The panel layout should be designed such that the seams between panels are orientated parallel to the line of maximum slope and that, in corners and other geometrically complex locations, the number of seams is minimized to prevent stress concentrations and complex welding layouts. The type of sheet used will depend on the engineering constraints of the site, particularly in relation to maximum slope angles, as HDPE material has an inherently low surface friction. The maximum size of rolls varies depending on the type and thickness of the sheet, and the manufacturer. Sheet length typically varies from about 100 m to 200 m, with widths between 5.7 m and 10.5 m.

On delivery to the site, a surface inspection of each roll should include verification of thickness of the sheet and a check for any visible defects or damage. The area used for storage of the liner should be chosen to minimize on-site transportation and handling and should have adequate protection against dirt, impact damage and other sources of damage. Each roll should be delivered with a quality control (QC) certificate from the liner manufacturer which includes the roll number/ identification code and results of the manufacturer's QC tests. Independent conformance testing should be undertaken on the rolls as delivered, to verify the data supplied by the manufacturer, with, as a general guide, one sample tested for every 10 000 m^2 of liner installed.

Prior to placement of the liner, inspections need to be undertaken to ensure that the sub-grade is suitable. To avoid stress concentrations on the liner, the sub-grade material (i.e. the surface that will directly underlay the liner) should be free from such things as irregularities, protrusions, loose soil, abrupt changes in grade, stones which may puncture to the geomembrane and areas which may be excessively softened by high water content.

Anchor trenches constructed to secure the liner at the edge of the site should be similarly prepared and have rounded corners where the geomembrane contacts the trench, so as to avoid sharp bends in the liner.

Placement of the liner also requires careful consideration. The equipment used and method of deployment should not damage the liner in any way. Immediately after placement of the geomembrane, temporary loading in the form of sand bags or tyres is advisable in order to minimize exposure to the wind. Panels should, where possible, be placed one at a time, with each panel seamed immediately after placement. Overlaps on the downslope direction of the base of the facility should be 'shingled' to facilitate rapid drainage following any precipitation.

Prior to seaming, the panel should be visually examined; any tears, punctures, holes or thin spots should be marked and the liner either condemned or repaired, depending on the extent of the damage. A simple field identification code is marked onto the sheet at this stage, corresponding to the panel layout drawing and cross-referenced with the manufacturer's identification code.

Seaming and testing

Two principal seaming techniques are employed for joining sheets of HDPE geomembrane: fusion welding and extrusion welding. To ensure weld efficacy, no welding should be undertaken during periods of high humidity, rainfall or strong winds. Geomembrane temperatures should normally be within the range 5–35 °C. Prior to welding, the seam area should be cleaned to ensure that it is free from moisture, dust, dirt or debris of any kind. Rollin *et al.* (1989) demonstrated that the bond strength for a weld could decrease by up to 35% with the introduction of clay particles onto the sheet being welded. A more practical point is that any debris may obstruct the welding machine and could result in localized over-heating and damage to the geomembrane.

The most important factors influencing the quality of the geomembrane field seam relate to the calibration and operation of the welding equipment which includes the welding speed, quantity of heat transferred to the sheets and pressure applied, together with the alignment of the sheets relative to the welding machine. Rollin & Fayoux (1991) state that other parameters influencing the quality of the seam are the geomembrane composition, thickness and surface neatness (presence of greases, water or soil particles), together with atmospheric conditions. However, they identify the most important factors as those related to the welding calibration set by the trained operators, such as welding speed, quantity of heat transferred to the sheet, pressure applied and alignment of the sheets relative to the welding machine.

It is necessary for the welding equipment to be calibrated at the beginning of each welding shift and trial seams carried out to verify weld efficacy.

On-site destructive testing should be undertaken on a test weld in shear and peel modes. In the UK, this site test is usually qualitative. Test strips (tabs) are taken and subjected to shear and peal using a field tensiometer; samples are deemed to pass if failure occurs outside the weld area.

In fusion welding, heat is necessary to melt the polymer at the sheet interface. 'Hot wedge welding' is the most commonly used fusion technique for seaming in the UK. In this technique the energy is generated by electrical elements placed directly between the geomembrane sheets. Rollers are used to drive the machine and to apply pressure to the heated strip of the geomembrane. Average temperatures of the heating element vary, depending on sheet properties and atmospheric conditions, but are generally about 375 °C. Welding speeds are variable, but in good conditions, 2–3 m min^{-1} can be achieved. The main advantage of hot wedge welding is that, with the use of two rollers, a double fusion weld is created. In this way an air gap is formed along the middle of the seam which facilitates simple non-destructive testing using the air pressure test method.

Prior to compliance testing of the seam, samples are cut from either end for on-site destructive shear and peel tests, similar to those used to check the welding equipment calibration. When undertaking an air pressure test, both ends of the seam are sealed by clamps and a pressure-feed device, usually a needle, is inserted into the air channel created by the double fusion weld. The seam is pressurized and the pressure is observed, generally for a minimum of 2 min. Any significant pressure drop indicates a leak in the system, and once identified it is clamped off and further air pressure tests are conducted on the remainder of the seam.

With extrusion welding a hot polyethylene fillet (identical in composition to that of the geomembrane sheet) is extruded onto the edge of the upper sheet to be welded. The heat energy produced from this fillet, together with pre-heating of the sheets with hot air, is sufficient to melt the polymer of the upper and lower sheets. The weld is secured with downward pressure from the shoe of the welding equipment. Pre-heating is essential with this welding technique to avoid thermal shock that will weaken the polymeric structure along the edge of the weld. Welding speeds for extrusion are generally of the order of 0.5–1 m min^{-1}. However, time-consuming preparation is involved with extrusion welding, as the geomembrane surfaces have to be abraded to improve bonding and then lightly tack-bonded together using hand-held hot air blowers. Consequently, this technique is generally only used for patch repair works and at locations where other welding techniques are impractical.

The most widely employed non-destructive testing technique for extrusion seams is based on electrical conductivity. A copper wire is placed at the junction of the sheets prior to welding with extrusion carried out over the top of the wire. On completion of welding, a high-voltage electric current is applied over the weld using a spark-gun, and where the weld is incomplete (i.e. where the copper wire is exposed) or where the weld is

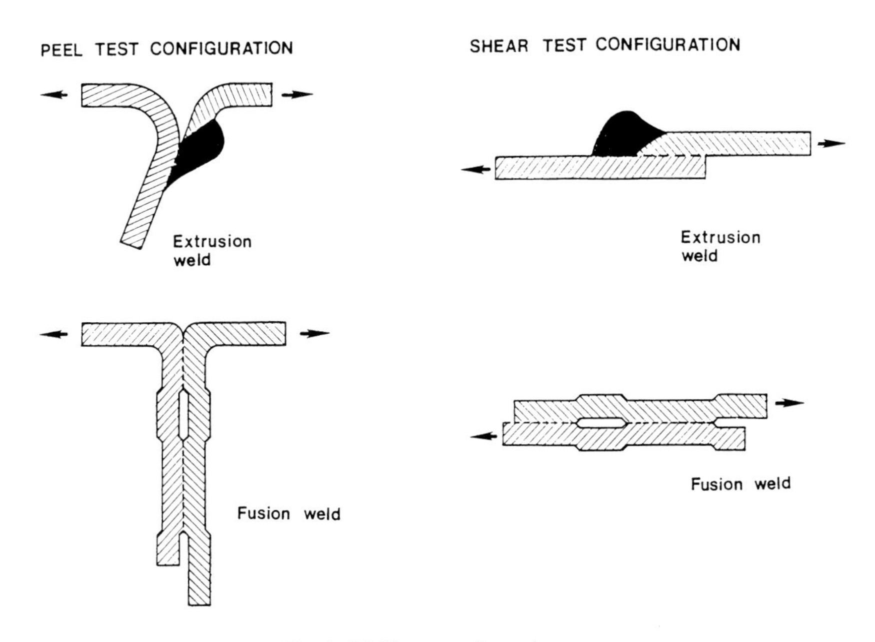

Fig. 1. Weld test configurations.

Fig. 2. Hot wedge welding.

thin, a spark is produced between the copper wire and the gun. Bennetton (1990) suggests that only unbonded areas greater than 1.5 mm in width can be easily detected with this method. The testing technique only identifies areas where the extrudate is thin or missing, and does not identify areas of poor bond in the weld.

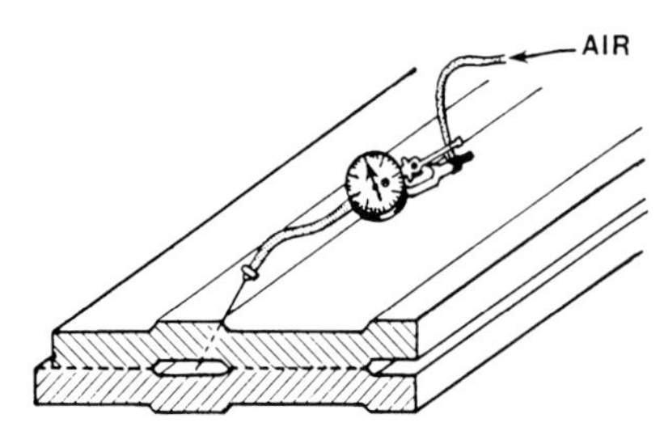

Fig. 3. Air pressure test.

A more effective and much less damaging method, although not widely used in the UK, is the vacuum chamber test technique. This is performed by applying a vacuum to a soaped section of an extruded seam. Any unbonded areas are identified from the formation of bubbles, caused by air being drawn up from beneath the liner.

Defects identified by on-site testing should be marked and then repaired and retested; this may involve either re-welding part of the seam or a further patch.

In addition to on-site destructive and non-destructive testing, samples should be taken periodically for quantitative laboratory compliance testing. Sample locations are selected randomly, unless prompted by suspicion of contamination, offset seams or obvious imperfect welding techniques; locations should not be agreed in advance with the installer. The frequency of sampling seam welds for laboratory destructive testing should be an average of one sample per 100–150 m of seam weld. This testing is to ensure seam strengths are within specified limits and tolerances.

Site documentation of all welding, sampling, testing and patching is vital. In this context, writing on the liner itself is to be recommended in addition to other documentation and construction record plans, as walking across the site will visibly reveal areas that have been completed and areas requiring further work. Prior to placement of cover materials, a final visual inspection and check of the record plans should be undertaken to ensure that all welding has been completed and tested and that all defects have been patched/repaired. Also requiring consideration at this stage are the likely effects of wrinkles and bridging, caused respectively by expansion and contraction of the geomembrane sheet. HDPE sheeting has a high coefficient of thermal expansion, due to the properties and colour of the material, and this can result in variations of as much as 10% of the sheet area.

An assessment of the wrinkles and bridging should be undertaken immediately prior to placement of cover materials, as trafficking or loading on these areas will result in uneven stresses on the geomembrane with possibly serious damage resulting. Large wrinkles, generally classed as those which can be folded against themselves, will also have an adverse effect on the basal drainage of the completed landfill. In both cases, repair is carried out by cutting and patching, together with suitable compliance testing.

Placement of cover

The end of welding operations is not, however, the end of the job. The placement and selection of cover materials and the first layers of selected waste are also processes which require careful consideration and control, to ensure continuing integrity of the geomembrane liner. If these operations are designed and carried

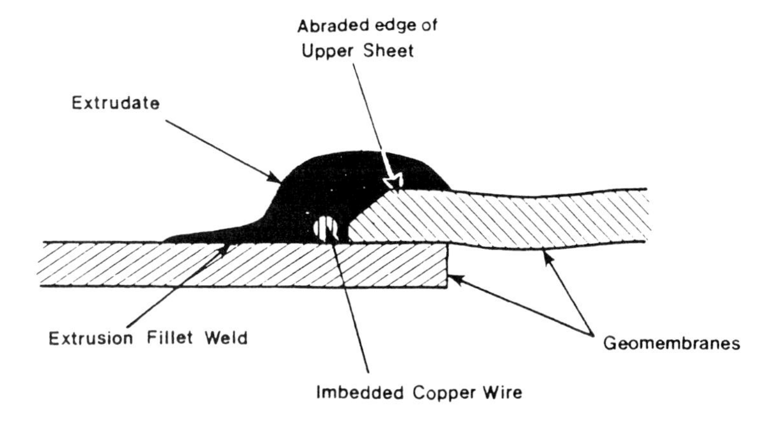

Fig. 4. Extrusion weld.

out without due respect to the relative fragility of the geomembrane, then all the hard work, expense, testing and monitoring employed in the HDPE liner construction could be negated.

Conclusions

It is to be assumed that all landfill liners will leak to some degree. However, the amount of leakage is not determined so much by the liner type as by the way it is installed and treated thereafter. HDPE geomembrane sheeting has an inherently extremely low permeability; a landfill liner is unfortunately never produced by a single sheet! High standards of workmanship and stringent quality control measures are therefore paramount at the installation stage.

Construction quality assurance (CQA) refers to the means and actions employed by an independent consultant ('third party') to assure the conformity of geomembrane preparation and installation to contractual/regulatory requirements, ensuring that the landfill liner is constructed in compliance with the project specification.

The CQA consultant is thus independent from owner, manufacturer, installer, designer and contractor and is responsible for observing and documenting activities related to the installation and repair of the components of the HDPE geomembrane. The CQA consultant is also responsible for issuing a final verification report.

The efficacy of properly implemented CQA is demonstrated by the work of Giroud & Bonaparte (1989) who noted that without CQA, one defect per 10 m of weld was typical, but with CQA, one defect per 300 m of weld

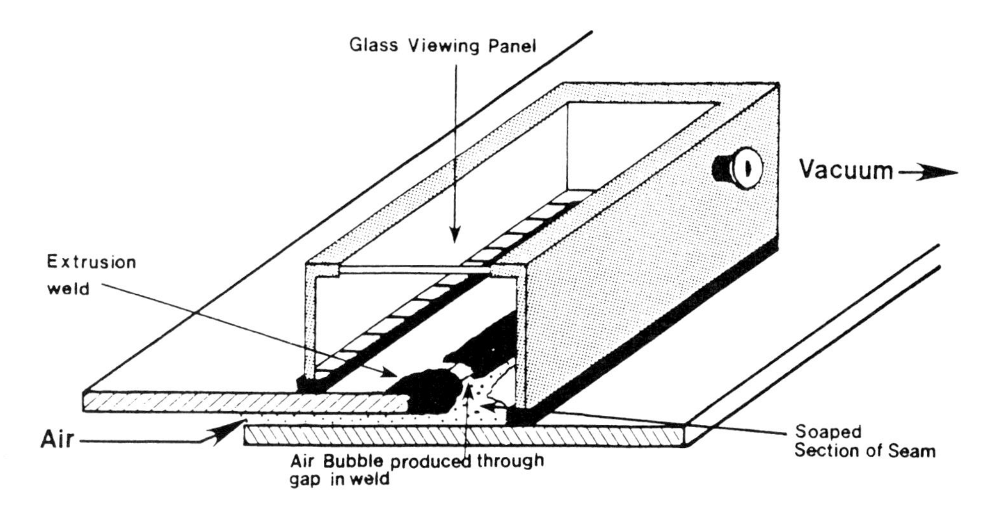

Fig. 5. Vacuum chamber test technique.

was characteristic. In terms of permeability, Attewell (1993) points out that CQA may lead to a final permeability performance two orders of magnitude lower than without CQA—a very significant improvement.

References

ATTEWELL, P. 1993. *Ground pollution.* Spon, London.

BENNETTON, J. P. 1990. *Evaluation de la qualite des soudures par essais non-destructifs.* CETE Report, Ministere de l'Equipement, Lyon, France.

GIROUD, J. P. & BONAPARTE, R. 1989. Leakage through liners constructed with geomembrane—Part 1. Geomembrane liners. *Geotextiles and Geomembranes*, **8**(1). 27–67.

TISSINGER, L. G., PEGGS, I. D. & HAXO, H. E. 1991. Chemical compatibility testing of geomembranes. *In*: ROLLIN, A. & RIGO, J. M. (eds) *Geomembranes Identification and Performance Testing*. Chapman and Hall, London.

ROLLIN, A. L. & FAYOUX, D. 1991. Geomembrane seaming techniques, *In*: ROLLIN, A. & RIGO, J. M. (eds) *Geomembranes Identification and Performance Testing*. Chapman and Hall, London

——, VIDOVIC, A., DENIS, R. & MARCOTTE, M. 1989. *Evaluation of HDPE geomembrane field techniques: need to improve reliability of quality seams. Proc Geosynthetics 89*, San Diego, 443–455.

The construction of clay liners for landfills

R. G. Clark & G. Davies

C L Associates, Prospect House, Prospect Road, Halesowen, West Midlands B62 8DU, UK

Abstract. Correct liner installation is fundamental to the construction of a containment landfill. The significance of material properties and the requirements of field placement and compaction are presented. The principle of demonstrating by means of construction quality assurance (CQA) that a mineral liner has been placed in accordance with the design requirements is discussed. Procedures are reviewed and the technical requirements of a landfill project that appertain to the CQA work are outlined.

Containment is now an established principle with regard to the design and construction of landfills. This containment is dependent on the effectiveness of the liner. In the case of clay liners considerable attention needs to be paid to the properties of the clay and the methodology of placing to ensure that a homogeneous material with the required limit on permeability is achieved. Construction quality assurance (CQA) can then be used to ensure that these requirements are met. The CQA should be carried out by an independent organization from the client and the contractor carrying out the works so that responsibilities are clearly defined and an independent compliance document can be prepared on completion of the works.

Material properties

In the case of a mineral liner there must be a suitable supply of low-permeability material available either on site or within an economically viable haulage distance. The suitability of a clay from a particular source can be determined by laboratory tests consisting of index, specific gravity (in order to calculate air voids lines), moisture content, particle size distribution, compaction and permeability determinations.

The NRA (1989) have published guidelines with regard to the suitability of materials. These are a liquid limit of less than 90%, a plasticity index of less than 65% and a clay content of greater than 10%. Murray *et al.* (1992) suggest that the plasticity index of the material should also be greater than 12% (see Fig. 1).

Permeability

As is the case for other types of earthworks, the maximum dry density and optimum moisture content of the liner material can be determined by laboratory

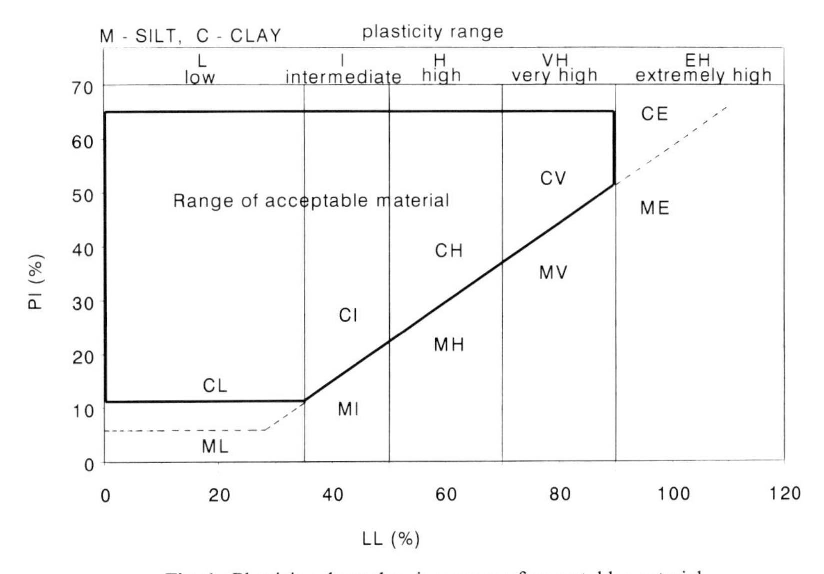

Fig. 1. Plasticity chart showing range of acceptable material.

From Bentley, S. P. (ed.) *Engineering Geology of Waste Disposal*,
Geological Society Engineering Geology Special Publication No. 11, pp 171–176

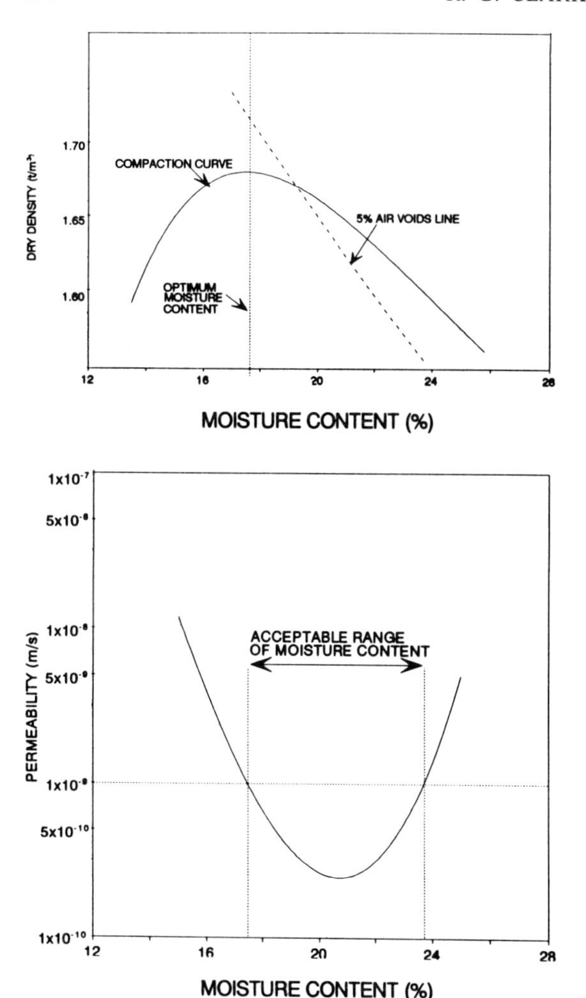

Fig. 2. Typical compaction, moisture content and permeability relationship.

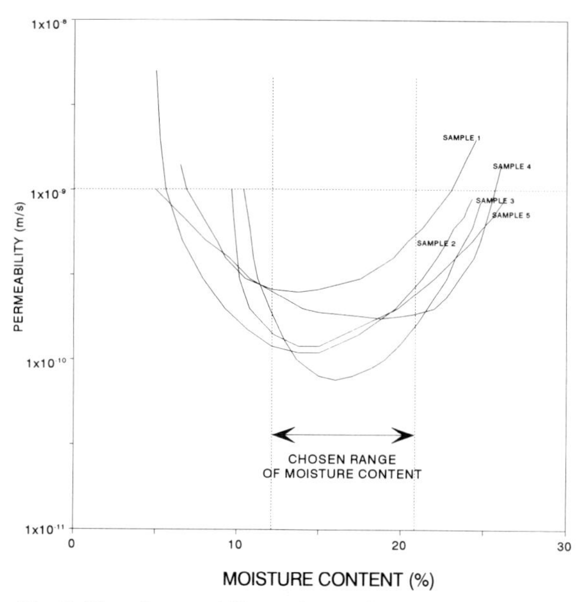

Fig. 3. Plot of permeability against moisture content for Pleistocene glacial clay.

testing. However, unlike other forms of earthworks where strength and settlement characteristics are the main criteria, the lowest permeability is likely to be at a moisture content which is greater than the optimum moisture content. Often the lowest permeability is obtained at some 2–4% above the optimum moisture content. A typical relationship between moisture content, compaction characteristics and permeability is shown in Fig. 2.

The generally accepted criterion is that a permeability of less than $1 \times 10^{-9}\,\mathrm{m\,s^{-1}}$ is to be achieved. As shown by Fig. 2, there will often be a specific range of moisture content within which this required permeability can be obtained. Figure 2 shows a typical curve for one sample. In order to check the variability of the material it is necessary to carry out tests on at least four samples. More may be required if the material is particularly variable. This will produce a family of curves (see Fig. 3 for an example showing results for a site in Pleistocene glacial clay) and the acceptable range of moisture content needs to be determined on the basis of considering all of these curves and making allowance for any uncertainty in the data.

It is impractical to attempt to measure continuously permeability in the field as the works proceed. Therefore, having carried out the laboratory tests and demonstrated that the clay is capable of achieving the required permeability, it is then necessary to reproduce in the field during the earthworks the same degree of compaction and range of moisture content. Also it is necessary to produce as far as possible a homogeneous clay liner rather than just a series of 'clods' of clay which would permit the possibility of flow paths between them.

Specification

A detailed specification should be prepared prior to commencement of the works so that all parties are aware of what is required. This specification should stipulate the preparation of the site, the drainage requirements, the procedure for selecting suitable material, the types of plant to be used, the densities and moisture content range that are to be achieved, the procedure for placing and compacting the clay, and full details of the monitoring and control of the works that will be exercised. The material properties that are required and the density and moisture content ranges that must be achieved are often referred to as the acceptance criteria. The specification should form part of the contract documents between the employer and the contractor for the construction of the landfill.

Field compaction

In general, the compaction of the clay should be by means of a tamping roller rather than a smooth drum roller. The feet on the tamping roller provide a kneading or remoulding action, thereby breaking down any clods of clay and producing as far as possible a homogeneous material when compacted. Daniel (1981) showed by laboratory tests that an increase in clod size in the finished liner could cause at least an order of magnitude increase in permeability. He also showed that kneading compaction produces a more homogeneous liner. Where problems have been experienced in the past, both in the UK and in the USA, in respect of poor-quality liners these can often be attributed to poor or inappropriate compaction.

It is important that the feet are of a correct shape and size so that the roller progressively applies compaction and does not lift up previously compacted layers. The roller should also possess effective scraper bars between the feet so that the efficiency of the roller is not impaired by clogging. Experience on a number of landfill projects has shown that in order to get the best results a vibratory tamping roller should be used rather than a static tamping roller. The only exception is where relatively high moisture content material is to be used and in these situations it may be found that a smooth drum roller is preferable and, even without the kneading action, can produce the required homogeneous material following compaction. A smooth drum roller should also be used at the end of each working day in order to seal the surface and prevent ponding of rainwater.

The plant used must produce good contact between successive layers. Otherwise the permeability parallel to the layers can be considerably greater than the permeability normal to the layers. This is particularly important where side seals are constructed in horizontal layers and horizontal permeability controls the effectiveness of this part of the liner.

Layer thicknesses after compaction must not be in excess of the specified thickness in order to ensure that adequate compaction and remoulding takes place throughout the depth of the layer. The authors usually specify a layer thickness of 150 mm after compaction in order to ensure that this is achieved, provided that roller weights are in excess of 3000 kg per metre width of roll.

Satisfactory compaction must be achieved at the interface between individual sections of seal, e.g. between base seal and side seals. This is often the most difficult to achieve and therefore should be specifically targeted for on-site testing.

Compaction trial

Before seal construction starts, a compaction trial should be conducted using the intended plant and methodology. *In situ* tests should be carried out to determine compliance with the acceptance criteria in respect of density (often above the 5% air voids line) and moisture content (as determined by the laboratory permeability tests). Often adjustments will have to be made to plant types and methodology as a result of these trials. The required number of passes of the compaction plant and any need for drying or adding moisture to the clay should also be determined together with ascertaining that the clay has been appropriately remoulded during placing and compaction.

Material conditioning

It is sometimes found that the natural moisture content of the clay does not fall within the required range of moisture content given by the acceptance criteria. The material will therefore need to be conditioned either by windrowing in order to dry it, which can be very time-consuming depending on weather conditions, or by the addition of moisture.

In adding water to the materials it is important that the method adopted imparts an even quantity of water over the whole area to be treated. It is imperative, therefore, that the water bowser or tanker being used either has a perforated bar, distributing the water over the full width of the tank, or an impact plate below the outfall of the tank which causes a powerful spray of water to be distributed over a wide area. The production of isolated pockets of moist and dry areas is unacceptable. It is also essential that, once the material to be treated has been bladed out, the material is either ripped or reworked using a plough prior to receiving water. Experience has taught us that it is not possible to add water evenly throughout the body of even a 150-mm-thick layer without the use of such plant. Indeed the process of ploughing and adding of water may need to be repeated, possibly a number of times, before adequate results are achieved.

Field testing and sampling

In situ density should preferably be measured by both sand replacement (SRD) and a nuclear density probe (NDP). This will enable cross-correlation between the

SRD and NDP results such that confidence can be gained in the subsequent use of the NDP. This will be advantageous in respect of the rapid determination of acceptability of particular layers of seal using the NDP so that corrective action or immediate removal and replacement of clay can be carried out where necessary as the works proceed.

The NDP can be sensitive, and in some cases very sensitive, to the type of surface on which the instrument stands when in operation. The surface must be completely flat and smooth without any voids or hollows, even small ones, being present at the ground surface immediately beneath the instrument. This can be a problem particularly when using a tamping roller because the feet of the roller will normally leave a series of indentations in the surface even when compaction is complete. An uneven surface can lead to considerable error in the readings given by the NDP. The reason is that for most of the instruments an average density is determined along an inclined path from the tip of the probe to a point on the base of the instrument. Therefore, if this path travels through a void over part of its length this will influence the density reading. Moisture content is determined over a curved path and is similarly affected.

This problem can sometimes be resolved by removing the top layer at the test location down to the maximum depth of the indentations and then forming a level smooth surface. However, the removal of the layer can in itself cause disturbance and hence affect the density which is then recorded. An alternative that has been found to be effective is to use a light smooth drum roller on the surface just at the test location following completion of the rolling using the heavy vibratory tamping roller. The smooth drum roller can provide a satisfactory surface for testing and, because it is only light in weight, does not alter the overall degree of compaction of the layer which, if it were to occur, would lead to false readings. In the case of stiff clays it may also be necessary to 'blade' the surface prior to the use of the smooth drum roller.

It should be noted that for safety reasons an NDP should not be used in wet weather. This may not be a problem if the works are also stopped by the wet weather. However, if this is not the case then the possibility of being unable to test the layers must be recognized when deciding whether to continue placing the liner, unless some form of shelter is provided.

As well as field testing, samples should also be taken as the liner material is being placed so that laboratory tests (index and particle size distribution) can be carried out to verify, on a progressive basis, that the materials used for constructing the liner have consistent properties to the samples used to establish the acceptance criteria.

If there is a requirement to carry out permeability tests on the placed materials this is usually achieved by taking 'undisturbed' samples and then carrying out permeability tests in the laboratory. *In situ* permeability tests in boreholes at such low permeabilities have been found to be unreliable, first because of the difficulty of sealing off test layers and ensuring that there are no leakages from casings, etc., and secondly because the mathematics behind the available analytical methods is dubious at these low permeabilities. B. S. 5930 (Anon. 1981) indicates that a permeability of 1×10^{-8} m s^{-1} is the lower limit of practicality for carrying out permeability tests in boreholes. Ring infiltrometers are one possible alternative. However, although there are a number of reports in the literature where this type of apparatus has been used in other countries, notably the USA, there does not appear to be much evidence of use in the UK, possibly because of the long duration of the test.

There must also be a means of establishing that the correct total thickness of seal has been obtained and that the correct side slope angles have been constructed. This would normally be achieved by means of surveying before, during and after the works. It is important to regularly survey the earthworks to ensure that construction is proceeding in accordance with the design drawings. Surveyors should liaise closely with the contractor so that surveying is carried out when needed and without delaying the construction of the liner. The alternative of either digging or boring a hole through the liner to verify thickness is not considered to be good practice because of the unreliability of the methods available for subsequently sealing the investigation hole.

Construction quality assurance

As well as a satisfactory design there must be complete confidence that the design has been properly implemented and that the required degree of containment has been achieved. CQA is therefore an important aspect of landfill construction in order to ensure that these requirements are met.

The staff carrying out CQA should be suitably qualified and should be experienced in this type of work. It is essential that they are appropriately briefed prior to the commencement of the works and are aware of both the design requirements and the client's brief. Their duties should be clearly defined in writing including types and frequencies of tests, range of responsibility and extent of authority on site, frequency and method of reporting, liaison and any other duties. There should be complete familiarity with the testing equipment to be used, including calibration checks and any factors that could lead to inaccuracy in the test results.

The personnel must be fully familiar with the acceptance criteria in respect of the materials to be used. If not already available, a chart should be prepared showing these criteria. Copies of this chart can then be used on site to plot the test results as the tests are being carried out. Any contravention of the acceptance criteria can then be noted at the time and any corrective measures implemented immediately without causing undue delay to the works. The results should also be presented in a tabular form and the location of each test recorded on a plan of the site.

Detailed daily records should be kept in the form of test results, site instructions and a diary which should include details of earthworks equipment being used, observations on the nature of fill materials, weather, etc. A photographic record should also be made of the site activities.

The staff carrying out CQA should be continuously monitoring the works to ensure that the specified procedures are followed. An appropriate amount of testing should be carried out on the placed layers and the staff should have authority to instruct further compaction or the removal of material if the test results do not conform to the acceptance criteria. Communication is important to ensure that the contractor is aware at the earliest opportunity that a section of liner is unacceptable.

Although staff carrying out CQA should ensure that no unacceptable 'shortcuts' are taken in the construction of the liner, they should, where possible, also use their experience and their knowledge of the project to facilitate the completion of the works as quickly and as economically as possible. Examples of this can be suggestions regarding improving the method of working or the type of plant, provided that these changes can be accommodated within the contractor's contract or that the client is agreeable to changes to this contract.

Test results

All test results should fall within the acceptance criteria. This is the only way of being able to ensure that all parts of the seal are acceptable and have a permeability of less than or equal to 1×10^{-9} m s^{-1}. Any placed material that is outside the acceptance criteria must be recompacted or removed and replaced at the time before the remainder of the works proceed. If the *in situ* results consistently fail then the works should be halted and samples taken for further laboratory testing including core cutter samples or some other form of 'undisturbed' sampling so that laboratory permeability tests can be carried out. These tests may result in a need to amend the acceptance criteria (providing this can be justified) or to reappraise either the suitability of the clay or the methodology of placing.

Considerable care is needed in the taking of so-called 'undisturbed' samples which are to be used for laboratory determination of permeability. Whichever sampling method is used, some disturbance is inevitable, the effect of which is to increase the permeability of the sample compared to the permeability of the placed liner.

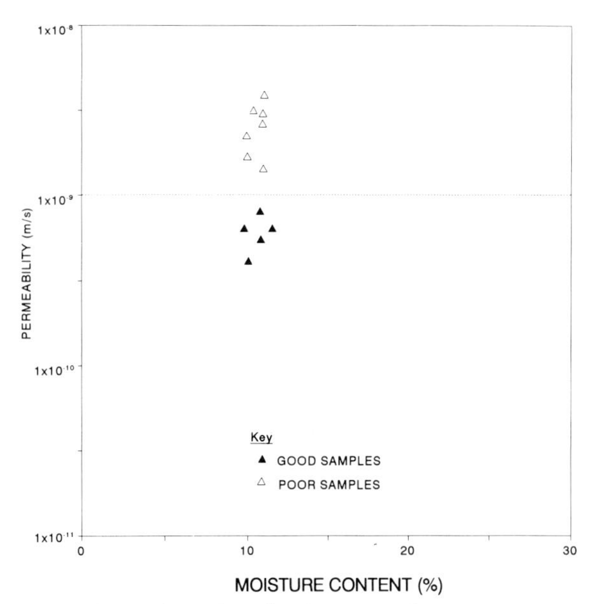

Fig. 4. Comparison of good and poor samples.

Figure 4 shows the laboratory permeability results for good and poor core cutter samples taken from the same liner. It can be seen that the difference amounts to almost an order of magnitude.

Many authors have indicated that measured field permeabilities on the completed liner can be as much as one or two orders of magnitude greater than permeabilities obtained in laboratory tests on recompacted samples prior to construction. It is not clear whether this difference is due to sample disturbance or an actual difference in permeability. It is quite possible that some of this difference can be attributed to sample disturbance and if so this emphasizes the need to take good quality samples.

Material quality

In our experience many sites have a consistent quality of low-permeability material available for use in seal construction. However, some sites, particularly those where the materials have not been naturally reworked (e.g. by rivers or glaciation), possess materials where the effects of weathering are quite limited. In such circumstances the moisture content and workability of the materials varies dramatically with the depth of excavation (e.g. some Mercia Mudstone or Coal Measures materials become drier and stiffer with depth). In such circumstances it may become necessary to spend significant (in terms of cost) periods of time working and reworking the materials before they conform with the requirements of the specification and the acceptance criteria established by the pre-construction laboratory testing. Even following this reworking they may still need to be mixed with other materials before being laid and compacted. All of this additional work has a cost implication and therefore needs to be taken into account at the feasibility stage, because it could have a bearing on the whole viability of the project.

Natural clay deposits may also contain geological features such as infilled erosion channels, granular layers or shear zones. Close examination of material both in the borrow pit and also as it is being placed is therefore necessary so that unsuitable material can be rejected before it is incorporated into the works. Such material would not necessarily be detected by on-site testing for density and moisture content. Unsuitable inclusions in the seal material, such as organic substances or rocks with a volume greater than $0.05\,m^3$ (NRA 1989), must be rigorously 'screened out' either by mechanical means or close visual inspection such that they can be removed prior to the layer in question being compacted.

Ingress of groundwater

Adequate compaction of the mineral liner is impossible if groundwater is allowed to ingress the site. The most likely source of water is from gravels overlying clay or other low-permeability strata. In general, such a situation will be detected by the pre-construction site investigation and dewatering provisions will be necessary. On two sites the authors experienced localized inflows of water that had not been expected. The flows were so great that all works had to stop until control measures could be installed. Impaired liner had to be replaced. In these circumstances it was the role of those carrying out CQA to determine the extent of the impaired liner and to ensure that appropriate remedial work was carried out.

Liner protection

Experience of managing the design and construction supervision of a considerable number of liner projects has indicated that a lot of time, effort and money can be wasted in constructing landfill liners and caps if protective measures are then not carried out closely on the heels of liner/cap construction. Desiccation cracks can form in dry weather and some materials are particularly prone to this phenomenon.

Compliance document

The works should be carried out and the CQA conducted in such a manner that a Compliance Document can be prepared upon completion. This document must be able to demonstrate that the liner has been placed in accordance with the specification, that it conforms to the acceptance criteria and that material placed in the field has at least the same degree of compaction and the same properties as the material that was shown in the laboratory to have a permeability of $1 \times 10^{-9}\,m\,s^{-1}$ or less. In summary, appropriate CQA is fundamental to achieving these requirements.

Conclusions

Pre-construction laboratory testing is required to establish acceptance criteria in respect of material properties and an acceptable range of density and moisture content.

Liner construction should not proceed without a specification for the works having been prepared. This should include the requirement for a compaction trial and state the acceptance criteria.

Correct compaction is imperative to achieving a satisfactory liner including the need in most cases for remoulding by means of a tamping roller. Wetting or drying of the clay may be necessary.

Field testing and sampling should be carried out by experienced staff with appropriate qualifications. The testing regime should be such that any unsatisfactory material can be immediately replaced or recompacted prior to other layers being placed on top.

CQA plays an important role in demonstrating that a satisfactory liner has been constructed which fulfils the design requirements. For an increasing number of sites the waste regulatory authorities require the submission of a compliance document.

References

ANON. 1981. B. S. 5930: *Code of practice for site investigations*. British Standards Institution, HMSO, London.

DANIEL, D. E. 1981. *Proc 4th Symposium on Uranium Tailings Management*. Fort Collins, California, 665–676.

——1987. Earthen liners for land disposal facilities. *In*: WOODS, R. D. (ed.) *Geotechnical Practice for Waste Disposal 87*. ASCE, 21–39.

MURRAY, E. J., RIX, D. W. & HUMPHREY, R. D. 1992. Clay linings to landfill site. *Quarterly Journal of Engineering Geology*, **25**(4), 371–376.

NRA 1989. *Earthworks to landfill sites*. National Rivers Authority, North West Region.

PARKINSON, C. D. 1991. *The permeability of landfill liners to leachate*. The Planning and Engineering of Landfills Conference, Midlands Geotechnical Society, University of Birmingham.

SECTION 3

GEOTECHNICS OF UNDERGROUND REPOSITORIES

Disposal of radioactive wastes in argillaceous formations

S. T. Horseman[1] & G. Volckaert[2]

[1] British Geological Survey, Kingsley Dunham Centre, Nicker Hill, Keyworth, Nottingham NG12 5GG, UK
[2] Studiecentrum voor Kernenergie, SCK/CEN, 2400, Mol, Belgium

Abstract. One option presently under investigation internationally for the disposal of the hazardous by-products of the nuclear fuel cycle is the construction of underground repositories in low-permeability argillaceous rock formations. The host-rock represents the most important barrier to the migration of radionuclides to the surface environment. The performance of this geological barrier will depend on a wide variety of factors, including the degree of compaction and diagenetic alteration of the rock during burial, affecting plasticity and response to deformation, stress history which determines some important structural attributes (faults, fissures, etc.), mineralogy and chemistry which are primary factors influencing radionuclide sorption and mobility, and the thickness and lithological variability of the low-permeability sequence. Our paper provides a brief overview of these considerations and identifies a number of areas of uncertainty, including our, as yet, incomplete knowledge of the precise mechanisms of groundwater movement in clay-rich media, possible departures from Darcy's Law, the occurrence of long-term transient flow, the role of 'coupled flow' phenomena, and the difficult problem of gas migration within the host-formation.

The multiple barrier concept is the cornerstone of all proposed schemes for the underground disposal of radioactive wastes. Based on the principle that uncertainties in performance can be minimized by conservatism in design, the concept invokes a series of barriers, both artificial and natural, between the waste and the surface environment, each successive barrier representing an additional impediment to the movement of radionuclides. Depending on waste category and disposal concept, the barriers are: (a) the chemical barrier (conditioned waste-form), (b) the physical barrier (waste containers), (c) the engineered barrier (buffer/backfill, lining, high-integrity seals, etc.), and (d) the geological barrier (low-flow geological environment). Waste-conditioning provides the first, or innermost, barrier to radionuclide migration. The waste may be incorporated in a stable and relatively inert matrix such as cement, bitumen, lead-alloy, polymer resin or glass. Due to the very low leach-rate of glass in groundwater, vitrification is internationally accepted to be the best method of immobilizing the aqueous products from the reprocessing of spent fuel. Although most waste containers will provide some form of physical barrier to groundwater, because of their relatively small total volumes, spent fuel, vitrified waste and other highly active wastes can be totally encapsulated in corrosion-resistant metal canisters which are designed to prevent groundwater entry for extended time periods. The buffer/backfill medium enclosing the waste will generally provide both a physical and a chemical barrier to radionuclide migration. Typically this would comprise compacted bentonite or reconstituted natural clay, giving a low permeability, an alkaline pH-buffered porewater which will limit the solubility and mobility of certain radionuclides (e.g. actinides), together with good retardation properties (high sorption and a capacity to filter colloids). The geological barrier is the final and most important impediment to radionuclide migration. Depending on details of the local geology, this may be considered to comprise the host formation itself (i.e. the clay stratum), extending above, below and laterally away from the repository, or the entire sequence of low-permeability rocks which may separate the repository from the surface and/or more permeable water-bearing strata. With appropriate site-selection, this barrier will ensure a low-flow environment for disposal, reducing the rate of degradation of the artificial barriers and limiting the release of radionuclides from the near-field. Furthermore, radionuclide migration within the barrier will be extremely slow and the flux will be attenuated along the migration pathway by sorption on mineral phases, radioactive decay and dilution. At some distance into the barrier, the flux will be sufficiently attenuated that the movement of radionuclides beyond this point, perhaps to the surface environment (the biosphere), will not at any time in the future constitute a hazard to man. The ideal, which may be achievable in the case of a clay host-medium, is total containment within the barrier.

From BENTLEY, S. P. (ed.) *Engineering Geology of Waste Disposal*,
Geological Society Engineering Geology Special Publication No. 11, pp 179–191

Argillaceous host-rocks

From the geological perspective, argillaceous rocks (mudrocks) encompass a wide variety of lithotypes ranging from soft clays, through more indurated rock types such as the mudstones, claystones and clayshales, to hard metamorphic rocks. The character and properties of these rocks depend, to a large extent, on their burial history and on the degree of diagenetic alteration suffered during burial. Mudrocks which have suffered minimal diagenetic alteration have the attributes of high water content and low strength and will tend to fail by plastic flow, rather than by brittle fracture. At the other end of the spectrum, highly altered and possibly metamorphosed mudrocks have very low water contents and high strengths and will display the brittle fracturing characteristics typical of all hard rocks deformed at shallow depths in the Earth's crust.

Degree of diagenetic alteration during burial is a key factor which distinguishes one potential argillaceous host-rock from another, and the position of a particular host-rock within this spectrum will determine its main characteristics as a barrier to radionuclide migration. The less-indurated rocks will deform largely in a plastic manner and will tend to self-seal by swelling if fractures do develop. While these characteristics are very beneficial in terms of their overall effect in limiting radionuclide migration, the very same properties of high plasticity, low strength and capacity to swell may create difficulties and constraints in repository engineering. At the opposite end of the spectrum, self-sealing and swelling are not characteristics of highly indurated argillaceous rocks and their overall performance, as host-media for disposal, will be similar to that of hard, fractured, crystalline rocks.

The character and properties of many mudrocks are not exclusively determined by burial diagenesis, since a number of geological processes occurring subsequent to burial can play an important part in determining the structural attributes of these rocks (e.g. faults, folds, joints and fissures). The most important of these are tectonic deformation, uplift and exhumation and, not infrequently, all three processes are closely interrelated as surface erosion strips away sediments thrown up by large-scale deformation, bringing the mudrock stratum closer to the surface. The combined effect of these processes is to impose a 'stress history' on the mudrock and it is this history which is the second major modifier of the character and properties of some potential argillaceous host-rocks. In some 'tectonized' mudrocks, this stress history may be exceedingly complex and almost impossible to unravel. In other mudrocks which have been subject to a simple cycle of burial and exhumation, it is possible to reconstruct the stress history and use it to predict some of the main attributes and properties of the rock. The term 'overconsolidated' is often applied to mudrocks which are presently at a depth that is less than the maximum burial depth, and the water content, strength, plasticity and rock-mass characteristics (joints and fissures) of these mudrocks are more or less quantifiable functions of stress history.

Stress history is not only an important modifier of mudrock properties, it may also have a dominant effect on the hydrogeology of low-permeability argillaceous sequences. In recent years an increasing amount of evidence has been presented to show that groundwater flow in some low-permeability environments *does not* occur under steady-state conditions. This has significant implications in hydrogeological modelling. The main reason for non-steady-state (transient) flow in thick sequences of low-permeability rocks is the phenomenon of hydro-mechanical coupling, whereby changes in pore pressure are occasioned by changes in stress. Expressed in simple terms, the groundwater flow in such environments is a function of the stress history. The enormous difficulty of predicting flow in mudrocks subject to a complex history of stress (e.g. 'tectonized' mudrocks) is immediately apparent.

The prediction of groundwater flow in an argillaceous formation may not be straightforward even in a simple geological setting, since there is also evidence to suggest that the flux of water in clay-rich media may not be exclusively a function of the hydraulic potential (head) gradient. The assertion is that the groundwater flux is driven, in part, by non-hydraulic potential gradients and that the phenomenon of 'coupled flow' is important.

Clay–water–solute interactions

The physical explanation of coupled flow, 'membrane effects' and many other important phenomena observed in clay systems lies in the generally large specific surface of clay minerals, the very small dimensions of the pores, and the complex interactions which occur between the clay mineral particles, water molecules and dissolved chemical species. These are of the utmost importance in determining the overall performance of an argillaceous medium as a barrier to radionuclide migration. Clay–water–solute interactions have a dominant effect on advective and diffusive mass transport in clay-rich media and are key considerations in specific mechanisms of radionuclide retention and retardation such as cation exchange and anion exclusion.

The surfaces of clay minerals within a mudrock have a negative charge which attracts cationic species and water to them to form adsorbed 'double-layers'. In the more compact mudrocks, most of the water present is strongly adsorbed in these double-layers and very little probably participates in normal advection. The interaction of double-layers in compact clays and mudrocks results in very large forces of repulsion between their constituent clay particles, and these are partly responsible for the marked swelling of these materials when they are de-stressed and exposed to fresh water. The forces of repulsion also act in a mudrock under *in situ* conditions, representing one of several forces acting between particles which enable the mudrock to support the weight of overlying formations. The sources of these forces of repulsion are the hydration of clay surfaces and possibly of adsorbed cations, and the osmotic pressure between the adsorbed aqueous solutions and free (macropore) water. An important conclusion is that any change of the local chemical and physical environment of the mudrock (e.g. stress, pore pressure, chemistry and temperature) will alter the relative magnitude of these interparticle forces, causing knock-on effects in other areas. This is at the heart of the very complex coupling between the thermal, chemical, hydraulic and mechanical responses of mudrocks. Osmosis represents only one facet of this complicated behaviour.

It seems probable that non-hydraulic mechanisms of groundwater flow will become more important as the ratio of the mass of adsorbed water to the mass of free water increases. Thus, although the groundwater flow velocities in more indurated and compact mudrocks (e.g. clayshales) are likely to be smaller than in the less-indurated clays, the mechanisms of groundwater movement are likely to be more complex in the more compact rock.

Radionuclide transport

The radionuclides which present a hazard in underground disposal will depend on the nature and origin of the waste. Based on the PAGIS (1988) and PACOMA (1991) safety studies of high- and medium-level waste disposal in clay, three groups of radionuclides have been defined: (i) critical radionuclides, (ii) possibly critical radionuclides and (iii) non-critical radionuclides. Critical radionuclides yielding the highest dose rates are ^{14}C, ^{99}Tc, ^{129}I, ^{135}Cs and ^{237}Np. Radionuclides which yield non-negligible dose rates and are possibly critical in risk assessment are ^{79}Se, ^{107}Pd, ^{93}Zr, and the U-, Pu-, Am- and Cm-isotopes. The main importance of these last three isotopes is that they are the precursors of the more critical radionuclides ^{237}Np, ^{235}U and ^{236}U in the actinide decay chains (De Preter *et al.* 1991).

Pathways to the biosphere

Calculation of radiological risk requires that all possible pathways to the biosphere be identified and that the complex processes governing the movement of radionuclides along these pathways be sufficiently well understood and quantified that reliable predictions of dose and risk can be made for extended time periods in the future. The potential pathways of radionuclide release from the disposal facility to the biosphere are: (a) the groundwater pathway, (b) the gas transport pathway, (c) the uplift/erosion/exhumation pathway, (d) the human intrusion pathway, and (e) the low-probability event pathway. It is generally agreed that transport in moving groundwater represents the single most important pathway by which radionuclides might enter the biosphere. This is therefore the first and foremost consideration in selecting the geological setting for the disposal facility. In examining the gas transport pathway, we are concerned mainly with the possible migration of active gases from the disposal facility to the biosphere. These gases may be produced directly by the waste or may be rendered radioactive by the process of isotopic exchange (e.g. ^{14}C for ^{12}C). However, non-active gases may also be produced and are also of concern since they may carry a charge of active gases and could, potentially, influence radionuclide migration along other pathways.

Transport in groundwater

Radionuclide transport in groundwater may be associated with the movement of a wide variety of chemical species: (a) ions (inorganic and organic charged species), (b) inorganic molecules (neutral species), (c) organic molecules and complexes, and (d) colloids, pseudocolloids and larger particles. Although the last two categories overlap somewhat, the important factors are the very broad range of sizes and the differences in electrical charge. Generally, the first three categories are transported in solution, whereas colloids, pseudocolloids (size 10^{-9}–10^{-6} m), together with larger particles ($> 10^{-6}$ m), are transported in suspension.

Radionuclide retention and retardation. Mechanisms of radionuclide retention and retardation in clay include: (a) filtration-type processes such as molecular and colloidal filtration, anion exclusion and ultrafiltration which are governed largely by the size and charge of the migrating species, (b) processes occurring primarily at mineral surfaces such as physical and chemical adsorption (including ion exchange) and substitution in mineral lattices (mineralization) which are dependent on radionuclide and clay chemistry, (c) complexation with immobile (mainly) organic phases, (d) mechanisms such as diffusion into dead-end pores which depend on rock fabric, and (e) indirect chemical processes such as precipitation, which are governed by changes in porewater chemistry. Biological organisms may also be able to 'fix' certain radioelements in the clay environment.

Factors affecting radionuclide migration. The marked nuclide retardation properties of argillaceous rocks are due, either directly or indirectly, to the clay mineral content and, in general, the higher the proportion of clay minerals in the host-rock, the better will be its performance as a barrier to radionuclide migration. The converse is also true, and silty/sandy layers will form preferential pathways for radionuclide migration.

One of the key characteristics of the clay minerals is their high surface area (specific surface) when contrasted with other rock-forming minerals. Total surface area (external plus interlayer) may be as high as 800 $m^2 g^{-1}$ in some smectites. Clay with high surface areas offer a large number of potential sorption sites, particularly for cationic species such as Cs^+ and Sr^{2+} which sorb by cation exchange. Thus mudrocks with a significant proportion of smectite or illite–smectite mixed-layer clays are particularly advantageous from the containment perspective. The correlation between radionuclide sorption behaviour and specific surface and/or ion exchange capacity has been discussed in Andersson *et al.* (1983). The sorption is generally increasing with increased available cation exchange capacity (CEC) for the non-hydrolysed spherical cations Cs^+ and Sr^{2+}; to some extent also for Am^{3+} at low pH, but to a lesser extent for species like NpO_2^+. A correlation between CEC and specific surface is observed in most silicate minerals (Allard *et al.* 1984) and is very evident in the clay minerals. Some of the main interactions between the properties of clay and the radionuclide transport parameters (effects of discontinuities not considered) are summarized in Fig. 1.

The non-clay mineralogy of the mudrock is also significant since certain minerals, notably the oxides and hydroxides of iron and manganese have high sorption capacities for certain radionuclides, in particular the actinides. Small quantities of pyrite and other minerals containing reduced iron and manganese such as siderite ($FeCO_3$) and rhodochrosite ($MnCO_3$) may influence the sorption of redox sensitive elements such as Np and Tc (Higgo *et al.* 1986). Iron oxides may also co-precipitate radionuclides as they 'age' over geological time-scales (Bruno & Sandino 1987). Carbonate such as calcite ($CaCO_3$) and dolomite ($CaMg(CO_3)_2$) are commonly present in argillaceous sediments and sorption behaviour in the presence of carbonates (or carbonate/radionuclide complexes) may be very different to that of pure clays (Higgo 1987). High molecular weight organic compounds (e.g. humic acids) in the mudrock may be immobile and could act as efficient scavengers for species prone to complexation.

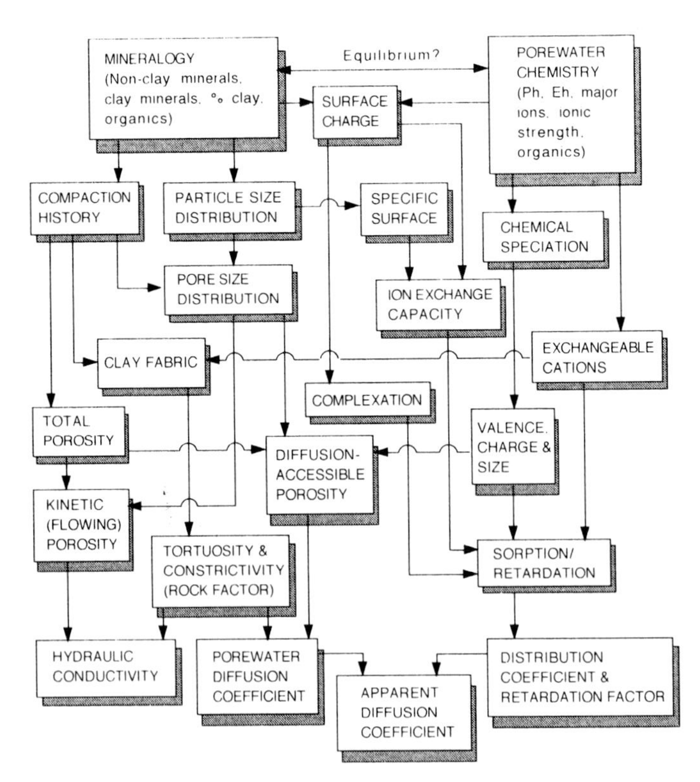

Fig. 1. Diagram showing the main interactions between clay properties and radionuclide transport parameters (effect of discontinuities not considered).

Degree of compaction is another key consideration. In mudrocks with low water contents, much of the water present may be structured water. Research on Bloom Clay suggests that only a fraction of the total water-filled porosity of this clay is available for diffusive transport and has prompted the use of a 'diffusion accessible porosity' term in transport modelling. Gillin *et al.* (1988) observed that anions were excluded from 50% of the total pore space in mudstones. A study by Henrion *et al.* (1991) suggests that the decreased mobility of small molecules and anions in the Boom Clay is due primarily to changes in this diffusion accessible porosity. It is anticipated that the largely physical mechanisms of radionuclide retention and retardation, including molecular and colloidal filtration and anion exclusion, would become more important as pore-throat sizes decrease under increasing levels of compaction.

Some of the organic matter in clays, notably humic acids, may be dissolved in the porewater. It has been suggested that the amount and the molecular sizes of such compounds found in a particular clay is indicative of the size selectivity of the clay filter (Henrion *et al.* 1991). Since these compounds can form complexes with radionuclides such as Tc, Am, Cm, U and Pu, the capacity of the host-rock to act as a filter is an important consideration. The mobility of organic material in Boom Clay is examined by Put *et al.* (1991) and the general importance of organics to migration is discussed in Carlsen (1989).

Radionuclide mobility is strongly dependent on the chemical speciation of the radionuclide in the clay porewater which, in turn, is critically dependent on the chemical composition and pH of the porewater and the redox state of the mineral–water system (Higgo 1987). The indications are that, at depths below the weathering zone, many natural clays exhibit neutral to moderately alkaline porewaters (pH 7–10). Bath *et al.* (1989) suggest that the pH of the London Clay at Bradwell in southeast England (generally pH 8–10) is the result of the overall hydrochemical evolution of the system, involving both the mass transfer and equilibria of exchangeable cations in the clay and solid-solution carbonate equilibrium. Reference groundwater chemistries for Swiss mudrocks reported by Wittwer & McKinley (1989) give pH 8.5 for the Opalinus clay and pH 7.5 for the lower freshwater Molasse. The general alkalinity of porewater in mudrocks is advantageous in terms of its overall effect on radionuclide mobility.

Discontinuities including faults, joints and fissures may represent conduits for the movement of groundwater and solutes. The hydraulic significance of these features will depend on the degree of diagenetic alteration of the rock during burial (which affects plasticity and strength), on genetic factors (shear or tensile origin), on the stress history of the rock (loading/unloading) and on the present-day *in situ* stress conditions. Since groundwater flow velocities will generally be greater for fracture flow and the surface area available for sorption will generally be smaller, the residence times of radionuclides migrating in a discontinuous rock mass may be substantially less than in the case of a more homogeneous and generally unfractured mudrock.

Radionuclide speciation. the importance of the chemical speciation of radionuclides under *in situ* conditions has already been emphasized. The following general observations on the speciation of 'critical' and 'possibly critical' radionuclides in mudrocks are drawn largely from work on the Boom Clay, which is summarized in De Preter *et al.* (1991, 1992). The *in situ* water content of this clay is in the range 18–29% by weight. The 'interstitial' porewater in fresh (unoxidized) Boom Clay has low ionic strength ($\approx 5 \times 10^{-3}$ M), the predominant ions are Na^+ and HCO_3^- and the water is alkaline with pH 9. The presence of pyrite is indicative of the reducing conditions (Eh -300 mV) at typical repository depths in this material (Henrion & Fonteyn 1984; Baeyens *et al.* 1985).

The speciation of ^{14}C in the clay environment is generally not well known, but possible vectors are HCO_3^-, which is not retarded by sorption, together with light organic molecules (CH_4, etc.). Technetium is known to exist in valence states from I to VII. In oxidizing groundwaters it is present as the pertechnetate ion TcO_4^-, which is very poorly sorbed (Higgo 1987). The possible Tc-species in the reducing environment of the Boom Clay are $TcO_2.2H_2O$ or $TcO(OH)_2$ and organic molecules. Tc, as well as Am, Cm, U and Pu, can be strongly complexed with organic molecules, emphasizing the requirement for detailed studies of the organic content of the clay and on the mobility or organic molecules. Iodine would be expected to form negative ions in the groundwater environment (mainly I^-). Thus ^{129}I may be expected to be poorly sorbed by most minerals, since anion sorption is usually low. However, due to the negative charge on the clay mineral surface, it is probable that anions will be largely excluded from the smaller (interparticle) pores of a mudrock (Bourke *et al.* 1989; Henrion *et al.* 1991). Iodine is exclusively present as I^- in the Boom Clay. As a poorly sorbed and long-lived species, ^{129}I constitutes a particularly important radiological hazard. Caesium is likely to be exclusively present as Cs^+ and its behaviour in clays is generally determined by ion-exchange processes and is therefore sensitive to changes in ionic strength and cation exchange capacity (Higgo 1987). The most selective sites for ^{135}Cs sorption are thought to be the charged frayed edges of clay minerals (Sawhney 1972; Cremers *et al.* 1988). Neptunium speciation in Boom Clay has yet to be fully elucidated. Np(IV) is a probable oxidation state in the clay environment and, although it forms only weak organic complexes, transport in this form cannot be completely ruled out. Overall ^{237}Np sorption in Boom Clay is probably controlled by a strong specific sorption on oxide-like surfaces (Henrion *et al.* 1990). No evidence of humic complexation has been found for selenium. The main species of ^{79}Se in Boom Clay are probably metallic Se and HSe^-. The importance of ^{107}Pd has recently been recognized and no speciation studies have not yet been undertaken.

Due to the complex chemistry of uranium, the speciation of ^{235}U and ^{236}U in the clay environment is poorly understood. The ratio of mobile (extractable) uranium in the clay to uranium in the associated porewater is suggestive of high sorption and low mobility in the Boom Clay, which is supported by measurements of uranium isotope ratios (Yoshida *et al.* 1991). Complex formation with low molecular weight organic compounds may be a factor in determining the mobility of the transuranic actinides Pu, Am and Cm in a mudrock.

Advection/dispersion model

The quantitative approach to the prediction of solute transport in porous media is based on the conventional advection/dispersion model. In this model, radionuclide migration is quantified by a number of hydrogeological and chemical parameters: (a) average linear velocity of groundwater, (b) porosity (appropriately defined), (c) coefficient of hydrodynamic dispersion, (d) retardation factor, and (e) radioactive decay constants. The transport equation is obtained by considering the mass balance within a volume element of the porous medium.

Transport equation including sorption. The one-dimensional transport equation for a sorbing radionuclide species i, including radioactive decay and in-growth from species j, is given by

$$R_i \frac{\partial C_i}{\partial t} = D \frac{\partial^2 C_i}{\partial x^2} - v \frac{\partial C_i}{\partial x} - R_i \lambda_i C_i + R_j \lambda_j C_j \qquad (1)$$

where C_i and C_j are the concentrations of the two radionuclide species in solution in the porewater ($kg\,m^{-3}$), x is distance along the flow path (m), D is the coefficient of hydrodynamic dispersion ($m^2\,s^{-1}$), R_i and R_j are the retardation factors, λ_i and λ_j are the radioactive decay constants (s^{-1}), and v is the average linear velocity of the groundwater ($m\,s^{-1}$). The decay constant can be calculated from the radionuclide half-life T (s) by $\lambda = 0.693/T$. The three-dimensional generalization of equation (1) is easily obtained.

Average linear velocity of groundwater. According to Darcy's Law, the average linear velocity of the groundwater v is given by

$$v = \frac{v_d}{n} = \frac{Q}{An} = -\frac{K}{n}\frac{dh}{dx} \qquad (2)$$

where v_d ($m\,s^{-1}$) is the specific discharge, n is the porosity (appropriately defined), Q ($m^3\,s^{-1}$) is the rate of flow of water through an area A (m^2) perpendicular to the direction of the flow, K is the hydraulic conductivity ($m\,s^{-1}$) and h is the hydraulic head (m).

Distribution coefficient and retardation factor. The distribution coefficient K_d ($m^3\,kg^{-1}$) may be defined as the mass of the radionuclide which is bound to the solid phase per unit mass of the mineral solids, divided by the mass remaining in solution per unit volume of groundwater.

$$K_d = (m_{rm}/m_m)/(m_{rw}/V_w) \qquad (3)$$

where m_{rm} is the mass of radionuclide sorbed on mineral solids of mass m_m and m_{rw} is the mass of radionuclides in volume V_w of groundwater solution. The mass of radionuclides sorbed on mineral solids per unit volume of the porous medium can then be expressed in terms of the groundwater concentration

$$(m_{rm}/V_t) = \rho_g(1-n)K_d C \qquad (4)$$

where ρ_g is the grain density of the mineral solids (m_m/V_m) and V_m is the volume of mineral solids. K_d is a valid representation of the partitioning between liquid and solids only if the reactions that cause the partitioning are fast and reversible and the isotherm is linear. The retardation factor R_i is related to the distribution coefficient by

$$R_i = \left[1 + \frac{(1-n)\rho_g K_{di}}{n}\right]. \qquad (5)$$

Dispersion. As a contaminant moves through a porous medium it tends to spread out from the path predicted by considering advective hydraulics alone. This spreading phenomenon is called hydrodynamic dispersion. The process occurs as a result of mechanical mixing and molecular diffusion. Mechanical dispersion and mixing is largely a consequence of the essentially inhomogeneous nature of flow in a porous medium. Contributions to the mechanical dispersion arise from (a) flow-velocity variations within individual pores due to the viscous drag exerted adjacent to the pore walls, (b) differences in the pore sizes along the flow paths, and (c) the tortuosity, branching and interfingering of pore channels. In fractured media, flow along paths of interconnected fractures results in very similar dispersion effects. Dispersion is an exceedingly complex and poorly understood process. Qualitatively, mechanical dispersion and diffusion have similar effects and dispersion is usually modelled using a diffusion-like term (see Equation (1) above), with the dispersive flux taken to be proportional to the concentration gradient. The physical constant which quantifies these effects is the coefficient of hydrodynamic dispersion D ($m^2 s^{-1}$):

$$D = \alpha|v| + D^* \quad (6)$$

where α is a property of the porous medium known as the dispersion length (m) and D^* is the 'porewater' diffusion coefficient ($m^2 s^{-1}$) for the particular radionuclide species and medium (Fried & Combarnous 1971).

It is generally observed that the dispersion length is related to the length-scale of the heterogeneities in a porous medium and increases somewhat for long-distance solute transport since the heterogeneities affecting the process become larger. Put *et al.* (1991) report a dispersion length $\alpha = 1 \times 10^{-3}$ m for Boom Clay, based on laboratory experiments on undisturbed clay cores.

Diffusion. In plastic clays, with low groundwater flow velocities, diffusion is the dominant mechanism of dispersion and the coefficient of hydrodynamic dispersion approximates to the coefficient of molecular diffusion. With the average linear velocity of the groundwater taken as zero and no radioactive decay ($\lambda_i = \lambda_j = 0$), the transport equation (Equation (1)) for a sorbing radionuclide species can be written as

$$\frac{\partial C}{\partial t} = D_{app} \frac{\partial^2 C}{\partial x^2} \quad (7)$$

which is Fick's second law of diffusion. The constant D_{app} is termed the apparent diffusion coefficient ($m^2 s^{-1}$) and is related to the porewater diffusion coefficient D^* by

$$D_{app} = \frac{D^*}{R} = \frac{D_0}{R \cdot R_f} \quad (8)$$

where R is the retardation factor and D_0 is the free water diffusion coefficient. Parameter R_f is the 'rock factor' used in the Belgian studies of radionuclide migration (Put 1984). Since the value of D_0 is generally well known or can be calculated with good precision from fundamental principles, Equation (8) provides a useful basis for the prediction of the apparent diffusion coefficients for a given chemical species and clay medium. By analogy with electrical resistivity studies, Henrion *et al.* (1991) show that a formation factor f^* may be defined for radionuclide migration in the Boom Clay

$$f^* = R_f/\eta \quad (9)$$

where η is the diffusion-accessible porosity. By analysing the results of flow-through diffusion experiments using non-sorbing I^- and HTO (tritiated water) on clay samples compacted to varying degrees, Henrion *et al.* (1991) have determined the following relationship for the product ηRD^*

$$\eta RD^* = 6.84 \times 10^{-10} (\eta)^{3/2} \qquad (m^2 s^{-1}). \quad (10)$$

De Preter *et al.* (1991) report the diffusion-accessible porosity of Boom Clay under *in situ* conditions to be 0.087 for I^- and 0.34 for HTO, which compares with a total porosity of about 0.4. This suggests that the decreased mobility of small neutral molecules and anions, under compaction, is solely the result of changes in diffusion-accessible porosity.

As has already been observed, in well-compacted mudrocks and bentonite backfills the larger part of the water present is probably structured water adsorbed to clay minerals. In these circumstances specifically bound cations in the inner Helmholz plane may have discernable mobility. This surface diffusion mechanism has been invoked to explain the unexpectedly high diffusion rates of ions such as Cs^+ and Sr^{2+} in compacted bentonite (Neretnieks 1982; Soudek *et al.* 1983). Thus, there may be two parallel paths for ion diffusion; the first through loosely held water generally more remote from clay surfaces, and the second through the adsorbed layers in close proximity to the clay surfaces (Charles *et al.* 1986; Muurinen *et al.* 1986). This remains a matter of controversy.

Hydrogeology

The main objective of hydrogeological studies, within the framework of site characterization and safety assessment, is the provision of data on the spatial (and possibly temporal) variations of groundwater flow velocity so as to quantify the advection terms in the radionuclide transport equation.

Although direct measurement of groundwater flow velocities and directions is feasible (tracer studies, etc.), it is not practical to characterize a regional hydrogeological system using direct methods of measurement. Thus, the universally accepted approach is to sample, by *in situ* testing, the spatial distribution of hydraulic potential (head) and hydraulic conductivity, to establish the geometry and boundary conditions of the problem (topography, geology, recharge, etc.), and to obtain the required flow velocities and directions by numerical modelling. In high-permeability environments (aquifers, etc.), it would appear that this approach can give satisfactory results provided that the heterogeneity and anisotropy of groundwater flow are adequately represented. This suggests that sampling and measurement errors do not unduly affect the calculations. Furthermore, it suggests that our 'conceptual model' of fluid flow in high-permeability environments does not diverge significantly from reality. The four implicit assumptions of this conceptual model are:

- Darcy's law is valid and local groundwater flux is determined *exclusively* by the negative gradient of hydraulic potential;
- the potential gradients arise because of elevation differences between the recharge and discharge areas;
- the hydrological system is in steady-state (i.e. no temporal variations in hydraulic potential or flux); and
- single-phase flow is occurring (no gas phase).

These assumptions are reasonable for groundwater flow, at depth, in most high-permeability rocks under natural conditions (not influenced by man), provided that recharge is moderately constant when averaged over appropriate time-periods.

In the first instance, the high-permeability conceptual model, assuming Darcy's law and steady-state, might be adopted for low-permeability media. If the measured hydraulic potentials can be reconciled with the predictions of the model, based on reasonably representative hydraulic properties, then the hydrogeological problem is more or less solved. Even in this case, the uncertainties are likely to be of sufficient magnitude to demand that the hydrogeological predictions be independently validated by geochemical studies of the groundwater chemistry. Unfortunately, it would appear to be a fairly common experience in radwaste disposal investigations that the measured hydraulic potentials cannot be reconciled with this simple conceptual model of groundwater flow. The growing volume of evidence for the occurrence of anomalous hydraulic heads in many low-permeability environments has been examined in an excellent review paper by Neuzil (1986).

Anomalous hydraulic heads

Based on the above discussion, a convenient definition of an anomalous head might be a head which cannot be reconciled with gravity-driven, single-phase, Darcy flow occurring under isothermal and steady-state conditions. Although the available data have yet to be fully compiled, hydraulic tests in a number of argillaceous rocks, under investigation as possible host-rocks for disposal, give hydraulic heads which are significantly lower than might be anticipated (e.g. Horseman *et al.* 1991). The common characteristic of these rocks is that they are fairly highly compacted clayshales with low water contents.

The general conclusion that such rocks invariably exhibit low heads is, however, far from true. The occurrence of anomalously high heads and pore pressures (overpressures) in oil-field shales is well known and there are a substantial number of case histories. By re-examining the assumptions of the conceptual model the possible causes of this phenomenon emerge as: (a) flow not steady-state (long-term transient flow occurring), (b) head measurements influenced by 'borehole effects', (c) flow not single-phase (gas present), (d) Darcy's law not valid (possible thresholds or nonlinearities), (e) potential gradients not exclusively topographically-controlled, (f) flow not exclusively associated with hydraulic potential gradients, and (g) flow not isothermal. One additional possibility is that geochemical processes such as clay mineral transformation (e.g. smectite to illite) could also cause anomalous heads.

Long-term transient flow

In many low-permeability environments it would appear that the commonest cause of anomalous hydraulic potentials is the occurrence of long-term transient flow. For the steady-state condition to be reached and maintained, groundwater flow must be occurring in a geological setting which offers fairly stable hydrogeological boundary conditions. This means, in effect, that geological processes such as burial, compaction, ice-loading, uplift, erosion, tectonic deformation, faulting, isostatic rebound and sea-level change, must occur at such slow and unvarying rates that collectively they cannot significantly affect groundwater movements. Furthermore, flow must have occurred under such stable conditions for some significant length of time up to the present day. This time period will depend on a variety of factors, in particular, the thickness and abundance of low-permeability rocks within the geological sequence (Neuzil 1986).

The fundamental cause of transient flow in most situations is hydro-mechanical coupling. In simple terms, if the stresses acting in a porous rock are altered (say, by some natural process), then the pore volume will also be altered. If advection is slow, as it will be in low-permeability environments, then the change in pore volume will cause a change in the pore pressure (Palciauskas & Domenico 1989). If the total stress increases, the pore pressures will increase, producing an overpressure situation; if this stress decreases, an underpressure situation will develop. The hydrogeological consequences of rapid burial, rapid uplift, surface erosion and neotectonic deformation are likely to be very significant in certain low-permeability environments. As an example, the Opalinus Clay of Switzerland has been subject to a complex history of burial, tectonic deformation and uplift, combined with unloading effects associated with both glacial and fluvial erosion. There seems little doubt that these geological and geomorphological processes have a major influence on present-day groundwater movement in this clayshale (Horseman *et al.* 1991). In a recent paper, Neuzil (1993) describes highly subnormal heads measured in the Pierre Shale of South Dakota which appear to be due to the lagging hydrodynamic response to rebound and cooling caused by erosion.

Non-Darcy behaviour

One basic requirement for the validity of Darcy's law is that the permeant should be non-reactive with the porous medium. Clay minerals generally exhibit exceedingly complex interactions with permeating solutions. Thus, only in the very special case of a permeant which is *chemically identical* (pH, Eh, ionic strength, major ions, etc.) with the clay porewater, does flow in a mudrock meet this basic requirement. Clay–water–solute interactions are at the heart of many of the anomalous hydrogeological responses of clays and mudrocks.

Nonlinearities and thresholds. Departure from Darcy's law in clays may take the form of nonlinearity between the water flux and the gradient of hydraulic potential, or a threshold hydraulic gradient may exist, which must be exceeded if flow is to occur. Laboratory evidence for non-Darcy flow through clays has been widely reported. Deviations from Darcy's law in pure and natural clays up to gradients of 900 were measured by Lutz & Kemper (1959) and experiments by Miller & Low (1963) led to the conclusion that there was a threshold gradient for flow in sodium montmorillonite. Apparent deviations from Darcy's law for flow in an undisturbed soft clay are also reported in Hansbo (1960). Possible explanations of the behaviour are the non-Newtonian behaviour of adsorbed water, electrochemical viscous drag, the occurrence of coupled flow (see below), hydrodynamic particle movement and clogging, and interactions with the permeant such as swelling. A number of suggestions have been made for alternatives to Darcy's law to describe the relationship between flow velocity and gradient (Hansbo 1960; Swartzendruber 1963). More recent investigations have failed to confirm the existence of threshold gradients (Olsen 1969; Chan & Kenney 1973). However, since most laboratory experiments have been performed on samples that are far less compacted than many naturally occurring argillaceous rocks, the possibility of non-Darcy flow in the more indurated rock types cannot be excluded. Intuitively, a number of the proposed mechanisms for non-Darcy flow would become more important as pore sizes decrease due to compaction.

Coupled flow phenomena. There is experimental evidence which suggests that the force driving groundwater movement in clays is not exclusively that associated with hydraulic gradient (Olsen 1969; Fritz & Marine 1983). There are many examples in nature of coupling in the flow of fluids, solutes, heat and electrical current and such phenomena are not exclusive to clays. The term 'Onsagerian coupling' is often applied to these transport phenomena to distinguish the concept from mathematical coupling (Carnahan 1987). Coupled flow phenomena which result in groundwater movement are given the general name 'osmosis'. In chemico-osmosis, the groundwater is driven by a force associated with the gradient of chemical potential, or loosely, by the solute concentration gradient. Similarly, in thermo-osmosis and electro-osmosis, it is a gradient of temperature and electrical potential which drives the water fluxes. For osmosis to be possible, the medium must act as a semipermeable membrane which allows water to pass but restricts the passage of solutes. Anion exclusion is thought to be the main source of these 'membrane effects' in clays (Fritz 1986).

A theory of one-dimensional transient flow in a non-ideal clay membrane under combined hydraulic and chemical gradients has been presented by Mitchell *et al.* (1973). Numerical simulations of coupled transport processes, including chemico- and thermo-osmosis, in saturated clays have been reported by Carnahan (1985), and Fargue *et al.* (1989) provide a detailed theoretical review of these 'non-dominant' transport mechanisms. Most work, to date, on the significance of the Onsager coupled processes to radioactive waste disposal has concentrated on mass transport in the near-field, where large thermal gradients in a buffer/backfill or clay host-medium may produce pore-water movements by thermo-osmosis and the movement of solutes by thermo-diffusion.

The large literature on the occurrence and possible effects of chemico-osmosis in sedimentary basins which has been examined by Alexander (1990) and the possible role of osmosis in groundwater flow in the mixed sedimentary formations beneath the AEA Technology Laboratory at Harwell in Oxfordshire is discussed in Brightman *et al.* (1987).

Although there is a reasonably large volume of evidence to support the existence and to define the general character of osmosis in clays, it is largely based on laboratory experiments on relatively pure, artificially compacted clays. Only electro-osmosis has been indisputably demonstrated to occur at the large scale in clay formations. It is worth noting, however, that no large-scale field measurements have yet been reported which are capable of identifying osmotic effects or of distinguishing them from other effects in clays. There is a clear requirement for such experiments to be undertaken.

Gases and gas migration

The principal mechanisms of gas generation within a repository are: (a) corrosion of metals, (b) radiolysis of water and other materials, (c) microbial and chemical degradation of organics, and (d) thermal release ($T > 100\,°C$) (Voinis *et al.* 1992). The anoxic corrosion of metals, in particular of ferrous metals used in waste packaging or repository engineering, is a major source of hydrogen. The rate of hydrogen production (mole year^{-1}) is directly proportional to the corrosion rate and the total surface area of ferrous metal subject to corrosion. Gamma radiation can penetrate the wall of a waste canister to produce radiolysis of the surrounding groundwater (Christensen & Bjerbakke 1985) producing H_2, O_2, CO_2 and CH_4, together with HCl, and Cl_2 in brines. Radiolysis effects on Boom Clay have been discussed by Put & Henrion (1990). Gases such as H_2 can also be produced inside the canisters largely due to gamma irradiation of the waste matrix, with a smaller contribution from beta irradiation. Where organic materials have been incorporated in the waste or have been used elsewhere in the repository, then the decomposition of these materials by chemical or microbial action will produce gases such as CO_2 under aerobic conditions and, primarily, CH_4 once all available oxygen has been consumed. Helium is produced by alpha decay in actinide wastes. Radioactive gases are also produced and are of two types: fission products such as ^{129}I, ^{3}H, ^{14}C, ^{85}Kr and the decay chains ^{220}Rn and ^{222}Rn. Isotopic exchange, such as ^{14}C for ^{12}C, will render other gases radioactive. Near-field temperatures significantly greate than 100 °C may also cause water to boil, releasing stea as an additional gaseous phase (thermal release) an naturally occurring gases such as C_4 may also present

The generation of gases in the repository is of concer for a number of reasons: (a) pressurization of wast containers, (b) perturbation of groundwater flux, (c effect on repository backfill and seals, (d) effect on host rock transport properties, (e) effect on thermal proper ties and heat dissipation, and (f) release of active gases The development of high gas pressures in a repositor could be problematic. Although venting has bee considered as a method of preventing pressure build up, it has the obvious disadvantages that active gase would be released and the venting arrangements coul provide an easy pathway for groundwater movement.

As has already been noted, under normal circum stances the flux associated with advective transport i clays is likely to be very small and the dispersio mechanism is likely to be dominated by Fickia diffusion. Calculations of the probable maximu diffusive flux of gas through a clay suggest that Fickia

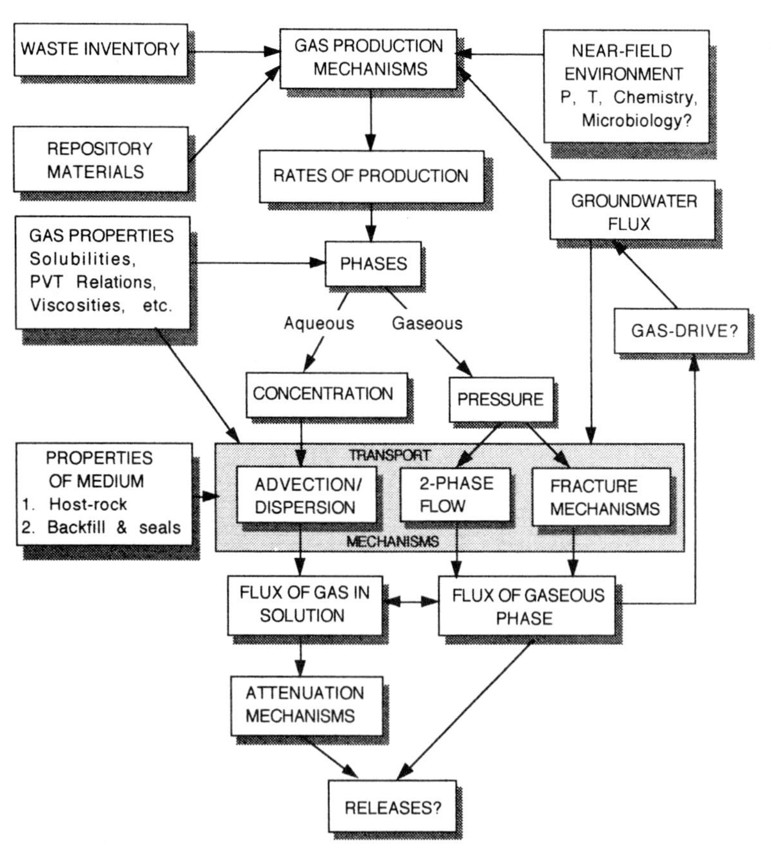

Fig. 2. System diagram showing some of the main interactions associated with gas production and migration (clay repository).

diffusion may be too slow a mechanism to accommodate the quantity (and rate) of gas produced in a repository (Lever & Rees 1987). This is, of course, highly dependent on specific details of the repository design and the waste inventory. Figure 2 summarizes some of the main interactions.

Possible mechanisms of gas transport are: (a) in solution by advection/dispersion (diffusion dominated), (b) as a gaseous phase in two-phase flow, and (c) as a gaseous phase flowing through natural or induced 'preferential pathways'. In fissured or jointed mudrocks, gas it likely to move through these discontinuities (Rodwell 1989).

Laboratory experiments on gas transport in bentonite have been reported by Pusch *et al.* (1985) and on intact clay samples by Lineham (1989). Some preliminary results are available for Boom Clay from the CEC Megas Project (Volckaert *et al.* 1993). Some general conclusions, to date, are (a) a gaseous phase will form in the repository, (b) during two-phase flow, the relative gas-permeability of the clay is a function of saturation and is effectively zero at water saturations exceeding about 90%, (c) if the gas pressure difference across the sample is raised, 'breakthrough' will occur which is apparent as a sudden increase in gas flux, (d) breakthrough pressures correlate well with hydraulic conductivity and are lower in a direction parallel to bedding than normal to it. Breakthrough pressures of 1–1.5 MPa above pore pressure are measured for flow parallel to bedding under typical *in situ* stress conditions.

Discussion

The choice of an argillaceous host-rock for the underground disposal of radioactive wastes has a number of distinct advantages. In the absence of fracturing, the required low-flow environment for disposal is more or less assured and, in the less-indurated mudrocks, the propensity for plastic deformation and for the self-healing of fractures by swelling are valuable attributes. Clay-rich media have a very marked capacity to retard the movement of radionuclides by sorption, cation exchange, filtration and other mechanisms. The chemical conditions prevailing, at depth, within an argillaceous formation are also favourable in so far as the generally alkaline pore water reduces the solubility of many of the radioelements and the speciation of these elements in a reducing environment serves to minimize their mobility.

A complicating factor is the intricate nature of groundwater and solute transport in argillaceous rocks. The usual hydrogeological assumptions that Darcy's law is valid, that the groundwater flux is determined *exclusively* by the negative gradient of hydraulic potential, that potential gradients arise because of elevation differences between the recharge and discharge areas, and that the hydrological system is in steady-state are all challenged when we examine low-permeability argillaceous formations. There is fairly conclusive evidence for long-term transient flow in many mudrock environments. There are also the possibilities of thresholds and nonlinearities in the relationship between flux and hydraulic gradient in clay-rich media. Furthermore, groundwater and solute fluxes may be driven by 'non-conjugate' thermodynamic forces (potential gradients) resulting in so-called 'coupled flow' phenomena.

Hydraulic testing in a number of the more indurated mudrocks (clayshales), under consideration as potential host-media for disposal, has revealed anomalously low hydraulic heads. This tends to confirm that one or more of the usual hydrogeological assumptions may not be valid for these rocks. This has important implications for site characterization and for the design of hydraulic testing programmes.

Finally we come to the issue of gas migration from the repository. It is perhaps ironic that the very same characteristics that make an argillaceous host formation ideal from the point of view of radionuclide transport in groundwater, become problematic when we examine the requirement to dissipate, without excess pressure build-up, the gas formed in the near-field of a repository. A detailed understanding is required of the mechanisms of gas movement in these tight rocks to that we can fully assess the impact, if any, on repository performance.

Acknowledgement. This paper is published by permission of the Director of the British Geological Survey (NERC).

References

ALEXANDER, J. 1990. *A review of osmotic processes in sedimentary basins*. Tech. Rept. of British Geological Survey (NERC), Keyworth, Nottingham, Report No. WE/90/12.

ALLARD, B., KARLSSEN, M., TULLBOURG, E-L. & LARSON, S. A. 1984. *Ion exchange capacities and surface areas of some major fracture-filling minerals of igneous rocks*. SKB Tech. Report TR 83-64, Swedish Nuclear Fuel and Waste Management Co., Stockholm.

ANDERSSON, K., TORSTENFELT, B. & ALLARD, B. 1983. *Sorption of radionuclides in geologic systems* SKB Tech. Report TR 83-63, Swedish Nuclear Fuel and Waste Management Co., Stockholm.

BAEYENS, B., MAES, A., CREMERS, A. & HENRION, P. N. 1985. In situ physico-chemical characterisation of Boom Clay. *Radioactive Waste Management and the Nuclear Fuel Cycle*, **6**(34), 391–408.

BATH, A. H., ROSS, C. A. M., ENTWISLE, D. C., CAVE, M. R., GREEN, K. A., REEDER, S. & FRY, M. B. 1989. *Hydrochemistry of porewaters from London Clay, Lower London Tertiaries and Chalk at the Bradwell site*. Nirex Safety Studies, Report No. NSS/R170.

BEAR, J. 1972. *Dynamics of fluids in porous media.* Elsevier, New York.

BONNE, A. & HEREMANS, R. 1981. Investigations on the Boom Clay, a candidate hostrock for final disposal of high level solid waste. *In* : VAN OLPHEN, H. & VENIALE, F. (eds) *Proc. 7th Int. Clay Conf.*, Bologna and Pavia, Italy, 799–818.

BOURKE, P. J., GILLING, D., JEFFERIES, N. L., LEVER, D. A. & LINEHAM, T. R. 1989. Laboratory experiments of mass transfer in the London Clay. *Proc. Mat. Res. Soc. Symp., Berlin*, **127**, 805–812.

BRIGHTMAN, M., ALEXANDER, J. & GOSTELOW, T. P. 1987. *Groundwater movements through mudrocks: measurements and interpretation.* Tech. Report of British Geological Survey (NERC), Keyworth, Nottingham, Report No. FLPU 87-1.

BRUNO, J. & SANDINO, A. 1987. *Radionuclide co-precipitation.* SKB Tech. Report 87-23, Swedish Nuclear Fuel and Waste Management Co., Stockholm.

CARLSEN, L. 1989. *The role of organics in the migration of radionuclides in the geosphere.* CEC, Nuclear Science and Technology Series, Report No. EUR 12024 EN, Luxembourg.

CARNAHAN, C. L. 1985. Thermodynamically coupled mass transport processes in a saturated clay. *Proc. Mat. Res. Soc. Symp.*, **44**, 491–498.

—— 1987. *Effects of coupled thermal, hydrogeological and chemical processes on nuclide transport.* Proc. of Geoval '87, Stockholm, Sweden.

CHAN, H. R. & KENNEY, T. C. 1973. Laboratory investigation of permeability ratio of New Liskeard varved soil. *Can Geotech. J.*, **10** 453–472.

CHARLES, R. J., COOK, A. J. & ROSS, C. A. M. 1986. *Solute transport in saturated clay.* Tech. Report of British Geological Survey (NERC), Keyworth, Nottingham, Report No. FLPU 86-11.

CHRISTENSEN, H. & BJERBAKKE, E. 1985. *Radiolysis of aqueous solutions.* 9th Int. Symp. Sci. Basis for Nuclear Waste Management, Stockholm, 9–11 September.

CREMERS, A., ELSEN, A., DE PRETER, P. & MAES, A. 1988. Quantitative analysis of radiocesium retention in soils. *Nature*, **335**, 247–249.

DE PRETER, P., P. PUT, M. & DE REGGE, P. 1991. Migration of radionuclides in Boom Clay: the interaction of safety assessment needs with experimental studies, pre-print of paper presented at Migration 91, Jerez de la Frontera, Spain, 21–25 October 1991.

——, PUT, M., DE CANNIERE, P. & MOORS, H. 1992. *Migration of radionuclides in Boom Clay: state-or-the-art report.* Belgian Agency for radioactive waste and Enriched Fissile Materials (Ondraf/Niras), Brussels, Report No. NIROND 92-07.

FARGUE, D., GOBLET, P. & JAMET, Ph. 1989. *Etudes sous l'angle thermodynamique des processus de transfert potentials non-dominant des radionucleides dans la geosphere.* Final Report No. LHM/RD/89/52, prepared for CEC by Ecole Nationales Superieure des Mines de Paris, Centre D'Informatique Geologique.

FRIED, J. J. & COMBARNOUS, M. A. 1971. Dispersion in porous media. *Advances in Hydroscience*, **7**, 169–282.

FRITZ, S. J. 1986. Ideality of clay membranes in osmotic processes: A review. *Clays and Clay Minerals*, **34**(2), 214–223.

—— & MARINE, I. W. 1983. Experimental support for predictive osmotic model of clay membranes. *Geochimica et Cosmochimica Acta*, **47**, 1515–1522.

GILLIN, D., JEFFERIES, N. L. & LINEHAN, D. J. 1988. *An experimental study of solute transport in mudstones.* Nirex Safety Studies, Report No. NSS/R109.

HANSBO, S. 1960. *Consolidation of clay with special reference to the influence of vertical sand drains.* Proc. Swedish Geotech Inst., 18, Stockholm.

HENRION, P. & FONTEYN, A. 1984. Chemical characterization of the Boom Clay. *In*: HEREMANS, R. *et al.* (eds) *R&D Programme on Radioactive Waste Disposal in Clay Formations.* CEC, Nuclear Science and Technology Series, EUR 9077 EN, Luxumbourg.

——, PUT, M., MONSECOUR, M. & DE REGGE, P. 1990 *Synthesis report on transport of radionuclides in Boom Clay, state-of-the-art.* CEN/SCK Report R-2863, Mol, Belgium

——, PUT, M. J. & VAN COMPEL, M. 1991. The influence of compaction on the diffusion of non-sorbed species in Boom Clay. *Radioactive Waste Management and the Nuclear Fuel Cycle*, **16**(1), 1–14.

HIGGO, J. J. W. 1987. Clay as a barrier to radionuclide migration. *Progr. in Nuclear Energy*, **19**, 173–207.

——, REES, L. V. C., COLE, T. & CRONAN, D. S. 1986 *Sorption of redox sensitive radionuclides to deep sea sediments in the absence of oxygen.* Report to UK Dept Environment, prepared by Dept. Chemistry, Imperial College, London.

HORSEMAN, S. T., ALEXANDER, J. & HOLMES, D. C. 1991. *Implications of long-term transient flow, coupled flow and borehole effects on hydrogeological testing in the Opalinus Clay: Preliminary study with scoping calculations.* Nagra Tech. Report NTB 91-16, Wettingen, Switzerland.

LEVER, D. A. & REES, J. W. 1987. Gas generation and migration in waste repositories. Paper presented at workshop on near-field assessment of repositories, Baden, Switzerland, 23–25 November.

LINEHAM, T. R. 1989. *A laboratory study of gas transport through intact clay samples.* Nirex Safety Studies, Report. No. NSS/R155, April.

LUTZ, J. F. & KEMPER, W. D. 1959. Intrinsic permeability of clay as affected by clay-water interaction. *Soil Sci.*, **88**, 83–90.

MILLER, R. H. & LOW, P. F. 1963. Threshold gradient for water flow in clay systems. *Proc. Soil Sci. Soc. Am.*, **27**(6), 605–609.

MITCHELL, J. K., GREENBERG, J. A. & WITHERPOON, P. A. 1973. Chemico-osmotic effects in fine-grained soils. *J. Soil Mech. and Found. Div. ASCE*, **99**, 307–322.

MUURINEN, A., RANTANEN, J. & PENTTILA-HILTUREN, P. 1986. Diffusion mechanisms of strontium, cesium and cobalt in compacted sodium bentonite. *Scientific Basis of Nuclear Waste Management*, **IX**, 617–624.

NERETNIEKS, I. 1982. *Diffusivities of some dissolved constituents in compacted wet bentonite clay—MX100—and the impact on radionuclide migration in the buffer.* KBS Tech. Report No. TR 82-27.

NEUZIL, C. E. 1986. Groundwater flow in low-permeability environments. *Water Resources Research*, **22**, 1163–1195.

—— 1993. Low fluid pressure within the Pierre Shale: a transient response to erosion. *Water resources research*, **29**, 2007–2020.

OLSEN, H. W. 1969. Simultaneous fluxes of liquid and charge through saturated kaolinite. *Proc. Am. Soil Sci. Soc.*, **33**(3), 338–344.

PACOMA 1991. *Performance assessment of the geological disposal of medium-level and alpha waste in a clay formation in Belgium.* MARIVOET, J. & ZEEVAERT, Th. (eds), CEC, Nuclear Science and Technology Series, Report No. EUR 13042 EN.

PAGIS 1988. Performance assessment of geological isolation systems for radioactive waste. *In*: MARIVOET, J. (ed.), *Disposal in clay formations*, CEC, Nuclear Science and Technology Series, Report No. EUR-11776 EN.

PALCIAUSKAS, V. V. & DOMENICO, P. A. 1989. Fluid pressures in deforming porous rocks. *Watwer resources research*, **25**, 203–213.

PUSCH, R., RANHAGEN, L. & NILDEN, K. 1985. *Gas migration through MX-80 bentonite.* Nagra Tech. Report NTB-85-36, Baden, Switzerland.

PUT, M. 1984. A uni-directional analytical model for the migration of radionuclides in a porous geological medium. HEREMANS, R. *et al.* (eds), *R&D Programme on radioactive Waste Disposal in a Clay Formation.* CEC, Nuclear Science and Technology Series, Report No. EUR 9077 EN, Luxembourg.

—— & HENRION, P. 1990. *Modelling of radionuclide migration and heat transport from a HLW-Repository in the Boom Clay.* SCK/CEN Report R-2910, Mol. Belgium.

——, MONSECOUER, M. & FONTEYNE, A. 1991. Mobility of the dissolved organic material in the interstitial Boom Clay water. Paper presented at: Migration '91, Jerez de la Frontera, Spain, 21–25 October.

RODWELL, W. R. 1989. *Modelling studies of gas injection into fractured rock: the interpretation of field results.* Nirex Safety Studies, Report No., NSS/R197

SAWHNEY, B. L. 1972. Selective sorption and fixation of cations by clay minerals: a review. *Clays and Clay Minerals*, **20**, 93–100.

SOUDEK, A., JAHNKE, F. M. & RADKE, C. J. 1983. *Ion-exchange equilibria and diffusion in an engineered backfill.* NRC Nuclear Waste Geochemistry, Report No. NUREG/CP-0052.

SWARTZENDRUBER, D. 1963. Non-Darcy behaviour and flow of water in unsaturated soils. *Proc. Soil Sci. Am.*, 491–495.

VOINIS, S., GAGO, J. & MULLER, W. 1992. Modelling of gas generation (Pegasus Project). *In*: HAIJTINK, B. & MCMENAMIN, T. (eds), *Proc. Prog. Meeting of* Pegasus Project, CEC, preprint volume, Brussels, 11–12 June.

VOLCKAERT, G., PUT, M., ORTIZ, L., DE CANNIERE, P., HORSEMAN, S., HARRINGTON, J., FIORAVANTE, V., IMPEY, M. & WORGAN, K. 1993. MEGAS: modelling and experiments on gas migration in repository host rocks. *In*: HAIJTINK, B. & MCMENAMIN, T. (eds), *Proc. Prog. Meeting of Pegasus Project*, CEC, pre-print volume, Cologne, 3–4 June.

WITTWER, C. & MCKINLEY, I. G. (eds) 1989. *Geochemical database for the sediment report.* Nagra Tech. Report NTB 89-02, Baden, Switzerland.

YOSHIDA, H., MONSECOUR, M. & BASHAM, I. R. 1991. Use of microscopic techniques in migration studies on Boom Clay. *Radiochemica Acta*, **52**, 133–138.

Geotechnical investigations for a deep radioactive waste repository: drilling

T. G. Ball,[1] A. J. Beswick[2] & J. A. Scarrow[1]

[1] KSW Deep Exploration Group, Soil Mechanics Ltd, Glossop House, Hogwood Lane, Finchampstead, Wokingham, Berkshire RG11 4QW, UK
[2] KSW Deep Exploration Group, Kenting Drilling Services Ltd, Trent Lane, Castle Donington, Derby DE74 2NP, UK

Abstract. Comprehensive geotechnical investigations of potential repository sites for deep disposal of low- and intermediate-level solid radioactive wastes at Sellafield, Cumbria and Dounreay, Highland Region included drilling and testing cored boreholes up to 2000 m deep. The drilling technology used combines geotechnical site investigation procedures with oil and gas drilling technology and mineral exploration wireline methods to produce high-quality continuous cores. The drilling programmes are closely linked to geophysical and hydrogeological and geochemical test schedules. At Sellafield, deep boreholes are drilled with standard diesel electric and mechanical oilfield land rigs modified with top-drive units for heavy-duty wireline coring and sensitive mud control, well control and automatic drilling systems. Lorry-mounted site investigation rigs are used for shallow boreholes which have been drilled for pump testing and *in situ* stress measurements. At Dounreay a hydraulic top-drive mining exploration rig was used for deep drilling. Triple tube corebarrels have been used throughout which include uPVC liners for core protection. By October 1994, 31 boreholes had been drilled, with over 19 km cored out of a total depth of 26½ km. A summary of the equipment and procedures adopted is presented together with some comments on the application of these experiences to future geotechnical investigations.

Investigations into the suitability of constructing a repository for disposal of low- and intermediate-level solid radioactive wastes by United Kingdom Nirex Limited (Nirex) included active consideration of two sites at Sellafield, Cumbria and Dounreay, Highland Region. These investigations included the drilling of boreholes to depths of up to 2000 m for geological, geotechnical and hydrogeological purposes. To date (October 1994), 29 boreholes at Sellafield and two at Dounreay have been drilled. A total of 26 585 m has been drilled, of which 19 146 m has been cored (Table 1). Preliminary results have led to Nirex selecting Sellafield as the site where they are concentrating their investigations. The drilling equipment and techniques used to recover continuous cores from these boreholes and facilitate high-quality testing programmes are described below.

Objectives

The conceptual design for a repository at a depth of at least 500 m at Sellafield involves a series of vaults up to 900 m long with a cross-section 25 m wide by 15 m high, and up to 600 m long with a cross-section 18 m wide by 26 m high, serviced by access, central and perimeter tunnels. Some of the geological factors affecting constructability that need to be considered are therefore rock mass quality, hydrogeology and *in situ* stresses.

The drilling investigation contractor is responsible for producing:

- high-quality continuous cores for detailed lithological and geotechnical logging and sampling, with as near as possible 100% real core recovery;
- a straight, smooth-walled borehole to optimize geophysical logging and imagery, hydrogeological, and hydrofracturing testing work;
- a variety of hydrogeological tests which are carried out both during and after drilling operations to obtain high-quality data from generally very low-permeability horizons; three main test programmes are being used comprising environmental pressure measurements, post-completion discrete extraction tests and Westbay multipacker multiport piezometer installations;
- measurement of *in situ* stresses to be carried out using the overcoring technique at depths of up to 250 m below ground level for input to the design of a proposed spiral decline from ground surface. The measuring method adopted had therefore to perform reliably under high hydrostatic heads and measure triaxial strain changes in a weak sedimentary rock mass.

From BENTLEY, S. P. (ed.) *Engineering Geology of Waste Disposal*,
Geological Society Engineering Geology Special Publication No. 11, pp 193–200

Table 1. *Borehole statistics*

Location	*Bore hole*	*Total depth (m)*	*Cored from (m)*	*Cored to (m)*	*Total cored (m)*	*Core recovery (%)*	*Average core run (m)*
Dounreay	D1	1327.40	8.05	1327.40	1319.35	99.05*	4.63
	D2	964.80	8.40	964.80	956.40	94.36*†	4.69
Sellafield	S2	1605.00	26.05	1605.00	1576.95	99.72*	4.76
	S3	1947.32	687.60	1947.32	1259.72	99.83*	5.12
	S4	1255.00	406.00	1255.00	849.00	100.09*	5.24
	S5	1255.00	152.00	1255.00	1103.00	100.43‡	5.46
	S7A	1004.00	455.00	1004.00	549.00	100.26‡	4.50
	S7B	465.00	58.00	465.00	407.00	99.62‡	5.50
	S8A	996.00	204.00	996.00	792.00	100.02‡	N/A
	S8B	244.95	15.15	244.95	229.80	99.32‡	1.44
	S9A	496.00	11.00	246.00	235.00 }	82.79‡	N/A
			246.00	496.00	250.00 }		
	S10A	1602.86	215.00	1602.86	1387.86	99.62‡	5.40
	S10B	252.29	22.18	252.29	230.11	92.67‡	1.42
	S10C	253.13	1.68	22.40	20.72	91.51‡	1.15
	S11A	1164.00	795.00	1164.00	369.00	98.78‡	4.15
	S12A	1144.00	105.50	348.38	242.88 }	99.97‡	5.02
			544.00	1144.00	600.00 }		
	S13A	1735.00	1096.00	1735.00	639.00	99.71‡	5.03
	S13B	295.70	34.85	295.70	260.85	99.55‡	1.45
	S14A	868.00	330.00	868.00	538.00	100.41‡	5.60
	RCF1	1144.00	374.00	1144.00	770.00	100.62‡	4.90
	RCF2	1144.50	420.50	1144.50	724.00	99.78‡	3.85
	RCF3	986.00	216.00	415.00	199.00 }		
			439.00	449.20	10.20 }	100.01‡	4.64
			462.02	986.00	523.98 }		
	RCM1	986.00	47.00	986.00	939.00	98.99‡	4.58
	RCM2	985.00	350.09	985.00	634.91	100.72‡	5.00
	RCM3	1029.00	398.90	569.76	170.86 }	99.66‡	4.99
			571.26	1029.00	457.74 }		
	PRZ2	555.00	235.00	555.00	320.00	in progress	N/A
	PRZ3	770.00	235.00	770.00	535.00	99.86‡	5.14
	SH1	17.19	1.72	17.19	15.47	100.13‡	1.29
	SH2	17.41	2.20	17.41	15.21	97.57‡	1.27
	SH3	15.16	0.52	15.16	14.64	101.84‡	1.46
	BH100	60.00	—	—	0.00	—	—
TOTAL		26584.71		19145.65			

* Measured core recovery (measured by client's consultants).
† Virtually all core losses occurred in a 20–30 m fault zone.
‡ Apparent core recovery, uncorrected during logging.
N/A, Not available.

A unique drilling system was required to meet the drilling and testing objectives (Beswick *et al.* 1992). It combined:

- the attention to detail associated with high-quality shallow geotechnical site investigations;
- drilling technology and engineering tolerances normally associated with oil and gas exploration;
- large-diameter continuous coring techniques which have become normal in some coal and other mineral exploration programmes.

In addition, a comprehensive Quality Management System in accordance with B.S. 5750 Part 2 (BSI, 1990) was commissioned. This project has extended the best geotechnical site investigation practices to a depth of 2 km; this is the first time that this type of investigation has been carried out in the UK, and it is understood that these are the first boreholes to be drilled to these depths under such rigorous quality controls (Ball & Tinkler, in prep.). Experience gained in meeting these requirements is described in a set of two papers: this paper describes the drilling equipment and procedures used; complimentary experience with *in situ* stress measurements are described in Christiansson *et al.* (1996). Whilst similar techniques have been used at both Sellafield and Dounreay, this paper concentrates on progress, procedures and results from the Sellafield boreholes.

Geology

At Sellafield, the generalized succession is principally a Permo-Triassic sequence comprising abrasive fine- to medium-grained sandstones, shales and poorly sorted breccias overlying Carboniferous Limestone and a basement of the Lower Palaeozoic Borrowdale Volcanic Group. The Permo-Trias includes the St Bees Sandstone, a formation noted for its abrasivity. Bit records from oil and gas wells in the adjoining Irish Sea Basin show abnormal gauge wear. The Borrowdale Volcanic Group comprises welded tuffs, with interbeds of volcanic breccia and some intrusive andesite sills.

At Dounreay, a Devonian sequence of interbedded siltstones and sandstones some 400–500 m thick overlies a basement complex of metamorphic Precambrian Moinian rocks. The Moinian rocks are very strong and abrasive with many quartz-rich feldspar veins.

Drilling rigs and associated equipment

Five rigs have been used to date for deep drilling, with a further two lorry mounted rigs for the shallow boreholes. The characteristics of the rigs for deep and shallow drilling are discussed separately below.

Deep drilling

Initially two oilfield land rigs were used at the Sellafield site. These are 3000 m capacity standard diesel electric land rigs with triples masts, modified with electric-driven 800 hp (540 kW) hydraulic power packs, hydraulic top-drive units and heavy-duty wireline coring equipment. The hydraulic top-drive units were originally used for the KTB super-deep pilot hole project in Germany (Chur *et al.* 1989). Both rigs incorporate main and secondary mud-mixing facilities to cater for the different open-hole drilling and coring mud requirements of this project. Other features include kinetic-energy monitoring and sensitive automatic drilling systems. Electromagnetic flow meters on the coring pump inlet and flow-out lines were used to monitor gains and losses during the coring operations. The flowmeter data were integrated with the mud logging and drilling data systems including the rig floor displays and alarms.

A similar third rig, with a doubles mast, was modified in comparable ways and commenced work on the project in June 1993. A purpose-designed diesel hydraulic rig was constructed to increase capacity for the project and commissioned at Sellafield in June 1994. This rig is equipped with a singles mast and state-of-the-art mechanized pipe handling and drilling monitoring systems and can be set up to drill holes at inclinations between vertical and 45° (this rig is fully described in Beswick & Hills, 1995).

At Dounreay an all-hydraulic, top-drive, heavy-duty, trailer-mounted mining rig with singles mast was used. This rig was capable of drilling to at least 2500 m with heavy drill strings. The rig incorporated a fine feed system for coring, a pipe-handling robotic arm to handle single lengths of drill pipe, and mud handling and flowmeter systems comparable to those used at Sellafield.

Heavy duty wireline coring system. The borehole specification required hole diameters in the continuously cored intervals of 159 mm principally to accommodate borehole imaging tools for fracture mapping. This diameter requirement could be accommodated with the largest size of heavy duty wireline equipment available. The string selected is one of a suite of wireline strings which has been widely used over the last 10 to 15 years in Europe. These strings were designed as a nesting set for drilling conditions where difficult and varied geology necessitates a system where the string can be used as casing if drilling cannot continue without casing due to hole conditions (Petersen 1987). These strings are rated on a strength criteria for use up to 4000 m. Noteworthy is that the drilling sizes allow the use of standard oilfield casing and tubing sizes. The string selected for this programme gave the desired 159 mm hole diameter and nominal 100 mm core.

With the objective of optimizing core quality and surface handling, a uPVC corebarrel liner was utilized in the inner tube of the corebarrel reducing the core size to 95 mm. The use of a liner results in an increased bit kerf thickness and a less advantageous area ratio which generally decreases coring penetration rates especially in strong, abrasive rocks. However, this was accepted as core quality was the foremost objective of the coring system. This type of liner is widely used in quality geotechnical and mining investigations to some 200 m depth where high-quality cores are essential for engineering purposes. With the exception of a modest trial at 800 m depth carried out by the Joint Venture, this particular material has not been used previously for holes deeper than 400 m. The liners were 2.65 mm thick and manufactured to fine tolerances and cut to a precise length so that insertion on site was simple. The manufacturing specification was as follows:

Length	6265.00 mm	±2.00 mm
Outside diameter	102.75 mm	±0.10 mm
Inside diameter	96.75 mm	±0.15 mm
Inner barrel ID	104.00 mm	nominal
	103.50 mm	minimum

Liner clearance	0.325 mm	minimum
Core diameter	95.00 mm	nominal
Core clearance	0.80 mm	nominal minimum

The liner material comprises Wellvic VB 167, a uPVC especially manufactured for stability on extrusion and known to be temperature stable to 60 °C. No liner distortion was noted with bottom hole temperatures approaching 60 °C and circulating temperatures up to 40 °C. To prevent orientation errors during core handling, the liners are marked with directional arrows during the manufacturing process.

A corebarrel length of 6 m was selected on the grounds that this was probably equivalent to the projected average run length and it facilitated surface handling without introducing especially heavy handling equipment and the risk of damage during handling.

The wireline coring system allows the sample barrel containing the core in the liner to be recovered from the borehole using an overshot, so significantly reducing the round trip time to recover core. A second inner barrel is returned to the corebarrel at the bottom of the string by free fall, with the restriction afforded by the small annular clearance providing the necessary braking effect to limit the landing impact on the landing ring. Tripping of the drill string was only required where a bit change was necessary or to facilitate testing and/or logging. Identical in-hole wireline coring equipment has been used for all the deep drilling.

The design and construction of wireline drill strings requires careful analysis. In contrast to the conventional drill strings usually used at these depths, wireline tool joints are internally flush, and hence relatively weak, to allow the passage of the corebarrel inner tube through the connection. During the planning stage of the KTB super-deep pilot hole in Germany, various wireline coring strings were reviewed for their suitability for drilling the 4000 m deep initial scientific borehole. The heavy-duty drill strings subsequently used for this project were included in that study which included finite-element analysis of the tool joints. The study concluded that these strings were capable of drilling to 4000 m on strength grounds.

The manufacturing process of the tube for wireline coring strings must also to be considered carefully. With tool joint strength so critical and the thickness of metal limited, good-quality, uniform wall thickness pipe is essential. Current welding technology, advanced processes and quality control during manufacture provide a high-quality tubing for pipe applications. Pipe failure with heavy-duty strings are very rare and no in-hole failures have been experienced to date with the 5½ in. (140 mm) string in use at Sellafield and Dounreay. However, a strict preventative inspection programme is in operation and all pipe run in the compression zone is inspected every 100 rotating hours and the pipe cycled in the string. The pipe is checked against a wear criteria similar to the American Petroleum Institute (API) classification for standard oilfield pipe. This classification does not currently set standards for this type of drill string.

A similar 124 mm heavy-duty wireline system was used where required by the particular borehole design. This produced cores with a nominal diameter of 76 mm.

Borehole design. The borehole design was dictated by the desire to provide isolation of certain formations for well control as well as for hydrogeological and geochemical evaluation together with the requirement for a 159 mm TD diameter and the need to accommodate multiple string completions and pumps for long-term testing. The drilling programme was developed between the client's specialists and the contractor.

In the initial holes, coring was carried out from near surface to final depth. The procedure was to core and carry out associated testing during drilling in the 159 mm cored hole as far as the next casing point. Sector logging and testing was carried out in the 159 mm hole after which the hole was opened to the appropriate diameter and standard oilfield casing in sizes from 508 mm (20 in.) to 178 mm (7 in.) installed and cemented. A typical design scheme for the most complicated geology is shown on Fig. 1; in practice no more than three casing strings have been used in any hole in addition to the short surface conductor pipe.

In some boreholes coring was commenced deeper. The initial drilling was carried out using conventional destructive drilling methods typically at 216 mm (8½ in.) to allow testing during drilling and high-resolution logging to be carried out. Holes were opened as for the coring sections prior to casing.

Wireline coring tool joints are relatively weak; consequently 178 mm (7 in.) technical casing was installed in hole sizes greater than 159 mm. In effect, the small mud-filled annulus between the pipe and casing or hole provides continuous support and reduces flexing and vibration in the string. The technical casing was hung free at the bottom of the larger casing and centralized as required.

Three boreholes (RCF1, RCF2 and RCM3) were steered using downhole motors and bent subs with deflection monitored by a positive pulse mud telemetry Measurement While Drilling (MWD) package. The boreholes were started vertically then deviated along predefined courses to intersect particular areas of geological interest. Final inclinations of up to 35° were achieved. The addition of the purpose-designed rig has enabled boreholes to be drilled at an inclination from the surface. Boreholes PRZ2 and PRZ3 have been drilled at a slant angle of 25.5° to the vertical using oilfield survey instruments to monitor and survey the trajectory. All these inclined boreholes have been completed within the specified directional tolerances.

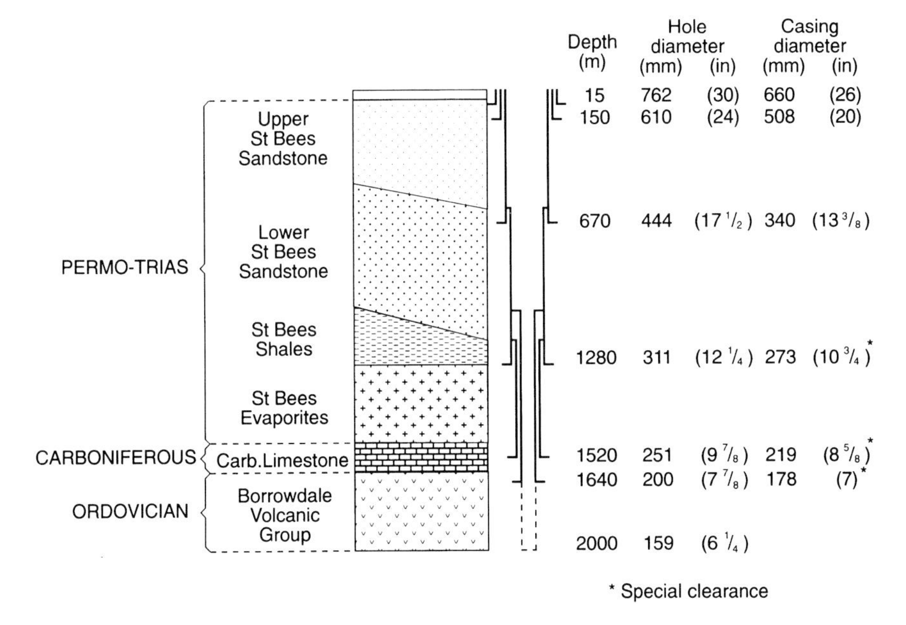

Fig. 1. Typical borehole design for the most complicated geology.

Drilling fluids. The drilling fluid specification was for a lightweight, non-particulate polymer with a tracer to assess the degree of invasion and contamination of the drilling fluid in the formation by analysis of the fluid produced during the various testing programmes. A bentonite gel mud system was used for some of the open holing intervals. The coring fluid used at Sellafield was a water-based dispersible xanthan-viscosifying biopolymer mud containing 1000 ppm (±2%) lithium ion (as lithium chloride) as the tracer. Drilling fluids control was generally conventional, with the obvious exception of the lithium tracer. Five or six flow line samples were taken each day and the tracer concentration checked. The high degree of control exercised over the mud system has greatly facilitated hydrogeological interpretation.

The properties of the drilling fluids at Dounreay and the control methods were very similar except that a lithium ion concentration of 250 ppm was considered appropriate.

Continuous coring. Continuous coring using the heavy-duty wireline coring system was carried out in all deep boreholes over part or all of the hole depth. Rotation was provided by the hydraulic top-drive units at speeds generally in the range of 150–220 rpm with corresponding weights on corebit of 1500–2500 kg. Drilling parameters were optimized throughout to ensure high-quality cores were recovered, and verticality and hole calliper were maintained. The coring was extremely successful, both in terms of overall recovery and core quality. As the figures in Table 1 illustrate, recovery has been consistently over 99% except in borehole D2 where losses occurred in a fault zone. Drilling was halted when potential losses were anticipated from the drill floor instrumentation and the core recovered immediately to minimize core erosion.

All nominally vertical boreholes were drilled well within the specified verticality tolerance of 10° at full depth. Controlled drilling with the wireline string resulted in actual maximum inclinations ranging from 0.9° to 6.4°.

Coring was interrupted each 50 m to allow testing to be carried out to determine the environmental pressure in the 50 m interval. This necessitated tripping the coring string and running a testing assembly on a testing string, and consequently slowed overall progress significantly but did allow the hydrogeological characteristics to be assessed at an early stage, assisting the design of both sector and long-term testing.

The initial design of corebits was based on the anticipated geology and previous experience from general geological descriptions so it was necessary to carry out some experimentation in the initial holes. In the first borehole at Sellafield, 19 bits were used with 12 different specifications, resulting in an average bit life of 83.0 m. Design changes were implemented by a constant feedback of information from site to the manufacturers and average bit life rapidly improved. It is so good now that it is difficult to quote figures on a hole-by-hole basis as many bits are used on more than one hole. The two bit specifications used most regularly now are:

- the Diamatec M7S impregnated bit, generally used for coring the Borrowdale Volcanic Group, with an average bit life of some 163 m;
- the Eastmann Christensen B9 impregnated bit, generally used for coring the Permo-Triassic sandstones, with an average bit life of some 443 m. One of these bits was still usable after coring 1022 m.

Penetration rates at Sellafield during coring ranged from 0.3 to 3.0 m h^{-1}, with an average of about 1.6 m h^{-1}. Penetration rates at Dounreay were slower and averaged about 1.0 m h^{-1}, but with only two holes drilled the opportunities for development of bit designs to optimize performance for the very hard crystalline rocks were limited. A full discussion of the evolution of bit design parameters for these projects is included in Beswick *et al.* (1992).

Well control. Perhaps the most important aspect in the application of wireline coring systems to deep drilling in potential hydrocarbon-bearing formations is the question of well control. The small annulus inherent with the wireline system necessitates attention to procedures for the identification of influxes and losses, tripping strings and corebarrel inner tubes and for the kill procedure itself in the event of a kick (influx of liquid or gas under pressure). In view of the relatively small annular volume compared with traditional pipe-hole systems, any gas influx will result in a gas column over an extended length of the borehole. High equivalent circulatory densities in the mud system arising from this drilling system may mask a kick until circulation is stopped. Furthermore, simulator studies illustrate that gas will take only a short time to reach the surface and killing a kick will take a relatively long time. Kick drills were held regularly to ensure that all crew members were familiar with the procedures and the need for rapid detection and response to kicks. To date, only very low-volume, low-pressure gas has been encountered; however, this is not allowed to lead to complacency.

The coring mud control system incorporated electromagnetic flowmeters to ensure early detection of small influxes and losses. The flowmeter data acquisition was incorporated in the mud logging unit and displayed as flow in, flow out and flow difference on the VDU screens in the mud logging unit and on the drill floor. The flowmeters were calibrated for flows from 0 to 6.3 l s^{-1} and flow differences to 0.6 ml s^{-1} were displayed with alarms set for flow differences of 6 ml s^{-1} (0.1 gal/min).

Swab and surge potential is greatly increased with small annulus systems, which also necessitates careful attention to tripping speeds. Swabbing increases the potential for influxes or breakout in highly stressed boreholes, and surge pressures causing fracturing of boreholes are important considerations. Simulator models suggest that surge pressures can increase from 6 Mpa to 25 Mpa from 500 m to 2000 m depth.

Shallow drilling

A total of eight shallow (up to 296 m deep) boreholes have been drilled to date using geotechnical site investigation rigs, a Boyles BBS 56HD and a Hydreq Gryphon 12T, and 1.5 m corebarrels. Borehole RCM2 was drilled to 250.00 m using a geotechnical rig, for operational reasons, and subsequently completed with an oilfield rig. Two of these shallow boreholes, S10B and S10C, were drilled adjacent to deep borehole S10A for *in situ* stress measurements (Christiansson *et al.* 1996) pump testing and long-term monitoring using a Westbay multipacker installation.

Borehole S10B was cored throughout using a Geobor S wireline coring string with a nominal 1.5 m corebarrel, producing a 102 mm core from a 146 mm borehole. As a prerequisite of the *in situ* stress measurements, potable water was the preferred drilling fluid for this borehole. However, as a result, initial core recoveries and borehole stability were poor and GS 550 polymer drilling mud was used below 40.79 m.

Borehole S10C was cored through the superficial deposits using the Geobor S wireline string. As cores were not required, it was drilled using conventional open-hole techniques to a depth of 253.13 m at a diameter of 108 mm, with potable water used as the flushing medium. Pump tests were then carried out and a Westbay multipacker string installed for long-term hydrogeological monitoring.

Core handling

Cores from both deep and shallow boreholes were handled in effectively identical manners, with procedures designed to ensure that all the core was handled in a controlled manner and eliminate orientation errors and minimize the possibility of induced fractures occurring during handling. The cores are handled several times in the site core facilities and accidental damage has been eliminated by designing the core-handling facilities such that lifting of unboxed core is unnecessary.

When the inner tube of the corebarrel is removed from the coring string, it is placed in a rigid 'scabbard' to protect it from bending and dynamic effects as it is lowered to the catwalk. The retrieval latching assembly, corebarrel head and core catcher and bit assembly are removed when the corebarrel is horizontal and the core within the uPVC liner extruded onto a purpose-designed cutting bench. The liner is cleaned and indelibly marked with reference letters across each cut position and the run number and additional orientation marks added. All marking is checked and the apparent recovery of the core in the liners measured and recorded for hole depth and recovery reconciliation before the core is cut into 1.5-m lengths for boxing. As each length is cut, tight-fitting end caps are fitted and the cut lengths are slid onto a trolley for transportation to the core store.

At the core store, the liner is cut longitudinally within a jig and all identification marks repeated on the lower part of the liner. The lengths are slid into boxes and photographed. The core is later logged and described, and specimens are selected for engineering evaluation. Core orientation is carried out using geophysical borehole imaging logs.

Quality management

A purpose-designed Quality Management System (QMS) has been commissioned which is in accordance with B. S. 5750 Part 2 and is tailored to cover the technical requirements for high-quality geotechnical investigations into sites such as those considered for radioactive waste disposal. Whilst the quality assurance policy was applied to all activities, coring, testing and data collection are obviously the prime considerations and were emphasized accordingly. The QMS builds on KSW member companies' experience in the application of BS 5750 (ISO 9000) through Europe over the last seven years.

The QMS defined KSW's organizational structure, the responsibilities of individuals and the company as a whole, and the procedures, processes and resources necessary for the provision of high-quality contract drilling services. The system was regularly audited formally by both internal auditors and the client; modifications were instigated wherever experience showed that improved procedures were desirable. The specific quality programmes for individual boreholes were detailed in a quality plan and associated inspection and test plans covering each discrete stage of the drilling programme. These documents were used to monitor progress and correlate these activities to the appropriate reference and verification documents.

The rigorous operation of such a comprehensive quality system has demanded a high level of communication between the various parties involved in the project, including sub-contractors and service companies. This has resulted in a marked lack of disagreement about operating, technical or financial responsibilities for particular actions in addition to the benefits of very strict control over quality.

Conclusions

The combination of geotechnical site investigation procedures with oil and gas drilling technology and mineral exploration wireline methods to produce high-quality continuous cores has potential benefits to all investigations where geological control data are not available and the geology at depth needs to be studied in detail. The combination of techniques used at Sellafield and Dounreay appears to provide a very efficient way of obtaining the highest quality cores and testing data from depths up to 2000 m, and there is no reason why the existing equipment could not extend this to at least 3 km or more with only minor modifications.

Rigs can be constructed that have the capability for considerable depth using small, lightweight strings provided that they are designed to give the flexibility required to cope with both surface hole drilling at sizes up to say 445 mm or even 508 mm and small-diameter high rotational speed wireline coring systems with hole sizes down to 76 mm, or even smaller in the lower sections of the hole. Whilst these rigs are obviously bigger than those used more conventionally for geotechnical site investigation, they can be significantly smaller than typical oilfield land rigs with consequential reductions in the cost of site preparation, rig moves and support services.

At depth, wireline coring has many advantages over conventional coring methods which include the following:

- core recovery to surface is rapid because it is not necessary to trip the drill string out of the borehole;
- it is possible to leave the coring string in place as a permanent casing if problems are encountered, and to continue drilling through the string with a smaller nesting size.

Some concerns remain about operational aspects of this work. Small-diameter systems provide less potential for successful fishing if the string parts in the hole, indeed wireline coring pipe can only be fished successfully with spears. Good quality control, preventative inspection, comprehensive well planning and improved drilling procedures have significantly reduced the risk of failure.

Kick detection and procedures will remain a key issue for small-annulus drilling systems. The use of electromagnetic flowmeters is a useful detection method which has worked well at Sellafield and Dounreay. However, well control methods will need to be fully investigated for individual projects where similar drilling techniques are going to be used. The approach of integrating the flowmeters with the mud logging unit points to one of the key areas that needs to be developed if this type of drilling is to become a useful cost-effective method for hydrocarbon exploration. The mud logging data, drilling parameter monitoring, kick detection and control, and all other data acquisition should be concentrated in one unit or service and provided to the driller on the appropriate displays. This approach lends itself to the introduction of automation and eventually artificial intelligence systems. The oil and gas industry is rapidly developing techniques for drilling what are in that industry's terms 'slim holes' and this project has illustrated that the geotechnical, mineral exploration and oil and gas drilling industries have much to teach each other and that there are many benefits in a joint industry approach to exploration drilling.

Acknowledgements. Acknowledgment is given to the Managing Directors of the three constituent companies of the KSW Deep Exploration Group Joint Venture, which comprises Kenting Drilling Services Limited, Soil Mechanics Limited and Bohrgesellschaft Rhein-Ruhr MbH (formerly Gewerkschaft Walter AS) for their support in publishing this paper. The authors are also grateful to project client, UK Nirex Limited, and their agents, British Nuclear Fuels plc and the UK Atomic Energy Authority, for permission to publish this information.

References

BALL, T. G. & TINKLER, J. in prep. A quality management system for deep core drilling.

BESWICK, A. J. & HILLS, D. L. 1995. *Low cost slim hole drilling system.* International Association of Drilling Contractors/ Society of Petroleum Engineers Paper Number 29388.

——, SCARROW, J. A. & MÜLLER-RUHE, W. 1992. *The integration of oilfield, mining and geotechnical exploration technologies for a deep scientific investigation to depths up to 2000 m.* International Association of Drilling Contractors/ Society of Petroleum Engineers Paper Number 23911.

BRITISH STANDARDS INSTITUTION. 1990. *BSI Handbook 22—Quality Assurance, 4th edn.* British Standards Institution.

CHRISTIANSSON, R., SCARROW, J. A., WHITTLESTONE, A. P. & WIKMAN, A. 1996. Geotechnical investigations for a deep radioactive waste repository: *in situ* stress measurements. *This volume.*

CHUR, C., ENGESER, B. & WOHLGEMUTH, L. 1989. KTB pilot hole—results and experiences of one year operation. *Oil Gas European Magazine*, **1/89**.

PETERSEN, G. 1987. Modern exploration by deep slim hole drilling and wireline coring. *Proceedings of the Third International Symposium on Observation of the Continental Crust through Drilling.* Mora, Sweden. Springer, Berlin 286–294.

Geotechnical investigations for a deep radioactive waste repository: *in situ* stress measurements

R. Christiansson,[1] J. A. Scarrow,[2] A. P. Whittlestone[2] & A. Wikman[3]

[1] VBB VIAK, Box 242, 791 62 Falun, Sweden.
[2] KSW Deep Exploration Group, Soil Mechanics Limited, Glossop House, Hogwood Lane, Wokingham, Berkshire RG40 4QW, UK.
[3] Vattenfall Hydropower AB, PO Box 800, S-771 28, Ludvika, Sweden.

Abstract. Comprehensive geotechnical investigations of potential repository sites for deep disposal of low- and intermediate-level solid radioactive wastes at Sellafield, Cumbria and Dounreay, Caithness included drilling and testing cored boreholes up to 2000 m deep. The investigation at Sellafield included the measurement of *in situ* stresses using the overcoring technique at various depths in a 250 m deep, vertical borehole drilled from the surface. The measuring instrument and method used for overcoring had, therefore, to perform reliably under high hydrostatic heads.

The equipment used to carry out these measurements was the Swedish Borre Probe. Previous work with this equipment has been carried out in slim holes and hard crystalline rocks. In contrast, this project required *in situ* stress measurements to be taken in a relatively large-diameter borehole, drilled in weak sedimentary rocks by wireline drilling techniques, and consequently, significant adaptations to the installation equipment.

The paper describes the adaption of the Borre Probe and its successful application in this project up to the end of 1992. Possible future applications of the technique are discussed.

Investigations into the suitability of constructing a repository for disposal of low- and intermediate-level solid radioactive wastes by United Kingdom Nirex Limited (Nirex) included active consideration of two sites at Sellafield, Cumbria and Dounreay, Highland Region. These investigations included the drilling of boreholes to depths of up to 2000 m for geological, geotechnical and hydrogeological purposes. Preliminary results have led to Nirex selecting Sellafield as the site where they are concentrating their investigations.

At Sellafield the generalized geological succession is principally a Permo-Triassic sequence comprising abrasive fine- to medium-grained, moderately strong to strong (Calder and St Bees) sandstones, shales and poorly sorted breccias overlying Carboniferous Limestone and a basement of the Lower Palaeozoic Borrowdale Volcanic Group. The Borrowdale Volcanic Group comprises welded tuffs, with interbeds of volcanic breccia and some intrusive andesite sills.

Objectives

The conceptual design of a repository at a depth of at least 500 m at Sellafield involves a series of vaults up to 900 m long, with a cross-section 25 m wide by 15 m high, and up to 600 m long, with a cross-section 18 m wide by 26 m high, serviced by access, central and perimeter tunnels. The *in situ* stress state is one of the primary pieces of geotechnical information that is required to assess the suitability of a location for such an underground construction. Hydraulic fracturing and stress relief are the two main methods by which successful *in situ* stress measurements have been obtained.

Hydraulic fracturing ('hydrofracture') can be carried out from the surface in deep boreholes and its relative simplicity is advantageous in the early stages of site assessment. However, evaluation of the measurement requires assumptions to be made about the vertical stresses; the method provides information on the stress components in the horizontal plane. This may be a valid assumption at great depth, but in situations where the stress tensor is oblique, such as at shallow depth or in areas of mountainous terrain, the technique is of limited value.

The stress-relief method involves the measurement of the displacements in a piece of rock when it is released from the rock mass. The *in situ* stresses are calculated using the measured strains and the elastic properties of the rock according to classical elastic theory (Leeman & Hayes 1966). The most common method for accomplishing stress relief is overcoring, which consists of coring a borehole at large diameter over a coaxial small-diameter pilot hole in which a strain-measuring instrument is located. A variety of instruments have been developed over the past three decades which permit the determination of the complete three-dimensional

From Bentley, S. P. (ed.) *Engineering Geology of Waste Disposal*,
Geological Society Engineering Geology Special Publication No. 11, pp 201–207

stress tensor from a single measurement. In general, the instruments have been developed for use in underground mining applications and have been designed for use from underground openings.

Several hydrofracture measurements have been made at the Sellafield site. To date, these have been concentrated in the Borrowdale Volcanic Group (Heath 1992) below 400 m below ground level (BGL). A programme of overcoring measurements was carried out at depths of up to 250 m BGL at Sellafield in late 1992. The objectives of these stress measurements were to better define the three-dimensional stress state in the shallow strata, using a more precise technique than hydrofracture, and to demonstrate that an overcoring technique was a viable method of stress measurement in deep boreholes in relatively weak materials.

Selection of measurement equipment

The overcoring stress measurements were specified to be carried out in a vertical borehole drilled from surface (Borehole S10B), at depths of between 50 and 250 m in the Calder and St Bees Sandstone strata. The borehole was also required to provide high-quality nominal 100-mm-diameter core for lithological and geotechnical logging, and to form a well for a pumping test on the completion of drilling. The drilling was carried out using a Boyles BBS 56HD rotary drilling rig using a Geobor S wireline coring system producing a nominal 146-mm-diameter borehole. Coring was carried out with a wireline double-tube corebarrel similar to that used in the other cored boreholes drilled at Sellafield (Ball *et al.* 1996). The instrument to be used for stress measurement, therefore, had to be capable of being installed and operated within a large-diameter wireline drilling system and of functioning under a significant hydraulic head, as the borehole would be water-filled.

A variety of instruments designed for overcoring stress measurement currently are commercially available worldwide. The two instruments used most commonly, both of which have been used in the UK, were operationally unsuitable for these measurements.

The USBM Borehole Deformation Gauge is a re-useable mechanical gauge which measures the stress relief by monitoring the changes in pilot-hole diameter (ISRM 1987). It is not bonded to the pilot-hole walls and can be installed readily underwater. However, its usage is limited practically by the effectiveness of the seals in the gauge body. The manufacturer states that these seals are only reliable up to a hydraulic head of approximately 60 m; which limits the depth at which the gauge could be used in a water-filled vertical downhole. Another disadvantage of the method for this project is that from the measured strains, the *in situ* stress state can be calculated only in a plane perpendicular to the direction of the borehole, i.e. the gauge only makes a two-dimensional stress measurement. To use the instrument correctly to determine the complete stress state, measurements in at least two orthogonal boreholes are required (Leeman 1967).

The CSIRO Hollow Inclusion Stress Cell is a strain cell that comprises three electrical resistance strain-gauge rosettes located inside its epoxy body which is bonded to the wall of the pilot hole (Worotnicki & Walton 1976). A single measurement allows the determination of the complete stress state (Duncun Fama & Pender 1980). Its disadvantage is that the adhesives suitable for use with the cell will not bond under any significant head of water; indeed bonding of the cell to damp pilot-hole walls without submergence of the cell is unreliable (ISRM 1987). The cell is not re-useable.

Further, both instruments require electrical connection to the surface during overcoring so that the strain changes can be monitored. In both cases this is achieved using an armoured cable passed through the drill pipe or rods. The significant cable lengths that would be required for measurements at 250 m added to the impracticability of these instruments. Despite these drawbacks, however, the value of these instruments is unquestioned in the situations appropriate to their design. The detailed investigations from the proposed Rock Characterization Facility (RCF) (Hutchins 1992) could include *in situ* stress measurement and both these instruments would be most appropriate for measurements in this situation.

The instrument found to be appropriate for the measurements during the present investigation was the Borre Probe (Hallbjörn *et al.* 1990). The Borre Probe has been developed in Sweden over the last 25 years to perform stress measurements in deep, water-filled boreholes from the surface. Development was carried out in the late 1960s by Hiltscher and Vattenfall, the Swedish State Power Board (Hiltscher *et al.* 1979). More recent development and commercial operation of the equipment has been carried out by Vattenfall Hydropower AB. Prior to the measurements at Sellafield, the Borre Probe had not been used in the UK.

The Borre Probe measures the complete stress state using electrical resistance strain gauges bonded directly to the walls of a pilot hole. Adhesives have been developed specially, capable of bonding the strain gauges to the rock successfully underwater. The instrument has been pressure tested at 110 bar (equivalent to a 1100 m head of water) over a ten-hour period and used successfully, submerged, on numerous occasions. The most recent development of the probe was the addition of a data logger to the body of the instrument, to record continuously and automatically the strain changes occurring during overcoring. This data logger operates down the hole without connection to the surface. To date, over 590 successful measurements have been made with the probe, of which 130 have used the downhole data logger.

Adaption of equipment

The Borre Probe was developed for use in slim holes (typically 76 mm) drilled in very strong, crystalline rocks using conventional drilling techniques. The nominal 12 mm annular overcore width produced by the typical system was considered inadequate for the weaker sedimentary rocks at Sellafield. Consequently the installation system was modified to operate in a 146 mm diameter borehole, with the wireline coring system used to produce an overcore with a nominal 33 mm annular width. New tools were manufactured specially that allowed the probe to remain unaltered. Adaption included the attachment of centralizing fins to the tool body to centre the equipment in the wireline casing and borehole during lowering and to guide the probe into the pilot hole.

Measurement equipment

The Borre Probe, described fully in Christiansson *et al.* (1989) and Hallbjörn *et al.* (1990), is cylindrical with a maximum diameter of approximately 54 mm and a length of about 550 mm (Fig. 1). It is lowered into the borehole on the wireline in a combined installation tool and weight. The total length of the probe, installation tool and weight used at Sellafield was about 2.6 m. A brief summary of the component parts of the probe follows.

Strain gauges

The instrument carries nine electrical resistance strain gauges mounted in three rosettes and located on three plastic cantilever arms at the lower end of the probe. The strain-gauge rosettes are bonded to the cantilever arms prior to the site work. The arms are located 120° apart at a known orientation to the main body of the instrument. The strain gauges are connected to the data logger up the inside of the probe.

Adhesive

The strain gauges are bonded to the pilot-hole wall using an epoxy or resin adhesive depending on the rock temperature and depth of the measurement location from the borehole collar. The composition of the adhesive is vital to the success of the measurement. Vattenfall have developed adhesives capable of providing acceptable bonding underwater. The time available for installation is limited by the pot life of the adhesive and, therefore, the adhesive used for any measurement is engineered carefully to provide adequate installation time as well as a good bond between the gauge and pilot-hole wall.

Data logger

Besides the nine strain gauges, the Borre Probe also contains a thermistor and one dummy gauge to assess the environmental effects on the readings during the measurement. The downhole data logger, in the main body of the probe (Fig. 1), records 11 channels of data at preset intervals from a preset start time. The logger is capable of storing 8 h of data recorded at 60 s intervals and is powered by a battery, also located in the probe's main body.

Prior to the installation of the probe, the data logger is connected to a portable computer and programmed with the measurement start time and recording interval. No further connection to the ground surface is required after this programming.

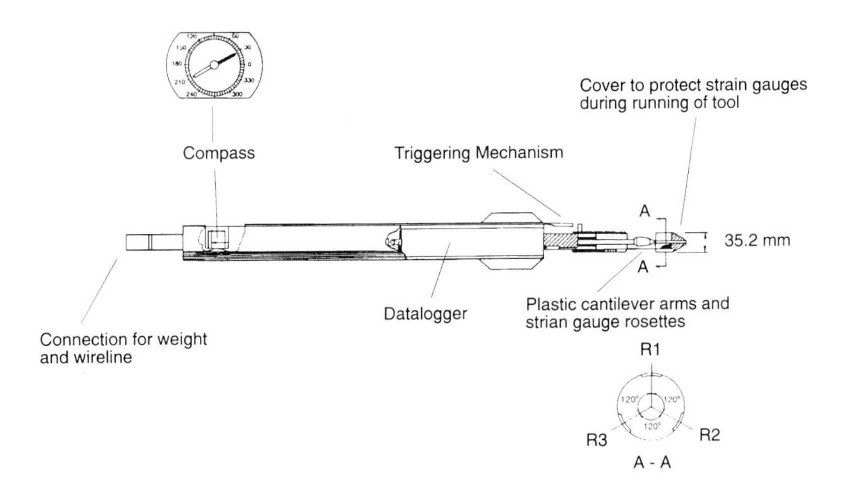

Fig. 1. Borre Probe in installation tool.

After overcoring, the probe is recovered with the overcore sample. The probe is again connected to the portable computer, before removal from the sample and disconnection of the strain gauges, and the data recovered using communications software.

Installation tool

The installation tool carries the probe down the hole and releases it into the pilot hole. As mentioned above, the installation tool was modified for the measurements at Sellafield. The tool contains a mechanical latch that is triggered when the base of the tool lands on the base of the main borehole. Triggering the latch releases the probe from the tool and forces the cantilever arms and strain gauges against the pilot-hole wall. The tool and weight, mounted above it, maintain the required pressure on the strain gauges and adhesive during hardening to ensure a good bond between the gauges and the pilot-hole wall.

The installation tool also carries a magnetic compass, connected to the latch, which is mechanically fixed in its orientation when the latch is triggered. This effectively records the orientation of the probe as it can only be set in and released from the tool in one orientation. To minimize the effect of the wireline drill string on the compass, the string was raised off the borehole bottom prior to the installation of the probe. The tool and probe have to be capable of passing through the core bit and have to maintain centralization in the open borehole to guide the probe into the pilot hole.

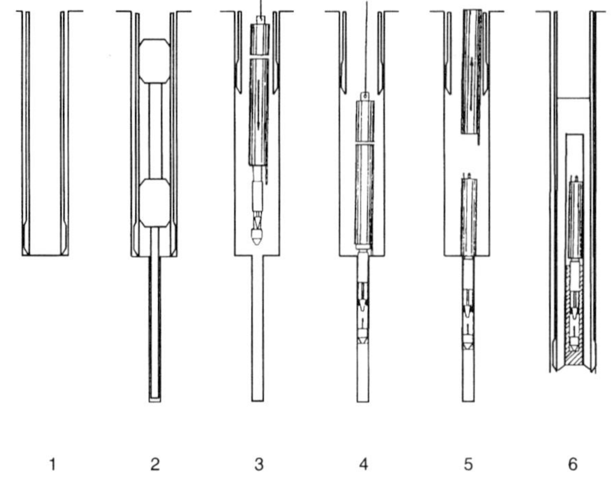
1
2
3
4
5
6
1 Advance main borehole to measurement depth
2 Drill 36 mm pilot hole and recover core for appraisal
3 Lower Probe in installation tool down hole
4 Gauges bonded to pilot hole wall
5 Raise installation tool. Probe bonded in place
6 Overcore Probe and recover to surface in core barrel

Fig. 2. Overcoring stress measurement in wireline drill string.

Measurement details

The measurements made using the Borre Probe and the overcoring technique at Sellafield were carried out in Borehole S10B which was drilled vertically from the surface. Fifteen measurements were attempted in the borehole at levels of approximately 100, 150, 200 and 250 m BGL in groups of three to six measurements. This configuration of measurements provided two groups of stress measurements in the Calder Sandstone and two in the St Bees Sandstone.

Measurement procedures

The procedures carried out before, during and after an overcoring measurement were similar to those carried out when performing overcoring stress measurements using the other 'strain cell' methods (ISRM 1987). The procedure for one measurement (Fig. 2), from commencement of drilling the pilot hole to recovery of the strain data from the logger after overcoring, took approximately 12 h.

Drilling the pilot hole

Stress measurement using the Borre Probe requires a centrally located, coaxial, clean, 36 mm diameter pilot hole. Diametral accuracy is necessary as the Borre Probe has a limited operating range during installation. The pilot hole needs to be centrally located in the borehole base to enable overcoring and biaxial testing of the recovered cylindrical rock sample. At the position of the probe the pilot hole should also be clear of drill cuttings to ensure a good bond between the gauges and the pilot-hole wall. It is preferable also to recover an intact core from the pilot hole. This core should be of sufficient quality to enable the quality of the rock in the large-diameter overcore sample to be anticipated so that the probe is not located in a position over or adjacent to pre-existing discontinuities; such locations will yield poor measurements.

The main borehole was advanced using standard wireline coring techniques as described by Ball *et al.* (1996). At the measurement depth, a coaxial 36 mm diameter pilot hole was drilled using the wireline string with an T36-size core bit and barrel attached, as an extension below the corebit, to an inner full-hole device. A specially manufactured planing tool, again run on the full-hole device, was used to prepare the base of the borehole prior to the drilling of the pilot hole. The core recovered from the pilot hole was examined for fractures or other discontinuities adjacent to the gauge position. If such fractures were evident, the pilot hole was overcored (without measurement) and another pilot hole drilled. If the recovered core indicated suitable conditions the Borre Probe was prepared for installation.

Installation of the Borre Probe

The connection of the strain gauges to the logger and other preparation and testing of the probe and installation tools were generally carried out during drilling to the measurement depth. To prepare the probe, the strain gauges were connected and checked, the data logger was programmed and the compass and probe were attached to the installation tool. Finally the latching mechanism in the tool was armed and the adhesive was applied to the strain-gauge rosettes. The installation tool was attached to the weight and the whole lowered down the borehole on the wireline.

As the installation tool reached the measurement level, the rate of descent was reduced and the release mechanism is activated as the tool touches the base of the borehole. The tool was then left in the hole for approximately 4 h until the adhesive was set completely.

Overcoring

The requirements for the overcoring of the Borre Probe are that the overcore is concentric to the pilot hole, drilled at a constant and steady rate, and that a suitable length of solid core is recovered at the position of the strain gauges. Concentricity of the overcore is both an operational requirement, to ensure the safety of the downhole equipment, and necessary for the calculations of the stress distribution and material properties from biaxial testing of the core sample. A constant drilling rate during overcoring ensures that stress relief on the core occurs in a controlled manner and that the strain-gauge response is not unduly affected by the drilling. Controlling the rate of penetration reduces the possibility of the overcore sample cracking in weaker strata. The recovery of solid core of a suitable length at the gauge position is required for subsequent biaxial testing. Cracking of the core in the proximity of the gauges can also influence the strain-gauges' response and possibly render the measurement incalculable.

After the epoxy setting period and as the downhole data logger was due to commence strain readings, the installation tool was recovered to the surface and the compass removed from the tool. The flush water was circulated for a 20–30 min period to stabilize temperatures before overcore drilling commenced. Overcoring was carried out at the same diameter as coring and when completed, the overcore sample and Probe were left at the base of the borehole for a 20 min period with the flush water circulating; the core and instrument were then recovered to surface.

Measurement success rate

Of the 15 tests attempted, 10 were successful. In the other tests, problems were of two kinds.

- Operational problems relating to the release and installation of the probe resulted in no measurement being made on four occasions. Twice the probe did not release from the installation tool and twice the pilot-hole diameter was oversize because of weakly cemented material at the measurement location. The increased pilot-hole diameter prevented the installation and bonding of the gauges.
- In one instance, two gauges malfunctioned, both oriented in directions non-parallel to the borehole axis. Such an occurrence renders the measurement incalculable; the *in situ* stress state cannot be determined.

This rate of success (10 out of 15, i.e. 66%) compares favourably with previous experience with overcoring in the UK in sedimentary rocks and is better than can usually be expected using other overcoring stress measurement techniques in vertical downholes (Worotnicki & Walton 1976). The success rate with the Borre Probe in hard crystalline rocks is, as is to be expected, usually somewhat better.

Biaxial testing

All suitable overcore rock samples were tested to determine the elastic properties using a biaxial cell. Testing was carried out in general accordance with the guidelines in ISRM (1987). During testing the strains induced in the core sample were monitored by the strain gauges installed by the Borre Probe, connected to a digital strain readout. A load range, which matched an estimated value of the *in situ* stress at the measurement depth, was applied to the core samples and released in a single loading cycle. Pressurization was stopped if the sample started to crack or creep. Some samples were reloaded.

The elastic properties determined from the biaxial testing were highly variable and atypical of sandstone strata. Some tests showed an unusual response to pressurization. These phenomena may have been caused by the presence of pore water in the sample, anisotropy of the sample or debonding of the strain gauges, as there was a delay of a few days from overcoring to biaxial testing. The phenomena may have been better understood if more comprehensive monitoring of the core samples had been carried out during testing; this is beyond the scope of the standard biaxial test carried out as part of *in situ* stress measurement.

Finally the overcore rock sample was described and the bond between the gauges and the rock was examined. The strength of the bond was consistently good, and the pilot hole was not significantly off-centre in any sample. The variation in the thickness of the overcore sample annulus was no more than 1 mm.

Assessment of results

The calculation of *in situ* stress state from the strain data and elastic properties followed the closed form solution described by Leeman (1968). The solution has been modified for the specific configuration of the Borre Probe which differs from the South African CSIR cell described in that paper. The calculations assume that the overcore sample is homogeneous and isotropically elastic. If the rock is neither isotropic nor elastic in its behaviour then it follows that errors could occur in both the calculated magnitude and orientation of the *in situ* stress field (Amadei & Goodman 1982). As up to nine strain readings were recorded by the Borre Probe for each measurement, statistical methods (least-squares analysis) were used to determine iteratively the *in situ* stress tensor. The methods described by Vreede (1981) formed part of the computer program used for the analysis and calculation.

The direction and magnitude of the principal stresses has been determined for each of the ten stress measurements made, assuming the rock mass to be both homogeneous and isotropically elastic. The results can be summarized as follows. The *in situ* stress state of the sandstone strata, calculated from the measurements made, are reasonable when considering the geological setting and structures in the vicinity of Sellafield. The results indicate that the major principal stress direction seems to be controlled by the direction of one of the major fault sets, parallel to the coastline. The horizontal principal stresses are generally in a plane near-parallel to the bedding direction of the strata. The magnitude of the near-vertical principal stress is relatively close to the theoretical lithostatic stress. The principal stress orientations scatter most in tests where steeply dipping cross-bedding was observed in the overcore sample. The results also seem to suggest that the anisotropy and porosity observed in the samples during the biaxial testing might have only a limited influence.

Future *in situ* stress measurements

Determination of the *in situ* stresses at the depth of proposed underground caverns or tunnels is the goal of a stress measurement programme performed as part of any site investigation. Initial studies provide both a general understanding of the state of stress at the proposed depth of construction and obtain the stress field for feasibility studies and preliminary underground design. Further stress measurements are generally carried out from underground facilities as construction commences to provide conformation and allow design adjustment if necessary. This sequence of events has occurred on site investigation programmes for deep underground excavations in other countries (Carlsson & Christiansson 1986).

In situ stress measurement by overcoring is the only method that can provide information on the complete state of stress from one borehole. The applicability of the overcoring technique using the Borre Probe in vertical boreholes drilled from surface has been demonstrated. With the techniques developed for the measurements at Sellafield it is considered that the equipment could be realistically used, with minor modifications, to make measurements at depths of up to approximately 1000 m BGL in conjunction with a wireline drilling system. The modifications of the equipment currently being considered to achieve this deep overcoring stress measurement capability include:

- further alterations to the installation tools to ensure better operation and centralization within the wireline drill string;
- addition of three extra strain gauges to the probe to give a 12 strain-gauge configuration measuring 10 independent strain directions, such a modification should increase the measurement success rate by improved redundancy and increase the confidence of the measurement;
- expansion of the data logger memory to allow a faster strain gauge sampling rate. An increased number of samples during overcoring will allow a more accurate strain-relief curve to be constructed and permit a better interpretation of measurement (Blackwood 1978). An increased memory will be required if the gauge configuration is changed from 9 to 12 gauges.

With these improvements and the procedures developed at Sellafield, the capability now exists for overcoring stress measurement within large-diameter wireline drilling systems. The capability also permits stress measurement to be performed in investigation boreholes drilled from the surface for a variety of purposes including the recovery of core for geological and geotechnical logging. Thus overcoring *in situ* stress measurement could form part of the comprehensive site investigations currently carried out for any underground projects. A set of closely spaced stress measurements could be carried out in the deep investigation boreholes drilled at proposed shaft positions, concentrated at the levels for proposed tunnels and caverns. Measurements could also be made in other boreholes in the vicinity and in the same tectonic unit to provide confirmation of the stress field. Additional measurements in other investigation holes in nearby tectonic units would also be of value for design and safety assessments. If underground construction commences, further stress measurements could then be carried out from the underground openings using other stress measurement instruments such as the CSIRO hollow inclusion cell or USBM borehole deformation gauge as well as the Borre Probe.

Conclusions

The equipment and techniques utilized and developed at Sellafield allow overcoring stress measurements to be carried out in deep vertical boreholes drilled from the surface using wireline drilling technology. It is believed that with some further modification to the equipment this capability could be extended to allow measurements at depths approaching 1000 m BGL. Such measurements would provide important geotechnical information for the feasibility assessment and design of proposed underground excavations at shallow or great depths prior to any underground development.

Acknowledgements. Acknowledgment is given to the Managing Directors of the three constituent companies of the KSW Deep Exploration Group Joint Venture which comprise Kenting Drilling Services Limited, Soil Mechanics Limited and Bohrgesellschaft Rhein-Ruhr MbH (formerly Gewerkschaft Walter AS) and Vattenfall Hydropower AB for their support in publishing this paper. The authors are also grateful to the project client, UK Nirex Limited, and their agents, British Nuclear Fuels plc, for permission to publish this information.

References

AMADEI, B. & GOODMAN, R. E. 1982. The influence of rock anisotropy on stress measurements by overcoring techniques. *Rock Mechanics*, **15**(4), 167–180.

BALL, T. G., BESWICK, A. J. & SCARROW, J. A. 1993. Geotechnical investigations for a deep radioactive waste repository: drilling. *This volume*.

BLACKWOOD, R. L. 1978. Diagnostic stress-relief curves in stress measurement by overcoring. *Int. J. Rock Mech. Min. Sci. and Geomech. Abstr.*, **15**, 205–209.

CARLSSON, A. & CHRISTIANSSON, R. 1986. Rock stresses and geological structures in the Forsmark area. *Proc. Int. Symp. Rock Stress and Rock Stress Measurement*, Stockholm, 457–465.

CHRISTIANSSON, R., INGEVALD, K. & STRINDELL, L. 1989. New equipment for in situ stress measurement by means of overcoring. *Proc. of Int. Cong. on Progress and Innovation In Tunnelling*, Toronto, 49–53.

DUNCUN FAMA, M. E. & PENDER, M. J. 1980. Analysis of the hollow inclusion technique for measuring in situ rock stress. *Int. J. Rock Mech. Min. Sci. and Geotech. Abstr.*, **17**, 137–146.

HALLBJÖRN, L., INGEVALD, K., MARTNA, J. & STRINDELL, L. 1990. New automatic probe for measuring triaxial stresses in deep bore holes. *Tunnelling Underground Space Technology*, **5**(1/2), 141–145.

HEATH, R. J. 1992. Deep discussions. A report on the British Geotechnical Society AGM meeting held on 24 June (1992). *Ground Engineering*, **25**(8) (October), 34–35.

HILTSCHER, R., MARTNA, J. & STRINDELL, L. 1979. The measurement of triaxial rock stresses in deep boreholes. *Proc. of 4th International Congress on Rock Mechanics*, Montreux, 227–234.

HUTCHINS, J. 1992. Nirex plans a rock laboratory at Sellafield. *Atom*, Nov/Dec.

ISRM 1987. Suggested methods for rock stress determination. *Int. J. Rock Mech. Min. Sci. and Geomech. Abstr.*, **24**(1), 53–73.

LEEMAN, E. R. 1967. The borehole deformation type of rock stress measuring instrument. *Int. J. Rock Mech. Min. Sci.*, **4**, 23–44.

—— 1968. The determination of the complete state of stress in rock using a single borehole—laboratory and underground measurements. *Int. J. Rock Mech. and Min. Sci.*, **5**, 31–56.

—— & HAYES, D. J. 1966. A technique for determining the complete state of stress in rock using a single borehole. *Proc. 1st Int. Cong. Rock Mechanics*, Lisbon, vol. 2, ISRM, 17–24.

VREEDE, F. A. 1981. Critical study of the method of calculating virgin rock stresses from measurement results of the CSIR triaxial strain cell. CSIR Report ME 1679, Pretoria.

WOROTNICKI, G. & WALTON, R. J. 1976. Triaxial 'hollow inclusion' gauges for determination of rock stresses in situ. *Symp on investigation of stresses in rock—advances in stress measurement*. Inst. of Eng. Aust. National Conf., 76/4, 1–8.

Geotechnical core and rock mass characterization for the UK radioactive waste repository design

C. G. Rawlings,[1]* N. Barton,[1] F. Løset,[1] G. Vik,[1] R. K. Bhasin,[1]
A. Smallwood[2] & N. Davies[3]

[1] Norwegian Geotechnical Institute, PO Box 3930, Ullevål Hageby, 0806 Oslo, Norway
[2] W. S. Atkins Consultants Limited, Woodcote Grove, Ashley Road, Epsom, Surrey KT18 5BW, UK
[3] UK Nirex Ltd, Curie Avenue, Harwell, Didcot, Oxfordshire OX11 0RH, UK
* Now at W. S. Atkins Consultants Limited.

Abstract. The NGI method of characterizing joints (using JRC, JCS and ϕ_r) and characterizing rock masses (using the Q-system) have been and are currently being used extensively in geotechnical consultancy projects. One such project recently completed for UK Nirex Ltd included the logging of 8 km of 100-mm-diameter drill core from boreholes up to 2 km in depth. Preliminary rock reinforcement designs were derived from the Q-system statistics, which were logged in parallel with JRC, JCS and ϕ_r. The data from the NGI method of characterizing joints and the Q-system for characterizing rock masses have also been used as the basis for UDEC-BB numerical modelling of the proposed cavern excavations for the disposal of solid, low- and intermediate-level radioactive wastes. The purpose of this numerical modelling was to investigate the stability of rock caverns and in particular the rock reinforcement requirements (giving predicted bolt loads and rock deformations), the extent of the disturbed zone (joint shearing and hydraulic aperture) with respect to cavern orientation, the effect of various pillar widths, and the effect of the cavern excavation sequence.

During the planning and construction of underground rock excavations for tunnels and caverns a large amount of engineering geological/geotechnical data are collected. These data arise from surface mapping, core logging or mapping in existing neighbouring underground excavations. These data can be used in the design of the proposed rock support for the underground caverns.

In major underground excavations, such as those proposed at Sellafield by UK Nirex Ltd for the disposal of solid, low- and intermediate-level radioactive wastes, the quantity of engineering geological/geotechnical data will be substantial. It is therefore of the utmost importance that the data are systematically recorded and stored in such a way to enable easy access and to enable easy use.

The NGI methods of characterizing joints (using JRC, JCS and ϕ_r) and characterizing rock masses (using the Q-system) are being used in a current geotechnical consultancy project for UK Nirex Ltd. For the purpose of recording, storing and manipulating the data a new method has been developed at NGI using geotechnical logging charts which combine the Q-system parameter histograms with more detailed joint and rock mass descriptions. In this method the geotechnical data are recorded on a logging chart displaying histograms which are subsequently punched onto a PC-based spreadsheet (Lotus and Impress) allowing the data to be manipulated.

Geology of the Sellafield area

The proposed repository rock at Sellafield comprises the volcanic rocks of the Borrowdale Volcanic Group. These rocks are Ordovician in age. Within the potential repository zone, the top surface of the volcanic rocks is at a depth of 400–600 m, occurring beneath and immediately overlying breccia known as Brockram. This is of Permian age and is overlain by a younger cover of Triassic sandstones. On moving west, towards the coast, the top of the volcanic rocks falls away to greater depth and an additional sequence of shales, evaporites and limestones of Permian and Carboniferous age occurs between the sandstones and the volcanic rocks. The rocks have been subjected to periods of folding and faulting during their geological history. It is this geological sequence of rocks which has been geotechnically logged in detail using NGI's geotechnical logging chart.

Geotechnical logging chart

The geotechnical logging chart can be used for engineering geological mapping underground, at the surface or during core logging. The chart used during

From Bentley, S. P. (ed.) *Engineering Geology of Waste Disposal*,
Geological Society Engineering Geology Special Publication No. 11, pp 209–214

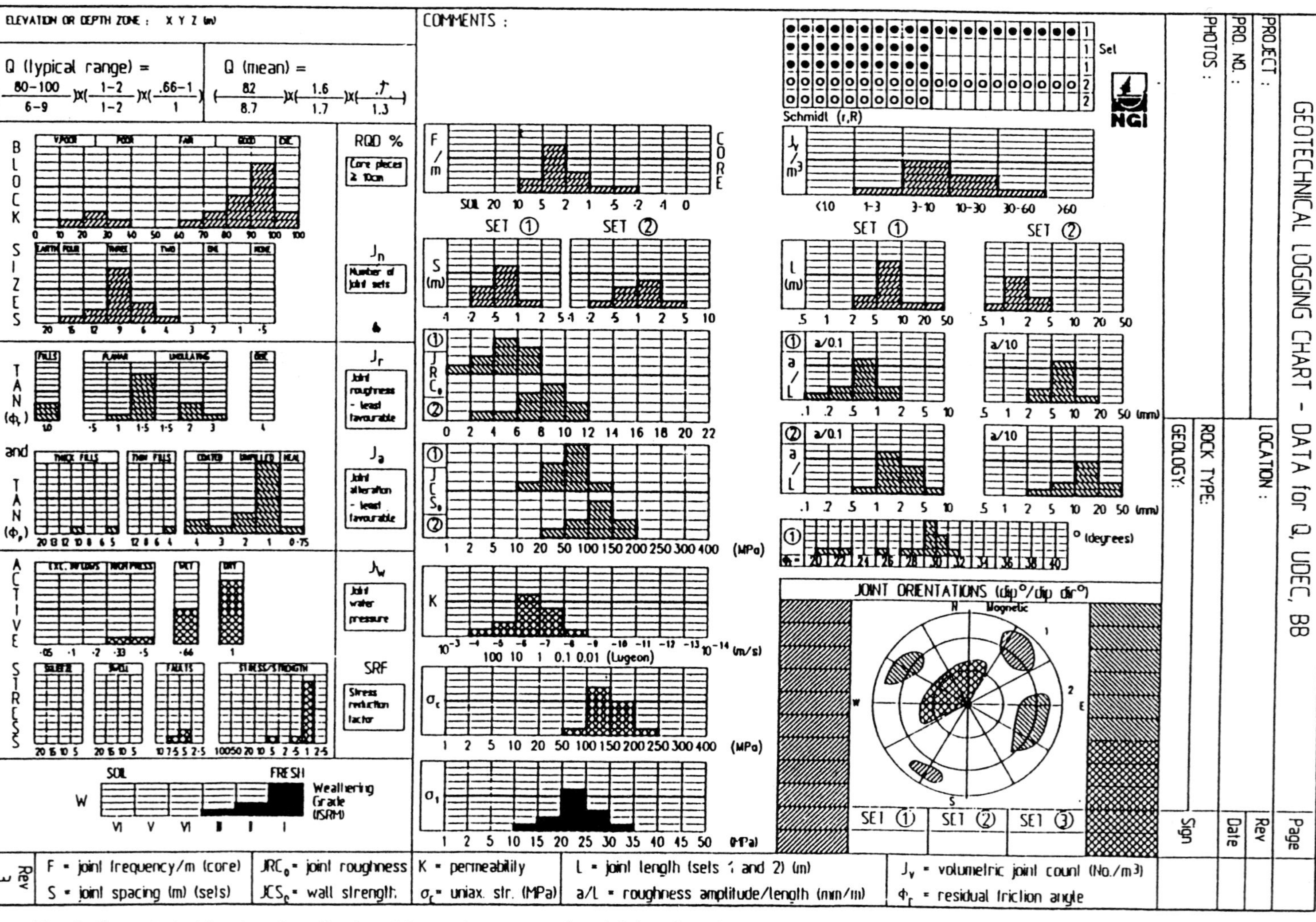

Fig. 1. Geotechnical logging chart for describing rock masses and rock joints for subsequent use in Q-system and UDEC-BB cavern design (field data).

mapping, is shown in Fig. 1. In this form the individual observations of the various parameters are recorded by hand in the rectangular graduated areas. The frequency of observations will depend on the type of mapping, e.g. during field mapping one rectangle may correspond to one exposure whilst during core logging one rectangle will correspond to a fixed length of core. On completion of this chart the histograms will show the range and frequency for each parameter. After core logging and mapping, the data are punched into the PC-based spreadsheet (Fig. 2) and the geotechnical logging chart is printed out. The Lotus spreadsheet automatically calculates mean values of the various parameters and can be used to combine several charts when information on specific depth intervals is required.

The basis of the engineering geological data is the six different parameters in the Q-system which are found on the left-hand side of the logging chart. The Q-value is a measure of the stability of excavations in a rock mass. The Q-value is based on six parameters (Barton *et al.* 1974).

$$Q = \mathrm{RQD}/J_n \times J_r/J_a \times J_w/\mathrm{SRF}.$$

Rock Quality designation (RQD) (Deere *et al.* 1967). The RQD is a modified core recovery percentage in which unrecovered core, fragments and small pieces of rock, and altered rock are not counted so as to downgrade the quality designation of rock containing these features. The RQD is the total length of core pieces over 100 mm expressed as a percentage of the core run length. The RQD is based on the assumption that if the joint spacing is sufficiently wide, then the deformation modulus of the mass will tend to that of the intact specimen. So as the RQD approaches 100% the field/laboratory modulus ratio approaches unity.

Joint set number, J_n. The joint set number is a measure of the joint sets. These numbers vary from 0.5 for unjointed rocks to 20 for completely crushed or earth-like rock masses. RQD and J_n together give a description of the joint pattern and block-size in the rock.

Joint roughness number, J_r. The joint roughness number is a measure of the roughness of the joint surface. These numbers vary from 4 for discontinuous joints to 0.5 for planar slicksided joints. For joints with soft infilling, $J_r = 1$ is used.

Joint alteration number, J_a. J_a describes the thickness and mineralogy (frictional properties) of the joint fillings or joint surfaces. These numbers vary from 0.75 for healed joints to 20 for discontinuities with thick fillings of swelling clay. J_r together with J_a gives an approximate indication of the friction angle along the joints or filled discontinuities. Other parameters which are supplementary to this are JRC, JCS and ϕ_r.

Joint water reduction factor, J_w. The joint water reduction factor varies from 1.0 for dry rock masses to 0.05 for rock masses with very high inflow, as recorded or anticipated at the proposed excavation depth.

Stress reduction factor (SRF). For the case competent rock, the SRF value is based on the ratio between rock strength (σ_c) and principal stress (σ_1) at the tunnelling depth. Special cases are swelling and squeezing rocks and fault zones. The factor may vary from 1 for hard rocks and moderate stress, to 20 or more for rocks under extremely high stress, or for cases of extreme squeezing or swelling. J_w together with SRF gives a description of the effective stress situation in the rock mass at the tunnelling depth.

In addition in the bottom left-hand corner of the logging chart the weathering (w) is recorded as six classes (I to VI) according to ISRM (1981). The typical minimum, maximum and mean Q-values are automatically calculated from the spreadsheet data and are shown in the top left-hand corner of the geotechnical logging chart.

In the middle section of the geotechnical logging chart the following parameters are recorded:

- *Joint frequency, F.* This parameter gives the number of joints (per metre) in drill cores or in line sampling of the rock mass mapping.
- *Joint spacing, S (in metres).* This value is the joint spacing (in metres) for the most prominent joint set (Set A) and the next most prominent joint (Set B) on the geotechnical logging chart.
- *Joint roughness coefficient (JRC).* These numbers describe the joint roughness and may vary from 0 for smooth, planar joints, up to 20 for very rough joints. These numbers may be measured in the laboratory by means of tilt tests or by profile gauges or estimated visually. JRC may also be calculated in the field by measuring the amplitude of the irregularities in relation to their length (a/L).
- *Joint wall compressive strength, JCS.* The strength (in MPa) is measured by a Schmidt-hammer on the saturated joint surface. In the laboratory and field, JCS is based on the best ten results of 20 Schmidt-hammer readings on each joint set at the natural moisture content conditions prevailing at the time.
- *Rock mass permeability, K* ($m\,s^{-1}$). Hydraulic measurements in the boreholes provide this information. Otherwise it must be estimated.
- *Uniaxial compressive strength,* σ_c. Measurements of compressive strength of the rocks are carried out on cylindrical samples prepared from drill cores.
- *Major principal stress,* σ_1. The magnitude of the major principal stress for the particular depth of interest is determined, so the ratio σ_c/σ_1 can be evaluated for purposes of choosing SRF in the Q-system.

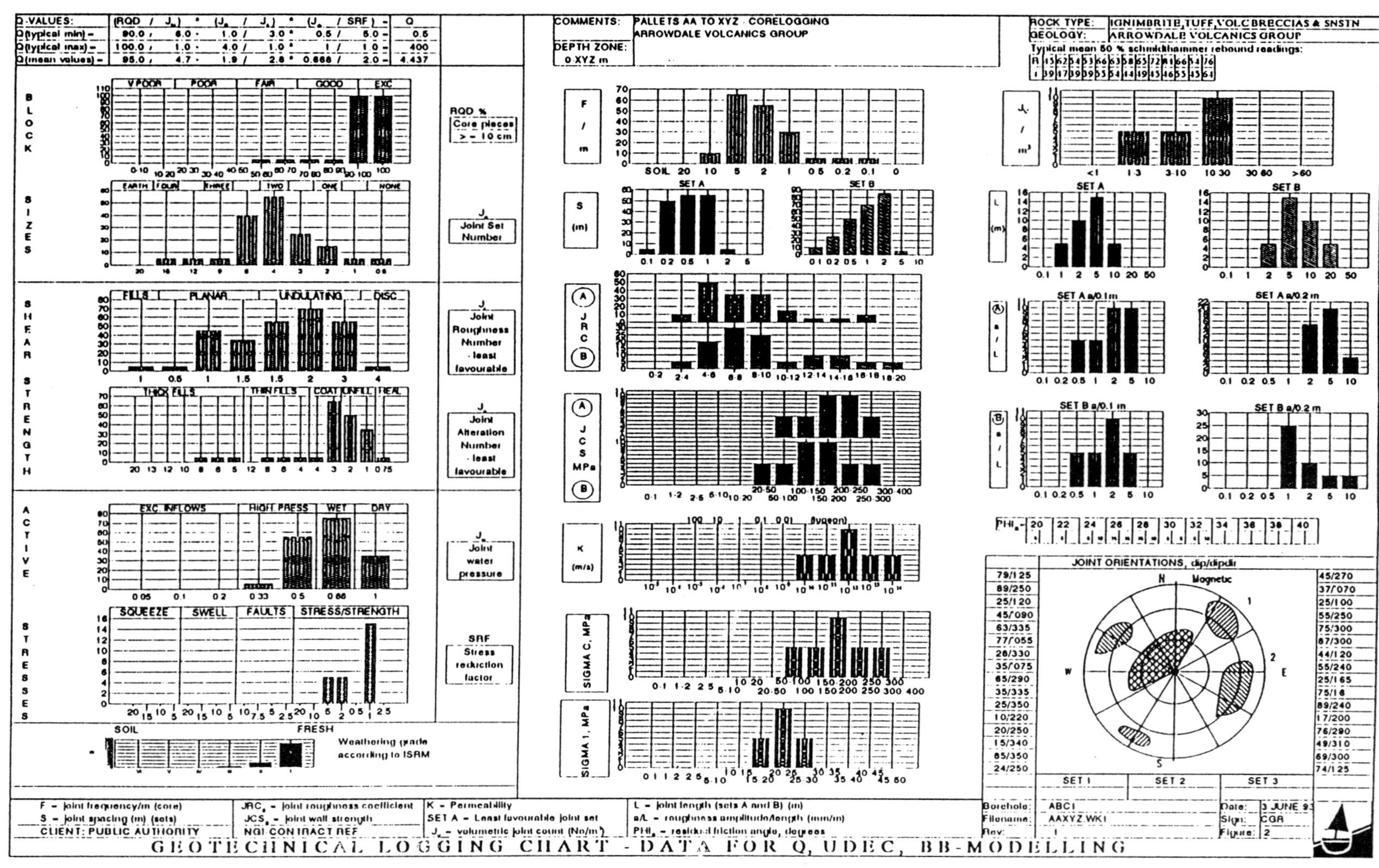

Fig. 2. Geotechnical logging chart for describing rock masses and rock joints for subsequent use in Q-system and UDEC-BB cavern design (core logging and/or field data).

On the right-hand side of the logging chart the following parameters are recorded:

- *Schmidt-hammer rebound number on fresh rock, R.*
- *Schmidt-hammer rebound number on the joint surface, r.*
- *Volumetric joint count,* J_v. This parameter gives the number of joints per cubic metre.
- *Joint length, L.* This value is the joint length (in metres) of individual members of the two prominent joint sets.
- *Roughness amplitude of asperities per unit length,* a/l *(mm m*$^{-1}$*).* The amplitude of asperities in millimetres are measured for two reference lengths along the joint plane, i.e. 0.1 m (laboratory) and 1.0 m (field). Measurement along the two most prominent joint sets (Sets A and B) are usually carried out in the dip direction (the presumed direction of sliding failure).
- *Residual friction angle,* ϕ_r. This value may be found using tests on smooth planar surfaces of the rock (ϕ_r) and Schmidt-hammer rebound tests on natural joints (r) and unweathered surfaces (R). A simple formula relates these parameters (Barton & Choubey 1977).

In the lower right-hand corner of the chart, space is allowed for joint orientation data to be added. The different joint sets are plotted on a stereo diagram. The lower hemisphere of the Schmidt net is used. The joint orientation is given as dip (0 to 90°) and dip direction (0 to 360°) related to magnetic north.

Joint characterization at Sellafield

To date some 8 km of 100-mm-diameter drill core has been logged and the results reported. Index tests to determine JRC (tilt tests, pull tests, and profiling), JCS (Schmidt-hammer tests), ϕ_r (tilt tests, pull tests and Schmidt-hammer tests) are carried out on joints recovered in the drill core. The index testing is performed in the NGI/WSA laboratory set up in BGS facilities at Keyworth. The NGI methods of tilt testing and Schmidt-hammer testing are described in detail by Barton & Choubey (1977) and by Barton & Bandis (1990). The extensive core logging performed during this project for UK Nirex Ltd is providing significant data with respect to the statistical variation of joint parameters and the potential variation of data from joint set to joint set. The index testing is supplemented by a lesser but significant number of laboratory tests such as direct shear tests (DST) and coupled shear flow tests (CSFT).

The Q-system data in the left-hand side of the chart are used to prepare preliminary designs for rock reinforcement for tunnels or waste storage caverns at appropriate depth. The recommended rock reinforcement (i.e. bolts of specific length, diameter and spacing) is subsequently modelled discretely (Cundall 1980) in UDEC-BB modelling.

The computerized data handling using the geotechnical core logging charts has enabled both the lateral and depth variation of the various parameters for the various jointed rock masses at the site to be carried out. An example of the depth related data is given in Fig. 3. The central black line represents the mean value of the parameter at a particular depth. to the left, the dark shaded area is plotted as far as the typical worst quality,

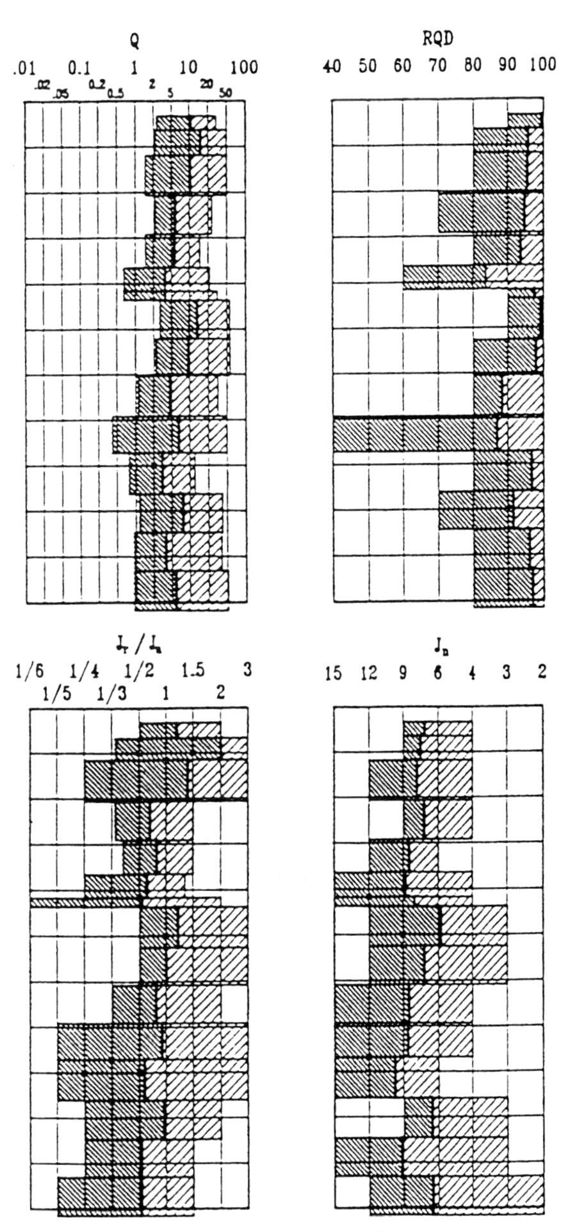

Fig. 3. Examples of depth logs for key Q-system parameters from a section of a Sellafield borehole.

while the lighter, shaded area is plotted to the right as far as the typical best quality. The parameter in question usually varies normally between these typical (but not extreme) limits.

Numerical modelling

The data from the NGI method of characterizing joints and the Q-system for characterizing rock masses have been used as the basis for UDEC-BB numerical modelling of the proposed cavern excavations for the disposal of solid, low- and intermediate-level radioactive wastes. The purpose of this numerical modelling has been to investigate the stability of rock caverns and in particular the rock reinforcement requirements (giving predicted bolt loads and rock deformations), the extent of the disturbed zone (joint shearing and associated changes in hydraulic aperture) with respect to cavern excavation, the effect of various pillar widths, and the effect of cavern excavation sequence.

Conclusions

- A systematic way, using the geotechnical logging chart developed by NGI, of recording engineering geological/geotechnical data during core logging and mapping has been detailed. The method enables both lateral and depth variation of the properties of jointed rock masses to be readily obtained.
- The NGI methods of characterizing joints (using JRC, JCS and ϕ_r) and characterizing rock masses (using the Q-system) have been incorporated into the geotechnical logging chart.
- Based on the geotechnical logging chart, data input files can be set up for UDEC-BB numerical modelling of the jointed rock mass behaviour.

References

BARTON, N. & BANDIS S. C. 1990. Review of predictive capabilities of JRC–JCS model in engineering practice. *Int. Symposium on Rock Joints, Loen. Proceedings*, 603–610.

—— & CHOUBEY, V. 1977. The shear strength of rock joints in theory and practice. *Rock Mechanics*, Springer, Vienna, No. 1/2, 1–54. Also NGI-Publ. 119, 1978.

——, LIEN, R. & LUNDE, J. 1974. Engineering classification of rock masses for the design of tunnel support. *Rock Mechanics*, **6**(4), 189–236.

CUNDALL, P. 1980. *A generalized distinct element program for modelling jointed rock*. Report PCAR-1-80, Contract DAJA#/-79-C-0548, European Research Office, US Army. Peter Cundall Associates.

DEERE, D. U., HENDRON, A. J., PATTON, F. D. & CORDING, E. J. 1967. Design of surface and near-surface construction in rock. *In*: FAIRHURST, C. (ed.), *Failure and Breakage of Rock*. Society of Mining Engineers of AIME, New York, 237–302.

ISRM 1981. *Suggested methods for the quantitative description of discontinuities in rock masses*. Rock Characterisation Testing and Monitoring, Pergamon, BROWN, E. T. (ed.)

Modelling *in situ* water uptake in a bentonite–sand barrier

H. R. Thomas,[1] S. W. Rees,[1] B. Kjartanson,[2] A. W. L. Wan[3] & N. A. Chandler[3]

[1] School of Engineering, University of Wales, Cardiff, PO Box 917, Cardiff CF2 1XH, UK
[2] Whiteshell Laboratories, AECL-Research, Pinawa, Manitoba, Canada R0E 1L0. Present address: Department of Civil and Construction Engineering, Iowa State University, Ames, Iowa, 50011-3232, USA
[3] Whiteshell Laboratories, AECL-Research, Pinawa, Manitoba, Canada, R0E 1L0

Abstract. An *in-situ* experiment being carried out to investigate the hydraulic behaviour of a bentonite–sand engineered barrier is described. Hydraulic interactions between the barrier, surrounding granite and a concrete sealing plug are being measured some 240 m below ground level at AECL's Underground Research Laboratory (URL). Observations from the test to date indicate that the buffer mass is taking on moisture. Pore pressures in the near-field rock appear to be low but generally continue to change with time. In contrast, pore water pressures in the far-field rock essentially remain constant. Supporting numerical investigations are also described. An axisymmetric solution of Richards' pressure head based flow equation is used. The model was calibrated against known experimental results and subsequently used to provide the anticipated response of the *in-situ* experiment. The importance of the unsaturated hydraulic properties of the granite are revealed, leading to the conclusion that further data on these properties, when subject to high suction values, are required.

The Canadian Nuclear Fuel Waste Management Program (CNFWMP) is investigating the feasibility of disposing of nuclear fuel waste in a vault, excavated at a depth of 500–1000 m in stable plutonic rock of the Canadian Shield. A system of natural and engineered barriers would be used to isolate the waste from the biota. The engineered barriers would include corrosion-resistant fuel-waste containers; bentonite–sand buffer material between the containers and the host rock; backfill, bulkheads and plugs for emplacement rooms, access tunnels and shafts; and seals for exploration boreholes.

One possible configuration of a repository is shown in Fig. 1(a). In this design, waste containers are placed in vertical boreholes drilled from the floor of a main gallery. Buffer material is required to (i) support the containers, (ii) provide an effective means of dissipating heat from the waste packages, and (iii) provide a seal against potential radionuclide migration.

A mixture of sodium bentonite and well-graded silica sand in a 1 : 1 dry mass ratio has been selected as the reference buffer material. It will be compacted *in-situ* to a minimum dry density of 1.67 Mg m^{-3}, corresponding to 95% modified Proctor maximum dry density, at a moisture content of 17–19%. The initial degree of saturation is between 75 and 85%. .This highly compacted material has been selected because of its well-known sealing properties, both in terms of its low permeability and its swelling characteristics.

During the waste deposition and backfilling/sealing operations of a vault, the underground excavations will obviously be dewatered, resulting in depressed hydraulic pressures in the rock immediately adjacent to the excavations. Following backfilling and sealing, the groundwater pressures will tend to recover to their pre-construction levels and the initially unsaturated clay-based barrier materials will take up water. This process is important for a number of reasons:

- container corrosion, fuel leaching and radionuclide transport cannot begin until a significant water phase is present, i.e. until the buffer material becomes virtually water-saturated;
- the mechanical, hydraulic and thermal properties of the buffer will depend on its moisture content.

An estimate of the time period required for the clay to become saturated is therefore of importance in clay barrier performance and safety assessment. To investigate this phenomenon, a large-scale *in situ* experiment has been installed at AECL's Underground Research Laboratory (URL). This paper describes the experiment.

To assist in the overall assessment of the system's performance a programme of numerical modelling work has also been commissioned by AECL. Such investigations can be performed at a fraction of the cost of large-scale *in-situ* experimental work and can be used to (i) provide an insight into the physical processes taking

From Bentley, S. P. (ed.) *Engineering Geology of Waste Disposal*,
Geological Society Engineering Geology Special Publication No. 11, pp 215–222

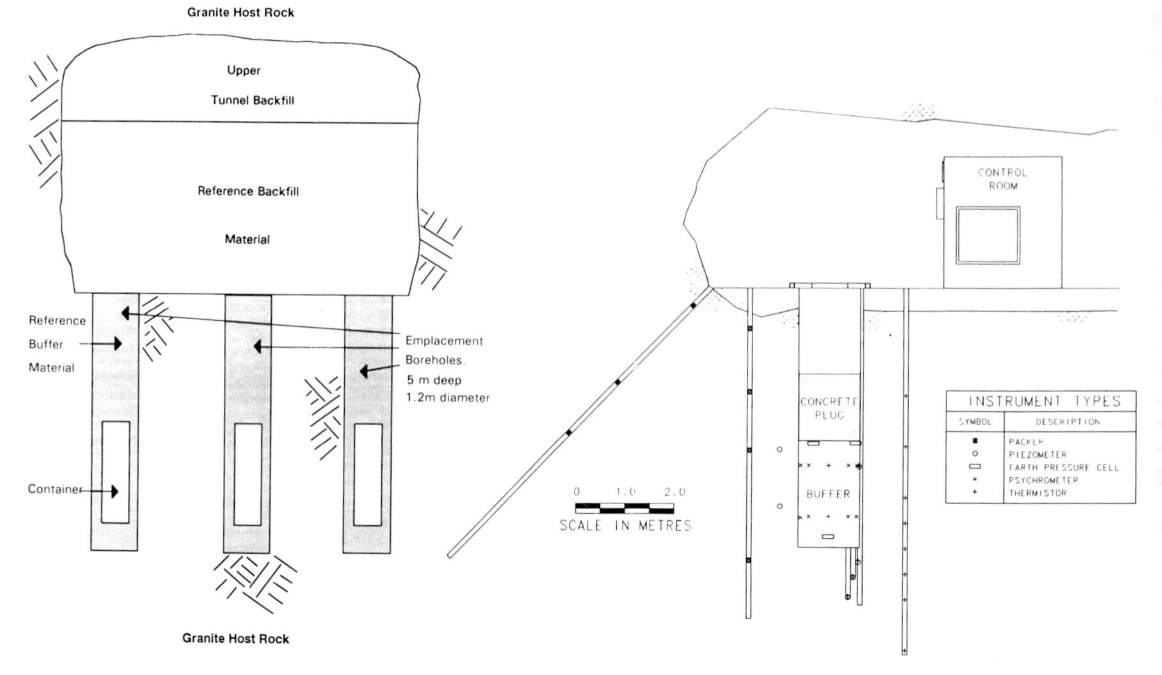

Fig. 1. (a) A possible repository configuration. (b) The experimental layout.

place, (ii) aid in establishing the relative importance of physical parameters via sensitivity analyses, and (iii) assist in the development of supporting laboratory experimental work that may subsequently be required, to define physical parameters.

The numerical modelling work performed consists of a finite-element simulation of the hydraulic interaction between the clay-based barrier and the surrounding rock. Since the buffer material is placed *in-situ* in an unsaturated state, in direct contact with the rock, it can be anticipated that suctions or negative pore pressures may be induced in the rock. Simulation of unsaturated flow in intact rock may therefore be required. Because of lack of experimental data related to unsaturated hydraulic behaviour of intact rock, the numerical results presented are based on assumed material parameters. Furthermore, the deformation characteristics of the buffer are not included in the analysis, as a first assessment of the problem. The numerical simulation approach adopted and the preliminary results achieved are presented here.

Buffer–rock–concrete plug interaction test

The URL, near Lac du Bonnet in Manitoba, Canada, is a geotechnical research and development facility constructed by AECL Research as part of the Canadian Nuclear Fuel Waste Management Program.

Geological conditions

The site is located in a previously undisturbed granite pluton, the Lac du Bonnet batholith, of the Canadian Shield. The URL consists of a 443 m deep shaft with two major experimental levels at 240 m and 420 m depths (240 Level and 420 Level respectively). Two low-dipping fracture zones intersect the shaft at the 130 Level and 240 Level respectively. The geology of the rock mass above the 240 Level is essentially pink heterogeneous granite characterized by extensive vertical jointing. Below the 240 Level, the rock is essentially unfractured grey granite with varying degrees of homogeneity.

Experimental design

The test is being carried out in a 5 m deep, 1.24 m diameter borehole at the 240 Level of the URL. The test arrangement, shown in Fig. 1(b), consists of 2 m of buffer material compacted into the bottom of the test hole, restrained by a 1.25 m thick, high-strength concrete plug and surrounding rock. Before installation of the buffer and the concrete plug, hydrogeological characterization of the borehole was carried out. This indicated that the intact granite adjacent to the hole has a hydraulic conductivity of about 10^{-11} cm s^{-1}. Flow into the hole was about 70 cm^3 day^{-1}, and water pressures in the rock adjacent to the test hole were drawn down as a result of the excavation of the openings. Instrumentation was installed in the buffer, the concrete plug and the surrounding rock to monitor changes in suction, pore water pressure, total stress, and deformation. The instrument layout design, shown in Fig. 1(b), was designed to provide information on the time-dependent changes in hydraulic gradients in the near-field rock and the buffer.

Pore water pressures in the rock were monitored by the hydraulic packers, pneumatic piezometers and hydraulic piezometers. The packers were used in the rock, at distances of at least 1 m from the test hole, with the piezometers located in the rock within 0.8 m from the wall of the hole. Thermocouple psychrometers were used to monitor changes in the total suction during water uptake by the buffer. A total of 24 psychrometers were located at four horizontal levels and two vertical planes within the buffer mass.

Placement of the buffer mass and installation of the buffer instruments began in September 1992, and was completed in four weeks. During this period, pore water pressures in the rock within 50 cm of the borehole wall were low, generally less than 50 kPa. In particular, the two hydraulic piezometers located 10 cm from the buffer showed sub-atmospheric pore water pressures in the rock. It should be noted that the pore water pressure measurements in the rock are limited by the inability of the piezometers to measure pore water pressures below −90 kPa. In contrast, piezometers located more than 50 cm from the borehole wall, and the hydraulic packers, showed relatively high pore water pressures ranging from 100 kPa (closer to the test hole) to 1200 kPa (6 m from the centre of the test hole).

The concrete plug was installed in November 1992. Following the installation of the plug, the packers farthest away from the test hole showed essentially no change in the pore water pressures in the rock, while the packers closest to the hole showed small but steady increases in pressure, as shown in Fig. 2. Data from the pneumatic and hydraulic piezometers indicated that the pore water pressures in the rock immediately adjacent to the hole continue to be small (less than 50 kPa), and in general showed little change with time.

At installation, the buffer had an initial total suction of about 4000 kPa. Data from typical psychrometers on the variations of the total suction in the buffer are shown in Fig. 3. As expected, the total suction in the buffer decreases with time as the buffer takes on moisture from the rock.

The data in Fig. 3 also show that, while the outer psychrometers which are closer to the rock generally show larger reductions in suction than the inner ones, no distinct suction fronts exist. This observation suggests that movement of moisture in dense buffer may be dominated by diffusion-type isothermal vapour transfer processes (Hillel 1980). Indication of moisture uptake is also provided by the total pressure cells mounted in the wall of the test hole at the buffer/rock interface and within the buffer mass. As shown in Fig. 4, the measured total pressures exerted by the buffer on the wall of the test hole are increasing steadily with time. This behaviour illustrates the tendency of the buffer to swell as it takes in water. Since expansion of the buffer is constrained by the granite host rock and the concrete cap, increased swelling pressures are recorded.

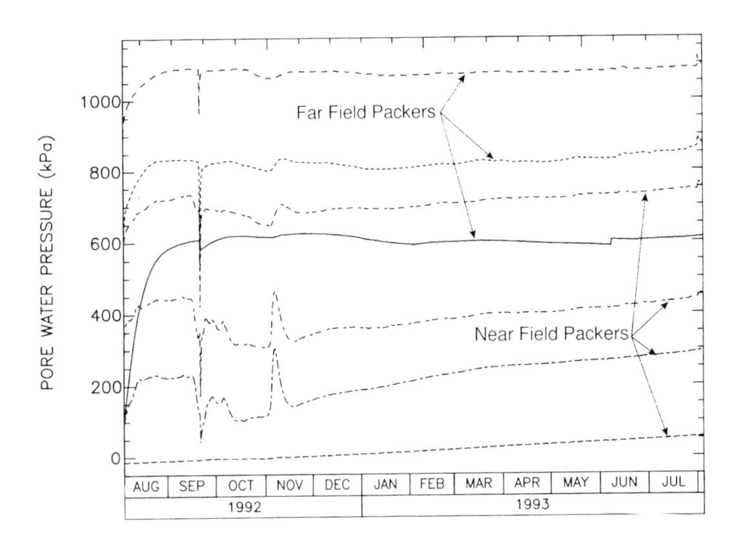

Fig. 2. Near-field and far-field pore water pressure responses in the granite.

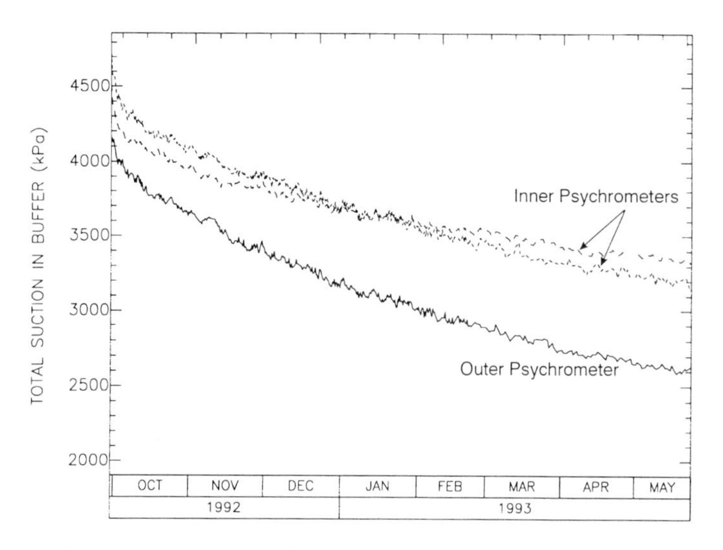

Fig. 3. Temporal and spatial variations of total suction in the buffer.

Modelling approach adopted

The numerical modelling study is based on an axisymmetric solution of Richards' (1931) pressure-based flow equation. The region modelled covers a depth of 6-m and since this is situated some 240 m below ground surface, the effect of the gravitational component of the total potential was considered to be negligible in comparison with the pressure head potential. The governing equation used to describe the flow of moisture is therefore:

$$\frac{\partial \theta}{\partial \psi} \cdot \frac{\partial \psi}{\partial t} = \frac{1}{R}\left[\frac{\partial}{\partial R}\left(KR\frac{\partial \psi}{\partial R}\right) + \frac{\partial}{\partial z}\left(KR\frac{\partial \psi}{\partial z}\right)\right] \quad (1)$$

where θ, ψ, K, R, z and t are the volumetric moisture content, capillary potential, unsaturated hydraulic conductivity, radius, elevation and time respectively.

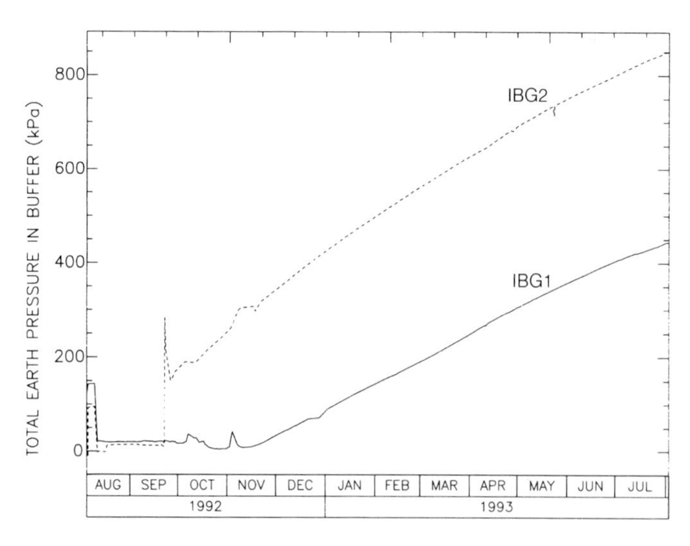

Fig. 4. Total earth pressure development in the buffer.

Equation (1) was solved by a combination of finite-element and finite-difference techniques, as described by Thomas & Rees (1990).

The domain to be modelled included a 2 m deep bentonite barrier capped with a layer of concrete 1.25 m thick. The finite-element mesh adopted for the problem is shown in Fig. 5. Numerical experiments were performed to check that this level of discretization produced converged results. Time stepping errors were also investigated and found to be insignificant provided that a maximum time step size of two weeks was not exceeded. In the first instance, it was assumed that the concrete cap was impermeable to moisture flow.

Fixed pore water pressure head boundary conditions were prescribed on the outermost vertical edge of the domain (side DE in Fig. 5). The exact values used were the steady-state values obtained previously for the granite (Chandler *et al.* 1992). Atmospheric pressure was applied to the boundary of the granite in the small region of 'open' borehole (sides AB and BC in Fig. 5).

Application of the model requires the specification of two material properties: the unsaturated hydraulic conductivity and the specific moisture capacity. The relationship between capillary potential and moisture content for the bentonite–sand barrier was determined experimentally using the psychrometric and the vapour equilibrium techniques. The results are shown in Fig. 6. Specific moisture capacity values were calculated directly from this relationship. It is also known from experimental work that the saturated hydraulic conductivity of the buffer material lies within the range 10^{-9} to 10^{-10} cm s^{-1}. An average value within this range was adopted here. No direct measurements of the unsaturated hydraulic conductivity were available. However, data provided by Radhakrishna *et al.* (1992) suggest that the magnitude of isothermal moisture diffusivity is approximately 6.3×10^{-5} cm^2 s^{-1} and remains relatively constant over the range of moisture content of interest. Therefore, this value was used, in conjunction with the specific moisture capacity values, to calculate the variation of unsaturated hydraulic conductivity with moisture content, for the bentonite–sand barrier. The relationship obtained is also shown in Fig. 6.

In the absence of any experimental data, the unsaturated hydraulic properties of the granite were assumed, as a first approximation, to be as shown in Fig. 7. The saturated hydraulic conductivity of the granite was previously measured to be 5×10^{-11} cm s^{-1}. The porosity of the rock was taken to be 0.0025 (Chandler *et al.* 1992). The variation of specific moisture capacity for the granite was calculated directly from the curve shown in Fig. 7.

Results and analysis

A simulation of the flow of water through saturated granite into the borehole, before emplacement of the buffer material and the concrete cap, was attempted first. This was performed so that the results could be

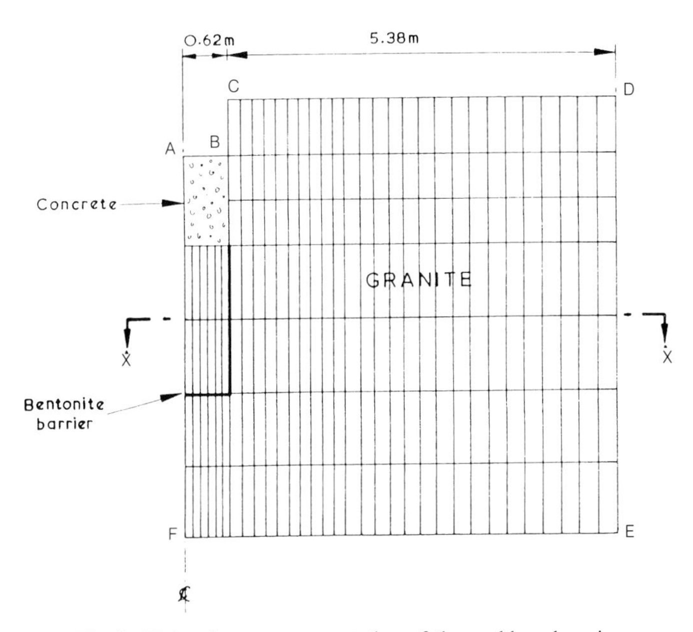

Fig. 5. Finite-element representation of the problem domain.

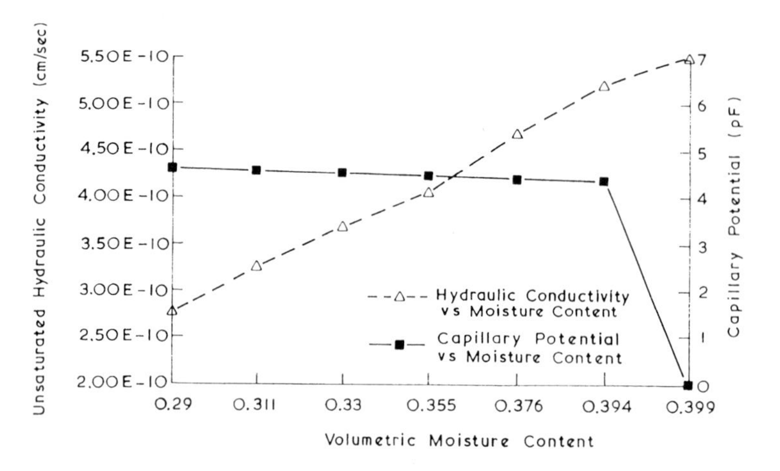

Fig. 6. Hydraulic properties of the bentonite–sand buffer.

compared with measured water inflows and independently calculated numerical inflows using FLAC (Fast Lagrangian Analysis of Continua), a commercially available, two-dimensional, finite-difference software package (Itasca 1989). In this way an assessment of the proposed simulation work could be achieved. The results achieved, in terms of inflow rates at various depths in the borehole, are presented in Table 1. Experimentally measured results and the numerical results obtained from FLAC are also shown. It can be seen that excellent correlations have been achieved between the two sets of numerical results, yielding confidence in the numerical solution techniques. Reasonable agreement can also be seen to have been achieved between numerical and experimental measurements although clearly greater differences are now apparent than observed when comparing numerical results alone. These discrepancies may be attributed to a range of factors. For example, the natural variability of rock may not be adequately represented by the single set of material properties adopted. A fuller investigation of these factors is outside the scope of this paper. However, for the exercise in hand, the correlations achieved are considered to be acceptable, leading to the conclusion that further simulation work can be performed.

The calibrated model was therefore used to investigate the effect of the installation of the bentonite–sand barrier. The initial distribution of pore water pressure in the granite was determined from the steady-state analysis described above. A uniform initial pressure head of −35 837 cm (about −3.58 MPa), corresponding to an emplacement moisture content of 18% (by weight), was specified throughout the barrier.

The results illustrated in Fig. 8 are for the second analysis in terms of predicted pore water pressure head

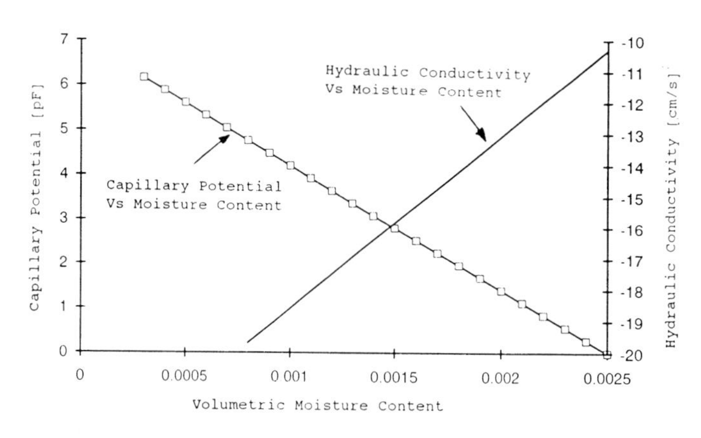

Fig. 7. Assumed hydraulic properties of the granite.

Table 1. *Predicted and measured inflow into an open borehole*

Height above bottom of borehole (m)	*Measure inflow ($cm\,s^{-1}$)*	*Predicted inflow-current ($cm\,s^{-1}$)*	*Predicted inflow (FLAC) ($cm\,s^{-1}$)*
1.0	1.28×10^{-8}	3.96×10^{-9}	3.60×10^{-9}
2.0	2.67×10^{-9}	1.90×10^{-9}	1.70×10^{-9}
3.6	8.25×10^{-10}	1.50×10^{-9}	1.45×10^{-9}

contours throughout the granite and the buffer at 200 days after emplacement of the concrete cap (23 May 1993). The results show that high suction values and steep suction pressure gradients exist in a narrow zone of rock (about 20 cm thick) immediately adjacent to the test hole. This pattern of behaviour is qualitatively consistent with measured data.

Presented in Fig. 9 are the results of the analysis in terms of the variation of pore water pressure head with time across section X–X at 0, 100 and 200 days (4 November 92, 2 February 93 and 23 May 93 respectively). The results of the analysis are compared directly with measured pore water pressure at each time. A good agreement between predicted and measured data can be seen to have been achieved at each time considered. The gradual resaturation of the barrier is evident.

Considerable differences exist, however, between experimental and numerical observations when considering the duration of the suctions in the rock. The numerical analysis indicates that significant negative pressure heads can be generated in the granite. The magnitude of such pressure heads and the extent of their penetration into the granite then influence the rate of saturation of the bentonite–sand barrier. The results obtained are strongly dependent on the assumed hydraulic properties of the granite. High suctions in the granite result in very low hydraulic conductivity creating a near-impermeable region.

It is clear that a more complete specification of the unsaturated hydraulic characteristics of the rock is required in order to investigate this problem further. While advances may be possible from numerical sensitivity analyses, some experimental work appears to be both necessary and justified to provide the form and magnitude of the relationships.

Conclusions

An experiment being carried out some 240 m below ground level at AECL's Underground Research Laboratory (URL) is described. Hydraulic interaction

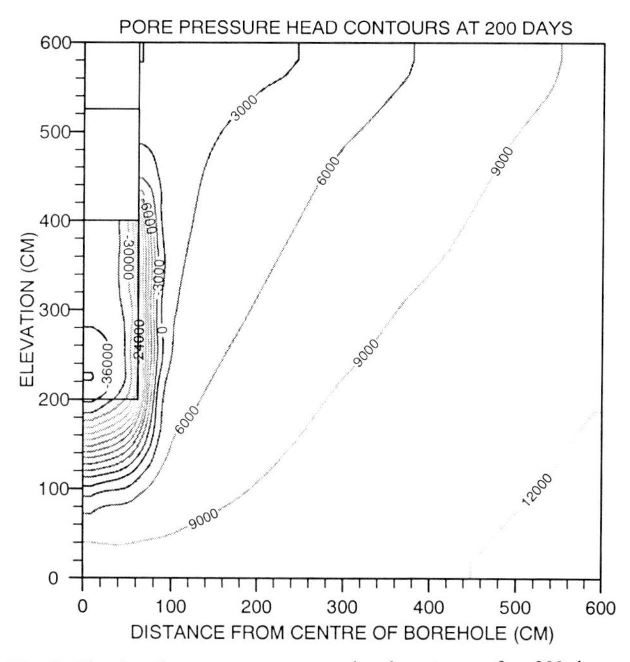

Fig. 8. Simulated pore water pressure head contours after 200 days.

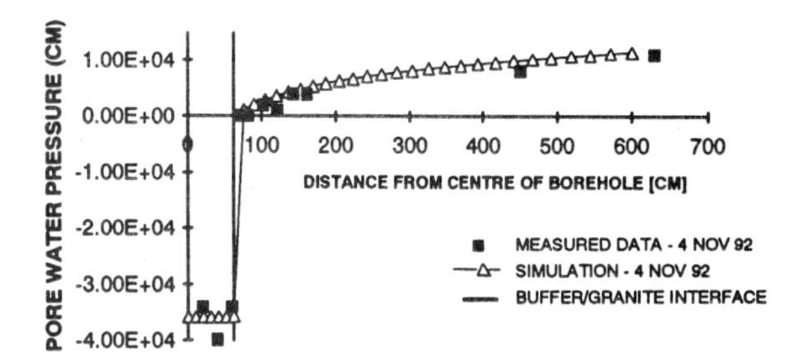

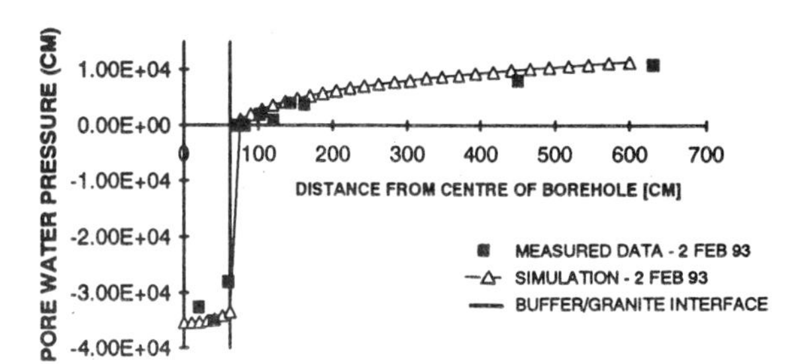

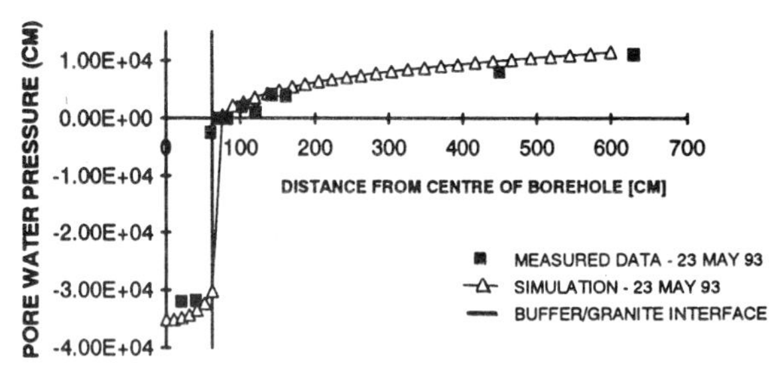

Fig. 9. Measured versus simulated pore pressure head profiles at section X–X.

between a bentonite–sand barrier, the surrounding granite and a concrete sealing plug is being monitored. *In-situ* measurements indicate that the buffer takes up water, with increasing total stresses and decreasing suctions. Hydraulic pressures within the rock are generally increasing. Pore pressures below atmospheric pressure were noted in the rock at the borehole wall for several months after installation of the buffer.

In support of the *in-situ* experimental work, the results of a numerical investigation of the hydraulic performance of a bentonite–sand barrier have been presented. An analysis of saturated groundwater flow through the intact granite host rock was performed to calibrate the modelling approach adopted. The results achieved were within one order of magnitude of the experimental results. This was considered to be acceptable.

The model then yielded results indicating that large negative pressure heads could arise in the granite due to the presence of unsaturated buffer material. Good agreement was achieved between predicted and measured pore water pressure profiles in the barrier and in the rock at each time considered.

Considerable differences emerged, however, between experimental and numerical results when considering the duration of the suctions in the rock. It is clear that to model more accurately the performance of the buffer, and in particular to provide reliable estimates of the times for saturation, more data on the hydraulic properties of unsaturated granite are required.

Acknowledgements. This research was jointly funded by AECL-Research and Ontario Hydro under the auspices of the CANDU Owners Group.

References

CHANDLER, N. A., KJARTANSON, B. H., KOZAK, E. T., MARTIN C. D. & THOMPSON, P. M. 1992. *Monitoring the geomechanical and hydrogeological response in granite for AECL-Research's Buffer/Container Experiment.* Proc., 33rd US Symposium on Rock Mechanics, Sante Fe, New Mexico, USA, June 1992.

HILLEL, D. 1980 *Fundamentals of Soil Physics.* Academic Press, New York

ITASCA CONSULTING GROUP, INC. 1989. *FLAC—Fast Lagrangian Analysis of Continua.* Suite 210, 1315 5th St. S., Minneapolis, Minesota.

RADHAKRISHNA, H. S., CRAWFORD, A. M., KJARTANSON, B. H. & LAU, K. C. 1992. Numerical modelling of heat and moisture transport through bentonite–sand buffer. *Canadian Geotechnical Journal*, **29**, 1044–1059.

RICHARDS, L. A. 1931. Capillary conduction of liquids through porous mediums. *Physics*, **1**, 318–333.

THOMAS, H. R. & REES, S. W. 1990. Modelling field infiltration into unsaturated clay. *J. Geotechnical Engng., American Society of Civil Engineers (ASCE)*, **116** (10).

Application of induced seismicity to radioactive waste management programmes

R. Paul Young

Applied Seismology and Rock Physics Laboratory, Department of Earth Sciences, Keele University, Staffordshire ST5 5BG, UK

Abstract. Induced seismicity can make a unique contribution to our understanding of rock mass response due to excavation and thermal-induced loads, during the site characterization and performance monitoring of a nuclear waste repository. Field data, with supplementary laboratory studies, are presented to show how induced high-frequency acoustic emission/microseismicity (AE/MS) can be used as a tool for passive volumetric remote sensing of failure processes. Case studies are described using results from experiments carried out at Atomic Energy of Canada Ltd's (AECL) Underground Research Laboratory (URL). This site has been operated for the last ten years to investigate the concept of safe disposal of nuclear waste fuel in a granitic rock mass at 420 m depth.

A significant proportion of radioactive waste from nuclear power stations is in the form of used nuclear fuel bundles. In Canada, AECL has been given the mandate to conduct research into developing and evaluating the technology for disposal of nuclear fuel waste in intrusive igneous rock and to prepare an assessment of the disposal concept for scientific, government and public review (Rummery 1992). The Canadian Nuclear Fuel Waste Management Concept is to immobilize used fuel in corrosion-resistant containers, then bury the containers in a disposal repository 500–1000 m deep in the rock of the Canadian Shield. Since the object is to isolate the waste from the biosphere, it is imperative that the waste is not dissolved and transported by groundwater. Long-term isolation is being studied using engineered barriers such as ceramic fuel pellets, canisters, backfill and sealing materials, as well as the natural barrier of rock at depth. This multi-barrier concept is being developed for either borehole emplacement, where containers are located in individual boreholes, or in-room emplacement, where the containers are stored in sealing and backfill materials within the room itself. These storage rooms form part of a disposal repository which in many respects resembles a deep mine. The proposed Canadian repository would be a network of tunnels and rooms over an area of 2 km by 2 km (500–1000 m deep), providing storage for Canada's anticipated used nuclear fuel production until the year 2035. Several countries including Sweden, Canada, Switzerland and more recently the United Kingdom have decided to study the problems of rock characterization, repository design and construction by building underground experimental laboratories. The role of these facilities is to help assess the feasibility and safety of deep geological disposal for nuclear waste.

Traditionally, the role of seismicity studies in nuclear waste management investigations has been restricted to hazard assessment and the influence of natural seismicity on the site selection and repository design process. Once a site is chosen for the possible storage of nuclear waste, the characterization of the site and the subsequent performance of the geological medium become critical issues. The role of seismicity studies in this regard has not been fully exploited.

The site characterization of a nuclear repository involves detailed knowledge of the response of the rock mass to excavating a network of rooms and tunnels. Seismicity studies can play an important role in this regard, because they provide a remote-volumetric and non-destructive way of monitoring the development of cracks, which can provide unanticipated pathways for radionucleide transport.

In situ studies of excavation-induced damage

Two examples of cases where induced seismicity has proven valuable to AECL during the *in situ* investigations of the URL are described below. The first case relates to the sinking of a vertical shaft created by drill-and-blast techniques. The AE/MS monitoring provided early information on the seismic response of the rock mass and the creation of an excavation damage zone. This allowed an optimization of the method for later experiments around an experimental tunnel. The second casc describes the 'mine-by tunnel' where AE/MS monitoring was used to quantify the damage zone caused by different excavation methods.

From BENTLEY, S. P. (ed.) *Engineering Geology of Waste Disposal*,
Geological Society Engineering Geology Special Publication No. 11, pp 223–230

Underground research laboratory (URL)

The URL is located near Pinawa, Manitoba, Canada and is situated within the Lac du Bonnet granite batholith, which is believed to be representative of many granitic intrusions in the Canadian Shield. The general geology of the site is shown in Fig. 1(a). Diamond drilling and *in situ* observations have shown that the grey granite, below a depth of 220 m, is much less fractured than the pink granite above this depth (Martin 1989). The pink granite is separated from the grey granite by a major fracture zone which is called Fracture Zone 2. Two other major fracture zones have also been identified, but outside of these zones the granite is of very good quality. Above and below Fracture Zone 2, different stress domains are recognized (Fig. 1(b)). Above the 240 m level the maximum stress is subhorizontal and trending N40 °E, whereas the minimum stress is subvertical. The stress below Fracture Zone 2 undergoes major changes, both in magnitude and orientation. The maximum horizontal stress is rotated by 90° and is oriented N130 °E, along the slip direction of Fracture Zone 2. The *in situ* stress field at 420 m depth is $\sigma_1 = 55$ MPa, $\sigma_2 = 48$ MPa and $\sigma_3 = 15$ MPa, with σ_1 and σ_2 subhorizontal and σ_3 oriented 14° from vertical (Martin 1990).

Seismicity induced by shaft excavation

An AE/MS network was installed at the URL in 1987 to monitor seismicity induced by shaft excavation between 324 and 443 m of depth (Young & Talebi 1988). Four inclined boreholes, drilled from the 300 m level, were used to install accelerometers and hydrophone sensors (Fig. 1(c)). The instrumentation recorded signals in the 100 Hz–10 kHz band and data were acquired at a sampling rate of 50 kHz per channel. Further details on the specific instrumentation used for this type of AE/MS monitoring can be found in Young *et al.* (1992). The excavation of a 4.6-m-diameter circular shaft was performed using a full-face drill-and-blast technique. After each blast a period of increased seismic activity was observed, lasting for about two hours and followed by a rapid decay of the number of events with time, leading to the normal rate of a few events per hour. In most cases the seismic events, monitored after each blast, clustered around the bottom and the walls of the shaft, near the newly created faces. They were located mainly within 5–10 m from the shaft bottom in the vertical direction and within 1–2 m from the walls in the horizontal plane. The accuracy of location is in most cases better than ± 1 m; confirmed by the source location of blasting caps used in the survey of seismic-wave velocities.

An interesting feature is the preferential clustering of the events in the northeast–southwest direction in the horizontal plane. This orientation is compatible with that of the minimum horizontal stress and orientation of overbreaks in the area. The trend can be seen in Fig. 1(c), where the horizontal distribution of seismic events following one blast is shown. The *in situ* stress at the URL has been extensively investigated using traditional methods, such as overcoring, hydraulic fracturing and borehole breakouts (Martin 1989). Figure 1(b) summarizes the *in situ* stress results and also shows the AE/MS interpretation of *in situ* stress orientation. The AE/MS results were obtained by computing a best-fit ellipse to the spatial distribution of source locations, following each round of excavation, and using the long-axis of the ellipse as the orientation of σ_3 (Talebi & Young 1992).

The results clearly showed that AE/MS monitoring could be used to delineate damage away from the shaft wall, provide information on the local *in situ* stress and confirm the nature of the borehole breakout failure process. Subsequent investigation of this data analysed the modes of failure of the seismic events and the distribution of source parameters (Gibowicz *et al.* 1991). These investigations confirmed the findings from source location results and indicated the complex nature of the failure process in the breakout zone. Shear and other modes of failure were interpreted to contribute to the process. The results from the shaft excavation experiment provided the design parameters for a more sophisticated seismic monitoring of the experimental tunnel at the 420 m level.

Seisimicity induced by tunnel excavation

AECL has completed an experimental excavation (mine-by tunnel) at the 420 m level in the URL. The site was instrumented with 16 triaxial accelerometers prior to the controlled excavation of a 3.5-m-diameter tunnel within the rock mass. Data were acquired at a sampling rate of 50 kHz with the AE/MS monitoring system (Feignier & Young 1992). The 48-channel seismic network envelops a $50 \times 50 \times 50$ m volume around the test tunnel and provides complete focal sphere coverage of the induced seismicity (Fig. 2(a)). A catalogue of 25 000 AE/MS ($M < 0$) events were recorded between October 1991 and July 1992. In order to determine the acoustic properties of the rock mass, velocity and attenuation surveys were carried out prior to the excavation of the tunnel. From the velocity survey it was found that the rock mass can be considered homogeneous, with $V_p = 5820 \pm 50$ m s^{-1}

Fig. 1. (a) Geology and stress field at the Underground Research Laboratory. (b) *In situ* stress orientation for σ_1 from conventional and AE/MS monitoring. (c) AE/MS source locations induced by shaft excavation. Note the ellipsoidal distribution of source locations and the orientation of principal stresses ($\sigma_{hmax} = \sigma_1$ and $\sigma_{hmin} = \sigma_2$).

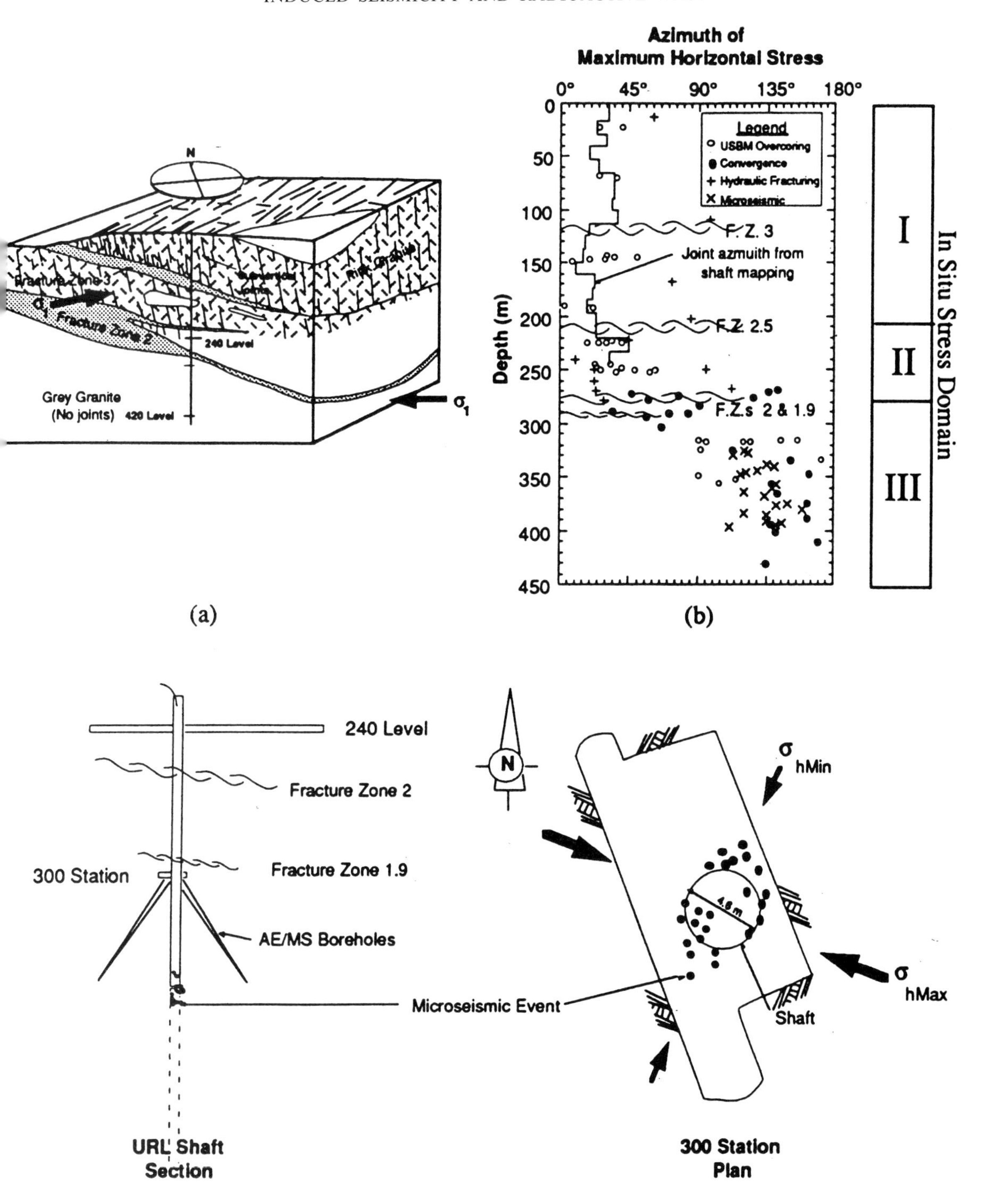
N
Fracture Zone 3
Fracture Zone 2
240 Level
Grey Granite
(No joints) 420 Level
σ_1
σ_1
(a)
Azimuth of
Maximum Horizontal Stress
0° 45° 90° 135° 180°
Depth (m)
0
50
100
150
200
250
300
350
400
450
Legend
USBM Overcoring
Convergence
Hydraulic Fracturing
Microseismic
F.Z. 3
Joint azmuith from
shaft mapping
F.Z. 2.5
F.Z.s 2 & 1.9
I
II
III
In Situ Stress Domain
(b)
240 Level
Fracture Zone 2
300 Station
Fracture Zone 1.9
AE/MS Boreholes
Microseismic Event
URL Shaft
Section
N
σ hMin
4.6 m
σ hMax
Shaft
300 Station
Plan
(c)

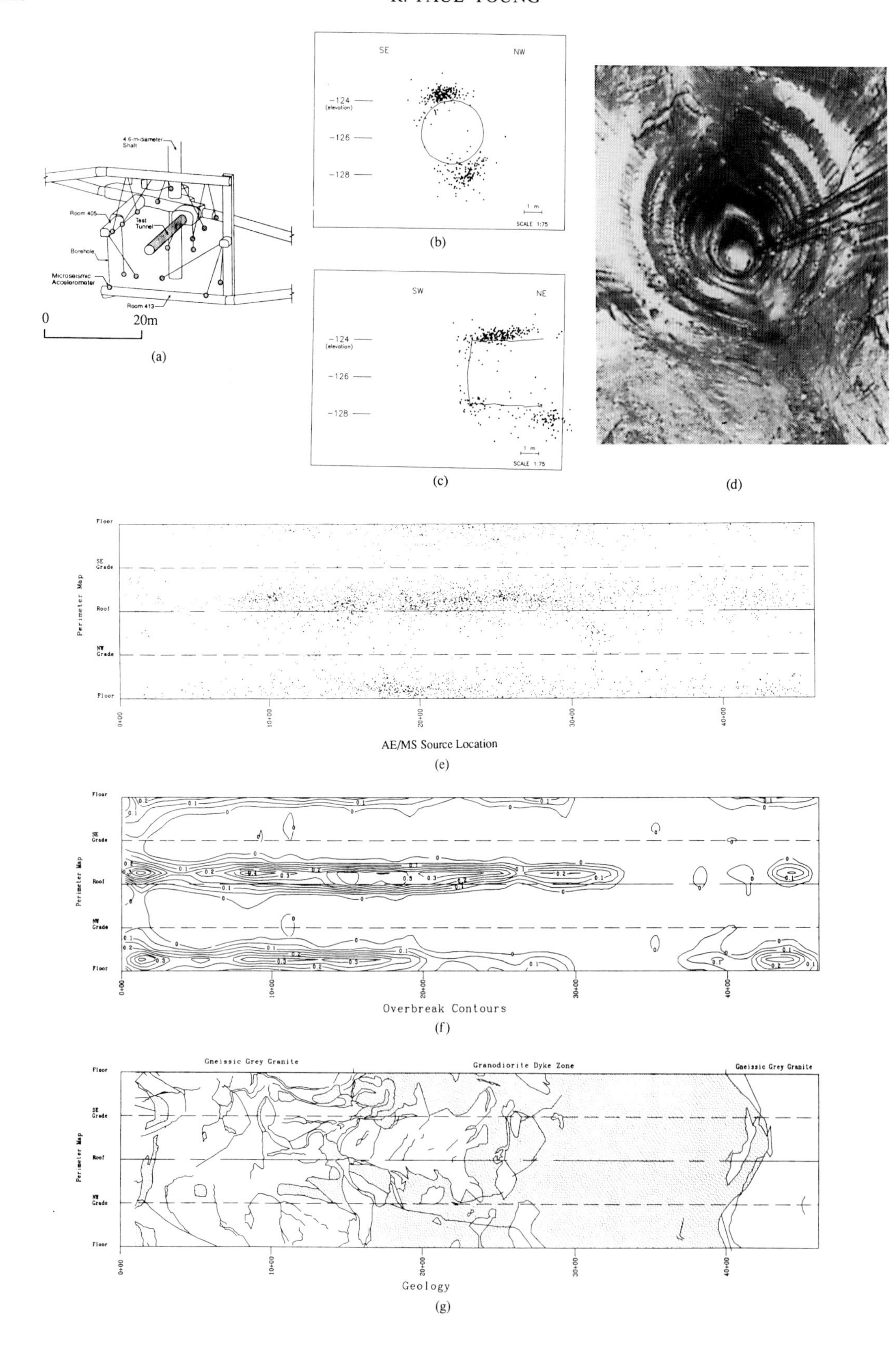
4.6-m-diameter Shaft
Room 405
Test Tunnel
Borehole
Microseismic Accelerometer
Room 413
0
20m
SE
NW
-124 (elevation)
-126
-128
1 m
SCALE 1:75
SW
NE
Floor
SE Grade
Roof
NW Grade
Perimeter Map
0+00
10+00
20+00
30+00
40+00
AE/MS Source Location
Overbreak Contours
Gneissic Grey Granite
Granodiorite Dyke Zone
Geology

(a) (b) (c) (d) (e) (f) (g)

and $V_s = 3360 \pm 50\,\mathrm{m\,s^{-1}}$. From the attenuation survey, estimates of the quality factor (Q) were obtained using the spectral ratio method (Feustel & Young 1993, 1994), with $Q_p = 223 \pm 37$ and $Q_s = 114 \pm 18$. The seismic data formed one part of an excavation response study (Read & Martin 1991). Detailed geological mapping, *in situ* stress measurements and continual deformation monitoring of the test tunnel by AECL provided unique data for validation of seismic model assumptions and interpretations.

A non-explosive rock-breaking technique was used for the tunnel excavation, to see if this would minimize the zone of excavation-induced damage. The tunnel was excavated approximately parallel to σ_2 so as to maximize excavation response. The excavation technique consisted of drilling 1-m-long holes (45 mm diameter) all around the perimeter of the tunnel. The holes were then reamed, up to 100 mm in diameter, to almost connect them. After completion of the reaming, microseismic data were collected. Then, hydraulic rock splitters were used to break out the interior of the tunnel stub. Once the rock breaking and mucking was complete, AE/MS monitoring was continued.

AE/MS monitoring data are now described from round 17, located in the middle of the tunnel where the system coverage was most optimum. The spatial distribution of source location results following excavation of the 1-m rock stub are shown in Figs 2(b) and (c). The monitoring period was 300 h following excavation and 359 events were recorded, with the number of events monitored showing a typical and rapid decay with time. The auto-locations were recomputed from manually picked arrival times, and the source location calibration gave a mean error of the order of 0.25 m. Figure 2(b) is a view looking into the test tunnel; the measured σ_1 orientation is 14° from horizontal, plunging to the southeast. The long axis of a best-fit ellipse to these data correlates to the average orientation of σ_3 (Talebi & Young 1992). Figure 2(d) is a photograph looking in the same direction as shown in Fig. 2(b). The well-developed stress-induced notch in the roof and floor can be seen clearly. Figure 2(c) shows a section view of the AE/MS events and highlights a different spatial distribution for events in the floor than in the roof of the tunnel. The notch in the floor clearly developed further back from the face than in the roof during the monitoring period. This is believed to be a result of gravity and additional confinement from waste rock at the face. However, lithological variations in the granitic rock at this location and the possibility of more saturated conditions in the floor also need to be investigated.

Figure 2(e) shows the spatial distribution of source-located AE/MS events from all 50 excavation rounds. The figure represents an unfolded perimeter map of the tunnel. The roof is shown as the centre line and the floor is shown as the top and bottom edges of the map. It should be noted that the first three excavation rounds, up to chainage 8.25 m, were carried out with drill-and-blast methods. The AE/MS system was set to be less sensitive and therefore only larger magnitude events, compared with later monitoring, were recorded for these rounds. However, from rounds 4 to 50 the sensitivity of the system and pattern of monitoring was standardized, and therefore the spatial distribution and density of source locations for these rounds is more meaningful. Figures 2(f) and (g) also show the mapped notch (overbreak) contours and the lithological variations in the tunnel, plotted on the same type of perimeter map for comparison with Fig. 2(e). The following important observations can be made from these data. First, the distribution of source locations map the distribution of breakout, and the apex of the notch is in the direction of σ_3. Second, the notch and seismicity are often more developed in areas of granite than granodiorite and, in general, the seismicity has a different spatial distribution and is more dominant in the roof than the floor. Third, the seismicity represents increased microcracking in the notch regions up to 1–2 m beyond the tunnel design line.

Laboratory investigations of thermally induced fracturing

In this experiment a 30-cm-diameter 22-cm-long cylinder of Lac du Bonnet granite was used to monitor thermally induced fracturing using ultrasonic imaging and AE. A near-surface sample was used to minimize stress-relief microcracking. A through-going 3 cm axial hole was bored down the centre of the core to accommodate the heater cartridge and three thermocouples (Fig. 3(a)). The cylinder mid-plane was chosen as a tomography plane. Average compressional velocities of $5400\,\mathrm{m\,s^{-1}}$ were measured, with an in-plane anisotropy of approximately 1.5%. Compressional velocities down the long axis of the core were somewhat faster, about $5800\,\mathrm{m\,s^{-1}}$.

Fig. 2. (a) Mine-by tunnel showing the location of microseismic sensors; (b) 359 source-located microseismic events following excavation of a 1-m section in the centre of the tunnel; (c) section showing the data in (b); (d) photograph showing the excavated mine-by tunnel and the breakout notches in the roof and floor (note the coincidence of microseismic events in the notch regions and the orientation of the measured σ_1); (e) perimeter maps of the mine-by tunnel for microseismic source locations following excavation of the 50 excavation rounds; (f) overbreak contours in metres; and (g) lithology.

A 1000 W electrical resistance cartridge heater served as a heat source and an internal thermocouple provided feedback for temperature control. A programmable temperature controller was used to provide controlled heating and cooling rates during the experiment. Temperatures were measured at five positions on the outer surface of the cylinder, at the level of the tomography plane and at three positions on the inner borehole. The temperature data were stored at 10 s intervals throughout both heating and cooling periods, and the tomography plane was positioned so as to bisect the heater cartridge. This plane was chosen to image the hottest portion of the rock where most AE activity was expected. Tomography data were collected prior to thermal loading and at the conclusion of each cooling phase. Collection of all tomographic data at room temperature ensured that the resulting images would only show structural changes due to the formation of the fractures in the rock, rather than changes due to thermal expansion effects. The sample was thermally loaded using six cycles to progressively higher temperatures, 125 °C, 150 °C, 170 °C, 180 °C, 200 °C and 245 °C. Ultrasonic tomography and acoustic emission data were collected using an array of 1 MHz piezoelectric transducers. Waveforms were recorded using Nicolet 440 digital oscilloscopes sampling at 10 MHz.

A simultaneous iterative reconstruction technique (Dines & Lytle 1979) was used to reconstruct the tomographic images and was modified to incorporate anisotropy (Stewart 1988). This algorithm reconstructed a base slowness plus an incremental elliptical anisotropy strength factor (slowness difference between the fast and slow directions). The orientation of the anisotropy was assumed to be uniform over the image region. A difference tomography technique was used in these experiments because the thermally induced changes in the slowness structure of the rock were fairly subtle (5–10%). To improve the visualization of these small changes, the differences in the arrival times, between successive heating cycles, were imaged directly to obtain a tomogram of the induced changes. These differences were computed using cross-correlation of the recorded waveforms for identical source–receiver pairs before and after each cycle (Falls *et al.* 1992).

Very rapid AE activity occurred during the fifth and sixth cycles. The acoustic slowness tomogram for the sixth thermal cycle is shown in Fig. 3(b). During previous cycles the slowness differences are broadly distributed over the whole tomography plane. However, the tomogram from thermal cycle 6 shows the fracture with remarkable clarity. The slowness difference tomogram shows the fracture as a localized band extending from the outer edge of the cylinder to the inner borehole. This band coincides with the dominant AE concentration and with visual observation of the fracture along the outer surfaces of the cylinder. Compressional wave slownesses increase by up to $4\,\mu s\,cm^{-1}$ or 20% along the fracture trace. Figure 3(c) shows the temperature, time, history and AE counts for thermal cycle 6. AE source locations for this thermal cycle are shown in Figs 3(d) and (e). Three orthogonal views are given for events occurring both early and late in the thermal cycle. Events during earlier cycles cluster around the borehole and are predominantly shear with dispersed tensile events in the heating cycle and compressive events in the cooling cycle. Locations for the sixth thermal cycle are shown divided into pre- and post-failure groups (Figs 3(d) and (e)). The vast majority of the events occur on the thermally induced fracture face. About 40% of the post-failure events are compressive and an equal number of shear events were also recorded.

It was possible to map the development of a macroscopic fracture in the granite sample extremely well using ultrasonic imaging and acoustic emission. The growth of the fracture from the outer edge of the sample inwards, is related to the stress pattern developed by strong thermal gradients in the sample. The thermal gradients generate compressive hoop stresses in the hotter portions of the rock and tensile hoop stresses in the cooler portions. The measured thermal gradients can produce tensile hoop stresses at the outer edge of the sample in excess of the tensile strength of the rock. AE source mechanisms appear to follow a consistent pattern during the experiment. The initial AE during each thermal cycle were low-amplitude shear events that clustered closely around the inner borehole surface. As higher temperatures were reached, AE activity was dominated by widely dispersed tensile events. A few shear events were detected along the incipient fracture plane during the sixth thermal cycle followed by abundant shear and implosional AE along the macroscopic fracture plane as the cylinder cooled. Further details of thermal studies in granite can be found in Richter & Simmons (1974), Carlson *et al.* (1993) and Jansen *et al.* (1993).

Fig. 3. (a) Sample geometry and transducer array for ultrasonic imaging and acoustic emission studies of thermally induced fracturing in granite; (b) slowness difference tomogram (units of $\mu s\,cm^{-1}$) and AE source locations for the sixth thermal cycle (the sketch indicates the position of the thermally induced macro-fracture); (c) cumulative acoustic emission events and temperature history for the sixth thermal cycle; (d) AE source locations and mechanisms of failure for events occurring pre-failure during the sixth thermal cycle, and (e) post-failure events during the sixth thermal cycle.

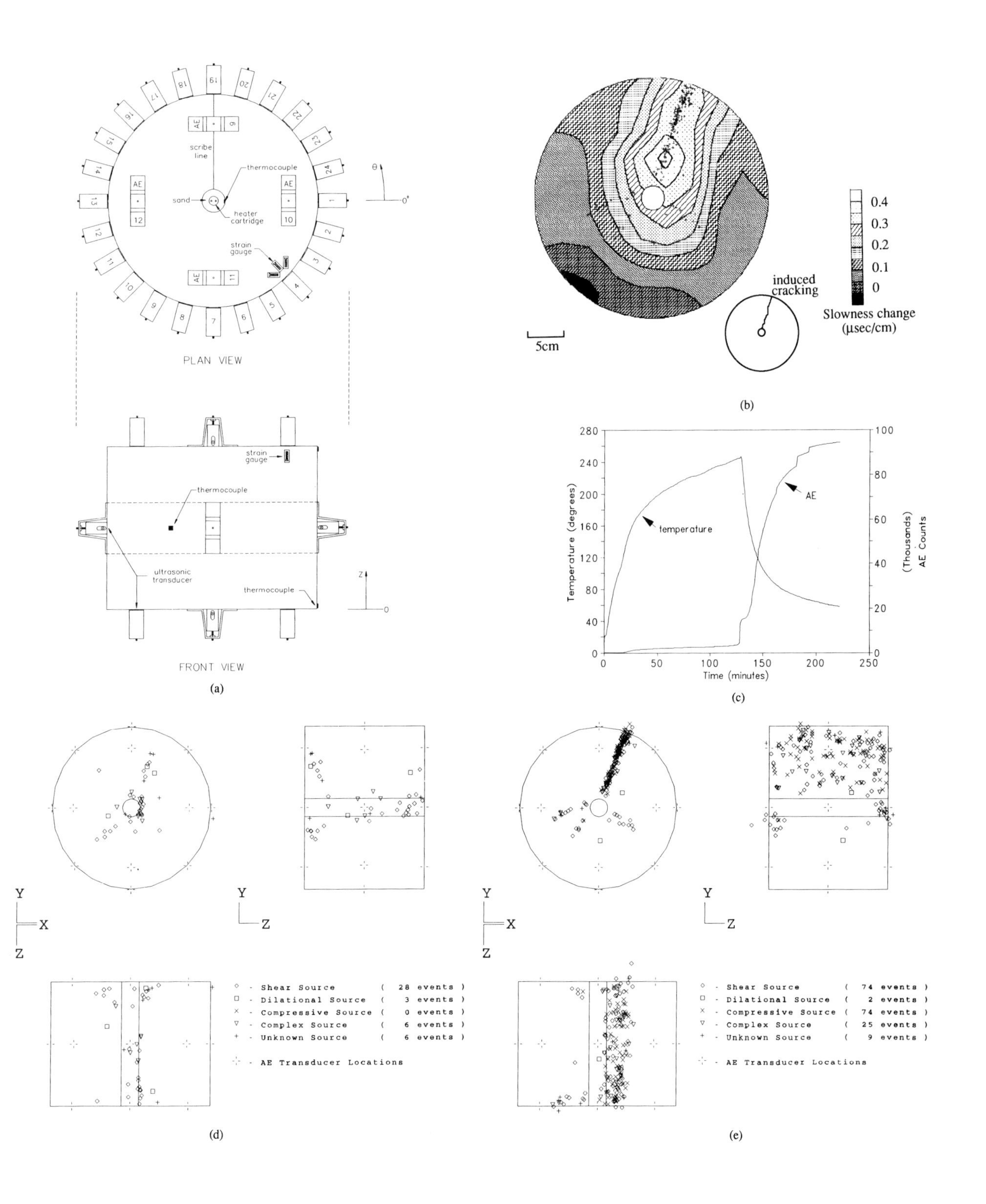
scribe line
thermocouple
sand
heater cartridge
strain gauge
AE
PLAN VIEW
strain gauge
thermocouple
ultrasonic transducer
thermocouple
FRONT VIEW
induced cracking
Slowness change (μsec/cm)
5cm
Temperature (degrees)
AE Counts (Thousands)
Time (minutes)
temperature
AE
Shear Source (28 events)
Dilational Source (3 events)
Compressive Source (0 events)
Complex Source (6 events)
Unknown Source (6 events)
AE Transducer Locations
Shear Source (74 events)
Dilational Source (2 events)
Compressive Source (74 events)
Complex Source (25 events)
Unknown Source (9 events)
AE Transducer Locations

(a) (b) (c) (d) (e)

Conclusions

Induced seismicity provides a unique approach to volumetric remote sampling of rocks, both in the laboratory at ultrasonic frequencies and in the field at seismic frequencies. At the URL, induced seismicity has been used to provide a valuable insight into the excavation damage zone around an underground opening. Source locations were shown to map the development of 'borehole breakout' on a large scale and delineate the excavation damage zone around the tunnel. A laboratory investigation in granite showed how tomographic difference imaging was sensitive to small physical property changes in samples subjected to thermal loading. In these experiments, the slowness difference images clearly showed the progressive damage in the sample created by cyclic loading. Acoustic emission data enhanced the interpretation by delineating the macroscopic fractures, determining the orientation of induced microcracks and classifying the mechanistic behaviour operative during the different phases of the experiment. Future *in situ* experiments will use induced seismicity in conjunction with conventional studies, to investigate how the borehole breakout failure process will be effected by additional thermal loading.

Acknowledgements. This work was supported by the Canadian Nuclear Fuel Waste Management Program with joint funding by AECL and Ontario Hydro under the auspices of the CANDU owners group. The author wishes to thank the Natural Sciences and Engineering Research Council of Canada and the Natural Environmental Research Council (UK) for additional funding. The author is also grateful to the staff of the Underground Research Laboratory and especially D. Martin for all his co-operation and insight. I also thank my former research staff and students at the Engineering Seismology Laboratory, Queen's University, Kingston, Canada for their contributions to different aspects of this work.

References

CARLSON, S. R., JANSEN, D. P. & YOUNG, R. P. 1993. *Thermally induced fracturing of Lac du Bonnet granite.* Report to Atomic Energy of Canada Ltd. RP020AECL, Queen's University, Canada.

DINES, K. & LYTLE, J. 1979. Computerized geophysical tomography. *Proc. IEEE.*, **67**, 1065–1073.

FALLS, S., YOUNG R. P., CARLSON S. & CHOW, T. 1992. Ultrasonic tomography and acoustic emission in hydraulically fractured Lac du Bonnet grey granite. *J. Geophys. Res.*, **97**(B5), 6867–84.

FEIGNIER, B. & YOUNG R. P. 1992. Moment tensor inversion of induced microseismic events: evidence of non-shear failures in the $-4 < M < -2$ moment magnitude range. *Geoph. Res. Lett.*, **19**(14) 1503–1506.

FEUSTEL, A. J. & YOUNG, R. P. 1993. *Attenuation anmalysis at the AECL Underground Research Laboratory using the spectral ratio method.* Report to Atomic Energy of Canada Ltd., RP022AECL, Queen's University, Canada.

—— & YOUNG, R. P. 1994. Q estimates from spectral ratios and multiple lapse time window analysis: results from an underground research laboratory in granite. *Geoph. Res. Lett.*, **21**(14), 1503–1506.

GIBOWICZ, S. J., YOUNG R. P., TALEBI, S. & RAWLENCE, D. 1991. Source parameters of seismic events at the Underground Research Laboratory in Manitoba, Canada: scaling relations for events with moment magnitude smaller than −2. *Bull. Seis. Soc. Am.*, **81**(4), 1157–1182.

JANSEN, D. P., CARLSON, S. R., YOUNG, R. P. & HUTCHINS, D. A. 1993. Ultrasonic imaging and acoustic emission monitoring of thermally induced microcracks in Lac du Bonnet Granite. *J. Geophys. Res.*, **98**(B12), 22 231–22 243.

MARTIN, C. D. 1989. Failure observations and in situ stress domains at the Underground Research Laboratory. *In*: MAURY, V. & FOURMAINTRAUX D. (eds), *Rock at Great Depth*, Balkema, Rotterdam, 719–726.

——1990. Characterizing in situ stress domains at the AECL Underground Research Laboratory. *Can. Geotech. J.*, **27**, 631–646.

READ, R. S. & MARTIN, C. D. 1991. *Mine-by experiment final design report.* Atomic Energy of Canada Report, AECL-10430, Whiteshell Laboratories.

RICHTER, D. & SIMMONS, G. 1974. Thermal expansion behaviour of igneous rocks. *Int. J. Rock Mech. Sci. Geomech. Abstr.*, **11**, 403–411.

RUMMERY, T. E. 1992. *Nuclear waste management.* Proceedings of the Sesquicentennial Symposium on Engineering and the Environment, Queen's University, 48–68.

STEWART, R. R. 1988. An algebraic reconstruction technique for weakly anisotropic velocity. *Geophysics*, **53**, 1613–1615.

TALEBI, S. & YOUNG, R. P. 1992. Microseismic monitoring in highly stressed granite: relationship between shaft-wall cracking and in situ Stress. *Int. J. Rock Mech. Sci. Geomech. Abstr.*, **29**(1), 25–34.

YOUNG, R. P. & TALEBI, S. 1988. *Microseismic monitoring and excavation damage assessment at AECL's Underground Research Laboratory.* Report to Atomic Energy of Canada Ltd., RP001AECL.

—— & MAXWELL, S. C. URBANCIC, T. I & FEIGNIER B. 1992. Mining-induced microseismicity and applications of imaging and source mechanism techniques. *Pure and Applied Geophysics*, **139**, (3/4), 697–720.

Prediction of groundwater flow around an underground waste repository

J. Zhao

School of Civil and Structural Engineering, Nanyang Technological University, Nanyang Avenue, Singapore 2263

Abstract. Leakage and transport of waste in a waste repository is generally through the circulation of groundwater flow. Waste repository design, therefore, requires the study of the hydraulic performance of the surrounding rock masses in the presence of groundwater. Rock mass permeability, conductivity and flow rate at various depths can be predicted by hydrogeological data of the joint frequency, joint aperture, stress–joint closure relationship and water-table, obtained at shallow subsurface exploration. Predictions using an improved hydraulic model show that the permeability of a rock mass decreases with increasing depth; the groundwater flow into and around the repository, however, increases with repository depth until it reaches a maximum rate. The groundwater flow stabilizes or decreases after the peak rate due to joint closure of the rock mass surrounding the repository under high *in situ* stress.

Underground repositories offer a possible site for the disposal of low- and high-level radioactive wastes and other hazardous wastes. The objective of underground disposal is to isolate waste from humans and their environment for such a period of time that any possible subsequent release of radionuclides and toxicant from the repository will not result in undue exposure.

Leakage and transport of waste in a waste repository is generally through the circulation of groundwater flow. Its design, therefore, requires a study of the hydraulic performance of the surrounding rock masses in the presence of groundwater. For those rocks normally considered for an underground repository, e.g. crystalline rocks, the hydrogeological properties are controlled by fluid flow within a highly impermeable but fractured medium since flow is concentrated in the joints.

Permeability of a single joint under normal stress

It has been shown that Darcy's permeability law can be applied to a rough surface joint in a rock mass using the smooth parallel plate theory (Witherspoon *et al.* 1980; Elliott *et al.* 1985; Zhao & Brown 1992). The permeability of a single joint, k_f (unit L^2), which is a property of geometry of the flow channel can be defined as

$$k_f = e^2/12 \tag{1}$$

where e is the equivalent hydraulic aperture of the joint.

The conductivity of a joint, K_f (unit LT^{-1}) can be related to the permeability by the equation,

$$K_f = k_f g/\upsilon \tag{2}$$

where g is the acceleration due to gravity, and υ is the kinematic viscosity of the fluid.

Laboratory hydraulic tests have been performed on single joints by a number of researchers (e.g. Louis 1967; Witherspoon *et al.* 1980, Gale 1982; Elliott *et al.* 1985; Zhao & Brown 1992). The relationship between joint permeability and the effective normal stress can be represented by a typical curve for a natural joint in the Cornish granite in Fig. 1. The joint permeability shows an asymptotic reduction with increasing effective normal stress. As the effective normal stress increases, the joint approaches a state of 'complete closure', with the flow rate approaching zero. The joint permeability–effective normal stress curve can be fitted by a simple logarithmic model (Walsh 1981; Zhao & Brown 1992),

$$k_f/k_0 = [1 - A\ln(\sigma'_n/\sigma'_0)]^2 \tag{3}$$

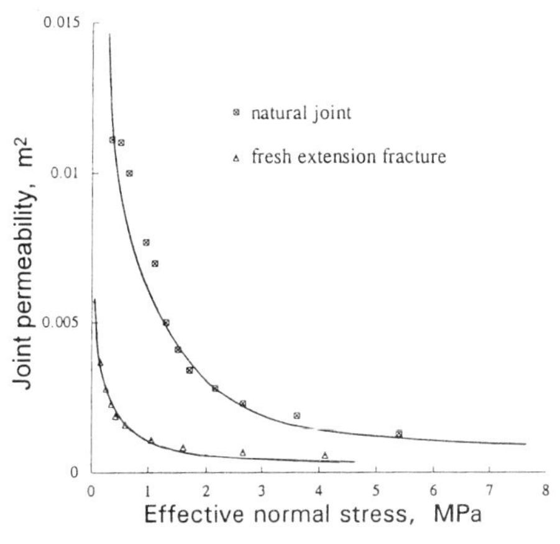

Fig. 1. Change of joint permeability with effective normal stress.

From BENTLEY, S. P. (ed.) *Engineering Geology of Waste Disposal*,
Geological Society Engineering Geology Special Publication No. 11, pp 231–236

where k_f is the joint permeability at effective normal stress σ'_n, k_0 is the joint permeability at a reference effective normal stress σ'_0, and A is a parameter having a value of between 0.14 and 0.22. Alternatively,

$$K_f/K_0 = [1 - A\ln(\sigma'_n/\sigma'_0)]^2 \quad (4)$$

or

$$e_f/e_0 = [1 - A\ln(\sigma'_n/\sigma'_0)] \quad (5)$$

where K refers to the hydraulic conductivity of the joint, and e refers to the equivalent hydraulic aperture of the joint.

Permeability of rock masses

For crystalline rocks, the rock mass permeability is controlled by the discontinuities since flow is concentrated in the joints. The rock mass permeability is a factor of (a) geometry of the joint system, and (b) hydromechanical characteristics of single joints. The geometry of the joint system (e.g. joint frequency, distribution and connectivity) can often be determined through detailed site investigation. Various model studies have been carried out to investigate the influence of joint system geometry on the rock mass permeability (Samaniego & Priest 1984; Kikuchi *et al.* 1991).

Goodman *et al.* (1965) has proposed a groundwater flow equation to predict water inflows into an underground excavation:

$$Q = \frac{2\pi KH}{\ln(2D/r)} \quad (6)$$

where Q is flow per unit length of the excavation, H is the hydraulic head (depth below groundwater table), K is the hydraulic conductivity of jointed rock mass, D is the average depth of the excavation below ground level, and r is the equivalent radius of the excavation cross-section.

It has been assumed that the hydraulic conductivity is a constant for excavation at various depths. However, experimental results show that the hydraulic conductivity of individual joints decreases with increasing effective normal stress. Furthermore, the joint intensity is also likely to decrease with increasing depth under normal circumstances. The coupled effects will reduce the rock mass conductivity at depth.

The above conventional formula can be improved by taking into account the variation of the rock mass conductivity with the depth. As illustrated in Fig. 2, groundwater flow into the excavation can be predicted by the geohydraulic data obtained from a conventional site investigation programme. The improved equation is

$$Q = \frac{2\pi K_i H}{\ln(2D/r)}\left(1 - A\ln D_i\right)^2 \quad (7)$$

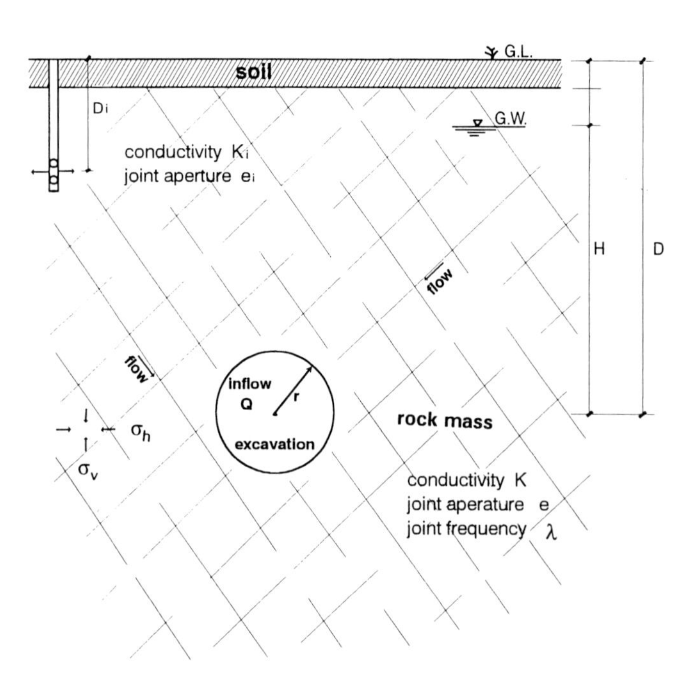

Fig. 2. Groundwater flow into and around an excavation in jointed rock mass (not to scale).

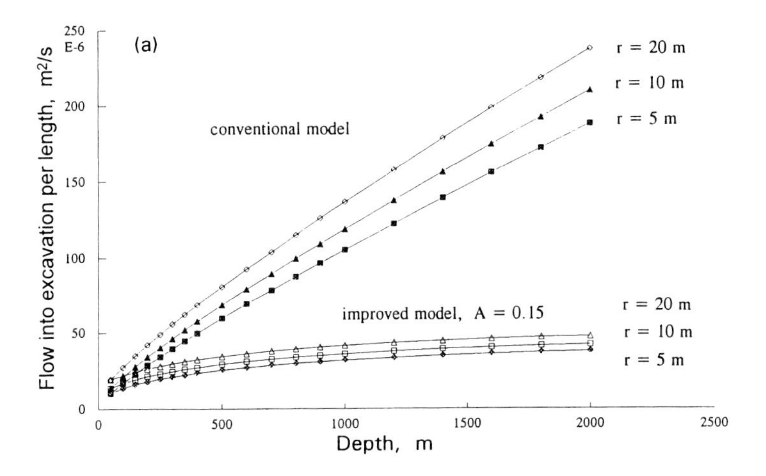

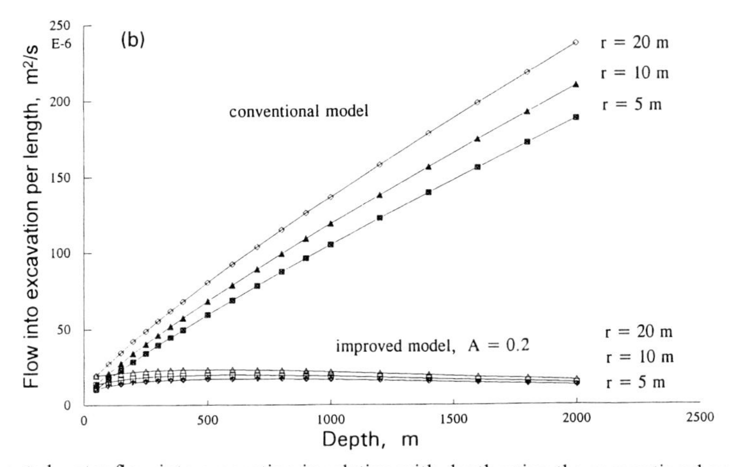

Fig. 3 . Computed water flow into excavation in relation with depth using the conventional and the improved models for different excavation sizes.

where K_i is the hydraulic conductivity measured from a shallow subsurface investigation at depth D_i.

The formula can be further modified to predict the water flow in a rock mass with basic geological data by replacing K with

$$K_i = gB\lambda e_i^2/12\upsilon \tag{8}$$

where λ is the joint frequency (fracture index) of the rock mass, and B is the ratio of conductive joints to the total joint numbers, since only a portion of joints are interconnected and conductive. Therefore,

$$Q = \frac{\pi g B\lambda e_i^2 H}{6\upsilon \ln(2D/r)}\left(1 - A\ln\frac{D}{D_i}\right)^2. \tag{9}$$

B usually has a value of between 0.1 and 0.3 for jointed crystalline rock mass and can be estimated through inspection. The joint hydraulic aperture, e, can be estimated from core logging or outcrop inspection, and usually varies from 10 to 100 μm.

Flow analyses are carried out using the conventional and the improved equations and the results are presented in Fig. 3. In the analyses, the hydraulic conductivity of the rock mass is assumed to be $1.0 \times 10^{-7}\,\mathrm{m\,s^{-1}}$ as obtained from field investigations of Swedish granite (Carlsson & Olsson 1977) and Singapore granite (Zhao 1993) rock masses at depths of 50 m, and groundwater level is assumed at the ground level. Groundwater flow into the excavation per unit length in relation to depth is calculated with various diameters and with different values of parameter A.

Comparison to field results

Carlsson & Olsson (1977) studied the hydraulic conductivity of Swedish crystalline rock masses with its relation to depth. Tests were conducted in 39 drill-holes at the Forsmark–Osthammar area in Sweden. They found that the hydraulic conductivity of rock mass decreases with increasing depth, as shown in Fig. 4.

Similar results were obtained by Bianchi & Snow (1968) who investigated the permeability of crystalline rock mass and interpreted the variation of joint aperture with depth. As shown in Fig. 5, the mechanical aperture of the joints in rock masses decreases with increasing depth. It should be noted that the mechanical aperture differs from the equivalent hydraulic aperture of a joint. However, at the stress range up to 8.0 MPa, the change of hydraulic aperture is proportional to the change of mechanical aperture (Zhao & Brown 1992).

By comparing both curves in Figs 4 and 5 with the logarithmic model presented in Equations (4) and (5), it is noted that the rate of reduction of hydraulic conductivity of the rock mass with depth (stress) is greater than that predicted by considering a single joint alone. This suggests a possible reduction of joint intensity with depth, and hence a greater reduction of rock mass permeability.

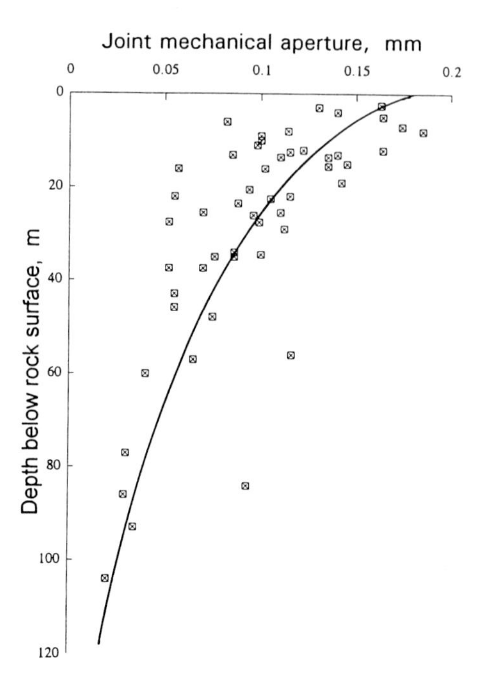

Fig. 5. Change of rock joint mechanical aperture with depth (after Bianchi & Snow 1968).

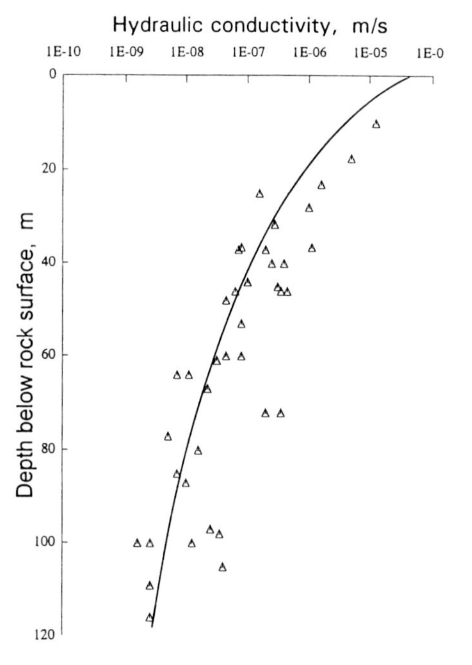

Fig. 4. Change of rock mass hydraulic conductivity with depth (after Carlsson & Olsson 1977).

Discussion

The conventional hydraulic relationship proposed by Goodman *et al.* (1965) predicts that the groundwater flow into an excavation increases with the depth, as Q is proportional to $D/\ln D$, for a defined excavation dimension, as shown in Fig. 3. The conventional model does not, however, take into account that the hydraulic conductivity of the joints in the rock mass decreases with increasing effective normal stress (and hence depth), and reduction of joint intensity with increasing depth. Therefore, the groundwater flow is largely controlled by the hydraulic head (i.e. the groundwater table).

Laboratory investigations carried out by various researchers have proven that the joint hydraulic conductivity is a function of the effective normal stress. Similar results have been obtained from *in situ* hydraulic tests indicating that the hydraulic conductivity of the rock mass decreases with increasing depth. Among the various numerical studies that have been conducted, Wei & Hudson (1988) suggested an exponential relationship between the rock mass conductivity and the effective stress.

The effect of decreasing hydraulic conductivity counterweighs the effect of increasing hydraulic head and therefore reduces the changing rate of water flow with depth. As with the results for coupled effects, the groundwater flow is usually greatest at shallow subsurface excavations, and is limited at deep excavations such as repositories and mines. However, when excavation cuts through major geological flow channels such as faults and karstic pipes, special studies should be carried out.

Experience in civil and mining engineering projects indicates that in deep excavations such as storage repositories and mines, inflow quantity is often dry and is much less than that predicted by the conventional water flow model. Generally, the quantity of water inflow into an excavation tends to stabilize or decrease with increasing depth. Observations usually indicate that the joints at the deep excavation surfaces are larger spaced and tighter, and water seeps through only some of the joints.

The improved water flow model takes into account the variation of hydraulic conductivity with stress, and therefore provides a better representation of the groundwater flow conditions with depth. The improved model, however, does not take into account the change in geological features, namely joint intensity, with depth. As shown in Fig. 3, groundwater flow predicted by the improved model first increases for a few hundreds of metres and then stabilizes or decreases with greater depth. The trends are the same for excavations of different sizes but are greatly influenced by the parameter A. For A having a value of 0.15, the flow rate increases but is stabilized after several hundreds of metres; the flow rate will eventually reach a peak and decrease. On the other hand, when A is equal to 0.2, the flow rate reaches a peak at about 500 m and then decreases gradually with depth. If A takes a larger value, the peak will be reached sooner and the rate of decrease of flow rate will be greater, and the flow rate will approach to zero at a few thousands of metres.

The groundwater flow into an excavation does not vary greatly with the dimensions of the excavation cross-section. By increasing the equivalent diameter from 5 m to 20 m, the flow rate increases by a maximum of 20%. The rate of increase is affected by parameter A; when A is larger, the rate of increase is reduced.

It should be noted that the hydraulic permeability and conductivity of the joints in a rock mass are influenced by the rock temperatures. If the repository is used for high-level radioactive waste disposal, a further 30–40 °C local temperature increment of the surrounding rock mass is expected due to the natural decay of the radioactive waste (CEC 1983). The increase in temperature will reduce the permeability of the rock joints due to joint expansion and healing (Zhao 1992), but will also reduce the viscosity of the water. The coupled effect is likely to increase the conductivity of the rock mass, and hence the flow rate into and around the repository.

Conclusions

Leakage and transport of waste in a deep underground rock repository are generally through the circulation of groundwater flow into and around the waste repository. Waste repository design, therefore, requires a study of the hydraulic performance of the surrounding rock masses in the presence of groundwater. Groundwater flow into and around the waste repository at depth can be predicted by hydrogeological data, such as conductivity, joint distribution and geometry, obtained in near-surface *in situ* exploration. The prediction is based on a stress–rock joint closure logarithmic relation verified in laboratory hydromechanical experiments and in field hydrogeological investigations. The improved model predicts the groundwater flow into and around the repository. This increases with repository depth until it reaches a maximum rate. The groundwater flow stabilizes or decreases after the peak rate due to joint closure of the rock mass surrounding the repository under high overburden stress.

The predicted groundwater flow by the improved model provides a better knowledge of the flow condition for the design and planning of repository excavation. Under normal circumstances, when no major conducting fissures and pipes are present, the rate of groundwater inflow tends to stabilize or, quite often, decrease with the depth of the excavation. The prediction matches well with the actual field observations and also yields realistic values of groundwater flow in the rock masses.

References

BIANCHI, L. & SNOW, D. T. 1968. Permeability of crystalline rock interpreted from measured orientation and apertures of fractures. *Annals of Arid Zone*, **8**, 231–245.

CARLSSON, A. & OLSSON, T. 1977. Hydraulic properties of Swedish crystalline rocks—hydraulic conductivity and its relation to depth. *Bulletin of the Geological Institutions of the University of Uppsala*, **7**, 71–84.

COMMISSION OF THE EUROPEAN COMMUNITIES (CEC) 1983. *Admissible thermal loading in geological formations—consequences for waste management methods, Vol. 2, crystalline rocks*. CEC 8179, Luxembourg.

ELLIOTT, G. M., BROWN, E. T., BOODT, P. I. & HUDSON, J. A. 1985. Hydromechanical behaviour of joints in the Carnmenellis granite, SW England. *Proceedings of International Symposium on Fundamentals of Rock Joints*, Bjorkliden. Centek Publishers, Lulea, 249–258.

GALE, J. E. 1982. The effects of fracture type (induced versus natural) on the stress-fracture closure-fracture permeability relationships. *Proceedings of 23rd US Symposium on Rock Mechanics*, Berkeley, California, 290–298.

GOODMAN, R. E., MOYE, D., SCHALKWYK, A. & JAVANDEL, L. 1965. Groudwater inflows during tunnel driving. *Engineering Geology*, **2**, 39–56.

KIKUCHI, K., MITO, Y., HONDA, M., MIMURO, T. & YOSHIDA, Y. 1991. In-situ experimental studies on groundwater flow analysis for jointed rock masses. *Proceedings of 7th International Congress on Rock Mechanics*, Aachen, **1**, 393–396.

LOUIS, C. 1967. *A study of groundwater flow in jointed rock and its influence on the stability of rock masses*. Doktor-Ingenieur dissertation, Universitat (TH) Karlsruhe, (in German). English translation in Imperial College Rock Mechanics Research Report, No. 10, September 1969.

SAMANIEGO, J. A. & PRIEST, S. D. 1984. The prediction of water flows through discontinuity networks into underground excavations. *Proceedings of ISRM Symposium on Design and Performance of Underground Excavations*, Cambridge. British Geotechnical Society, London, 157–164.

WALSH, J. B. 1981. Effect of pore pressure on fracture permeability. *International Journal of Rock Mechanics and Mining Sciences and Geomechanical Abstracts*, **18**, 429–435.

WEI, Z. Q. & HUDSON, J. A. 1988. Permeability of jointed rock masses. *Proceedings of International Symposium on Rock Mechanics and Power Plants*, Madrid. Balkema, Rotterdam, 613–625.

WITHERSPOON, P. A., WANG, J. S. Y., IWAI, K. & GALE, J. E. 1980. Validity of the cubic lafor w fluid flow in a deformable rock fracture. *Water Resources Research*, **16**, 1016–1024.

ZHAO, J. 1992. Coupled hydro-thermal cracking in granite fractures and application to radioactive waste repository. *Geotechnical Engineering, Journal of Southeast Asian Geotechnical Society*, **23**, 61–75.

——1993. Hydraulic properties of the Bukit Timah granite and potential inflow during cavern construction. *Proceedings of Seminar on Utilization of Underground Space in Singapore*, Singapore, 93–100.

——& BROWN, E. T. 1992. Hydro-thermo-mechanical properties of joints in the Carnmenellis granite. *Quarterly Journal of Engineering Geology*, **25**, 279–290.

Storage of hazardous waste at shallow depths

R. Christiansson[1] and R. Jernlås[2]

[1] VBBVIAK, Box 1902, 791 19 Falun, Sweden
[2] Vattenfall Hydropower AB, Box 800, 771 28 Ludvika, Sweden

Abstract. The possibilities of depositing environmentally hazardous waste in crystalline rock are evaluated for the special case where the waste is placed above the groundwater table. Layouts are proposed together with the siting process and costs. The environmental aspects are discussed for three types of waste. It is concluded that storage of hazardous waste could be done with good long time stability at many locations in Sweden. However, depending on the amount of dilution, some wastes might need improvements.

In the spring of 1990, SAKAB commissioned a study of the possibility of depositing environmentally hazardous waste in rock. SAKAB is a company that specializes in dealing with environmentally hazardous waste from local authorities and industrial enterprises in Sweden. The waste in this case was batteries containing mercury, ash produced by burning coal (pressurised fluidized bed combustion product, PFBC), flue gas products (flue gas cleaning by-product, FGCB) from SAKAB's own plant, and metal hydroxide sludge (MeOH).

SAKAB's requirements were that the waste should be placed above the groundwater table and that the functional service life of the repository should extend to the next ice age.

Waste and volumes

Types of waste

The metal hydroxide sludge (MeOH) is the remains from a surface treatment industry. About 10 000 tonnes are produced annually in Sweden. In addition, there was at the time of the study (1990) a temporary store containing 90 000 tonnes. SAKAB's own flue-gas cleaning product (FGCB) is produced at a rate of about 8000 tonnes a year and there was a temporary store containing 42 000 tonnes. No figures were available on the annual rate of dealing with alkaline batteries but 1000 tonnes were contained in a temporary store. Remains from coal combustion (PFBC) were also stated as a possible material for future deposition. These remains may consist of waste from gasification processes, ash and desulphurization products.

The quantity was said to be completely dependent on the scope of any new construction of combined heat and power stations.

Physical properties

The compaction properties are important for storage. Table 1 shows approximate bulk densities for materials compacted at optimum moisture content.

The grain size distribution of the cyclone ash may be described as a sandy silt, and the bed ash as a coarse sand. Only information on the consistency of the metal hydroxide is available: it was paste-like in the moist state.

Chemical properties

Table 2 provides a summary of the chemical contents of the various products. The contents of different metals in the metal hydroxide sludge are mean values from analyses from a repository newly completed by SAKAB. The contents of PFBC ash are given as mean values for cyclone ash and bed material from a plant in Malmö (Nilsson 1987). The value for the flue gas product from SAKAB is the mean of 20 or so samples taken during 1990–1991.

Table 1. *Bulk density on compaction at optimum moisture content*

Material	*Bulk density* ($kg\,m^3$)	*Moisture content* ($kg\,kg^{-1}$)
Flue gas products	1070	0.40–0.45
Metal hydroxide	1070	0.30–0.40
PFBC product, bed	1500	0.15–0.20*
PFBC product, cyclone	1600	0.15–0.20*
PFBC product, bed + cyclone	1850	0.15–0.20*

* Hartlén *et al.* (1989).

From Bentley, S. P. (ed.) *Engineering Geology of Waste Disposal*,
Geological Society Engineering Geology Special Publication No. 11, pp 237–244

Table 2. *Total contents (ppm)*

	FGCB	*MeOH*	*PFBC*
As		140	112
Cd	40	90	0.93
Co		760	
Cr	214		128
Hg		26 500	0.86
Cu	1 485	20 500	185
Mo		1 300	
Ni	342	33 100	200
Pb	3 903	2 800	210
Zn	9 634	9 700	

Layout

Main components

The main parts of the permanently drained repository for hazardous waste include (cf. Fig. 1).

- an access tunnel from the base of the hill; the tunnel is excavated at an upward inclination some 5‰ to allow drainage;
- a drainage tunnel system under and around the repository;
- a drainage gallery of boreholes drilled from the drainage tunnel to ensure permanent lowering of the groundwater table below the lowest part of the caverns;
- an excavated volume, such as caverns or silos in rock for the waste;
- sealing of the fractures exposed at the rock surface and/or other activities to increase runoff and decrease percolation of the precipitation down to the repository area.

Excavation methods and repository layout

Sweden has a long tradition in underground design. One of the objectives of the study was to assess traditional methods of underground excavation. These include excavation by conventional drill and blast methods; and reinforcement of the rock using by bolts and, shotcrete, if required. The amount of rock reinforcement required is a function of:

- the rock mass quality;
- the dimensions of the opening;
- excavation damage;
- the designed life span;
- the safety aspect for personnel working underground.

Experiences from Swedish underground excavation and mining in crystalline rock show that:

- small tunnels in old mines are still stable after hundreds of years, without any reinforcement;
- larger openings, such as rock caverns with a cross-section measuring some 20 × 20 m, are stable for at least decades with reasonably small amounts of reinforcement;
- underground openings in rock of good quality could be safe for operating personnel with a minimum of reinforcement, if the rock surfaces could be inspected and sealed approximately every third year.

There is a long tradition of building rock caverns and tunnels for various purposes in Sweden. The openings are normally built with a horseshoe-shaped roof/walls configuration.

With a rock mass of normally good quality, containing seldom more than three or four natural fractures per metre, an underground opening could be built with a rock cover towards the surface that is smaller than the span of the opening. The natural fractures divide the rock mass into irregular blocks, which often lock one another in position.

Rock caverns. Rock caverns could be built with various cross-sectional areas. The length of the cavern is only

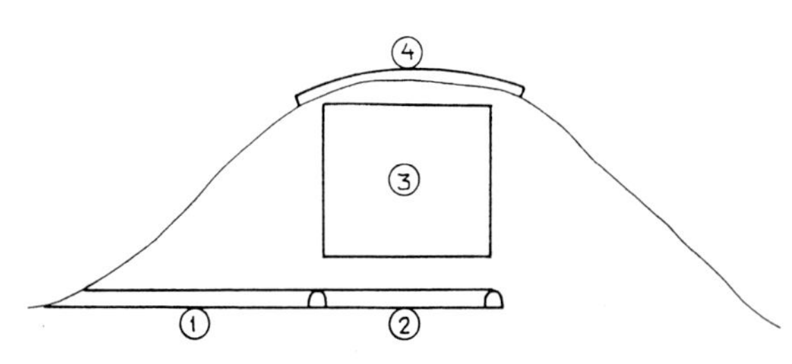

Fig. 1. The main components of a drained repository. 1, Access/drainage tunnel; 2, drainage tunnels and boreholes; 3, repository volume; 4, sealing or other activities to reduce percolation.

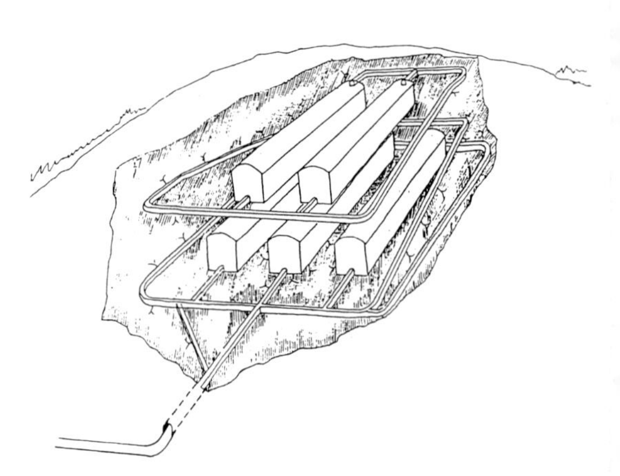

Fig. 2. Tentative layout of a rock cavern repository.

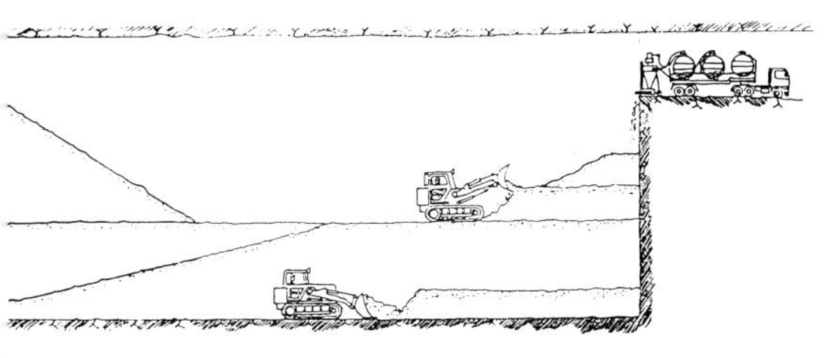

Fig. 3. Handling of waste in a rock cavern.

limited by variations in the geological conditions. It is, however, advantageous for drainage of the rock surrounding the repository to employ a reasonably square or circular repository area.

Rock caverns for large volumes could be built as parallel galleries. If the available height of the drained rock volume is sufficient, rock caverns could be built at two levels. Rock mechanics analysis indicates that for good long-time stability, the distance between caverns with a square cross-section should be at least close to the span of the cavern.

A possible layout for a rock cavern repository is shown in Fig. 2. Access is gained to the caverns through the drainage tunnel. The access tunnels to the top headings of the caverns, excavated as a spiral around the caverns, are a good base for drainage of the rock mass around the repository. Boreholes for drainage are required, however. If the lowest drainage tunnel is located at least 5 m below the lowest parts of the repository, a borehole gallery with boreholes at approximately 3 m centres is sufficient for drainage. The boreholes must be drilled at a small downward angle, so that the drainage system is water-filled. If oxidation occurs, precipitation of natural solutes in the groundwater could plug the boreholes and reduce drainage in the long term.

The waste could be filled into the caverns from environmentally controlled areas in the access tunnels to the top headings. Bulldozers could be used for handling and compacting the waste in the caverns (Fig. 3). It is estimated that after late consolidation, some years after final filling of the caverns, approximately 90% of the caverns will be filled with waste.

If an inspection programme in the future were to show problems with the concept, the waste could be retrieved after excavation through the concrete plugs that seal the cavern entrances.

Silos. Silos in bedrock could be built with various dimensions. The largest silo built in rock so far in Sweden is a part of the Final Repository for Low and Intermediate Reactor Waste in Forsmark, and it is 31 m in diameter and 69 m high. Under geological conditions like those normally found in Swedish crystalline rock, the diameter is more critical for reinforcement demands than the height.

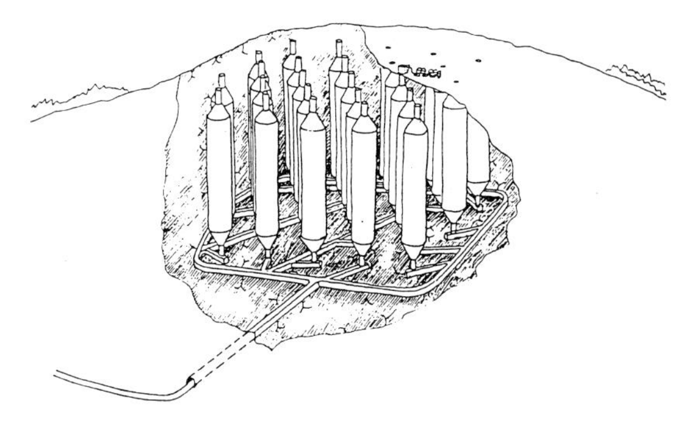

Fig. 4. Tentative layout of a silo repository.

The optimum layout for a number of silos is to excavate them in a hexagonal pattern. The stability of such a set of circular openings is controlled by the secondary stresses induced around the silos, and the strength of the rock where the stresses are induced. In a probable geological situation the risk of blocks being divided by natural fractures involves the greatest stability problem.

For a given ratio between the silo diameter and the smallest distance between the silos, the same repository volume could be excavated under a given area of a hill, whatever the diameters of the silos. There is therefore no benefit in building large silos. A preliminary silo layout is shown in Fig. 4. The diameter of the silos is 16 m and the smallest distance between the silos is 20 m. The high circular walls can be made stable with a very small amount of reinforcement. A possible and cost-effective excavation method is illustrated in Fig. 5.

The shaft under the silo is plugged before operation. Waste is filled from the surface down into the silos. Dust can be controlled by simple measures. The waste can be compacted using a falling weight. Settlement caused by consolidation can easily be monitored. The silos can be completely filled with waste before the upper shafts are plugged and sealed.

If an inspection programme in the future were to show problems with the concept, the waste could be retrieved after opening up the plugs in the shafts under the silos.

Hydraulic principles for a drained repository

The annual precipitation in central Sweden is 650–700 mm a^{-1}. The average percolation is about 150 mm a^{-1}. The percolation in the upper part of a hill of crystalline rock, covered with a thin layer (0.2 m) of glacial till, is probably smaller. Relatively simple activities, such as sealing larger fractures on the rock surface and smoothening the surface to increase runoff will strongly reduce the percolation into the top of the hill. The percolation can easily be reduced to 10 mm a^{-1}. It is possible to reduce percolation to 1–2 mm a^{-1} by systematically sealing the surface.

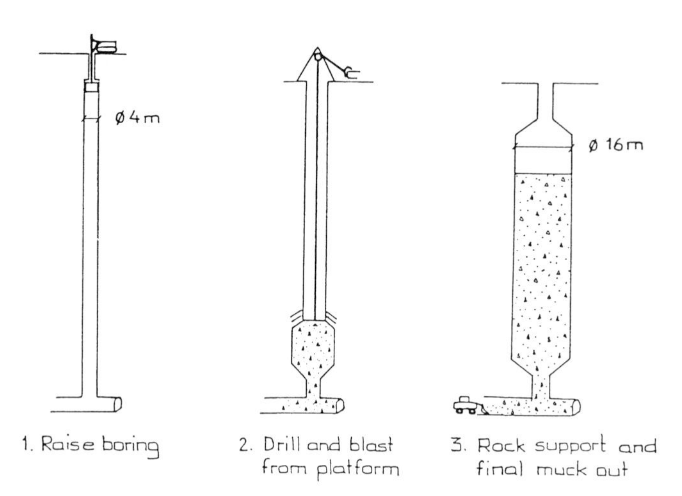

Fig. 5. Example of the excavation sequences for a silo.

The groundwater balance in a drained hill is illustrated in Fig. 6. For a hill of geometry similar to that shown in Fig. 6, the drainage system will drain about 15 000 $m^3 a^{-1}$ from the rock mass outside the repository.

The more the percolation is reduced and the lower the groundwater table under the repository, the smaller will be the volume of polluted water that will seep from the repository to be mixed with freshly drained water.

The water flow in and around the waste is unsaturated. In the safety analysis for the concept it is assumed that all water that percolates passes through the waste.

Monitoring and long-term stability

The groundwater draining from the hill and the more or less contaminated unsaturated flow that seeps out from the repository can be separated underground and can be controlled before mixing and discharge to a recipient.

Maintenance of the drainage/access tunnel will therefore be required for an (as yet) unknown time. The monitoring period must be defined on the basis of predicted seepage, experience of operation and post-closure time and legal aspects.

The stability of underground openings can be monitored by normal rock mechanics measures during operation. After the caverns of silos have been filled up, the long-term stability conditions for underground opening are established. Due to the geotechnical properties of the waste, minor rock displacements can be foreseen. This could at most extend the fractured zone by some metres from the surfaces of the caverns or silos. No rock mechanical problems of significance could be foreseen after closure, unless excavation is made for other purposes in the vicinity, or unless a glacier erodes the landscape.

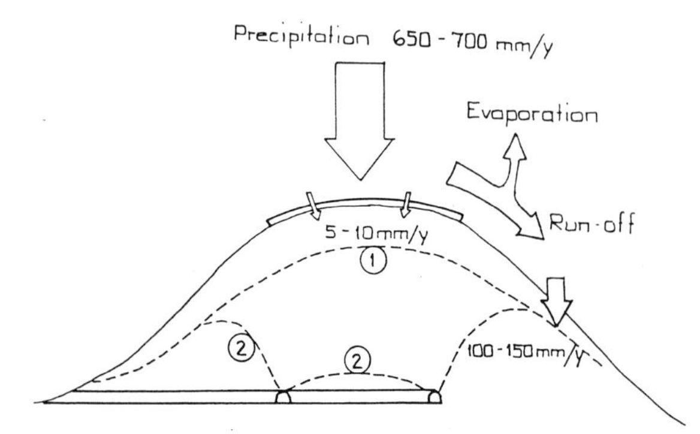

Fig. 6. Groundwater balance for a drained repository. 1, Original groundwater table; 2, drained groundwater table.

Siting process

Geological and topographical conditions in central Sweden

The bedrock in Sweden is mainly of crystalline origin. Except from the Fennoscandian mountain ridge and some smaller areas of sedimentary rock, the majority of the bedrock is 1400 Ma to 1800 Ma old.

The rock types in central Sweden are mainly:

- old metasediments;
- intruded granites of various ages;
- intruded greenstones;
- relatively young dykes of various rock; dykes of diabase intersect the region in a north-northwest direction;

- mineralizations are common in the metasediments, especially close to younger intrusions.
- migmatization occurs in the metasediments, but is rare in intrusions.

The intruded rocks are normally homogeneous, like in the old Stripa mine, where a granite was found from a depth of 360–400 m. This granite has been studied within nuclear waste management programmes for about 15 years.

Many intrusions that today reach the surface contribute to the broken landscape. Within the forested regions hills with a relative height of about 100 m are commonly found. The hills increase in height towards the north of the Swedish mainland.

Site studies

Seven topographical maps were randomly selected (Fig. 7). Each map covers 25 × 25 km. The following topographical and geological conditions were used as the criteria for selecting which sites would be studied:

- the uppermost 5 m of the hill must cover an area of at least 40 000 m^2;
- the height from the top to the foot of the hill must be at least 75 m;
- the hillside must be steep, to require as short an access/drainage tunnel as possible;
- a nearby road must be available close to the foot of a steep hillside.

On the seven map sheets, 31 hills were identified. The average dimensions are shown in Fig. 8. The geology of the hills was studied. The majority of the hill were intrusions, mainly of granite. Mineralizations were known at the foot of a few hills. Tectonic information was compiled, mainly from three mines in the area (cf. Fig. 7). Map studies and field inspections of three of the hills indicated that only minor fractured zones could

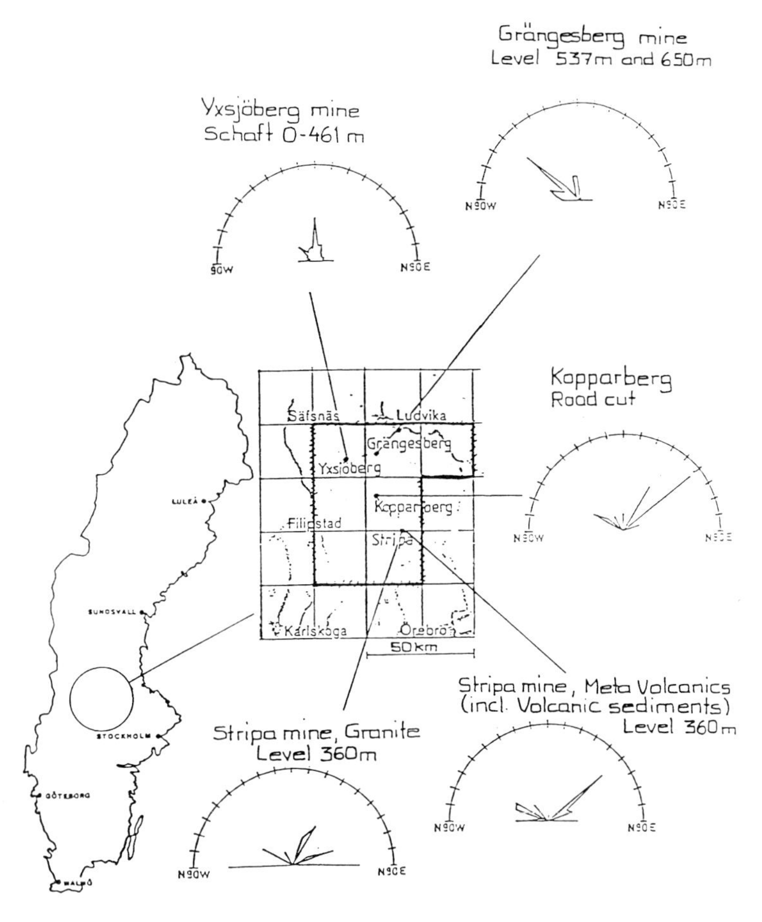

Fig. 7. Study area and orientation of fractures at some locations in the area.

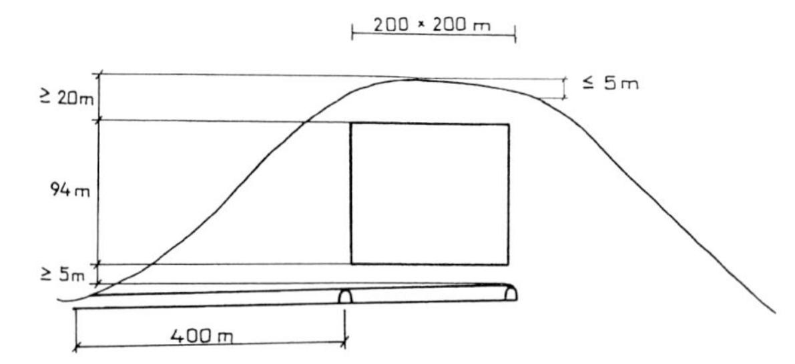

Fig. 8. Summary of average dimensions of 31 hills within the study area.

intersect the hills. The larger fracture zones are located in valleys between the hills.

It was concluded that suitable hills for siting a drained repository for hazardous waste could easily be found in central Sweden. This finding is probably valid for other hilly areas of Sweden as well.

Other factors that control the siting of a repository

It was concluded that geological conditions could be feasible at many locations. Other factors of greater importance include:

- distance from waste producers to the repository;
- possibilities of transporting waste to the site;
- nearby recipient for drained water;
- location for dumping excavated rock;
- isolated location with respect to the change in scenery that sealing or activities increasing runoff on top of the hill could cause;
- environmental and legal aspects, for example, public land use for recreation;
- local opinion.

Some of these aspects were studied as well. It was concluded that although the legal process could be extensive, the concept is feasible. Local opinion could be a big issue as well. It is therefore beneficial that the site location is not primarily dependent on specific geological conditions.

Environmental aspects

Time aspects

The requirement on the functional service life time of the repository is that it must extend up to the next ice age. Beyond this time no requirements are made on its function. Unfortunately the possibility of verifying the properties of building materials over such a long time are limited. Even natural processes such as clogging of drains as a result of rust deposits or the formation of secondary minerals is both possible and probable during such long time spans. The layout and materials selected must, as far as possible, be responsible for ensuring that the function of the repository is not affected detrimentally by such processes. It must be possible to check the function of the repository from time to time and, if necessary, restore it.

Contents and effects on recipients

It has been estimated that the sealed top surface covering about 4.5 ha can permit about 50 m^3 of water a year to penetrate. This corresponds to somewhat over 1 mm a^{-1}. Parts of this will pass through the waste and be able to leach out various contaminants. This leachate will then be mixed with the unpolluted groundwater that enters the drainage system. The drainage flow has been estimated at some 15 800 $m^3 a^{-1}$, with an assumed permeability of rock of 1.0×10^{-8} m s^{-1} (average for superficial rock in Sweden).

In a conservative calculation it may be assumed that all water that passes through the sealed top surface will also pass through the waste. Table 3 presents estimates of the contents of materials in the water leaving the repository, that is, the mixture of drain water and leachate. For this estimate, data on the waste types obtained from batch and column tests were used.

When this discharging water reaches the first small river, the concentrations will be further reduced by dilution. The concentrations after dilution in the river may be calculated on the basis that such a river drains a small runoff area covering 5 km^2, with a runoff rate of 12 l s^{-1} km^{-2}. These concentrations are tabulated in Table 3 together with the assumed background contents of respective substances.

Table 3. *Contents in discharge water (ppb)*

Substance	*Concentration*		
	FGCB	*MeOH*	*PFBC*
Al	14	6.7	—
As	2.2	1.4	1.1
Cd	<0.05	<0.05	0.03
Cr	0.03	0.005	0.14
Hg	0.14	1.3	0.007
Cu	2.9	7.6	1.1
Mn	7.1	64	—
Ni	5.5	9.7	5.1
Pb	4 900	7.3	0.09
Zn	22	21	—
Cl^-	84 000	15 000	5 500
SO_4^{2-}	17 000	32 000	18 000

Table 4. *Concentration after dilution in the first recipient (ppb)*

Substance	*Surface*	*Waste*		
		FGCB	*MeOH*	*PFBC*
Al	70	65.67	65.11	—
As	0.3	0.45	0.39	0.36
Cd	0.03	0.03	0.03	0.03
Cr	1	1.17	1.37	0.97
Hg	0.003	0.01	0.10	0.00
Cu	0.7	0.87	1.23	0.73
Mn	10	9.78	14.17	—
Ni	1	1.35	1.67	1.32
Pb	0.4	379.10	0.93	0.38
Zn	0.3	4.47	4.93	—
Cl^-	4 000	68 616.27	4 850.21	4 115.94
SO_4^{2-}	20 000	19 768.12	20 927.12	19 845.42

The contamination factor (*Kf*) concept (SNV 1990) may be used as the starting point for assessing the effect of metals in the surface water. This factor is defined as follows:

Kf = present content/original content

The original content is considered to be the local background. In the event of there being no known value, standard values are used. The National Environment Protection Board (SNV 199) has proposed standard values for the metals arsenic, cadmium, chromium, copper, nickel, lead and zinc. These values have been used in Table 4 to describe the probable composition of the surface water. Values considered reasonable have been used for other substances.

On the basis of the *Kf* value, the degree of influence can then be graded on a four-value scale as shown in Table 5. In this case, the concentrations assumed for river water may be regarded as local background values. The influence on the water leaving the repository and river water after mixture with the discharging water are shown in Table 6.

If a strong influence (2) is considered acceptable, the present influence on the metal contents of the river are unacceptable as regards mercury in a repository containing metal hydroxide sludge, and as regards the lead

Table 5. *Degree of influence on water (SNV 1990)*

Kf	*Degree of influence*	
<1.5	0	None or insignificant
1.5–3	1	Clear
3–10	2	Strong
>10	3	Very strong

Table 6. *Degree of influence*

Substance	Discharge			River water		
	FGCB	MeOH	PFBC	FGCB	MeOH	PFBC
Al	0	0	—	0	0	—
As	2	2	2	0	0	0
Cd	1	1	0	0	0	0
Cr	2	2	0	0	0	0
Hg	3	3	1	2	3	0
Cu	2	3	1	0	1	0
Mn	0	2	—	0	0	—
Ni	2	2	2	0	1	0
Pb	3	3	0	3	1	0
Zn	2	2	—	0	0	0

content in a repository containing the flue gas product. There are several means of achieving acceptable contents. One method is to improve the quality of the waste. Another is to deposit the various types of waste at the same installation and use the fact that the different types of waste have different contamination profiles. A third alternative is to increase the dilution. This can be done both in direct connection with the deposit by design of the drainage or by localization of the deposit so that discharge water is diluted in a larger watercourse than that considered in this example. Of these alternatives, the first and third should give the best effect.

Costs

The following assumptions were made.

- The excavation costs assume all excavation continuously under one contract, including 400 000 m^3 repository volume in silos or 420 000 m^3 repository volume in rock caverns for 400 000 m^3 waste.
- The financial costs assume stepwise excavation of either four silos of 17 000 m^3, or two rock caverns of 84 000 m^3 at a time. Average costs for construction are used. Expansion of the repository in steps could offer the possibility of disposing of different wastes in separate rooms.
- The excavation and operational times are estimated at 30 years.
- The study was based on 1990 prices. Costs have been upgraded to take into account inflation in Sweden to May 1993 (14%), and converted to UK£ (£1.00 = SEK 11.40).

This method gives the average cost for the 30-year life span. In reality, the initial costs will be higher due to, for example, investments in access roads and tunnels. The results are summarized in Table 7.

The cost per tonne of different wastes depends on the optimal density after compaction. The optimal density

Table 7. *Costs*

Costs of repository	*Silos*	*Caverns*
Excavation and sealing (£)	19.6×10^6	16.3×10^6
Excavation and sealing costs (£ m^{-3})	48.95	38.80
Financial costs 6% (£ m^{-3})	7.30	12.80
Operating costs (£ m^{-3})	4.70	11.80
Total (£ m^{-3})	60.95	63.40

also depends on the water content. Considering the water contents of some of the waste (cf. Table 1) and the practical moisure contents after dewatering, the disposal cost for some wastes in silos will range from £25 to £59 per tonne, excluding transportation costs.

Conclusions

It is concluded that:

- disposal of hazardous waste could be done in permanently drained underground openings;
- the geological, topographical and hydrological conditions are suitable at many locations in, for example, central Sweden;
- factors other than the geology will probably control the location of such a repository;
- the proposed layouts will permit simple monitoring during operating and post-closure phases;
- correction activities could easily be performed, for example, improving the sealing of the surface or expanding the drainage;
- the waste could easily be retrieved;
- the long-term safety assessment could be made within a reasonable time after closure;
- it is reasonable to assume, that the repository would be safe up to the next ice age, say some 5000 years;
- although rock cavern excavation costs are lower, the overall cost of the silo concept is lower.

References

HARTLÉN, J., ROGBECK, J., LINDAU, L. & NILSSON, C. 1989. *Kolförbränningens restprodukter*. Stiftelsen för värmeteknisk forskning, Rapport nr 3144.

NILSSON, C. 1987. *Restprodukter från förbränning i fludiserande bädd—egenskaper vid deponering och återanvändning*. Stiftelsen för värmeteknisk forskning, Rapport 276.

SNV 1990. Bedömningsgrunder för sjöar och vattendrag. Klassificering av vattenkemi samt metaller i sediment och organismer. *Naturvårdsverkets allmånna råd*, **90**, 4.

A geochemical data management system for radioactive waste repository feasibility investigations

C. P. Nathanail

Centre for Research into the Built Environment, The Nottingham Trent University, Nottingham NG1 4BU, UK

Abstract. The geochemistry and hydrogeochemistry of the ground is a prime influence in determining the suitability of a site for the underground storage of radioactive waste. Geochemical testing is carried out during and after drilling of investigation boreholes to determine the nature of the groundwater regime and the connectivity between the repository and the biosphere.

The aim of this paper is to describe a data management system for the entry, validation and reporting of geochemical testing using dBase IV version 1.5 running under MS-DOS 5.0 on an IBM-compatible 486DX personal computer.

Data entry is in four phases for sample register details, tracer sample analyses, hydrochemical sample inventory details and hydrochemical sample analyses. Each record is committed on entry and then validated by a senior chemist. A six-stage validation scale was developed to track the progress of the analysis for individual determinands.

Results are reported in full for each sample and as a series of summary sheets for rapid assessment. The summaries display the QA status of each analysis from 'not required' through 'pending' to 'fully documented procedures'. Output is to a laser printer and a series of Lotus 1-2-3 spreadsheet files for further analysis and interpretation. Reporting is both interim on a daily and fortnightly basis during testing and final upon completion of testing.

The conflicting requirements of security, quality assurance, ease of use and a rapidly evolving specification mean that communication between geochemists and the database programmer have to be effective and response times minimized.

The aim of this paper is to describe a database management system for the storage, validation and reporting of geochemical analyses.

Geochemical testing

The aim of geochemical testing of potential nuclear waste repositories is to determine the hydrogeological regime of the area. The age of groundwater and the rate and direction of groundwater flow are of particular interest (Ireland 1992). Boreholes are sunk to determine the geology of the site and to obtain samples for geochemical testing.

The hydrogeological interpretation is primarily based on ^{14}C isotope determinations. Much effort is put into ensuring the reliability of samples selected for ^{14}C testing. The drilling fluid is spiked with lithium of known concentration. The lithium concentration can then be used to measure the extent of groundwater–drilling-fluid mixing. Field measurements of temperature, pH, Eh and conductivity are made at the time the sample is taken. Labile determinands are analysed within 24 hours of the sample being taken. Labile determinands comprise pH, conductivity, total carbonate alkalinity (TCA), total alkalinity (TA), total inorganic carbon (TIC), total organic carbon (TOC), total and ferric iron, nitrate, nitrite, ammonia, phosphate and total dissolved solids (TDS). Stable determinands are analysed over a longer time-scale. Stable determinands are sodium, potassium, calcium, magnesium, lithium, manganese, strontium, barium, aluminium, silica, sulphate, chloride, bromide, iodide and fluoride. Tracer samples are analysed for only a small sub-set of all possible determinands. Hydrochemical samples are analysed for a wider range of determinands which is decided by the location from which the sample was taken. Some of the hydrochemical samples are analysed for ^{14}C.

System overview

The database system is designed to store, check and report geochemical analyses undertaken as part of the investigations of a potential underground radioactive waste repository (Fig. 1). It was developed using dBase IV version 1.5 running on a 486DX IBM-compatible personal computer under MS-DOS 5.0. The system files take up approximately 3 MB and the data for an individual borehole take up to 5 MB.

From BENTLEY, S. P. (ed.) *Engineering Geology of Waste Disposal*,
Geological Society Engineering Geology Special Publication No. 11, pp 245–248

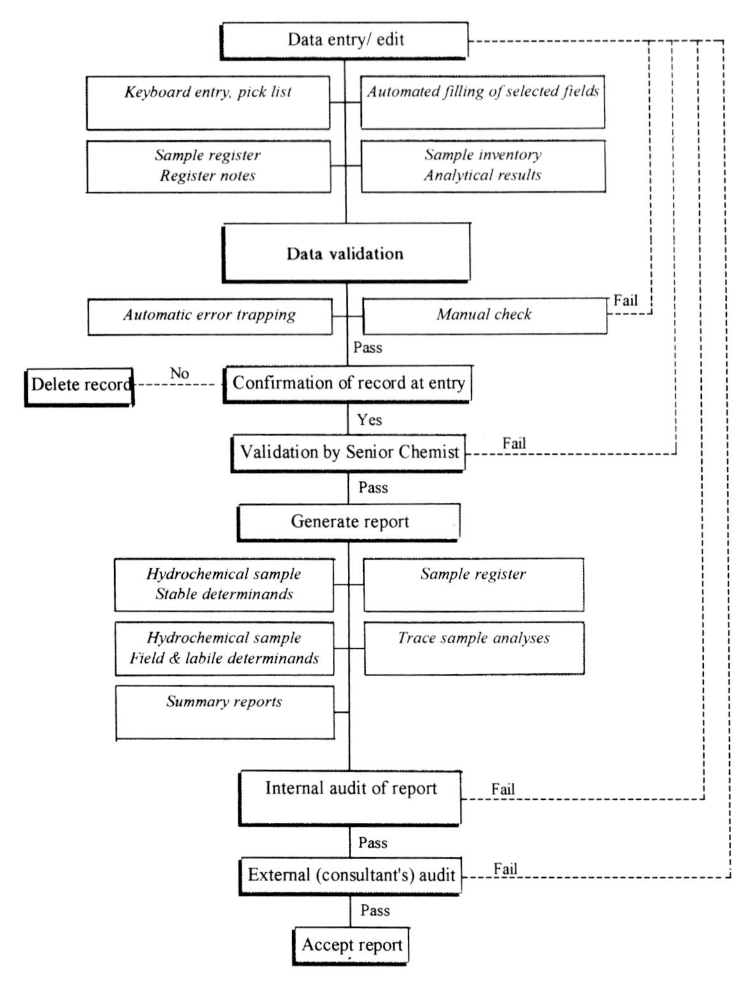

Fig. 1. Flow diagram of database operations.

Temporary files require up to a further 5 MB. The system is menu-driven and uses either a mouse or the cursor keys to move within menus or screens. The 'active' system was designed to be used to produce interim reports on a daily basis thereby replacing an earlier system developed solely for the production of final reports upon the completion of testing of an individual borehole.

Program, screen, report form and control database files are held in the *c:\active* sub-directory. Datafiles for each borehole are held in separate sub-directories. For example, borehole 3 in *c:\bh3*; post-completion testing for borehole 7 in *c:\bhpct7* and so on. On-off programs, needed to assist in setting up data structures, to transfer historical data from different file formats and to respond to urgent requests, are stored in a separate sub-directory.

File-naming conventions

A file-naming convention was developed to assist in programming and maintenance. The function of most of the program, screen and report files can be gleaned from their name. The convention splits the eight-character DOS prefix of the name into four bigrams. Most files begin with a bigram for the project although some program (*.prg) files begin with 'QB' to indicate that they were derived from dBase query-by-example

(*.QBE) files. The second bigram is usually one of the following:

AD	adding data
CH	checking/authorizing data
RE	reporting
RS	reporting summaries
BR	browse
MA	maintenance

The third and fourth bigrams are sometimes combined to give:

REGI	sample register
TRAC	tracer sample
FULL	full sample (a synonym for hydrochemical sample)
FIEL	field determinands
LABI	labile determinands
STAB	stable determinands

The chemical symbol (e.g. Li for lithium) or accepted acronyms (e.g. TOC for total organic carbon) of individual determinands were used to name relevant fields and files.

So, for example, CPADTRAC.PRG allows the ADdition of TRACer sample analyses; CPCHREGI allows the CHecking of sample REGIster information, and the LI_CONC field was used to store the CONCentration of lithium.

Data files

Sample register information is stored in one table while analyses for tracer and hydrochemical analyses are stored in two separate tables. The reason for this is that tracer samples are reported in groups of up to six samples and are only tested for a small number of determinands. Hydrochemical samples are reported individually and can be analysed for the complete suite of determinands. Separate tables are used to store borehole details, sample type descriptions and allowable values. The tables are related using key fields. The borehole number, sample number and sample type were used as key fields. During system testing it was found that index files (*.MDX) used to filter data and speed up searching operations were easily corrupted and needed to be rebuilt at frequent intervals.

Security

Access to the database is controlled by a user name and password. Once a user has logged in, only those menu options available to their security level are displayed. Access is allowed at five levels:

1. data entry only
2. data entry and validation
3. deletion of records and creation of new borehole directories
4. modification of form layout
5. supervisor/programmer

Data entry

Data are entered in the following sequence:

1. sample register
2. register notes
3. sample inventory (for hydrochemical samples only)
4. analytical results

Samples are entered on the sample register table (CPREGIST.DBF) before analytical results can be entered in the table for hydrochemical (CPFULL.DBF) or tracer samples (CPTRACER.DBF). Records have to be committed at the time of entry by setting the CONFIRM flag to 'Y'. Analytical results can only be entered once a sample has been entered onto the register. If an attempt is made to enter results for a sample not on the register the user is given the option of going to the register entry screen or to change the sample number being worked with.

'Pick lists' are used to restrict the entry of borehole number, sample type and sample location codes. Pick lists are stored in a DBF file and are activated when the user attempts to enter information into a field. Only the options within the pick list may be entered.

The fields which identify the sample number, type and location have to be entered before the user can continue. The sample number is unique and identifies which borehole the sample was taken from and whether the sample was a tracer or hydrochemical type. The borehole number and sample number are automatically cross-checked for consistency.

Data validation

The aim of validation is to ensure that the analytical results have been correctly transcribed from the analysis sheet onto the computer. Validation is a five-stage process involving:

1. error trapping by database software;
2. confirmation of a record at data entry;
3. validation of a record by a senior chemist;
4. internal audit of report;
5. external audit of report.

Error trapping involves the use of pick lists of allowable responses and ranges of values within which the analysis must lie.

Each record has to be committed at data entry stage. If a record is not committed, it is automatically deleted when exiting the system. For each determinand, six possible validation stages are available:

A analysis not requested
B analysis requested but will not be possible
C result of analysis pending
D analysis entered but not validated
E analysis validated
F analysis and supporting documentation validated

The sample type usually determines which determinands are to be analysed and the status flags are automatically set to either C or A as appropriate. Analyses can only be entered for a determinand at status C and edited at status D. Records have to be validated from status D to E by a senior chemist, or higher. This is done in one go for field determinands and field and laboratory environment measurements and individually for each labile and stable determinand. Once a result has reached status E it should be *locked* and can then only be changed with the intervention of an off-site programmer.

An internal audit compares the database hard-copy output with the original manuscript analytical results sheets. An external audit checks a percentage of each batch of transmitted results. The detection of any errors should result in the batch being rejected.

Reporting

Digital reports comprise dBase (*.DBF) and Lotus 1-2-3 (*.WKS) format files of sample register and analytical results for use in interpretation. Routine hard-copy reports, output from the database to a Canon LBP III laser printer, comprise:

- sample register
- tracer sample analyses
- hydrochemical sample
- stable determinand reports (one for each determinand)
- summary reports

Additional report formats or selections of the data are produced by the on-site database programmer in response to requests from the geochemists.

System testing

The database system was tested over a period of several months using live data. The number of bugs and modifications to the specification was found to decrease rapidly as time progressed.

Conclusions

The database system was designed to allow the input, validation and reporting of geochemical analyses by site staff with minimal training and no assumed knowledge of computers or databases. The principles in the database design have broader application in other areas of data storage for waste disposal or contaminated groundwater studies.

The conflicting requirements of security, quality assurance, ease of use and a rapidly changing specification mean that communication between geochemists and the database programmer have to be effective and response times kept to a minimum.

Acknowledgements. The author acknowledges the contribution of former colleagues at Wimpey Environmental in the work leading up to this contribution. The opinions expressed in this paper and any errors in it however are the sole responsibility of the author.

Reference

IRELAND, T. J. 1992. Characterisation of the Sellafield area for the development of an underground radioactive waste repository. *In*: HUDSON, J. A. (ed.) *ISRM EUROCK '92: Rock characterisation*. British Geotechnical Society, London, 474–481.

SECTION 4

CONTAINMENT PROPERTIES OF NATURAL CLAYS

Evaluation of clays as linings to landfill

E. J. Murray,[1] D. W. Rix[1] & R. D. Humphrey[2]

[1] Murray Rix Geotechnical, 5A Regent Court, Hinckley, Leicestershire LE10 0AD, UK
[2] Lincs Laboratory, Lincolnshire County Council, St Georges Lane, Riseholme, Lincoln LN2 2LQ, UK

Abstract. The use of remoulded natural clays as low-permeability barriers or as part of multiple-layered lining systems to landfill cells is widespread. Geotechnical considerations in the evaluation of possible sources of clay and the design and construction of clay liners are addressed.

A study made of individual geological deposits within Lincolnshire is outlined and use of the term 'material suitability' to define the potential of a deposit to form a low-permeability barrier is discussed. A scheme for classifying deposits as unsuitable, marginal or suitable based primarily on plasticity data and variability is discussed and examples from the classification study are presented.

The term 'acceptability in earthworks' is also introduced and relates to the requirement to place acceptance limits on the proposed lining material to achieve the desired low-permeability barrier and to ensure the practicality of constructing the lining. The use of the moisture condition value (MCV) test in the assessment of acceptability and control of earthworks is discussed. The requirement of a maximum allowable permeability for a clay lining dictates the upper limit to the acceptable MCV range while the shear strength would usually dictate the lower limit.

Landfill comprises a mélange of waste materials, including a high percentage of decomposable wastes, which give rise to the production of various potentially troublesome gases along with a leachate containing both chemical and biological pollutants. It is common practice to construct cells (as illustrated in Fig. 1) to contain the possible pollutants. The linings to these cells generally consist of a compacted layer of low-permeability clay possibly in conjunction with a synthetic liner or liners. Guidance on containment methods can be obtained from the National Association of Waste Disposal Contractors (NAWDC) *Codes of practice for landfills* (1989), the Department of the Environment (DOE) Waste Management Paper No. 26 (1990) and the North-West Waste Disposal Officers (NWWDO) *Guidance on the use of landfill liners* (1986).

There is a recognized need to protect the environment surrounding landfill sites. To this end the proposed EU Landfill Directive, along with the Environmental Protection Act (1990), the Control of Pollution Act (1974), and water-protection legislation as detailed in the National Rivers Authority's (NRA) *Policy and practice for the protection of groundwater* (1992), emphasize the legislative and environmental obligation to prevent pollution of the ground and groundwater and obviate the dangers to future generations.

The overriding requirement in the evaluation of a clay for use as a lining material is that it should be capable of providing a low-permeability barrier following emplacement. In order to achieve this it is usual to specify the use of a clay with suitable 'material characteristics' (B.S. 5930: Anon. 1981) as defined by its plasticity, material variability and clay content. This paper outlines the use of a classification scheme to identify possible sources of suitable natural materials and is based on a study carried out in Lincolnshire.

For individual landfill sites it is also important to consider the strength, structure and other 'mass characteristics' of the clay proposed for use as a lining as it must be possible to excavate, segregate, handle and compact the deposit to an acceptable state. Thus in addition to the term 'material suitability', the term 'acceptability in earthworks' is introduced. The definition of acceptability is comparable with that adopted by the Department of Transport in the *Specification for highway works* (1991), this specification often being used as a guide to the compaction requirements for a clay lining. It is apparent that a material that is unsuitable is also unacceptable but a material that is suitable will not necessarily be acceptable.

In general highway earthworks, the use of the moisture condition value (MCV) apparatus is widespread. The MCV compaction test (Parsons & Boden 1979) was developed at the Transport and Road Research Laboratory as a control on material acceptability. The test yields reproducible results and the apparatus can be used on site to yield a rapid appraisal of material properties. On landfill sites a high degree of site control is essential and the MCV test lends itself readily to the determination of the 'acceptability' of clays forming linings. The use of the MCV test is illustrated by examination of the results for a glacial till.

From BENTLEY, S. P. (ed.) *Engineering Geology of Waste Disposal*,
Geological Society Engineering Geology Special Publication No. 11, pp 251–258

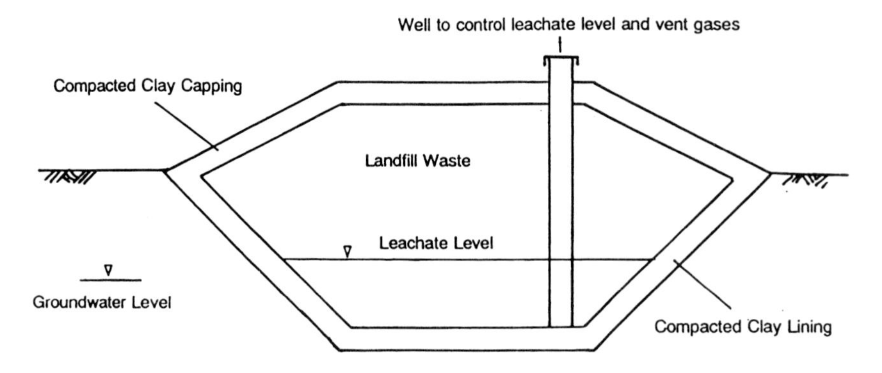

Fig. 1. Typical landfill cell.

Material suitability

Material suitability relates to the material type and whether it could potentially form a low-permeability barrier. A permeability of 10^{-9} m s^{-1} or less is usually specified. This degree of permeability is often defined as 'practically impervious' and taken as distinguishing clays from higher permeability silts (e.g. Somerville 1986). In terms of plasticity the division between clays and silts is known as the A-line, as shown in Fig. 2. In the study within Lincolnshire, data on individual strata were obtained from a large number of sources, but as might be expected, relatively few permeability results were obtained. It thus became necessary to identify potentially suitable deposits using the more readily available results for soil plasticity. However, permeability is related to a large number of factors and materials classed as silts based on plasticity data could achieve the required low permeability if adequately compacted within an acceptable moisture content range. Nevertheless, because of the lack of permeability test data and in accordance with NRA requirements, materials plotting below the A-line were, in general, defined as unsuitable in the study, as were materials with greater permeabilities, e.g. sands and gravels. Conversely, clays which plot above the A-line were deemed suitable or marginal.

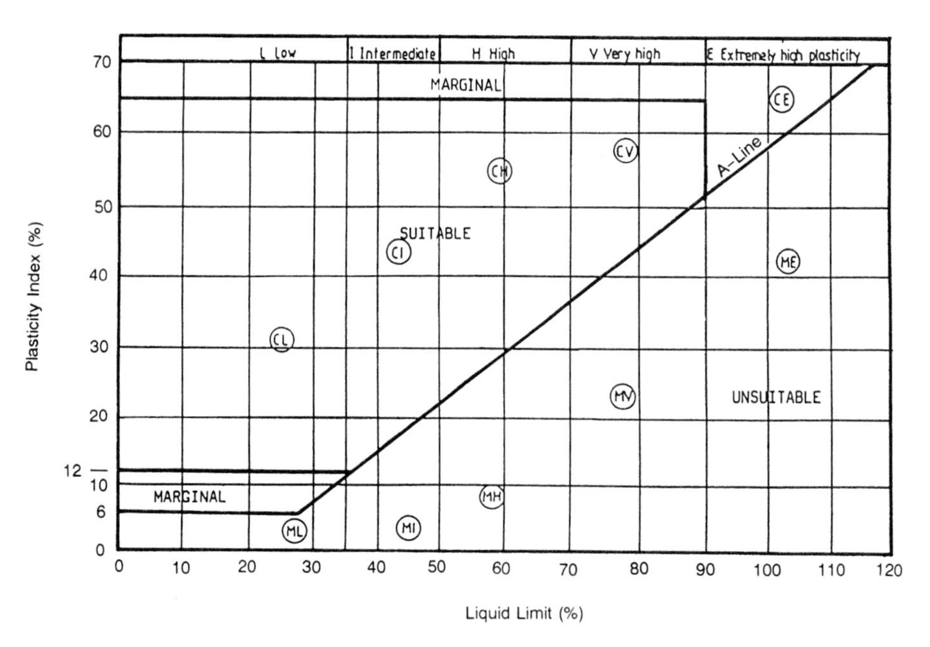

Fig. 2. Plasticity chart showing ares defined as 'suitable', 'marginal' and 'unsuitable'.

The NRA (1989) define suitable materials as those clays with a liquid limit (LL) of less than 90% and a plasticity index (PI) of less than 65%. These upper limits to LL and PI are based on criteria defined by the Department of Transport (1991) for modern earthworks plant. Materials in excess of these limits can give rise to problems with stability, deformation and compaction in earthworks. However, these limits preclude the use of extremely plastic clays. Such clay would exhibit very low permeability characteristics and if used with care may be adopted as a lining material. Sources of such materials are scarce and have not been identified within Lincolnshire. Further study of the use of such deposits would be warranted. However, for the purpose of this study, materials above these limits were defined tentatively as marginal.

The permeability of clays varies with the PI, and Murray *et al.* (1992) suggest that there appears to be a notable increase in permeability when the material has a PI of <12%. This suggests that clays with a PI of less than or equal to this value would best be defined as marginal in terms of suitability. This is considered to correspond reasonably with the NRA requirement that suitable materials should have a clay content (particle size less then 0.002 mm) of greater than 10%.

The variability of a deposit also influences its suitability; for example, glacial till whilst predominantly clay may exhibit significant variations in plasticity over short distances and contain pockets of sand, silt or other unsuitable material. Care must be taken when collating laboratory test results on a deposit to ensure that preferential sampling and testing is taken into account and the influence of material variability is fully assessed.

Based on the foregoing, deposits may be defined as 'suitable', 'marginal' or 'unsuitable' for use as linings to landfill sites and the relationship of the plasticity of clays to suitability as a lining material is illustrated in Fig. 2.

Mapping of material suitability

From published maps and geological memoirs, it is possible to identify those deposits which would obviously fall into the category of unsuitable materials, e.g. rocks, sands and gravels. The predominantly cohesive deposits require more detailed assessment. The results of the collation of plasticity data on glacial till and Lower Lias Clay from various ground investigations within Lincolnshire are presented in Figs 3 and 4. The results are interpreted as illustrating the glacial till as being marginal in terms of suitability due to the inherent variation in plasticity results, including a proportion of material with a PI of less than 12%, and the presence of unsuitable materials. The actual proportion of unsuitable materials is not readily identifiable from a cursory examination of the data in Fig. 3 and a careful examination of excavation and borehole logs is necessary to appreciate fully the influence of preferential sampling and testing.

The Lower Lias Clay results of Fig. 4 are considered to indicate the deposit as suitable (mainly CI and CH clays with a small proportion of MI and MH silts) as there is a relatively close grouping of plasticity data and little

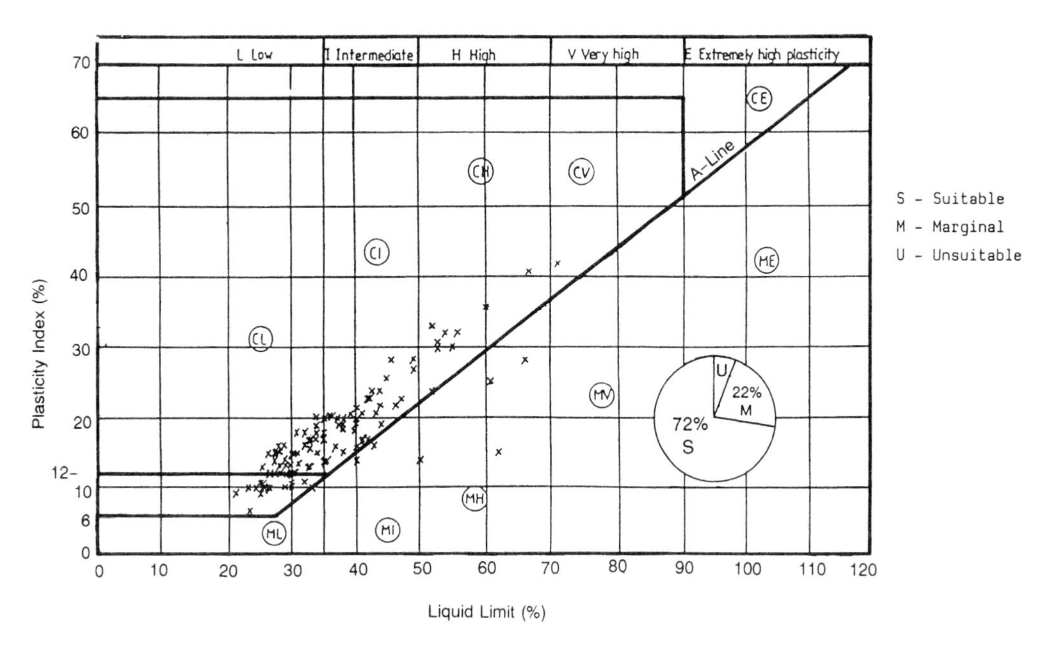

Fig. 3. Plasticity chart: glacial till.

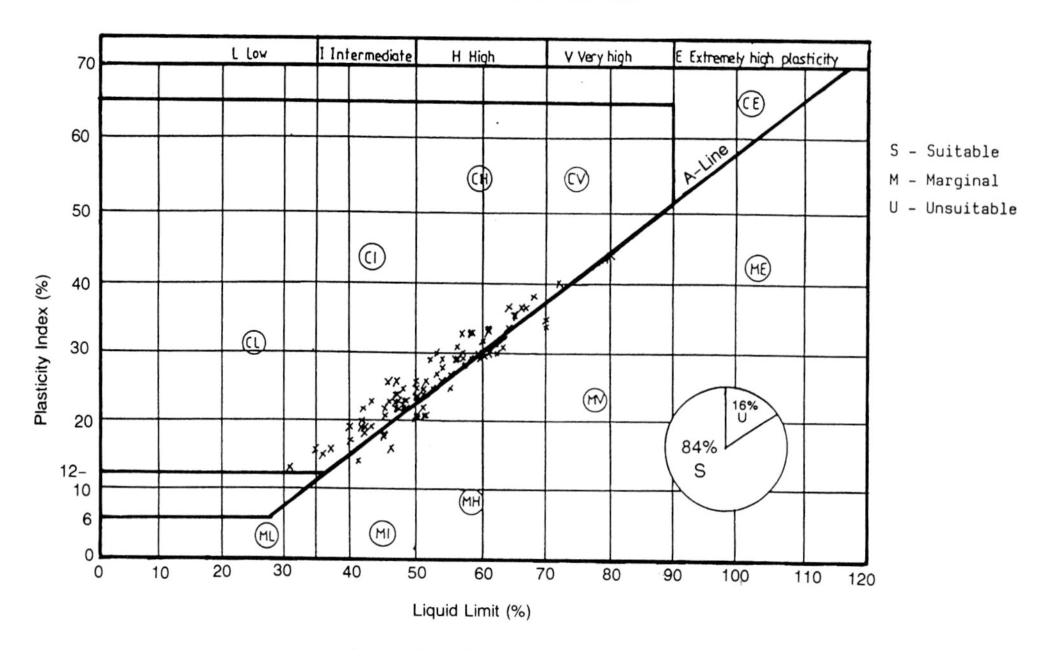

Fig. 4. Plasticity chart: Lower Lias clay.

variation in material characteristics. Those samples yielding results below the A-line were generally described visually as silty clays suggesting a significant clay content. These are thus likely to achieve the overriding requirement of a permeability of less than $10^{-9}\,\mathrm{m\,s^{-1}}$ on recompaction provided they are within an acceptable moisture content range. In addition, the MI and MH soils are likely to be readily mixed with the CI and CH soils on excavation, handling and recompaction. However, this does not preclude the need to examine carefully the material from a selected source to ensure its suitability as ground conditions can vary unpredictably.

Having established a map of potential sources of suitable and marginal clays the planning of landfill sites and an appraisal of the economics of development and potential sources of clay for linings may be made. However defining a material as suitable does not mean that it is acceptable in the earthworks and this requires further analysis. Jones *et al.* (1993) present a further discussion on suitability of materials for landfill lining.

Acceptability in earthworks

A clay may have suitable material characteristics but the variation in permeability with moisture content, degree of compaction and soil structure must also be taken into account. Acceptability relates to the excavation, handling, trafficability and compaction of the materials required to achieve the desired low permeability.

Adequate compaction of materials depends on a number of factors including strength, moisture content, type of plant used, etc. If a material is too soft, stability under earthworks plant may lead to trafficability problems. If the material is too stiff it may not be possible to compact it sufficiently to achieve the low degree of permeability required. It is difficult to adequately remould very stiff clays and remove discontinuities which can lead to the establishment of seepage paths. A material defined as suitable may thus be deemed unacceptable for use in the earthworks.

Compaction and material acceptability

In determining acceptability for use as a lining material it is normal practice to carry out permeability tests on samples recompacted in the laboratory and on samples recovered during compaction trials on site as well as from the lining during construction. However such testing is time-consuming and earthworks are often controlled by moisture content and density determinations. Alternatively, the moisture condition value (MCV) test (Parsons & Boden 1979) provides a rapid means of determining the acceptability of clay soils (Murray *et al.* 1992) which, coupled with a maximum air voids requirement to monitor the degree of compaction, may be used to aid assessment of the adequacy of the completed lining.

Cobbe & Threadgold (1988) discuss the use of the MCV test in general earthworks. As shown in Fig. 5, the

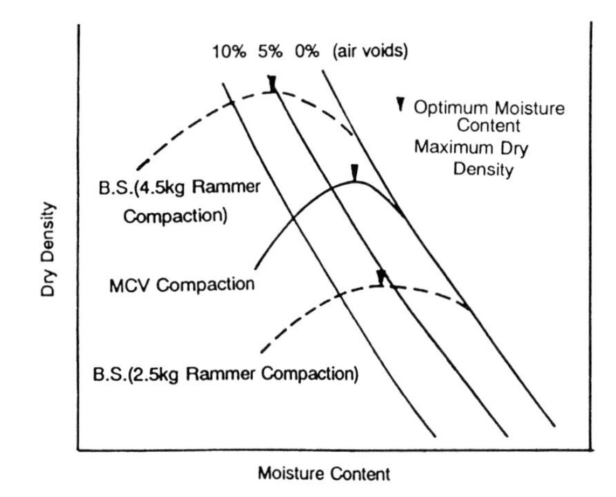

Fig. 5. Typical compaction curves.

degree of compaction achieved during the MCV test generally lies between that achieved by the 4.5 kg rammer and the 2.5 kg rammer methods of B. S. 1377 (Anon. 1990). As limiting permeability is an overriding requirement of a clay lining, there is a need to ensure a thoroughly compacted, uniform, homogeneous lining. This will necessitate detailed site monitoring and compaction probably in excess of normal earthworks levels. However, the degree of compaction achieved in a laboratory 4.5 kg rammer test is unlikely to be realized within a clay lining. Equally the low-permeability requirement would normally necessitate compaction in excess of that obtained using a 2.5 kg rammer. Densities obtained using the MCV test are thus considered more likely to represent desirable site compaction levels.

The MCV test is based on compacting a soil sample until no further change in density occurs, whereas the B. S. 2.5 kg and 4.5 kg tests are based on applying a given amount of compactive effort to a soil sample. However, the general forms of the B. S. compaction curves are similar to the MCV compaction curve but because of the different compaction criteria the optimum moisture content in the MCV compaction test tends to be closer to the zero air voids line as illustrated in Fig. 5. The differences in the compaction tests do not limit the use of the MCV test as all laboratory testing should be correlated with on-site compaction trials which may highlight the need for modifications to proposals based on laboratory testing alone. Experience suggests that more consistent results are obtained using the MCV test than the B. S. compaction tests and it is, therefore, recommended for use in the design and construction control of clay linings.

Figures 6 (a) and (b) show results for four series of tests (indicated by the different symbols) on a glacial till comprising a low-plasticity clay (PI of 15% and a clay content of 26%). These results indicate an increase in

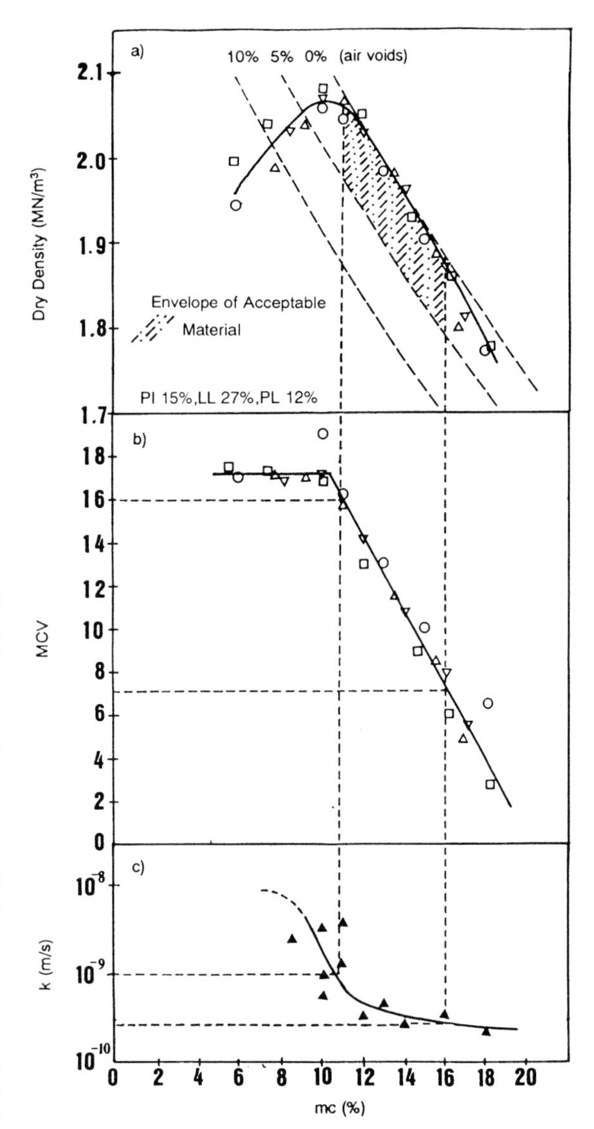

Fig. 6. MCV compaction and permeability results for low plasticity clay.

MCV with decreasing moisture content until the optimum is reached. Thereafter, for further reductions in moisture content the MCV is essentially constant. The increase in MCV is matched by an increase in remoulded shear strength as shown in Fig. 7.

Permeability requirements

In most earthworks it is the strength and degree of compaction that are the controlling parameters but in

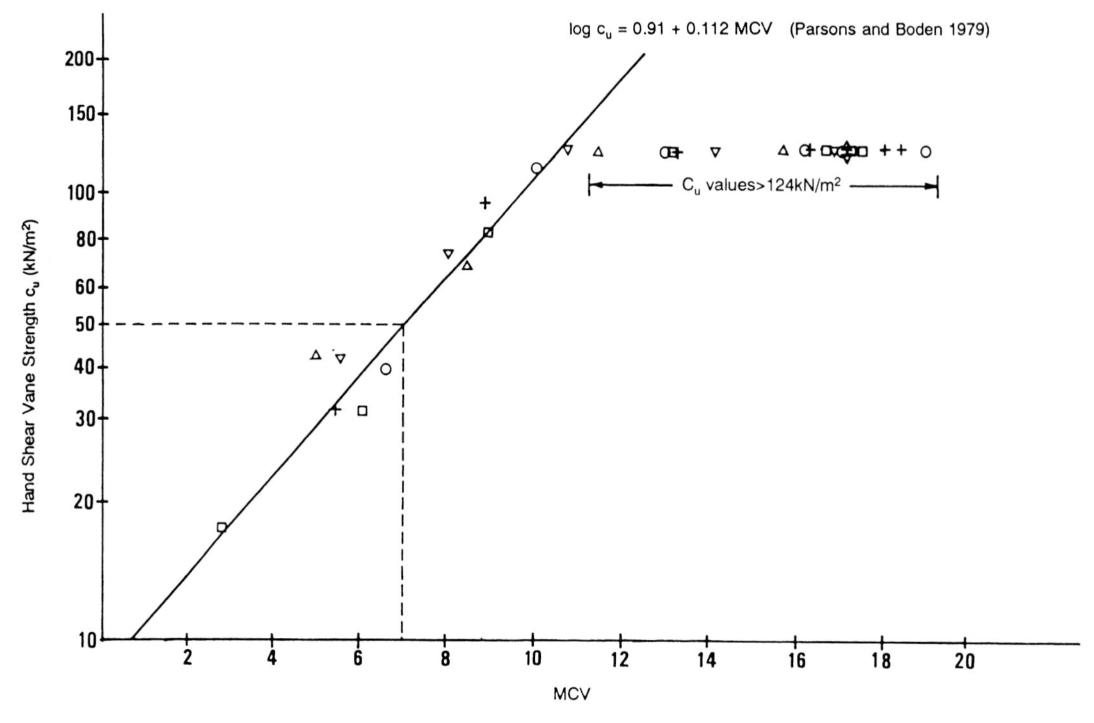

Fig. 7. Hand shear vane strength against MCV for clay of Fig. 6.

the construction of a clay lining to a landfill site a requirement for low permeability presents an additional burden on the selection and emplacement of clays.

The results of Figs 6 (a)–(c) show an increase in permeability with decreasing moisture content even though there is an increase in dry density of the clay. The results were obtained from both constant and falling pressure tests as outlined by Head (1985) and the behaviour pattern is consistent with the results reported by Needham (1991), Parkinson (1991) and Seymour (1992) amongst others. At around the optimum moisture content there is a more rapid increase in permeability, reflecting the lack of remoulding of the clay at moisture contents below the optimum value and the presence of fissures resulting in preferential seepage paths. Obviously, greater compactive effort at these relatively low moisture contents would produce a greater degree of compaction and a reduction in permeability, but as discussed previously this may not be readily achievable on site. Observations during laboratory compaction operations on the drier samples suggest the presence of discontinuities which are also likely to exist during *in situ* emplacement of the fill. The acceptable lower limit to moisture content should therefore be greater than the optimum moisture content as determined from the MCV compaction series.

Control of construction

It is necessary to establish relationships between the permeability and other soil parameters such as MCV (or moisture content) and density and use these to control acceptability and compaction. However, this should not be taken as precluding the need for further permeability testing on the compacted lining material as an assurance that the control criteria are adequate.

For a clay of suitable plasticity, test results suggest that the lower limit for the moisture content should be dictated by the permeability requirement. However, the upper limit to the moisture content is likely to be dictated by the shear strength of the clay because although the permeability requirement may be met, handling, compaction and trafficking become more difficult. This, in conjunction with stability considerations, makes a minimum shear strength requirement essential. Typically an undrained shear strength (c_u) of no less than 40–50 $kN\,m^{-2}$ is required in earthworks.

Figure 7 presents results of hand shear vane strength tests carried out during the four series of tests reported in Figs 6(a) and (b) and indicated by the different symbols. It can be seen that for a strength of 50 $kN\,m^{-2}$ an MCV of 7 would be required. Thus in order to achieve a permeability of less than $10^{-9}\,m\,s^{-1}$

and to satisfy the emplacement requirements, the laboratory test results suggest it appropriate to ensure the clay of Fig. 6 has an MCV of between 7 and 16. Compaction trials should be undertaken to ensure the required densities (and permeability) are achievable using a practical layer thickness and number of passes of the selected roller. Should the requirements not be met, the results of the trials may be used to restrict further the acceptable MCV range. A controlled compaction trial prior to the main earthworks is considered essential to alleviate potential problems in the construction stage.

Parkinson (1991), amongst others, has noted that permeabilities from field tests are perhaps 10–100 times greater than those from laboratory tests. It may be possible to reduce the discrepancy by careful site control but if necessary this may be allowed for by reducing the acceptable upper limit of MCV.

It is also worth noting that the lower limit to moisture content of 11% (MCV of 16) is close to the plastic limit of 12% for the glacial till. The plastic limit is the moisture content at which a cohesive soil stops behaving as a plastic material and starts behaving as a brittle material. At the plastic limit cracking starts to appear in thin threads (3 mm) of cohesive soil rolled by hand. These discontinuities, which can also be expected to be present in clay linings, would tend to increase permeability. This appears to be reflected in Fig. 6 where there is a more rapid increase in permeability for a given change in moisture content below a moisture content of 11–12%. There also appears to be a greater scatter of permeability results below the plastic limit and this may be a function of the frequency, orientation and continuance of fissures which will vary from sample to sample. On this basis, an MCV of 14 corresponding to a moisture content of 12% (the plastic limit of the glacial till) may be considered a more appropriate limit to ensure the permeability requirement is met.

Based on the laboratory results it is possible to set upper and lower bounds on material acceptability. If this is added to a compaction requirement of no more than 5% air voids, the envelope of acceptable material and compaction as detailed in Fig. 6(a) is obtained. It would be necessary in practice to ensure that with 5% air voids the permeability requirement is still met. However, the uptake of free water by the clay forming the lining would mean a reduction in air voids, due to suction and percolation effects, and the clay complying closer with the fully compacted state and an acceptable permeability. This would result in a corresponding softening of the lining, the consequences of which may have to be taken into account.

Consideration should also be given to the type of compaction plant to be employed on site. It is recognized that a discontinuity often exists between clay layers compacted using a smooth roller. Such discontinuities may well present seepage paths. For this reason it would be considered preferable to adopt tamping rollers in the construction of clay linings. These rollers knead and remould the soil, providing a more homogeneous material and reducing the risk of major discontinuities (and thus potential seepage paths) remaining. A smooth wheeled roller may then be used to provide an acceptable surface finish to the lining.

Conclusions

A study carried out in Lincolnshire to identify possible sources of 'suitable material' to form landfill liners has shown that materials may be classed as unsuitable, marginal or suitable based on the collation of plasticity data and an appraisal of variability.

For a material to be designated as suitable does not mean that it is acceptable for use in earthworks. In order to achieve the desired low permeability and to facilitate excavation, handling and compaction, the material must have acceptable mass characteristics such as moisture content, structure and strength.

The acceptability of a material for use in a clay lining is dictated by a maximum allowable permeability. Although this will require permeability determinations during construction of the lining, earthworks operations are generally controlled by moisture content, density and strength determinations. The alternative use of the moisture condition value and air voids determinations is recommended as a control on material acceptability and as an indicator of the degree of compaction achievable on site.

In control of the earthworks by the MCV test the lower limit to moisture content should be greater than the optimum moisture content (as defined by MCV compaction series) and is controlled by the maximum permeability requirement of 10^{-9} m s^{-1}. This dictates the upper limit to the acceptable MCV range.

The upper limit to moisture content is likely to be controlled by the strength of the clay and its handling, traffickability and compaction requirements. This would dictate the lower limit to the acceptable MCV range.

Acknowledgements. The authors would like to thank Linc Waste for their co-operation and permission in allowing this paper to be published and to Mr S. Brooks and Mr J. Race for the preparation of the drawings.

References

ANON. 1981. B.S. 5930: *Code of practice for site investigation.* British Standards Institution, HMSO, London.

—— 1990. B.S. 1377: Part 5: *Methods of tests for soils for civil engineering purposes, Code of practice*: British Standards Institution, HMSO, London

COBBE, M. I. & THREADGOLD, L. 1988. Compaction control by the moisture condition test–a new approach. *Journal of the Institution of Highways and Transportation*, Dec., 13–18.

DEPARTMENT OF THE ENVIRONMENT 1990. *Landfilling wastes.* Waste Management Paper No. 26. A Technical Memorandum for the Disposal of Wastes on Landfill Sites, HMSO, London.

DEPARTMENT OF TRANSPORT 1991. *Specification for highway works*, HMSO, London.

ENVIRONMENTAL PROTECTION ACT 1990. *Waste management, the duty of care: a code of practice.* HMSO, London.

HEAD, K. H. 1985. Effective stress tests. *Manual of Soil Laboratory Testing* vol. 3. Pentech Press Ltd.

JONES, R. M., MURRAY, E. J., RIX, D. W. & HUMPHREY, R. D. 1993. Selection of clays for use as landfill liners. *Green '93 an International Symposium on Geotechnics Related to the Environment, Waste Disposal and Landfill 2*, Session 4, June–July, Bolton UK, 77–82.

MURRAY, E. J., RIX, D. W. & HUMPHREY, R. D. 1992. Clay linings to landfill sites. *Quarterly Journal of Engineering Geology*, **25**, 371–376.

NAWDC LANDFILL COMMITTEE, August 1989. *Codes of practice for landfills.* NAWDC.

NEEDHAM, A. 1991. Clay liners—the answer lies in the clay. *Surveyor Special Supplement to Wastexpo International '91.* 18–19 Sept, Cannock, Staffs, 11–13.

NORTH-WEST WASTE DISPOSAL OFFICERS 1986. *Guidance on the use of landfill liners.* Waste Disposal Authority, Preston.

NRA 1989. Earthworks to landfill sites. National Rivers Authority, North-West Region.

—— 1992. *Policy and practice for the protection of groundwater.* National Rivers Authority.

PARKINSON, C. D. 1991. The permeability of landfill liners. *The Planning and Engineering of Landfill.* Conference Proceedings, University of Birmingham, 147–151.

PARSONS, A. W. & BODEN, J. B. 1979. *The moisture condition test and its potential application in earthworks.* Transport and Road Research Laboratory Supplementary Report 522.

SEYMOUR, K. J. 1992. Landfill linings for leachate containment. *The Institution of Water Management*, **6**, August, 389–396.

SOMERVILLE, S. H. 1986. *Control of groundwater for temporary works.* Construction Industry Research and Information Association Report 113.

Factors affecting the containment properties of natural clays

Jonathan Arch,[1] Emma Stevenson[2] & Alex Maltman[2]

[1] Aspinwall & Company, Walford Manor, Baschurch, Shrewsbury SY4 2HH, UK
[2] Institute of Earth Studies, University of Wales, Aberystwyth SY23 3DB, UK

Abstract. Natural clays are often relied upon to provide containment for the safe disposal of wastes, either as an *in situ* material or as an engineered fill material. It is generally assumed that these materials will permanently exhibit a low permeability, and that they are isotropic and homogeneous. However, anisotropies and discontinuities are often present in clays, or may be induced by deformation which can occur during or after construction of waste disposal sites. Anisotropic fabrics may also be developed or enhanced within clays as they are compacted and consolidated. Many landfill sites are dynamic environments where deformation will occur both within the waste and in the material surrounding the waste. When clays deform, for example due to slope instability or differential settlement, the deformation will generally be accommodated by the development of discrete microscopic shear zones which have the potential to significantly change the containment ability of clays. This paper presents the results of a study to investigate the variation in the permeability of clays as a result of their deformation. Natural examples, theoretical studies and experimental data are combined to demonstrate that fluids will preferentially flow along shear zones or induced fabrics. The data are used to demonstrate that, in a dynamic environment such as a waste disposal site, the permeability of clays can increase substantially in materials which become heavily deformed.

Most new landfills are now developed as containment sites, where natural and/or synthetic materials are relied upon to provide containment and management of leachate and landfill gas. In many sites, containment is afforded by the provision of a clay barrier surrounding the waste, either as an *in situ* material or as an engineered fill material. A standard specification, which is now widely used for this clay barrier, is that it should have a hydraulic conductivity no greater than $1 \times 10^{-9}\,\mathrm{m\,s^{-1}}$. This specification is obtainable with many naturally occurring clays in the UK, and has been adopted by most regulatory authorities as one which results in an acceptably low rate of leakage. However, anisotropies in clays (e.g. fissures, laminations, shears) result in the *in situ* material rarely attaining the appropriate specification in all locations in all directions. For this reason, clay is often reworked in a controlled earthworks operation to produce a homogenous material, and tested on a pre-specified grid to ensure that the required specification is attained. If this testing proves satisfactory, it is often assumed that the clay has attained the appropriate standard and will retain that same standard for the design life of the site.

It has previously been recognized that this assumption may not be valid. Clay can be subject to chemical changes which may alter its permeability (Peters 1993). Similarly, physical changes (for example, those associated with the poor compaction of the material) may cause subsequent desiccation which can alter the clay's permeability. This paper presents evidence in relation to another example of a mechanism whereby the permeability of clay may alter after its placement, namely deformation.

Dynamic environments in waste disposal sites

Many sites where the low permeability of clays is relied upon to provide the effective containment of wastes, are actually dynamic environments where deformation of the waste and of the surrounding material is taking place. Such deformation can occur in at least three locations, namely:

- the base of the site due to foundation settlement;
- the side-slopes of the site due to slope instability;
- the top of the site due to the settlement of the waste.

In each of these situations, any clays which are present around the waste, either as an *in situ* material or as a placed fill material, will deform in response to the stress regime to which they are subjected, which may be extensional or compressional. Each of these situations is described below.

Foundation settlement

Waste disposal sites are commonly situated in areas where the ground conditions are such that they provide a poor foundation. Mine workings or other cavities may

From BENTLEY, S. P. (ed.) *Engineering Geology of Waste Disposal*,
Geological Society Engineering Geology Special Publication No. 11, pp 259–265

be present, and the stress unload–reload cycle which often occurs, for example due to excavation followed by landfilling, may trigger collapse of these voids. Alternatively, any compressible ground will be subject to settlement when a large vertical stress in the form of a considerable volume of waste material is placed above it. These factors have the potential to cause differential settlement at the base of a landfill site which would typically subject any clay material present, either as an *in situ* material or as a placed base liner, to an extensional stress regime (see Fig. 1(a)). It is worth noting that this deformation would normally be expected to occur once waste had been placed in the site; observation of such deformation is therefore blinded by the presence of the waste.

Side-slope instability

The economics of landfilling are such that there is an incentive for operators to attempt to maximize the void capacity of a waste disposal site. This is often achieved by utilizing side-slopes which are as steep as possible. A justification often used for this practice is that the slope is only temporary as it will effectively be buttressed when waste is placed against it. This ongoing practice, and the utilization of slopes which often have factors of safety perilously close to 1.0 has led to a large number of side-slope failures in landfill sites. In addition, slope failures are often present in old quarries which are subsequently infilled with waste. If slope faliures occur within *in situ* clay, this material will be subjected to an extensional stress regime at the top of the slip and a compressional regime at its base (see Fig. 1(b)). Instability will similarly affect any clay material that has been placed on the slope as a liner. When slope failures are observed, these are often simply tidied up, or waste material is hurriedly placed against the affected area.

Waste settlement

Clay capping layers have been constructed on many landfill sites in an effort to effectively seal the top of the waste, but these capping layers have often not attained their anticipated performance as a low-permeability layer above the waste. The top of a landfill site can represent a highly dynamic environment as the waste settles, often by as much as 20% of the total thickness (Watts & Charles 1990) which on deep sites can represent settlement of several metres. This settlement is often differential due to the heterogeneous nature of the waste. This can lead to the development of depressions on the landfill surface (see Fig. 1(c)) where

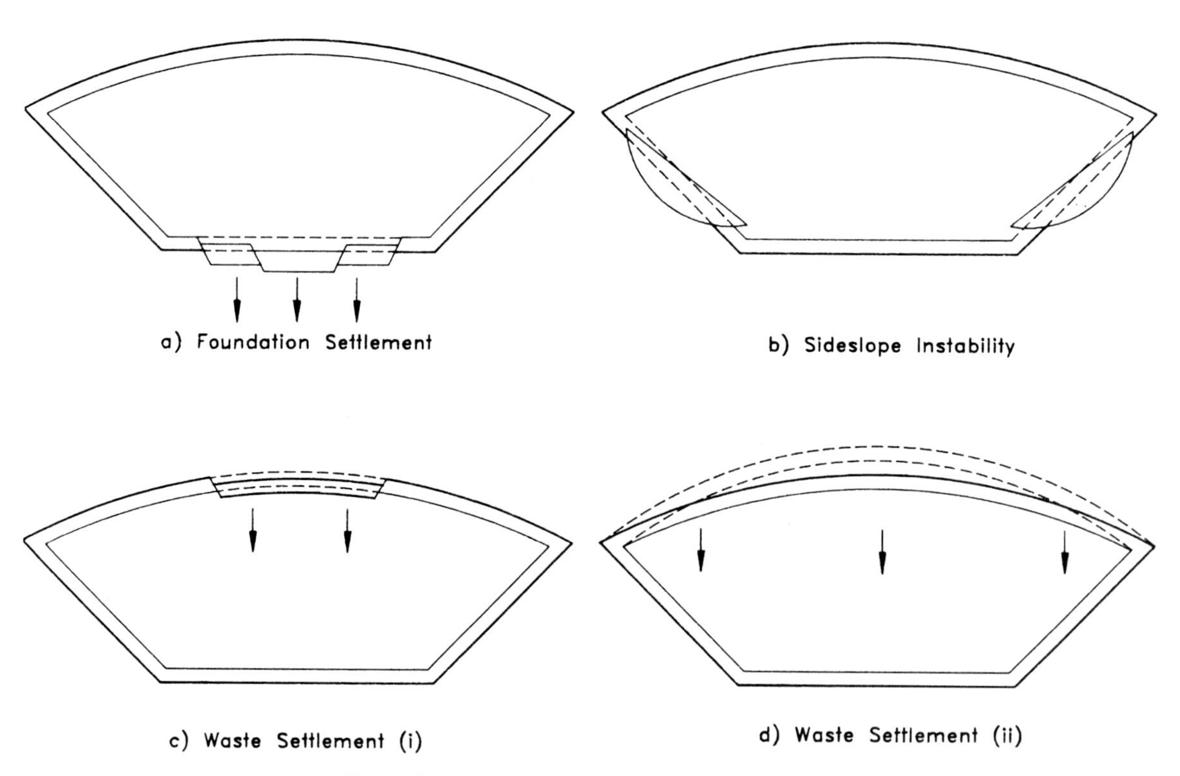

Fig. 1. Dynamic environments in waste disposal sites.

the stress regime is extensional. However, in a typical domed landfill site, settlement will lead to a net shortening in the length of the capping layer and the clay cap will be subject to an overall compressional stress regime (see Fig. 1(d)).

What happens when clays deform?

It has been argued above that waste disposal sites are dynamic environments with any clay material which is present surrounding the waste liable to deformation in response to either an extensional or a compressional stress regime. It is often assumed that this deformation is accommodated by homogeneous ductile flow within the clay, but there is now a wealth of evidence to suggest that this is not normally the case. Observations in the geological record (Maltman 1988) and in engineering soils (Morgenstern & Tchalenko 1966), and several programmes of laboratory testing (Arch 1988) have all indicated that the deformation of clays is typically accommodated by the development of discrete microscopic shear zones. These shear zones represent areas of preferred orientation of clay particles which align themselves parallel to the sides of the zone. Microscopic observations suggest that these shear zones often anastomose and splay to form wide shear zone arrays, particularly in specimens deformed to high strains (Arch *et al.* 1988).

Because of the relatively weak bonding and friction between clay particles, the development of shear zones takes place by inter-granular sliding. This is confirmed by electron microscope observations of the zones which show the individual clay particles to be intact, closely packed and aligned parallel to the sides of the zone. It is important to note that these features are non-brittle, with cohesion being maintained across the zones both during and after deformation. Given the observed geometric relationship between the clay particles in the shear zones and those outside it (assumed to represent the clay fabric prior to the development of the shear zones), any mechanism for the formation of the shear zones must involve initial dilation; it would not be geometrically possible for the particles to rotate into their ultimate orientation without initial dilation. Particles must then rotate to define the preferred orientation within the zones and collapse to form the observed closely packed structure. The observed closely packed structure suggests a final collapse associated with the expelling of fluid from the shear zone.

Fluid zones along shear zones

Natural examples

The geological record (rocks, recent sediments and soils) contains many examples of clays which have deformed by the development of the shear zones referred to above, and there is often evidence that these zones have acted as preferential pathways for the migration of fluids; for example , the zones often show mineralization, sediment entrainment and enhanced diagenesis (e.g. Shipboard Scientific Party 1988). The slowness of some of these chemical processes means that flow along the zones must be prolonged, persisting long after the zones have formed.

Theoretical considerations

The deformation processes described above have implications for the migration of fluids within the clays both during and after deformation. For example, if a simple relationship between permeability and porosity is assumed (e.g. Terzaghi & Peck 1967), the dilation essential to the development of shear zones will locally increase the porosity of the material and hence increase its permeability whilst the zones are forming. Subsequent collapse of the structure will result in a reduction in the porosity, and hence permeability, to a level probably lower than that of the host structure. Competing against this effect will be changes in permeability brought about by the development of a preferred particle orientation from a less-well-defined clay fabric. Intuitively it would be expected that the permeability parallel to shear zones would be higher than that perpendicular to them (Arch & Maltman 1990). In order to consider these factors together, Pouisselle's Law (England *et al.* 1987) can be used which relates permeability, k, to the tortuosity, T (defined as the length of the tortuous fluid flow route divided by the length of the direct route), of the fluid flow path and the mean pore radius, r:

$$k = \frac{Dr^2}{T^2} \quad (1)$$

where D is a constant. As the porosity of the clay changes, the mean pore radius will change accordingly causing an associated isotropic change in permeability. As preferred orientation develops parallel to the orientation of shear zones, the tortuosity parallel to the shear zone will decrease (thus increasing the permeability) and that perpendicular to the zone will increase (with an associated decrease in permeability). Given the square relationships in Pouisseille's Law, small changes in either porosity or tortuosity will result in similar larger changes in permeability.

In the case of fluid flow perpendicular to the shear zone, the final permeability should always be lower than that of the original undeformed material as the increase in tortuosity and decrease in porosity will both contribute to this effect. However, in the case of fluid flow parallel to the shear zone, the ultimate permeability will depend on the relative changes in porosity and tortuosity, since the changes which occur will have conflicting influences. The tortuosity variation will depend on the intensity of any primary fabric in the clay and the orientation of any shear zones relative to this fabric. The magnitude of porosity reduction would be expected to be less, for example, in a clay which was initially over-consolidated than one which was initially normally or under-consolidated. As such, it would be expected that, in a normally consolidated clay which contained a fabric sub-parallel to the shear zones, the effect of porosity would dominate and the overall permeability would decrease, whereas in an isotropic over-consolidated clay, the influence of tortuosity would dominate and the overall permeability would increase. Since most clays used in landfills in the UK are over-consolidated (e.g. London Clay, glacial till), and rarely show significant fabric development, it would be expected that the overall permeability due to deformation would increase.

The type of permeability variations described above will also apply to clays which are not deforming but simply consolidating. Anisotropic permeability in undeformed clays has been known for some time (e.g. Al-Tabbaa & Wood 1987); for example, it has been well documented that laminated clays exhibit a horizontal permeability higher than their vertical permeability. Consolidation theory in clays (e.g. Hedberg 1936) defines a 'house of cards' structure which develops a horizontal preferred orientation of clay particles and a reduced porosity as the material consolidates. Referring to Equation (1), the mean pore radius would reduce for both vertical and horizontal orientations. However, the tortuosity would increase with regard to vertical flow and decrease with regard to horizontal flow. Both these effects would result in a reduction in the vertical permeability during consolidation. The horizontal permeability may increase or decrease depending on the relative magnitudes of the porosity decrease and the tortuosity increase.

Experimental work

It follows from the above arguments concerning particle realignment that the fluid transfer properties of a clay will change during periods of deformation. As a first step, a series of experiments has been designed to explore the relationships between deformation and clay permeability. The experimental system is based on combining a triaxial cell with a constant head method of permeability determination. Pressure head across the specimen is recorded by a differential pressure transducer, with the downstream permeant being automatically and continuously monitored by means of a sensitive electronic balance. Data concerning stress, strain, volume and pressure are displayed and recorded on a PC. From Darcy's relationship, the permeability during accumulated strain (referred to here as the 'dynamic' permeability) is then calculated. Potential errors could arise in this arrangement from incomplete saturation of the permeant, seeing as the measurements are carried out at atmospheric pressure rather than under some back-pressure, and through evaporation from the effluent receptacle. These effects have been assessed and found to be negligible, with the permeability results being highly reproducible and consistent with those obtained by other measurement modes in other laboratories. The apparatus produces rapid and accurate results, and is routinely used to determine the permeability of core samples, often from landfill sites.

In the experiments reported here, a constant pressure gradient was maintained at some specified value between 100 kPa and 400 kPa, along the length of a cylindrical sample 54 mm diameter and, typically, 100 mm in length. The specimen was encased in a latex sleeve and confined in the triaxial cell at a specified pressure between 300 kPa and 700 kPa. Strain rates employed in the tests were of the order of $10^{-4}\,s^{-1}$. The permeant was in all cases distilled water. A variety of soils and sediments have already been investigated, including, for example, laboratory-generated sands, and silts cored from the deep-ocean floor, as well as the clays reported here. The general patterns of behaviour discussed below have now been recognized in a variety of test materials, but the results presented here are from ball clay, pure kaolinite (in some cases with added fine sand) and samples of clays currently being employed as landfill linings. The former two clays were obtained from commercial ceramic supplies, mixed with distilled water and consolidated under loads ranging from self-weight to 1000 kPa. The tests demonstrate that there is some variety in the behaviour of different materials, but it seems that this can be simplified into three main types.

Type 1 behaviour: permeability decrease with deformation. An example of this type of behaviour is shown in Fig. 2 which shows the permeability changes in a kaolinite-fine sand (KFS) sample generated in the laboratory as strain progresses. The rate of permeability change generally decreases as stress increases and strain accumulates. The initial rapid loss of permeability is lessened at a point close to that of yielding in the clay. This kind of behaviour is thought to be associated with under-consolidated and normally consolidated soils; for example, 'self-weight' consolidated samples all show this affect well. The inference is that strain is largely accommodated by consolidation and improved packing of the clay particles with an associated porosity reduction leading to lower permeabilities.

Type 2 behaviour: permeability increase with deformation. Figure 3(a) shows an over-consolidated (OCR = 2.7) KFS specimen with an initial permeability of $4 \times 10^{-9}\,m\,s^{-1}$, increasing to $6.5 \times 10^{-9}\,m\,s^{-1}$ after 30% strain, representing an increase of over 60%. Early stress increments are accompanied by some permeability decrease, which may reflect some initial consolidation.

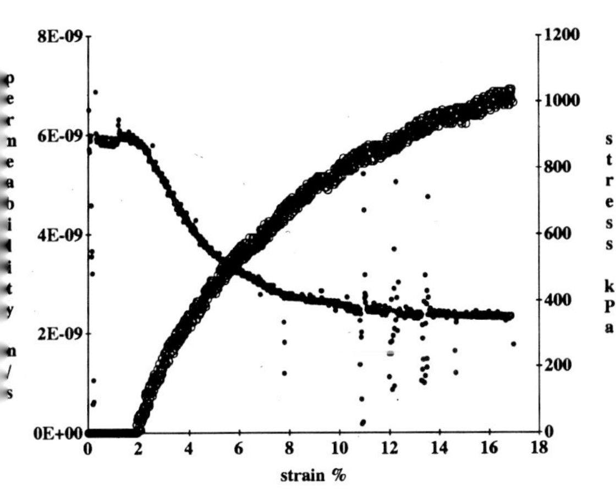

Fig. 2. Permeability reduction with deformation.

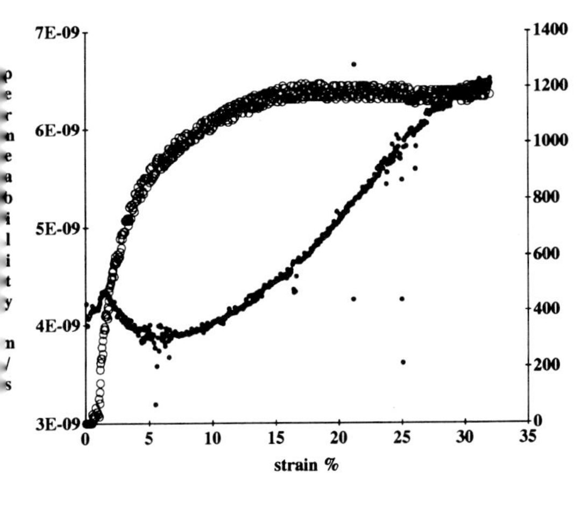

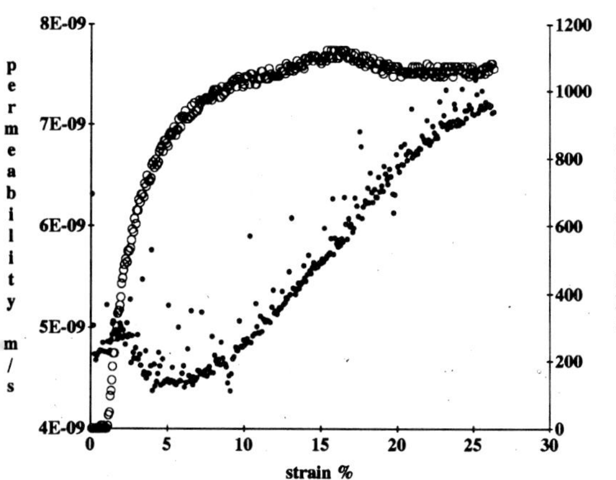

Fig. 3. Permeability increase with deformation.

However, continued strain is then accommodated by the development of shear zones. As argued above, because no cataclasis is observed in the zones, the preferred alignment of clays within the shear zones can only be achieved through dilation. It is this increase in porosity and interconnectivity, perhaps accompanied by diffuse, bulk dilation, that is thought to be responsible for the observed permeability increase. Such dilation effects are likely to be transient, and the levelling of permeability increase at 30% strain may indicate that sufficient zones of aligned clays have been established to allow continued low-friction sliding. The final permeability is then higher due to the lower tortuosity within the shear zones. The interference of the role of dilation associated with tortuosity reduction is supported by the slightly more consolidated (OCR = 2.94) KFS specimen shown in Fig. 3(b). The pattern of permeability behaviour is similar, but the increase in permeability is greater, as reflected by the slightly steeper gradient of the curve, presumable because of the additional dilation. This phenomenon is of particular concern in the context of waste containment, because the observed increase in permeability comes about in over-consolidated materials, the very soils often employed as clay barriers.

Type 3 behaviour: complex permeability-deformation interplay. Some specimens have shown much more complicated changes in permeability with progressive strain, often with a remarkably complimentary stress level. An example is given in Fig. 4(a), a lightly over-consolidated ball clay synthesized in the laboratory. Superimposed on a slight overall reduction of permeability with deformation are marked intervals of greater permeability which correspond to periods of lower stress level. These intervals of strain softening may reflect the continued development of shear zones, with corresponding increases in fluid flow. Subsequently the zones lock up and cause strain hardening, further strain-induced consolidation, and reductions in permeability. A similar though less clearly defined behaviour is shown by the clay used as a landfill liner in mid-Wales, illustrated in Fig. 4(b). Note that the data scatter in these two examples is similar; it appears greater in Fig. 4(b) because of the expanded scale of the permeability axis.

Discussion of the experimental results. Two lines of evidence substantiate the above interpretations. First, electron microscopic analysis has documented the bulk compaction and shear zone generation inferred in the above summary. Secondly, by accurately monitoring the volume change needed to maintain a constant pressure head in the experiments, it becomes possible to compare the ingress of permeant with the measured effluent, and hence detect any change in pore volume within the specimen. Initial results from this testing programme have broadly confirmed the compaction and dilation effects suggested above.

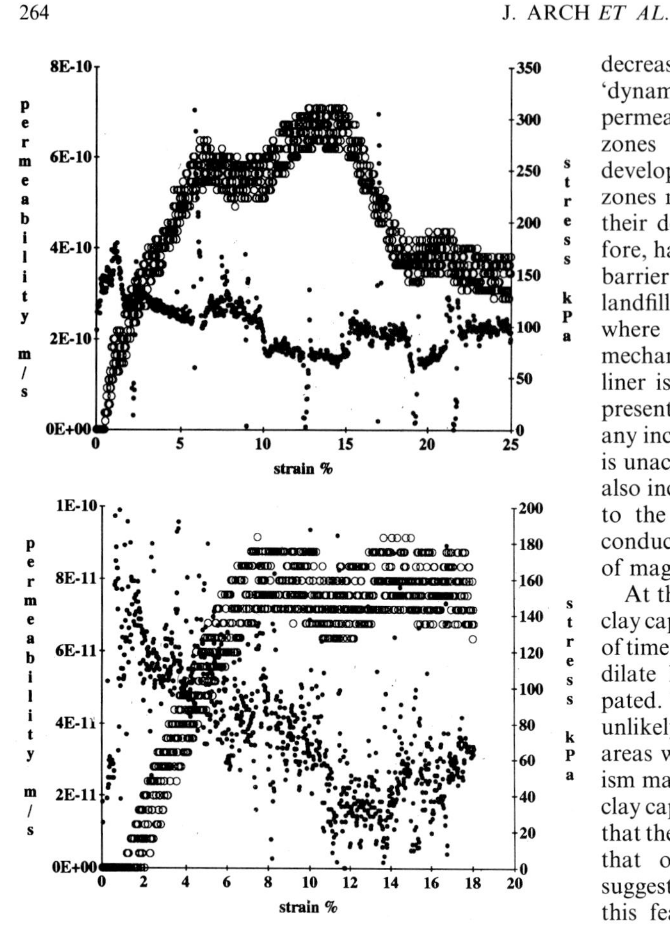

Fig. 4. Complex permeability–deformation interplay.

Detailed investigations are continuing into these various inter-relationships and the factors controlling them. It remains unclear, for example, exactly what factors determine whether an over-consolidated clay will adopt the relatively simple type 2 behaviour as opposed to the more complex type 3. Nevertheless, the overall implication for clays as barriers is clear. The experiments suggest that an over-consolidated clay undergoing strain is likely to increase in permeability, at least temporarily and possibly permanently. Moreover, the experiments have so far dealt only with lightly over-consolidated clays. With the heavily over-consolidated clays often used in landfill sites, the effects could be even more marked.

Implications for waste disposal sites

In a uniform 'static' situation, clay which is present at the base of a landfill site will tend to consolidate due to the weight of the overlying waste, with an associated decrease in vertical permeability. However, in more 'dynamic' situations it has been demonstrated that the permeability of a clay can increase parallel to shear zones which develop, particularly whilst they are developing. The permeability perpendicular to shear zones may significantly decrease, particularly following their development. The presence of shear zones, therefore, has the potential to affect the performance of a clay barrier if areas of deformation coincide with parts of a landfill where a significant head of leachate exists or where landfill gas is present under pressure. This mechanism is of particular concern where a clay base liner is placed, or where thin layers of *in situ* clay are present, on areas of soft or unstable ground, and where any increase on the contaminant loading to groundwater is unacceptable. Any shear zones which are present may also increase the potential for off-site gas migration (due to the lower viscosity of landfill gas, the pneumatic conductivity of a clay liner is approximately two orders of magnitude higher than the hydraulic conductivity).

At the top of a landfill site, active deformation of any clay capping layer is virtually inevitable over a long period of time whilst the waste is settling, and the development of dilate high-permeability shear zones would be anticipated. The permeability of a clay cap is therefore highly unlikely to retain its original permeability, particularly in areas where deformation is concentrated. This mechanism may contribute to the frequent poor performance of clay capping layers. One of the conclusions of this study is that the permeability across shear zones will be lower than that of the undeformed material. It is, therefore, suggested that there may be scope to take advantage of this feature by the inducement of shear zones in an appropriate orientation (e.g. parallel to the landfill base) in order to reduce the vertical permeability. However, this should be done with caution as shear zones may be induced in less favourable orientations, and the geotechnical properties of the material (at its residual as opposed to its peak strength) may lead to subsequent slope stability problems.

Conclusions

This paper has presented the results of a study aimed at investigating the way in which the permeability of clays varies as the material deforms, and has discussed the relevance of these results for waste disposal sites. The main conclusions are listed below.

- Waste disposal sites are often dynamic environments where any clay barriers which are present are liable to be subject to deformation.
- The deformation of clays will be accommodated by the development of discrete microscopic shear zones where a preferred orientation of clay particles parallel to the edges of the zone is developed.

- Fluid flow along shear zones will increase, particularly during their development. Fluid flow across shear zones will tend to be inhibited, particularly in the time following their genesis.
- Depending on the orientation and density of shear zones which develop in clay barriers surrounding landfill sites, there is the potential for the permeability of the clay barriers to significantly increase. Leakage from the landfill would then be over and above that designed for.
- Clay capping layers will inevitably deform, often to considerable strains, and shear zones will continue to develop for a considerable length of time as the waste settles. This is likely to have a significant detrimental effect on the performance of clay capping layers.
- In the over-consolidated clays often used as landfill containment layers, active physical movement will increase the permeability substantially, at least temporarily and possibly permanently.

References

AL-TABBAA, A. & WOOD, D. M. 1987. Some measurements of the permeability of kaolin. *Geotechnique*, **37**, 499–503.

ARCH, J. 1988. An experimental study of deformation microstructures in soft sediments. PhD thesis, University of Wales, Aberystwyth.

ARCH, J. & MALTMAN, A. J. 1990. Anisotropic permeability and tortuosity in deformed wet sediments. *Journal of Geophysical Research*, **95**, 9035–9045.

——, —— & KNIPE, R. J. 1988. Shear-zone geometrics in experimentally-deformed clay: the influence of water content, strain rate and primary fabric. *Journal of Structural Geology*, **10**, 91–99.

ENGLAND, W. A., MACKENSIE, D. M., MANN, D. M. & QUIGLEY, T. M. 1987. The movement and entrapment of petroleum fluids in the subsurface. *Journal of the Geological Society of London*, **144**, 327–347.

HEDBERG, H. D. 1936. Gravitational compaction of clays and shales. *American Journal of Science*, **31**, 241–287.

MALTMAN, A. J. 1987. Shear zones in argillaceous sediments—an experimental study. In: JONES, M. E. & PRESTON, R. M. F. (eds) *Deformation of sediments and sedimentary rocks*. Geological Society Special Publication No 29, 77–87.

MALTMAN, A. J. 1988. The importance of shear zones in naturally deformed wet sediments. *Tectonophysics*, **145**, 163–175.

MORGENSTERN, N. R. & TCHALENKO, J. S. 1966. Microstructural observations on shear zones from slips in natural clays. *Geotechnical conference, Oslo, Proceedings*, **1**, 147–152.

PETERS, T. 1993. Chemical and physical changes in the subsoil of three waste landfills. *Waste Management and Research*, **11**, 17–25.

SHIPBOARD SCIENTIFIC PARTY 1988. Synthesis of shipboard results: leg 110 transect of the northern Barbados Ridge. *Proceedings of the Ocean Drilling Program Initial Report*, Part A, 110, 577–591.

TERZAHGI, K. & PECK R. B. 1967. *Soil Mechanics in Engineering Practice*. John Wiley and Sons.

WATTS, K. S. & CHARLES, J. A. 1990. *Proceedings of the Institution of Civil Engineers*, Part 1, **88**, 971–93.

The engineering of a cementitious barrier

P. A. Claisse[1] & H. P. Unsworth[2]

[1] School of The Built Environment, Coventry University, Coventry, UK
[2] Department of Civil Engineering, Universitee of Dundee, Dundee, UK

Abstract. When cementitious materials are used for the containment of waste they act in two different and sometimes conflicting ways. The first is physical containment in which the waste is physically isolated from the environment. The second is chemical containment in which the water passing through the barrier is buffered to high pH thereby very substantially reducing the solubility of many harmful species and promoting sorption onto the cement matrix. Chemical barriers have been extensively researched for nuclear waste containment and this paper explores the possibility of using them for non-nuclear waste.

Initial results are presented which show that cementitious barriers containing cement replacement materials may be well suited to barrier construction. It is, however, concluded that cementitious barriers would have to be used as part of a composite system in order to comply with current legislation.

The cementitious chemical barrier is one of the main engineering features of the current plans for a UK repository for medium and low level nuclear waste. The concept has been developed in response to a requirement for a barrier which will have a predictable performance in a deep saturated geological environment over a time-scale of up to a million years. The barrier is built out of conventional engineering materials but its method of operation is far from conventional for an engineering structure because it is essentially sacrificial. The main function of the barrier is to condition the chemistry of the repository to high pH by dissolving alkalis in the groundwater. The alkalis are free sodium, potassium and lime and subsequently the calcium silicate hydrate which forms the structure of the hardened cement.

This paper outlines the main features of a cementitious chemical barrier and explores the possibility of applying this technology to non-nuclear waste. Figure 1 shows schematic a representations of a repository for nuclear waste. The grout contains cement, lime, bentonite and other components, it is a soft material with a strength of about 4 MPa and is the main chemical barrier. Figure 2 shows a typical arrangement of a landfill for non-nuclear waste. The waste may have been solidified using cement or some other alkaline material but this is not normally the case and the chemical barrier could work by establishing a local alkaline region while the bulk of the waste remained acidic.

Objectives for waste containment

Short term

In this discussion the short term is considered to be the working life of the landfill and the early post-closure phase until the first deposited waste has been in place for about 50 years. This is the time when the 'landfill reactor' is working most effectively on the organic component of the waste. The objective for landfills with a substantial organic loading will therefore be to provide complete containment and a leachate balance which provides sufficient moisture to promote the reactions but controls the leachate head on the liner.

Long term

In the very long term the contents of a landfill will disperse into the environment from which they came.

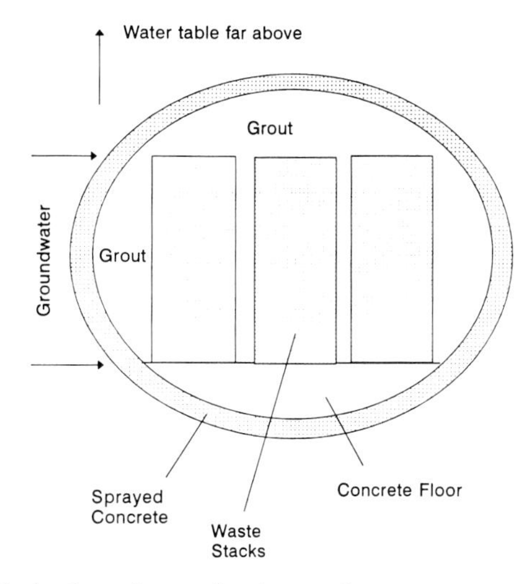

Fig. 1. A scheme for a radwaste repository.

From BENTLEY, S. P. (ed.) *Engineering Geology of Waste Disposal*,
Geological Society Engineering Geology Special Publication No. 11, pp 267–272

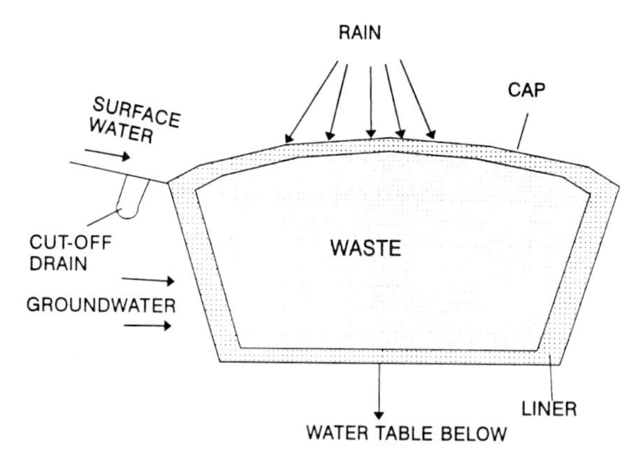

Fig. 2. A landfill.

For nuclear waste the objective is to contain the waste until the activity has substantially decayed but, once the organic degradation is complete, there is no further reduction in toxicity for non-nuclear waste. The long-term objective for non-nuclear waste containment may therefore be to provide an environmentally acceptable transition between the aim of absolute containment in the short term and the certainty of dispersion in the very long term. The absolute physical certainty that all landfills will eventually have to dilute and disperse their inventory of heavy metals and other stable toxins does not appear to be recognized by current legislation. It is of note that some waste materials (e.g. pulverized fuel ash) have a sufficiently low permeability that in normal deposition they do not generate leachate for about 30 years so short-term containment is irrelevant.

Properties of cementitious materials in barriers

Physical containment

Physical containment with concrete is well understood and documented. The degree of containment will depend on the permeability of the barrier. The permeability of concrete is relatively easy to measure and this has been achieved in a wide variety of different ways. One of the better methods is the use of a Hoek cell (Anon 1994).

Chemical containment

Chemical containment has been studied in detail for nuclear waste (UK Nirex 1993). In the type of repository for which a chemical barrier would be used, the main mechanism of loss of radionuclides is caused by flowing groundwater. This flow may be present in the area before the repository is built or it may be caused by the heat generated in the repository. In order to operate for a long time a chemical barrier depends on other barriers to limit the flow of groundwater through it. This is normally achieved by positioning the repository in a geology with a very low permeability. In this situation the permeability of the repository itself can be shown not to have a significant effect on the flow of water through it.

Thus water will enter the repository very slowly and the chemical barrier works by conditioning it before it reaches the waste and also after it leaves the waste but before it leaves the repository. Before it reaches the waste the barrier will raise the pH of the water, reduce the Eh, and remove many dissolved ions such as sulphates. In this way the barrier will ensure that the solubility of the radionuclides in the waste is as low as possible. For example, raising the pH from 8 to 12.5 will reduce the solubility of uranium by an order of magnitude, plutonium and protactinium by one and a half, and americium by three and a half. After the water leaves the waste the barrier will provide a high capacity for sorption to remove radionuclides from it.

It may be seen that, unlike a conventional engineering structure, the method of operation of a chemical barrier is sacrificial. As it operates the cement matrix carbonates and reacts with sulphates and other materials to an extent which would indicate failure in a conventional structure.

Calculating the rate of loss of radionuclides from the repository is made relatively simple by assuming that the chemical barrier is in complete equilibrium. Thus if the sectional area of the repository perpendicular to the direction of groundwater flow is A (m^2), the Darcy velocity of the flow is V ($m\,s^{-1}$) and the solubility of the radionuclide in the pore solution of the barrier is C ($mol\,m^{-3}$) the loss rate is CAV ($mol\,s^{-1}$). It should be noted that only a small fraction of the radionuclides in a cementitious barrier can be lost by this mechanism since most of them are sorbed onto the matrix. The ratio of those in solution to the total inventory in the barrier is P/α where P is the porosity and α is the capacity factor and typical values for this are 3×10^{-5} (Uranium) and 6×10^{-6} (Plutonium).

Measurements of the buffering and sorption capacity of concrete have been carried out using 'batch' experiments in which a solution with a fixed contaminant loading is mixed with ground concrete and permitted to come to equilibrium. The solution is then filtered off and analysed for pH and loss of contaminants. An alternative system is to use a 'through flow' cell in which the solution is pumped through an intact sample of concrete. The requirements of the apparatus are similar to those for permeability measurements and a modified Hoek cell is again very suitable for this.

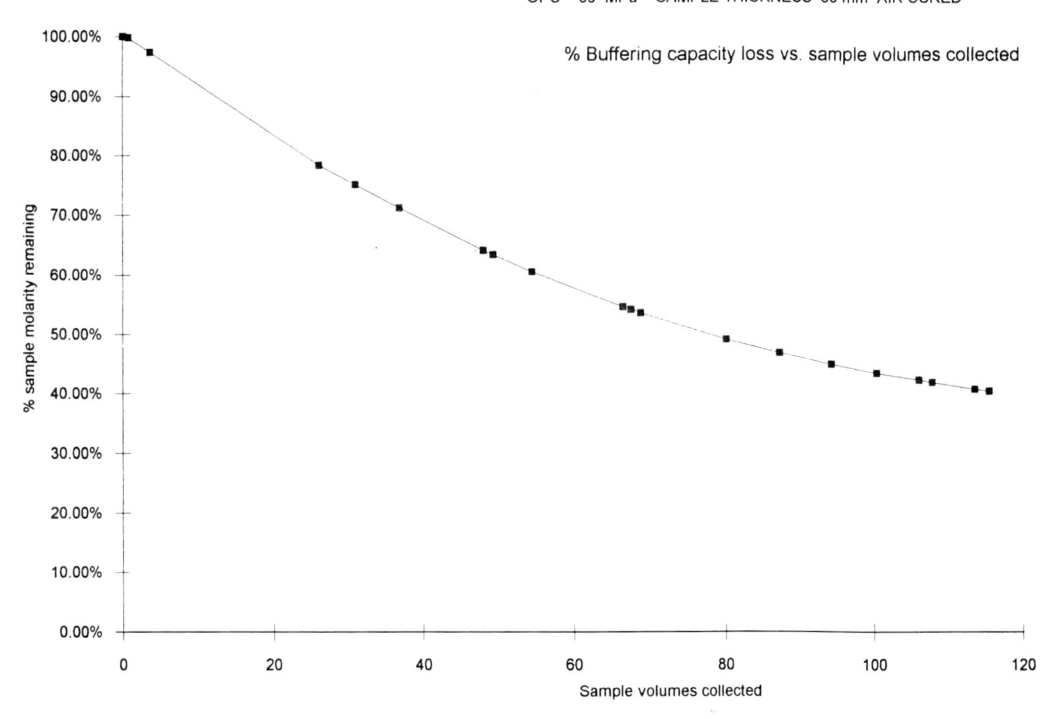

Fig. 3. Typical result from initial experiments.

Initial experimental observations

When a barrier material is tested it will act both physically and chemically as it would in service. Both types of containment are therefore measured in the same experiment. In the experiments which are reported here, samples of cementitious materials approximately 20 mm thick were exposed to water pressures up to 10 MPa in a 100-mm-diameter modified Hoek cell. To measure the physical containment the flow rate was measured and this was used to calculate a coefficient of permeability.

To measure the ability of the barrier to buffer the leachate, the buffering capacity of the water flowing from the cell was measured by titration. The theoretical buffering capacity of the sample was calculated and the proportion of this that was remaining was plotted against the number of sample volumes that had flowed through the cell. A typical plot is shown in Fig. 3. The results from the initial observations are summarized in Table 1, these show the effect of permeating a volume of water equal to 30 times the sample volumes.

Table 1. *Initial experimental results*

*Mix type**	*% buffering remaining after 30 volumes*	*Initial intrinsic permeability (m^2)*	*Intrinsic permeability after 30 volumes (m^2)*
OPC 35 MPa	94	2×10^{-16}	9×10^{-17}
OPC 20 MPa	70	9×10^{-16}	4×10^{-16}
60/40 OPC/PFA	90	9×10^{-18}	4×10^{-18}
55/45 OPC/PFA	98	8×10^{-17}	1×10^{-17}
45/55 OPC/GGBS	98	7×10^{-19}	5×10^{-17}

* OPC: Ordinary Portland cement; PFA: pulverized fuel ash; GGBS: ground granulated blastfurnace slag.

The results showed considerable variations between samples but it is clear that the blended cements release their buffering more slowly and would therefore appear to be more suitable for barriers.

Modes of failure

Exhaustion of buffering

The principal mode of chemical failure of the barrier is the exhaustion of its buffering and sorption capacity. Fortunately it is possible to carry out accurate calculations of the design life of the barrier to this mode of failure. The relationship between groundwater flow rates and the volume of material in the barrier indicates a life in excess of a million years for current design concepts for nuclear waste.

Cracking

Cracking is the obvious mode of physical failure, and has been the reason why concrete barriers have not been common in the past. Cracking could be caused by drying or thermal effects or the imposed stresses on the barrier. The problem is overcome by using composite barriers which are discussed in the last section of this paper (Fig. 5).

High pH 'boulder' formation

A possible cause of premature chemical failure which is the subject of a research programme is the formation of impermeable 'boulder-like' pieces with preferential flow paths for water around them. These boulders could develop impermeable surface layers through the formation of carbonates, chloroaluminates or magnesium compounds in a similar manner to that observed at the surface of existing concrete structures in hostile environments. If this occurred the alkaline buffering and sorption capacity of the interior of the boulders would be lost. In this way the total buffering and sorption capacity of the repository would be substantially reduced.

In current plans for nuclear waste it is envisaged that almost all of the cementitious material will be in the form of a soft grout. This material has been chosen to comply with various operational criteria including being readily pumpable into small spaces between the packages and having a low strength. These requirements have the effect that the formation of hard impermeable boulders will be strongly inhibited.

Action of sulphates

Sulphates react with hardened concrete and cause expansion of the matrix, which leads to a significant loss of strength. This effect may be prevented by the use of sulphate-resisting cement or blastfurnace slag cement. It is of note, however, that in a deep nuclear repository the effect is harmless because the expansion is contained by the surrounding rock. For non-nuclear applications where the containment pressures are insufficient it would be necessary to use sulphate-resisting cement but the aluminate phases of the cement which are omitted in order to give it sulphate resistance have substantial capacity for sorption so this method should not be used if it is avoidable. Cement replacement with blastfurnace slag probably represents a better option.

Gas transmission through the barrier

Gas generation

Gas is generated in significant quantities in most forms of waste. Organic materials are responsible for a number of gases including methane, carbon dioxide, nitrogen and hydrogen and this will be the main source of gas in a normal near-surface landfill. In deep nuclear waste disposal the main source of gas is hydrogen from the anaerobic corrosion of steel.

Transmission in saturated barriers

Concrete has a low permeability to water but if it is dry it will transmit gas very easily. This will be the situation if a concrete cap is used on a landfill and represents a major potential advantage for concrete barriers. If the concrete is saturated, however, the permeability will be very low. A major study has been carried out in this area for nuclear waste disposal (Harris *et al.* 1992) and the results of this are summarized in Fig. 4. In this study four different cementitious formulations were tested and for each there is a horizontal line on the graph drawn at a stress level corresponding to its tensile strength. The four mixes were 'PFA', a pulverized fuel ash concrete; 'BFS', a blastfurnace slag concrete; 'LIMESTONE', a grout with limestone dust in it; and 'LIME', a grout with lime in it. The saturated and dry coefficients of permeability of each mix were measured and are represented on the graph. Numerical modelling of the effect of releasing gas at the estimated generation rate for the repository into voids of different radii was carried out. These voids were assumed to be present above the vents which will be placed in the top of each drum of waste and the resulting stresses generated by the gas are shown by the curves on the graph for three different radii. The solid line is for a radius of 0.1 m which will be the radius of the vents. It may be seen from the graph that the calculations indicated that the grouts would survive but the concretes would fail. The backfill in the UK repository will be a soft grout so the results are satisfactory.

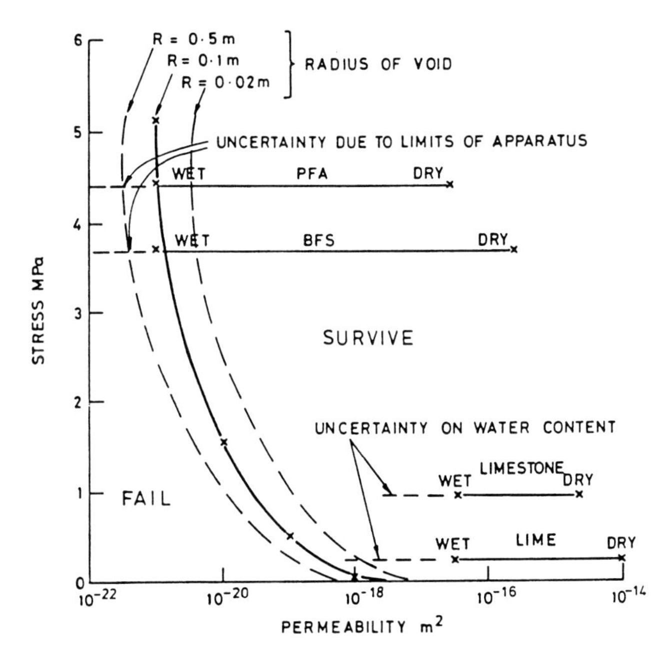

Fig. 4. Gas transport in a repository.

Pore water expulsion

It may be seen from Fig. 4 that there was some uncertainty with the measured saturated permeability of the grouts because they may not have been fully saturated. The samples were kept under water after casting and held in sealed containers over a flat water surface during testing but they still lost some weight. These grouts have substantial porosity with maximum radii in the 1 μm range. Classical calculations indicate that a pore with this radius should remain full at a relative humidity of 99.9% and have a capillary height of 14 m, and should thus have been full. As expected, pore water expulsion at pressures over about 150 kPa was observed. The observation that the pores tended to empty at pressures below this indicates, however, that barriers made with materials of this type might drain into gas-filled areas beneath them, permitting easy escape of the gas.

Design of cementitious barriers

It is clear that a single cementitious barrier will not provide sufficient short-term containment to satisfy current regulations for a liner. In order to take advantage of this type of barrier it will be necessary to use a composite system with several cementitious layers or layers of other materials. Figure 5 shows a schematic diagram of a composite barrier with several cementitious layers which has been proposed by the authors. The two concrete layers are different in that one is reinforced and the other is not. They will therefore crack in different places and through pathways will not be formed. The mix designs for the concrete layers are intended to ensure that they have unreacted PFA in them, so if cracks do form, lime leaching from the grout layer will react pozzolanically

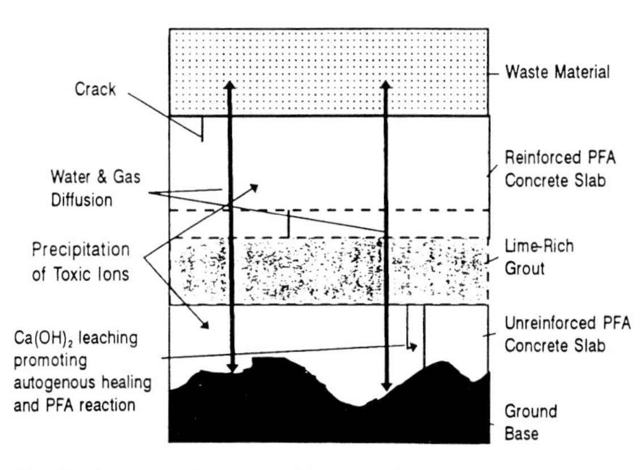

Fig. 5. A composite cementitious barrier.

with the PFA and seal them. Alternatively, high-density polyethylene (HDPE) liners could be used in the cementitious barrier. The HDPE would be protected by the concrete and help to cure it. Using a layer of bentonite in a concrete barrier might be possible but calcium bentonite reacts with the free lime in concrete and destroys both, so the type of bentonite to be used would have to be carefully selected and tested.

References

ANON. 1994. Technical literature on Hoek Cell and drainage platens. Robertson Geologging, Conwy, Gwynedd LL31 9PX.

HARRIS, A. W., ATKINSON, A. & CLAISSE, P. A. 1992. Transport of gases in concrete barriers. *Waste Management*, **12**, 155–178.

UK NIREX 1993. Report No. 525, Scientific Update. United Kingdom Nirex Ltd. Harwell, Didcot, Oxon, UK, 25.

A model for adsorption of organic species by clays and commercial landfill barrier materials

M. C. R. Davies, L. M. R. Railton & K. P. Williams

School of Engineering, University of Wales College of Cardiff, Newport Road, Cardiff CF2 1XH, UK

Abstract. The adsorptions of four carboxylic acids and phenol onto three substrates have been examined. It has been demonstrated that modifying montmorillonite with the diethyl ammonium ion increases the adsorption of the acids at the higher concentration. The amount of adsorption cannot be explained simply by consideration of adsorbate solubilities in water, but depends on more complex interactions between adsorbate–adsorbent and solvent. A model to predict the adsorptive lifetime of a barrier has also been developed.

It has long been recognized that clay minerals, in particular montmorillonite, have the ability to adsorb chemicals from aqueous environments and that this could be a valuable asset in designing barrier systems for landfill sites. Griffin *et al.* (1976) used column leaching methods to demonstrate the potential of clays for attenuating the constituents of landfill leachate. They found that heavy metals were adsorbed most efficiently followed by K, NH_4, Mg, Si, Fe, Cl, Na and finally organic compounds. The adsorption was affected by pH, cation exchange capacity of the clay, exchangeable cations and leachate composition. Similar conclusions were reached by Frost & Griffin (1977). Bagchi (1987) investigated the mechanisms of attenuation in the leachate–soil system, concluding that cation exchange, biological uptake and precipitation reactions were the controlling mechanisms. More recently, several papers dedicated to the studies of adsorption onto clays and soils were presented at Geoconfine 93. Of particular note in this respect are the papers by Dearlove (1993), Hockley *et al.* (1993), Sergeev *et al.* (1993), Wagner *et al.* (1993) and Pousette & Johnson (1993).

Organo-clay reactions have been studied extensively over the years and important reaction mechanisms have been discovered. Some of the main reviews include those by Greenland (1965), Mortland (1970), Brindley (1970) and Theng (1974), and these reviews should be consulted for a fuller discussion of the bonding mechanisms between the clay minerals and organic cations.

Certain properties of organic cations set them apart from metal ions and influence their adsorption onto clay surfaces. These include hydrogen bonding, ion-dipole and physical forces, and their importance depends on factors such as molecular weight, nature of the functional groups present and the configuration of the molecule (Mortland 1970). The primary bonding mechanisms between the organic cations and the clay surface are thought to be electrostatic, but it is thought that physical non-coulombic forces also contribute to adsorption (Raussell-Colom & Serratosa 1987). Van der Waals' forces control the adsorption process when large cations are involved and hydrogen bonding is dominant if the organic compound contains ammonium ions.

Adsorption onto the clay mineral depends on pH and dissociation constants. If the pH is too low adsorption will be hindered due to competition with the H^+ ions or metal cations. The solubility of the organic base in water also affects the rate of adsorption. Mortland *et al.* (1986) examined organic clays for their ability to adsorb phenol and chlorinated phenols. Clay–organic complexes containing the long-chain alkyl groups are the most hydrophobic and Mortland *et al.* found that phenols were adsorbed from water in proportion to their hydrophobicities, which increase with chlorine addition. Boyd *et al.* (1988) used similar clay–organic complexes to investigate the adsorption of pentachlorophenol. They found that generally the more hydrophobic the cation on the clay, the greater the uptake of PCP from water.

Experimental

Materials

Three clays were used in this study. One was a commercial sodium montmorillonite (NaMo) obtained from Steetly Minerals Ltd., UK. An organically modified clay (OMC) was prepared according to a method proposed by Mortland (1970). This involved hydrating 1 kg of sodium montmorillonite with 1.2 litres of distilled water in a mixer, before 200 g of dimethyl ammonium chloride was added. The organically modified clay was stored in slurry form.

The third clay was an industrial mixture of sodium montmorillonite, ordinary Portland cement, a retarder

From BENTLEY, S. P. (ed.) *Engineering Geology of Waste Disposal*,
Geological Society Engineering Geology Special Publication No. 11, pp 273–278

and a dispersant. The mixture is one that is typically used for slurry wall construction. It was received from a contractor engaged in an actual construction project, in solid form. It was slightly oxidized on the surface and was dried slowly in an oven at 70 °C and ground to a powder. All other chemicals were of Analar grade and all water was either distilled or demineralized. X-ray diffraction studies were performed on the three substances using a Philips Analytical PW 1729 X-ray generator. The XRD trace for sodium montmorillonite was found to coincide with that given in the *Mineral Diffraction File* (Anon. 1980). The OMC still had the sodium montmorillonite structure but the industrial material is mineralogically different, and is obviously not a clay. The XRD trace for this material shows more similarities with traces obtained by Johnson (1991) for OPC.

The acids used are as shown in Table 1, together with their chemical formulae.

Adsorption isotherms

A 2 g sample of each clay was placed in 250 ml of the appropriate solution at the specified concentration of the component of interest. Initial concentrations were chosen to reflect concentrations found in leachate from landfill sites. That is, concentrations of 20, 100, 200, 600, 1000, 2000, 5000 and 10 000 ppm for the organic acids and 2, 5, 10 and 20 ppm for phenol solutions. After 30 min of agitation by magnetic stirrer the solutions were centrifuged at 3000 rpm on an MSE centrifuge (Model No. GF8). The supernatant was collected and analysed. An ion chromatograph (Dionex 4000i) was used to analyse for the organic acids. The method used for the determination of phenol was taken from *Methods for the Examination of Waters and Associated Materials* (Anon. 1981) and a Unicam SP600 Series 2 Spectrophotometer was used for the analysis.

In order to allow comparisons with other data for inorganic species (Davies *et al.* 1993) the adsorption experiments were performed for 30 min. The authors note, however, that this may not be long enough for the organic acids to reach equilibrium with the clays. Kinetic studies were performed and it was found that although there was considerable adsorption in the first half hour, adsorption was still apparent after 48 hours.

Results

Carboxylic acids

The adsorption isotherms for acetic acid, lactic acid, malic acid and tartaric acid are presented in Figs 1, 2, 3 and 4, respectively. All of the curves show a similar trend in that adsorption is weak at low concentrations and then becomes stronger. These type V isotherms (Gregg & Sing 1982) are typical of weak adsorbent–adsorbate interactions, the adsorption increasing at the higher concentration as the result of co-operative bonding. At these higher concentrations it is clear that the adsorption is consistently greater on the organically modified clay (OMC) than on the sodium montmorillonite (NaMo), the data for the commercial barrier material being rather inconclusive because of curtailment of the studies at the lower concentrations due to a

Table 1. *Chemical formulae and solubilities of the adsorbates*

		Solubility in water (parts per 100 at 20 °C)*
Acetic acid	CH_3—COOH	∞
Lactic acid	CH_3—CH(OH)—COOH	∞
Malic acid	HOOC—CH_2—CH(OH)—COOH	0–140
Tartaric acid	HOOC—CH(OH)—CH(OH)—COOH	120–140
Phenol	C_6H_5OH	8.2

* Range of solubilities corresponds to the isomeric form of the compound. Source of data: *Chemical Engineers Handbook*, Perry and Chilton, McGraw-Hill.

lack of samples. However, the adsorption is suitably encouraging to warrant extension of the study.

Examination of Figs 1 and 4 shows that the adsorptions of acetic and tartaric acids are very similar but much lower than the corresponding behaviour for malic and lactic acids (Figs 3 and 4). These latter adsorbates again show similar behaviour, with the malic acid exhibiting slightly greater affinity for the OMC at the highest concentrations.

Phenol

Figure 5 demonstrates that the isotherms for phenol are also likely to be of type V, albeit that the concentration range is much smaller than that used for the acids to reflect the lower levels commonly encountered in landfill leachates. The adsorption is relatively strong and is similar on the two clays.

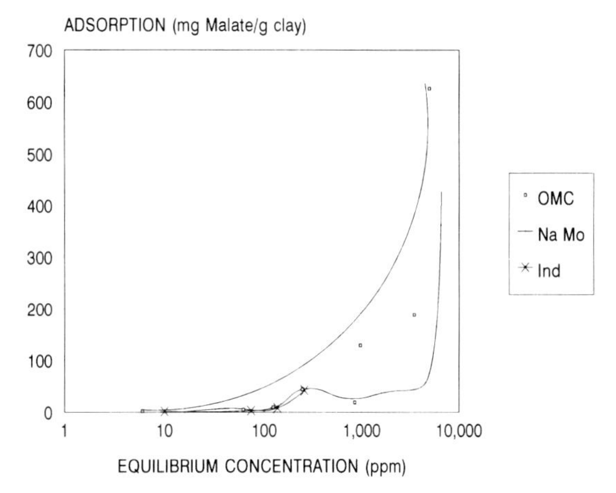

Fig. 3. Adsorption isotherms for malic acid.

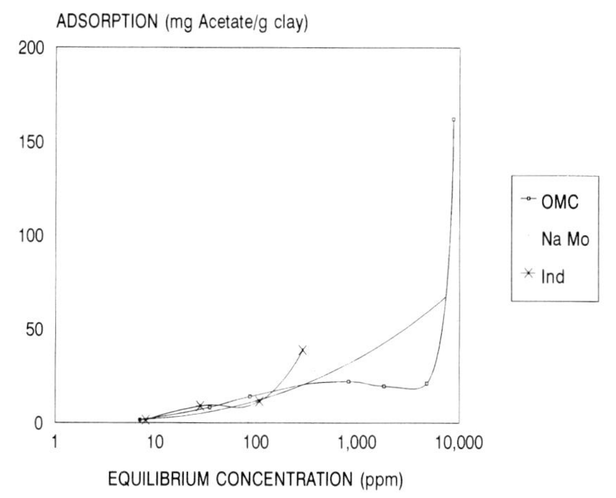

Fig. 1. Adsorption isotherms for acetic acid.

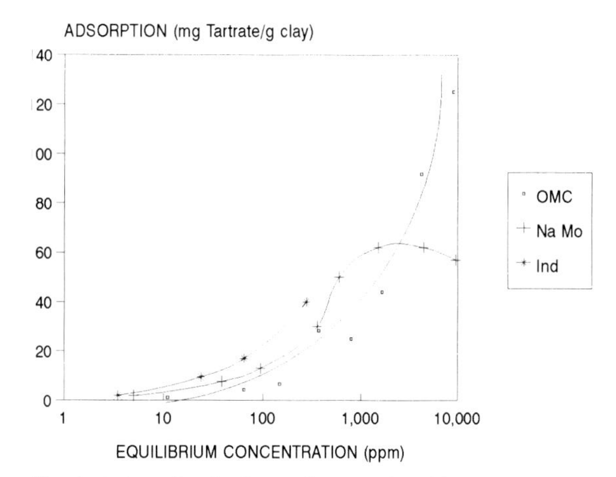

Fig. 4. Adsorption isotherms for tartaric acid.

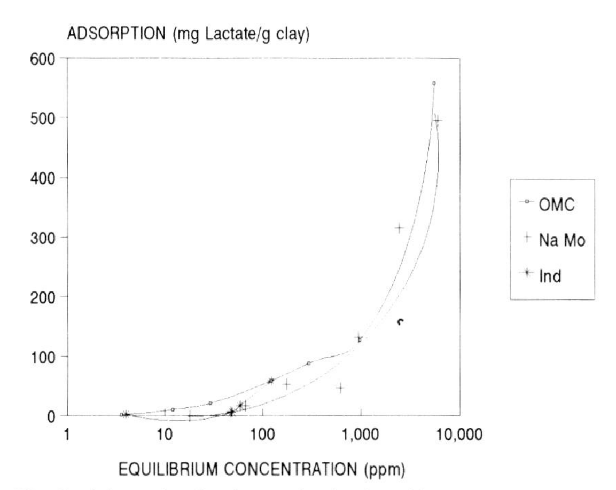

Fig. 2. Adsorption isotherms for lactic acid.

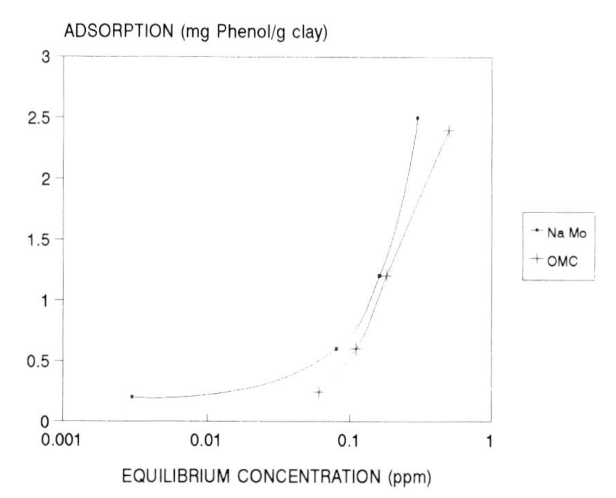

Fig. 5. Adsorption isotherms for phenol.

Discussion

Mortland *et al.* (1986) and Boyd *et al.* (1988), whilst studying the adsorption of phenol and chlorophenols onto organo-clays, concluded that there was a good correlation between the hydrophobic character of the adsorbate and the amount of adsorption obtained. For these materials it was found that the adsorption increased as the solubility of the component in water decreased. On this basis it would be predicted that the order of adsorption onto OMC for the compounds studied in the present work would be:

phenol > malic > tartaric > lactic = acetic

Allowing for the fact that phenol was only considered at low concentrations it was seen that the actual order was:

phenol > malic > lactic > tartaric > acetic

indicating that solubility in water, whilst allowing a reasonable assessment to be made of the likely adsorption, can be misleading. It is obviously necessary to consider other factors such as hydrogen bonding, coulombic interactions, the geometry of the molecule and dipole moments. Dipole moments and solubility are, however, directly related and solute insolubility in water is the primary factor affecting the soil organic matter–water partition coefficient, K_{om} (Chiou *et al.* 1983).

Prediction of adsorptive lifetime of a barrier wall

Despite much research activity it is still not possible to predict exactly what will happen to the components of a leachate as they pass through a clay barrier. Sergeev *et al.* (1993), whilst appreciating this difficulty, attempted to predict the lifetime of a barrier system from a knowledge of readily measured parameters. The following equation is a modified form of that presented by Sergeev *et al.* (1993) but with the insertion of the density term which was omitted in error in the original publication:

$$t = \frac{Ad\rho_d N_{max}}{qc} \quad (1)$$

in which t is the lifetime of the barrier for complete adsorption, A is the area of the barrier wall, d is the thickness of the barrier, ρ_d is the mass of dry clay per unit volume of barrier (i.e. dry density), q is the volumetric flow rate of leachate, c is the mass of component per unit volume of leachate, and N_{max} is the maximum value on the static adsorption isotherm (mass of component adsorbed per unit mass of clay).

It can be shown that,

$$\rho_d = \left(\frac{G_s \rho_w}{1+e}\right) \quad (2)$$

in which G_s, ρ_w and e are the specific gravity of the soil particles, the mass density of water and the void ratio of the barrier, respectively. In addition, the flow of water through the barrier is governed by Darcy's law, viz:

$$q = Aki \quad (3)$$

where k is the permeability of the barrier and i the hydraulic gradient (i.e. the head drop across the barrier, Δh, divided by the thickness of the barrier, d).

The dry density and area terms are eliminated by substitution of Equations (2) and (3) into Equation (1). This yields an equation for predicting the lifetime of the barrier, t, in terms of readily obtainable geometric and material parameters, i.e.

$$t = \frac{d^2}{\Delta h} \cdot \frac{G_s \rho_w}{k(1+e)} \cdot \frac{N_{max}}{c} \quad (4)$$

Equation (4) may be used to estimate the adsorptive capacity of a barrier wall, using the N_{max} values obtained from the adsorption isotherms. To illustrate its application an example using the model is presented below. A site in the Cardiff Bay Development area will be used for this example; more details of the site can be found in Railton (1994). The adsorptive life span of the clay barrier will be calculated considering the ionic species individually rather than as in a leachate mix. This is an oversimplification because the interaction of other components in the leachate system will cause a reduction in the adsorptive capacity of the clays as opposed to the individual adsorption of the species (Railton 1994). The width of the slurry wall constructed at the site was $d = 0.6$ m. Consolidation tests (Davies *et al.* 1993; Railton 1994) indicated that typical values of void ratio, e, for the sodium montmorillonite and organically modified clay are 15 and 2.5 respectively, and that a representative permeability for these clays is $k = 1.5 \times 10^{-9}$ m s^{-1}, which is of the order normally specified in practice. A head drop across the barrier, Δh, of unity and a specific gravity of the individual particles in the barriers $G_s = 2.65$ are selected as typical values for use in this analysis.

Table 2 illustrates the length of time it would take to saturate a sodium montmorillonite and an organically modified slurry wall with the considered contaminants. An initial concentration was chosen for the acids to reflect concentrations found in leachate produced in a landfill site.

From the results presented in Table 2 it can be seen that the time to saturate the organically modified barrier is much greater than that of the sodium montmorillonite. This is due to the fact that organically modified clay will usually adsorb more organic species than the sodium montmorillonite, and also the organic clay is denser because it has a much smaller voids ratio and thus more sites available for adsorption. In both cases

Table 2. *Length of time for total saturation of a sodium montmorillonite wall, with individual species*

		N_{max} (mg g^{-1})		*t (years)*	
Component	*c* (mg l^{-1})	*NaMO*	*OMC*	*NaMO*	*OMC*
Lactic acid	6000	495	558	104	536
Malic acid	6000	426	625	89	600
Tartaric acid	6000	57.1	125	12	120
Acetic acid	6000	54.8	162	12	156
Phenol	2	2.5	2.4	1576	6915

the results presented in Table 2 are only for individual pollutants, not as combined in leachate. Nevertheless, the data indicate that clay barriers are capable of adsorbing chemical species from leachates over considerable periods of time. Modifying the clay, for example by organic substitution in the present work, can increase the adsorptive lifetime very significantly for reasons of change in geotechnical properties as well as surface chemistry. It is interesting to note that seepage velocities in barriers constructed from either of the clays considered in this work would mean that contaminated leachate would pass through the barrier in less than 10 years. The adsorptive capability of the barrier is, therefore, of obvious environmental benefit. However, if clay barrier systems are to remain effective for periods of several decades, then the structural properties must also remain, and this is a subject requiring further investigation.

Acknowledgements. The work described has been financed by the Science and Engineering Research Council and Cardiff Bay Development Corporation under the Total Technology Scheme. Thanks are also extended to the Welsh Development Agency for their support in this work, and to the staff at Celtic Technologies Ltd for useful discussions.

References

ANON. 1980. *Mineral Powder Diffraction File Search Manual.* International Centre for Diffraction Data, Swarthmore, USA.

ANON 1981. *Methods for the examination of waters and associated materials. Determination of monohydric phenols in effluents and wastewaters, 4-aminoantipyrine (pH 10) spectrophotometric method.* HMSO, London.

BAGCHI, A. 1987. Natural attenuation mechanisms of landfill leachate and effects of various factors on the mechanisms. *Waste Management and Research*, **5**, 453–464.

BOYD *et al.* 1988. Pentachlorophenol sorption by organo clays. *Clay and Clay Minerals,* **36**(2), 125–130.

BRINDLEY, G. W. 1970. Organic complexes of silicates. *An. Reun. Hisp. Belga Min. Arci.*, Madrid, 55–56.

CHIOU, C. T., PORTER, P. E & SCHMEDDING, D. W. 1983. Partition equilibria of nonionic organic compounds between soil organic matter and water. *Environ. Sci. Technol.* **17**(4), 227–231.

DAVIES, M. C. R., RAILTON, L. M. R. & WILLIAMS, K. P. 1993. Adsorption and consolidation properties of modified bentonites. *In*: ARNOULD, M., BARRES, M., & COME, B. (eds), *Geology and confinement of toxic wastes*, vol. 2, 671–678.

DEARLOVE, J. P. L. 1993. Geochemical interaction processes between landfill clay liners and leachate. *In*: ARNOULD, M., BARRES, M., & COME, B. (eds) *Geology and confinement of toxic wastes*, vol. 1, 330–39.

FROST, R. R. & GRIFFIN, R. A. 1977. Effect of pH on adsorption of copper, zinc and cadmium from landfill leachate by clay minerals. *J. Environ. Sci. Health*, **A12**(4, 5), 139–156.

GREENLAND, D. J. 1965. *Soils Fert.* **9**, 457.

GREGG, S. J. & SING, K. S. W. 1982. *Adsorption, surface area and porosity.* Academic Press, New York, 303.

GRIFFIN, R. A. *et al.* 1976 Attenuation of pollutants in municipal landfill leachate by passage through clay. *Environmental Science and Technology*, **6**, 1262–1268.

HOCKLEY, D. E. *et al.* 1993. Processes affecting contaminant release from waste—soil interfaces. *In*: ARNOULD, M., BARRES, M., & COME, B. (eds) *Geology and confinement of toxic wastes*, vol. 1, 73–79.

JOHNSON, A. G. 1991. *Cement stabilisation of tailings*, PhD thesis, University of Wales College of Cardiff, Department of Materials and Minerals.

MORTLAND, M. 1970. *Advances in Agronomy*, **22**, 75–117.

——, SHAOBAI, S. & BOYD, S. A. 1986. Clay–organic complexes as adsorbents for phenol and chlorophenols. *Clays and Clay Minerals*, **34**(5), 581–585.

OWESIS. I. S. & KHERA, R. P. 1990. *Geotechnology of waste management.*

POUSETTE, K. & JACOBSON, A. 1993. Liner-test—permeability and fixation of leachate compounds in natural barrier materials. *In*: ARNOULD, M., BARRES, M., & COME, B. (eds) *Geology and confinement of toxic wastes*, vol. 1, 251–255.

RAILTON, L. M. R. 1994. *Adsorption and consolidation properties of modified bentonites.* PhD thesis, University of Wales College of Cardiff, Division of Materials and Minerals.

RAUSSELL-COLOM, J. A. & SERRATOSA, J. M. 1987. Reactions of clays with organic substances. *In*: NEWMAN, A. C. D. (eds) *Chemistry of clays and clay minerals*. Elsevier Applied Science, London and New York.

SERGEEV, V. I. *et al.* 1993. Procedure of investigating a subsoil layer as a geochemical barrier for heavy metals. *In*: ARNOULD, M., BARRES, M., & COME, B. (eds) *Geology and confinement of toxic wastes*, vol. 1, 115–121.

THENG, B. K. G. 1974 *The chemistry of clay organic reactions*. Adam Hilger, London.

WAGNER, J-F. *et al.* 1993. Migration of heavy metals in natural clay barriers. *In*: ARNOULD, M., BARRES, M., & COME, B. (eds) *Geology and confinement of toxic wastes*, vol. 1, 129–132.

Diffusion of contaminants through a clay barrier under acidic conditions

A. M. O. Mohamed & R. N. Yong

Geotechnical Research Centre, McGill University, 817 Sherbrooke St West, Montreal, Canada H3A 2K6

Abstract. In this study, the characteristics of contaminant–soil interaction and retention regarding soil and influent pH values were examined. Leaching column tests were used to investigate the retention capacity of a natural clay soil. The laboratory tests showed good attenuation for Pb^{2+} and Zn^{2+} and lesser attenuation for Na^{+}, K^{+}, Ca^{2+} and Mg^{2+} salts under acidic conditions.

To calculate the diffusion parameters, a phenomenological approach was developed. The technique is based on an analytical solution via a Fourier series, a non-dimensional analysis and the measured experimental data. The calculated diffusion parameters were similar to values reported in the literature. The developed technique can be used to calculate diffusion parameters for both adsorption and desorption processes.

Evaluation of contaminant migration through clay barriers has been the focus of recent investigations by many researchers. For example, Elliott & Liberati (1981) and Yong *et al.* (1986, 1990a) among others have reported that soils have a high capacity to retain heavy metals. The studies showed that the capacity of soil to retain heavy metals could be influenced by several factors depending on soil constituents, leachate and waste characteristics. However, little attention has been given to the migration of contaminants through clay barriers under acidic conditions. Acidic leachate can be generated from sulphide ore operations of mining waste, for example. The sulphide minerals are unstable and oxidize when exposed to oxygen and water. The resultant reaction yields sulphuric acid which increases the solubility of heavy metals and promotes their mobility. Rainfall and snow melt flush the toxic solutions from the waste sites into the downstream environment. As the generated toxic solutions move downward, various contaminants interact with soil components, resulting in such processes as physi-sorption, chemi-sorption, precipitation or complexation (Yong *et al.* 1992a). The result is the retention of contaminants by soil constituents.

It has been shown by Harter (1983); Yong *et al.* (1990b); Yong & Phadungchewit (1993), and Mohamed *et al.* (1994) that soil pH is an important factor in heavy metal retention and movement in soils. Hence, the ability of soils to retain heavy metals depends on the resistance of soils to a change in pH. The migration of low-pH leachate through soil would affect the soil pH and the retention of heavy metals in the soil. This effect might be small or large depending on the capacity of soil to resist a change in pH. Furthermore, if an acidic leachate was to pass through the soil continuously, the capacity of the soil to resist a change in pH would decrease. Hence, the soil's ability to retain heavy metals would decrease.

Many available models for transport and fate of various contaminant species in the subsurface environment are based on the following assumptions:

(1) the material is homogeneous, isotropic and non-deformable;
(2) the material has low permeability;
(3) contaminant transport is governed by a steady flow of an incompressible fluid;
(4) sorption reaction is linear and reversible; and
(5) coupled flow processes are neglected.

From a practical viewpoint all the above assumptions can be justified except assumption (4) which is highly dependent on soil pH. For example, for both high and low values of pH, precipitation and dissolution occur respectively. This in turn could invalidate assumption (4). Furthermore, the predictive models require knowledge of the spatial and the temporal variability of parameters used as input data to calculate the diffusion parameter, yet these are difficult to provide. Therefore, a technique based on collected effluent concentrations as a function of time should be developed to calculate the diffusion parameters.

In this study, the characteristics of contaminant–soil interactions and retention regarding soil and leachate pH values are examined. Natural clay soil was used in this investigation. To calculate the diffusion parameters, an analytical technique is developed. The technique is based on: (1) a Fourier series; (2) non-dimensional analysis, and (3) measured experimental data. Furthermore, a root-time method is developed to calculate the diffusion parameters.

Material and methods

The natural micaceous soil used in this investigation contains illite, phlogopite, hydrobiotite and vermiculite

From Bentley, S. P. (ed.) *Engineering Geology of Waste Disposal*, Geological Society Engineering Geology Special Publication No. 11, pp 279–289

as basic minerals. The detailed mineralogical analysis of the soil used was reported by Mohamed *et al.* (1994). Specific surface area (206 $m^2 g^{-1}$) was determined by using the ethylene glycol monoethyl ether (EGME) adsorption method (Carter *et al.* 1965). The cation exchange capacity of the soil was determined using two methods: (1) the batch equilibrium test (ASTM D4319: 1984), which gave a CEC of 14.89 meq per 100 g; and (2) the silver–thiourea method (Chhabra *et al.* 1975), which gave a CEC of 13.2 meq per 100 g. The low CEC values indicate that the soil is mainly illitic and micaceous in composition. The engineering properties of this soil are: (1) a maximum dry density of 1.81 $Mg m^{-3}$; (2) an optimum moisture content of 16.1%; and (3) a permeability of $2.3 \times 10^{-9} m s^{-1}$.

Test procedure

Leaching column tests were conducted to study the adsorption characteristics of the soil under investigation. Sample dimensions were 0.11 m in diameter and 0.12 m in height. Soil was compacted at its optimum moisture content and maximum dry density using the static compaction test method. Leaching was conducted using municipal solid waste leachate spiked with Pb^{2+}, Zn^{2+}, Na^+, K^+, Mg^{2+} and Ca^{2+} in the form of chlorides. The pH of the reconstituted leachate was also lowered by adding concentrated HCl. The chemical composition of the reconstituted leachate is as follows: (1) heavy metal concentrations of 1372.2 ppm Pb^{2+} and 1141.6 ppm Zn^{2+}; (2) cation concentrations of 346 ppm Na^+, and 164.8 ppm K^+, 43.8 ppm Mg^{2+} and 95.4 ppm Ca^{2+}; (3) pH 1.33; and (4) a specific conductivity of 16.833 $mS cm^{-1}$.

Leaching was carried out under a constant pressure of 103.5 kPa, i.e. an equivalent water head of 10.6 m, resulting in a hydraulic gradient of 87.2. During the leaching process, effluent was collected as a function of time and analysed for chemical composition. After the test, samples were extruded and cut into 10-mm-thick slices. The soils in each slice were analysed for the pore fluid concentrations (i.e. soluble and exchangeable) using the batch equilibrium test (ASTM 1984). Cations and heavy metal concentrations were determined using atomic adsorption spectrophotometry (AAS) after various appropriate dilutions. Chloride was determined using titration with $AgNO_3$. The pH and specific conductivity were determined using a pH meter and the electrophoretic mass transport analyser respectively.

Results and discussions

Migration of heavy metals

In a clay soil system, heavy metals may be present in the following forms: (1) ion-exchange sites; (2) incorporated into or on the surface of crystalline or non-crystalline inorganic precipitates; (3) incorporated into organic compounds; or (4) in soil pore solution (Yong *et al.* 1992a). Most investigators have recognized that heavy metals occur predominantly in a sorbed state. Because of their low solubility, movement of heavy metals in soil has generally been considered to be small.

The effluent relative concentration, C_e/C_o, for Pb^{2+}, and Zn^{2+}, as a function of the collected pore volumes is shown in Fig. 1, where, C_e is the concentration of a

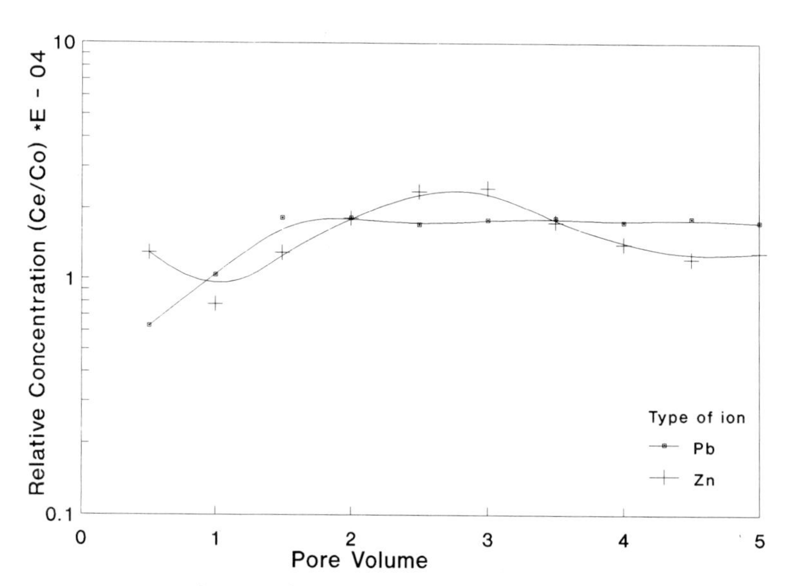

Fig. 1. Variations of Pb^{2+} and Zn^{2+} relative concentrations with accumulated pore volumes.

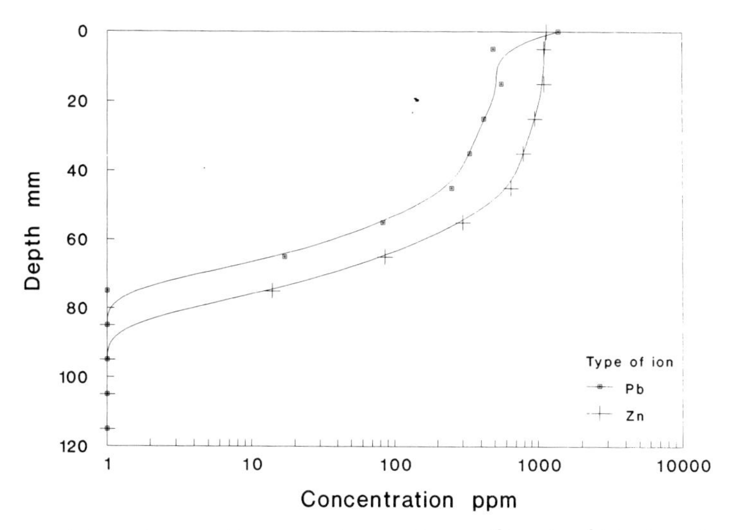

Fig. 2. Variations of pore fluid concentration profiles of Pb^{2+} and Zn^{2+} in the soil column.

particular ion in the effluent, and C_o is the original concentration of a particular ion in the influent. It can be seen that very little Pb^{2+} and Zn^{2+} was collected in the effluent, indicating that most of the Pb^{2+} and Zn^{2+} was retained by the soil. The results show that significant amounts of heavy metals were retained in the top portion of the soil samples, as seen in the concentration profiles depicted in Fig. 2. Due to the low pH of the influent, it is expected that the retention capacity of the soil in the top part of the column will be reduced with time. However, this depends on the buffering capacity of the soil to any change in pH. It is known that heavy metals would generally precipitate if the solution pH is high, e.g. Pb^{2+} precipitates at $pH > 5$. Since soil pH was initially about 6.5, therefore, Pb^{2+} precipitates in the soil at the start of leaching. Further leaching decreases soil pH and enhances the mobility of Pb^{2+} in solution (Mohamed *et al.* 1994). After five pore volumes the top 25% of the soil pH ranges from 1.33 to 5, which enhances the mobility of Pb^{2+}. In the top part of the soil column, Pb^{2+} was retained by cation exchange. For the rest of the soil column with a $pH > 5$, the Pb^{2+} retention mechanism is due to precipitation in a hydroxide form. It can also be seen from Fig. 2 that the amount of Zn^{2+} retained in the soil column is less than the amount of Pb^{2+} retained. This can be explained by the ease of exchange or the strength with which cations of equal charge are held. Adsorption is generally inversely proportional to the hydrated radii, or proportional to the unhydrated radii (Bohn *et al.* 1979). Therefore, the predicted order of soil retention based on unhydrated radii is Pb^{2+} (0.120 nm) > Zn^{2+} (0.074 nm) (Elliott *et al.* 1986); and based on metal ion softness, which is function of ionization potential, charge of metal ion and ionic radius (Stumm & Morgan 1981), is Pb^{2+} (3.58) > Zn^{2+} (2.34). Both predictions are in agreement with the experimental data.

Migration of cations

The effluent relative concentration, C_e/C_o, in the leachate for Na^+, K^+, Mg^{2+} and Ca^{2+} is shown in Fig. 3. With an increasing number of pore volumes permeated, the relative concentrations of cations increases. For Na^+, Mg^{2+} and Ca^{2+}, the relative concentrations are greater than 1.0 due to desorption of cations from the solid particles. The high relative concentrations of Ca^{2+} and Mg^{2+} in the collected effluent could be attributed to cation exchange, i.e. replacement by Pb^{2+} and Zn^{2+}. Due to the adsorption and incorporation of K^+ into the interlayer lattice of micaceous soils, the recorded concentrations of K^+ in the effluent are low. Also, Na^+ relative concentrations reached steady-state conditions after the passage of approximately four pore volumes, whilst for K^+, Mg^{2+} and Ca^{2+}, steady-state conditions arrived after approximately two pore volumes.

The pore fluid cation migration versus soil column depth profiles are shown in Fig. 4. The migration profiles depict how a particular cationic species migrates through the soil column with increasing permeation by the leachate. The initial concentration of Na^+ in the influent was 345 ppm, while the measured concentrations in the pore fluid were greater than 500 ppm which indicated Na^+ desorption. Similar results are obtained for K^+, Mg^{2+} and Ca^{2+}. As discussed, this behaviour is attributed to cation exchange, i.e. replacement by Pb^{2+} and Zn^{2+} in the top part of the soil column; whilst, at

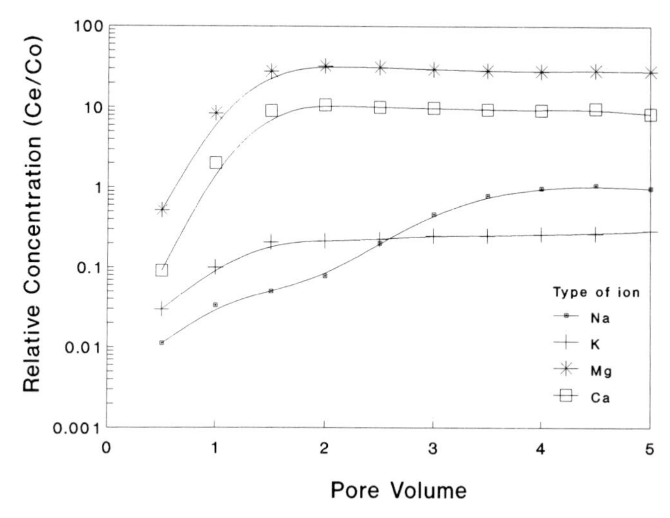

Fig. 3. Variations of cation relative concentrations with accumulated pore volumes.

the bottom part of the column, the interaction mechanism is due to cation exchange process between Ca^{2+}, Mg^{2+}, Na^{+} and K^{+}, i.e. the replacement of Na^{+} and K^{+} ions in the exchangeable sites by Ca^{2+} and Mg^{2+} to balance the charge deficit that occurs by the desorption of Na^{+} and K^{+}, as shown in the results (Fig. 5) depicting exchangeable cations versus depth of the soil column. These observations are similar to those made by Yong *et al.* (1986) for results of Na^{+} and Ca^{2+} migration in their studies, suggesting a requirement to balance the charge deficit that occurs by the desorption of sodium. Similar results were obtained by Crooks & Quigley (1984) for results of Ca^{2+} and Mg^{2+} migrations, suggesting a requirement to balance the charge deficit that occurs by the desorption of Na^{+} and Cl^{-} ions.

The experimental profiles of the total exchangeable cations versus depth of the soil column are shown in Fig. 5. As discussed earlier, the concentrations of the exchangeable cations in the top part of the soil column were decreased due to low pH values (1.33 to 5). The range of corresponding soil column depths is 0–35 mm. With the depth increase, the concentrations of exchangeable cations are increased, hence the retention capacity of the soil is increased.

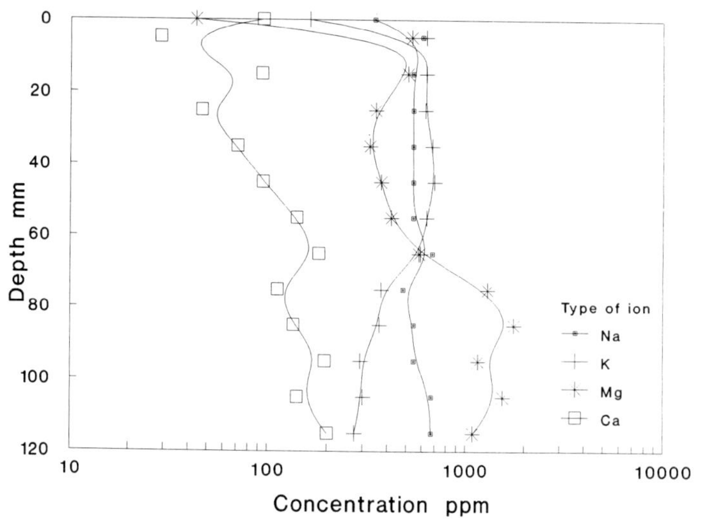

Fig. 4. Variations of pore fluid concentration profiles of cations in the soil column.

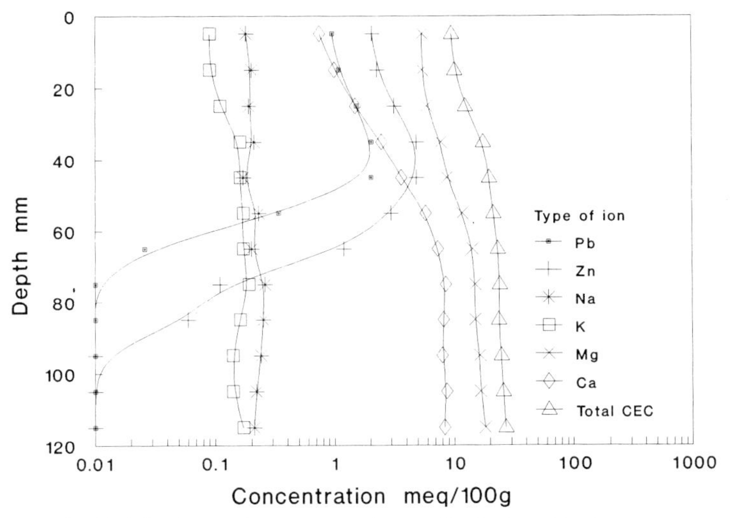

Fig. 5. Adsorbed cation concentration profiles in the soil column.

Theoretical analysis

Flow of solutes due to diffusion in one-dimension takes the following form:

$$J_c = -D \frac{\partial C}{\partial x} \tag{1}$$

where C is the solute concentration in the pore fluid, x is distance; t is time, and D is a diffusion parameter.

If there is an external force, F, acting in the x-direction upon the dissolved molecules or colloidal particles, and if u be the mobility of the particles under consideration, the steady velocity of these particles will be $F \cdot u$, and the resulting flux is:

$$J_h = CFu = Cv. \tag{2}$$

Consequently, the total current, J, due to diffusion and the action of the eternal force (i.e. hydraulic) will be:

$$J = J_c + J_h = -D \frac{\partial c}{\partial x} + Cv \tag{3}$$

and the rate of change of concentration with time, due to this current

$$\frac{\partial C}{\partial t} = D \frac{\partial^2 C}{\partial x^2} - \frac{\partial}{\partial x}(Cv). \tag{4}$$

If the force, F, and, consequently, the steady velocity, v, are independent of x, Equation (4) is reduced to:

$$\frac{\partial C}{\partial t} = D \frac{\partial^2 C}{\partial x^2} - v \frac{\partial C}{\partial x}. \tag{5}$$

Equation (5) is the diffusion–convection equation in its one-dimensional form. Furthermore, Equation (5) can be reduced to the ordinary diffusion equation by the following transformation (Furth 1931):

$$C = C_T \exp\left[\frac{v}{2D}(x - x_0) - \frac{v^2 t}{4D}\right]. \tag{6}$$

Substitution into Equation (5) gives the differential equation for C_T

$$\frac{\partial C_T}{\partial t} = D \frac{\partial^2 C_T}{\partial x^2}. \tag{7}$$

The boundary conditions corresponding to the experiments are given by

$$C_T = C_1 \quad \text{at} \quad x = 0; \qquad C_T = C_2 \quad \text{at} \quad x = L. \tag{8}$$

The initial conditions corresponding to the experiments are given by

$$\begin{aligned} C_T &= C_1 \quad \text{at} \quad x = 0 \\ C_T &= C_2 \quad \text{at} \quad 0 < x < L. \end{aligned} \tag{9}$$

Equation (7) can be presented in a non-dimensional form by using the following representation:

$$C^* = \frac{C_T - C_2}{C_1 - C_2}; \qquad \tau = \frac{Dt}{L^2}; \qquad \xi = \frac{x}{L} \tag{10}$$

where C^* is the non-dimensional solute concentration; τ is a non-dimensional time factor; L is the length of the soil column, and ξ is the non-dimensional distance.

The governing equation in non-dimensional form can be written as follows:

$$\frac{\partial C^*}{\partial \tau} = \frac{\partial^2 C^*}{\partial \xi^2}. \tag{11}$$

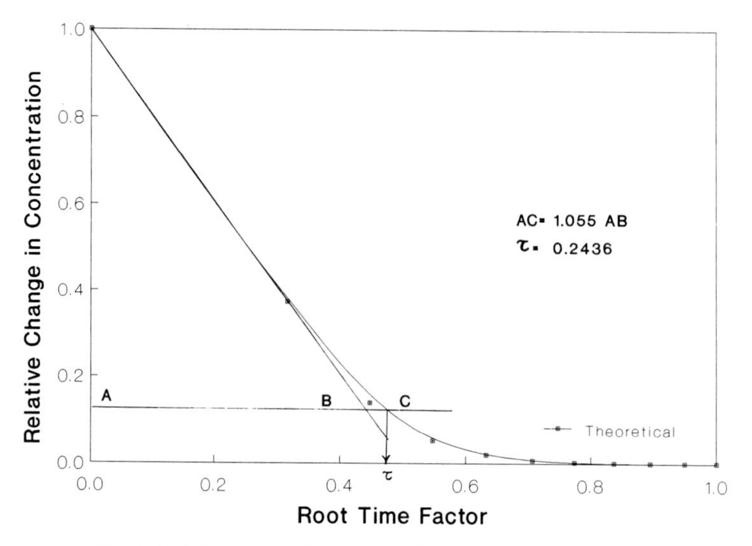

Fig. 6. Relative change in concentration with root time factor.

The boundary conditions in non-dimensional form are given by:

$$C^* = 1 \quad \text{at} \quad \xi = 0; \qquad C^* = 0 \quad \text{at} \quad \xi = 1. \tag{12}$$

The initial conditions in non-dimensional form are given by:

$$C^* = 1 \quad \text{at} \quad \xi = 0; \qquad C^* = 0 \quad \text{at} \quad 0 < \xi = 1. \tag{13}$$

The analytical solution for this initial-boundary value problem can be obtained by using a Fourier series as follows:

$$C^*(\xi, \tau) = (1 - \xi) + \sum_{n+1}^{\infty} a_n(\tau) \sin(n\pi\xi). \tag{14}$$

Equation (14) shows that the boundary conditions given by Equation (12) are automatically satisfied. Substituting Equation (14) into Equation (11), one gets

$$\dot{a}_n(\tau) = (n\pi)^2[-a_n(\tau)]. \tag{15}$$

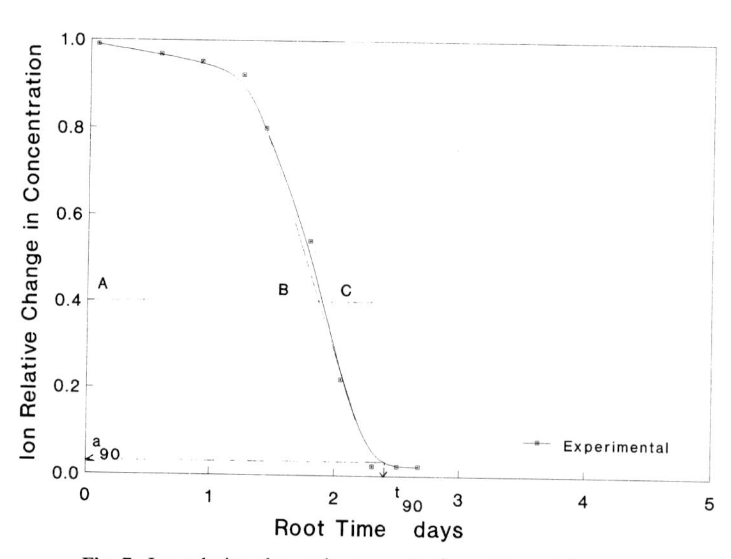

Fig. 7. Ion relative change in concentration with root time.

Hence

$$a_n(\tau) = A_n \exp(-(n\pi)^2\tau). \quad (16)$$

Using the initial conditions represented by Equation (13), A_n can be determined. After lengthy derivations, A_n takes the following form:

$$A_n = -\frac{2}{n\pi}. \quad (17)$$

Substituting Equations (16) and (17) into Equation (14), the following expression for non-dimensional solute concentration can be obtained

$$C^*(\xi,\tau) = (1-\xi) - \sum_{n-1}^{\infty}\left(\frac{2}{n\pi}\right)\exp(-n(n\pi)^2)\sin(n\pi\xi). \quad (18)$$

If τ is large enough (i.e. steady-state condition), Equation (18) can be approximately written as

$$C^*(\xi,\tau) \approx (1-\xi) - \frac{2}{\pi}\exp(-\pi^2\tau)\sin(\pi\xi). \quad (19)$$

Determination of the diffusion parameter

The diffusion parameter is calculated based on the theoretical relationship between $C^*(\xi,\tau)$ given by Equation (19) and the root-time method which will be explained in this section. The procedure will be as follows.

For $\xi = \frac{1}{2}$, Equation (19) is reduced to

$$C^*(\tfrac{1}{2},\tau) = \tfrac{1}{2} - \frac{2}{\pi}\exp(-\pi^2\tau). \quad (20)$$

The relative change in non-dimensional solute concentration, C^*_{RC}, may be expressed as:

$$C^*_{RC} = \frac{C^*(\frac{1}{2},\tau) - C^*(\frac{1}{2},\infty)}{C^*(\frac{1}{2},0) - C^*(\frac{1}{2},\infty)} = \exp(-\pi^2\tau). \quad (21)$$

The relationship between C^*_{RC} and given by Equation (21) is shown in Fig. 6. The theoretical curve is linear up to $C^*_{RC} = 0.2$ (80% equilibrium), and at $C^*_{RC} = 0.1$ (90% equilibrium) the abscissa (AC) is 1.055 times the abscissa (AB) of the production of the linear part of the curve. This characteristic is used to determine the point on the experimental curve corresponding to $C^*_{RC} = 0.1$.

The experimental data reduced to relative change of concentration of specific ion in the collected effluent versus root time generally consists of a short curve representing initial increase in concentration (in the effluent), a linear part and a second curve. The point D (shown in Fig. 7) corresponding to the initial condition is obtained by producing back the linear part of the curve to the ordinate at zero time. A straight line (DE) is then drawn having an abscissa 1.055 times the corresponding abscissa on the linear part of the experimental part. The intersection of the line (DE) with the experimental curve locates the point (a_{90}) corresponding to $C^*_{RC} = 0.1$ and the corresponding value t_{90} can be obtained. From Fig. 6, the value of s corresponding to $C^*_{RC} = 0.1$ is 0.2436 and the diffusion coefficient, D, is given by:

$$D = \frac{0.2436L^2}{t_{90}}. \quad (22)$$

Application

The experimentally measured cation (Na^+, K^+, Mg^{2+} and Ca^{2+}) and anion (Cl^-) concentrations in the effluent are used to calculate the diffusion parameters based on Equation (22). The data are shown in Figs 8 and 9 regarding the relative change of concentrations versus root time. For example, Fig. 8(a) shows the calculated relative change of concentrations versus root time for Na^+ ion. The relative change of concentrations $(C - C_f)/(C_0 - C_f)$ is calculated as follows:

(1) C is assigned to the measured Na^+ concentration values in the effluent as a function of time;
(2) C_0 is assigned zero, since the initial concentration of Na^+ in the soil pore fluid was zero; and
(3) C_f is assigned to the concentration of Na^+ in the influent (i.e. 346 ppm).

Note that the relative change of concentrations could be positive or negative, depending on whether the ions are adsorbed or desorbed. For example, for Na^+, K^+, and Cl^-, the relative change of concentrations is positive, whilst for Mg^{2+} and Ca^{2+}, the relative change of concentrations is negative due to ion desorption.

Using the procedures outlined above and the experimental data shown in Figs 8 and 9, the calculated diffusion parameters for the case of adsorbed cations (Na^+ and K^+), desorbed cations (Mg^{2+} and Ca^{2+}) and chloride (Cl^-), movement are shown in Table 1. Also, the diffusion parameters reported by Yong & Warith (1990) are shown in the table. The soil used by Yong & Warith (1990) was natural clay from Montreal with a CEC of 60 meq per 100 g; a permeability of 7×10^{-9} m s^{-1}; a dry density of 1.39 Mg m^{-3}; and a specific surface area of 80 m^2 g^{-1}. A hydraulic pressure head of 2.5 m of water was used. The finite-difference technique was used to calculate the diffusion parameters, which requires concentration data as a function of space and time. It shows that the difference in the calculated diffusion parameters are of one order of magnitude. This difference could be attributed to: (1) the initial and boundary conditions used in both studies; (2) the hydraulic pressure used in this study is being three times higher than that used by Yong & Warith (1990).

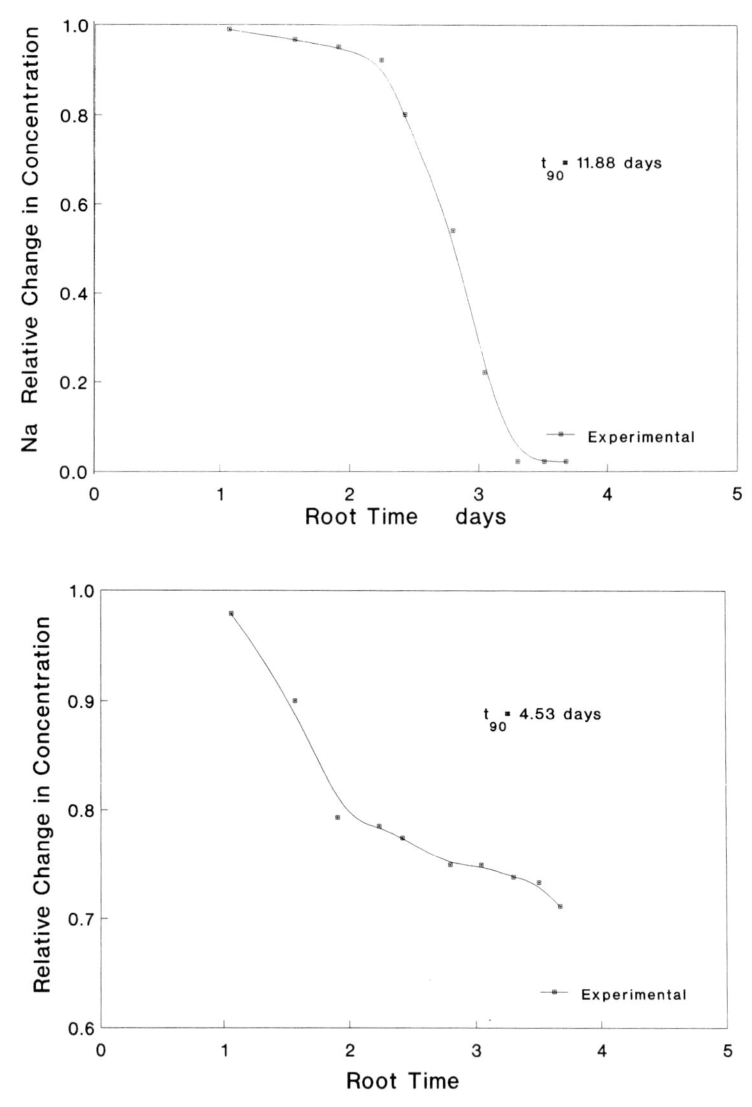

Fig. 8. (a) Na^+ relative change in concentration with root time. (b) K^+ relative change in concentration with root time. (c) Mg^{2+} relative change in concentration with root time. (d) Ca^{2+} relative change in concentration with root time.

This, in turn, can be attributed to the calculated high diffusion parameters.

In this study, the calculated diffusion parameter for anion Cl^- is almost the same as those calculated for cations (Mg^{2+} and Ca^{2+}). This could be attributed to desorption of Mg^{2+} and Ca^{2+} with time. Hence their movements were similar to anion Cl^-. It should be emphasized that data reported in literature showed that diffusion parameters of anions are higher than cations due to adsorption of cations. This is not so in this study due to changes in soil pH during the process of leaching.

Conclusion

Leaching column tests were conducted to study the adsorption characteristics of natural clay under acidic conditions. The recorded effluent pH values were 5.7–6.9 in spite of the low pH (1.3) of the acidic influent. This is indicative of the high buffering capacity of the natural soil. The pore fluid pH values of the soil column changed form 1.3 to 5 in the top part of the soil column (approximately 25% of the soil column), while the rest of the soil column had a pH of around 6. The pore fluid

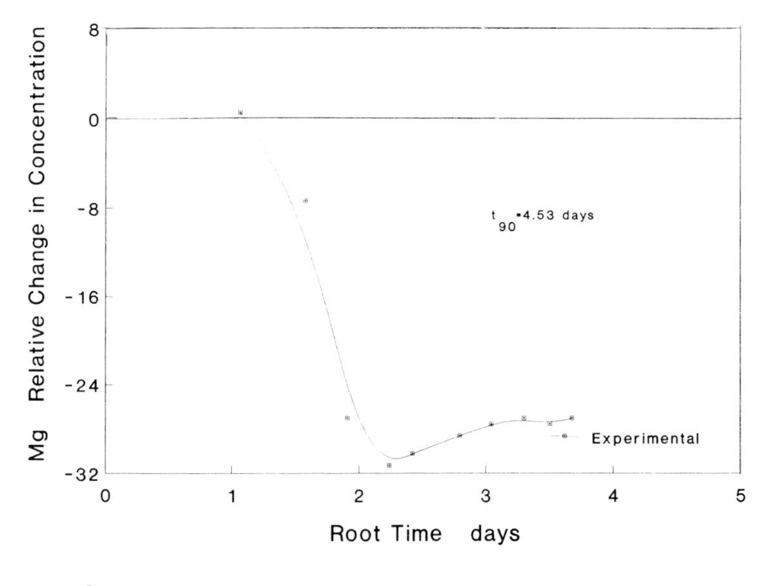

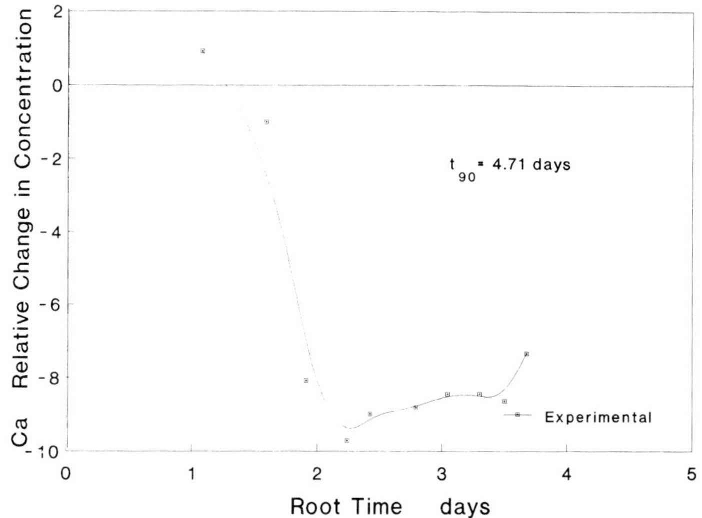

concentration profiles indicated that breakthrough ($C_e/C_0 = 0.50$) occurred for Na^+, Mg^{2+}, Ca^{2+} and Cl^- below the maximum five pore volumes of leaching conducted, with desorption of cations from the solid particles. On the other hand, the natural soil has a high attenuation capacity for K^+, Pb^{2+} and Zn^{2+}, as indicated from the concentration profiles of the soil pore fluid with depth after leaching.

The exchangeable cation variations with depth indicated high cation exchange or retention at the upper portions of the soil column, in particular Pb^{2+} and Zn^{2+}. Retention of Pb^{2+} and Zn^{2+} is mainly through precipitation (high pH values of >5) and cation exchange, the latter being confined to the upper portions of the soil column only (pH <5).

Furthermore, a phenomenological approach is proposed in this study to evaluate the diffusion parameters. The approach is based on an analytical solution for the differential equation of solute transport in a clay barrier. The diffusion equation is first cast in a non-dimensional form and the Fourier series is used to solve the differential equation for specified initial and boundary conditions. Data reduction based on the proposed technique and the experimental data are presented. The calculated diffusion parameters are average values to reach steady-state conditions. The technique can be used to calculate the

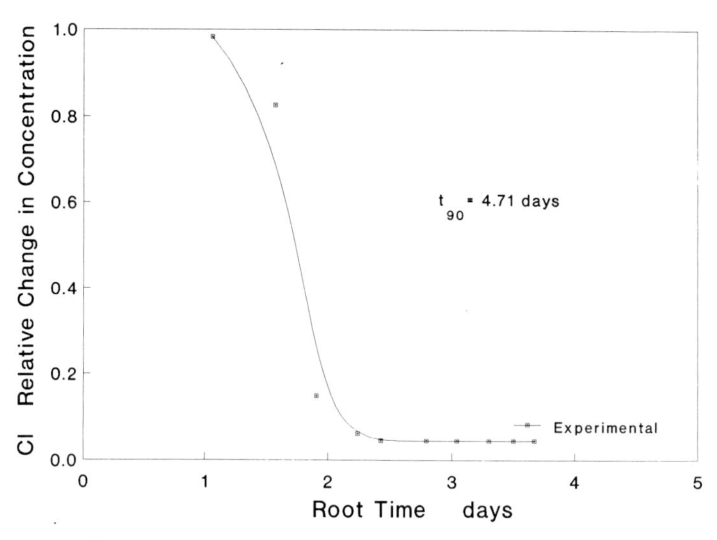

Fig. 9. Cl^- relative change in concentration with root time.

Table 1. *Calculated diffusion parameters*

	Ion				
	Na^+	K^+	Mg^{2+}	Ca^{2+}	Cl^-
Diffusion parameter × (10^{-4} m^2 day^{-1})					
This study	3.002	7.878	7.878	7.573	7.573
Yong & Warith (1990)	—	0.4	0.14	0.12	0.33

diffusion parameters for both adsorption and desorption processes. The calculated diffusion parameters for various cations and anions were one order of magnitude different from those reported by Yong & Warith (1990) using the finite-difference technique. The difference was mainly attributed among other things to the high hydraulic gradient used in this investigation.

Acknowledgements. This study was supported by grants in aid of research from the Natural Sciences and Research Council (NSERC), Grant No. OGP0046418 for the first-named author and Grant No. A-882 for the second-named author.

References

ASTM 1984. ASTM D4319-83: *Standard test method for distribution ratios by the short-term batch method.* Annual ASTM standards, **4.08**, Soil and Rock; Building Stones, American Society for Testing and Materials, 766–773.

BOHN, H. L., NCNEAL, B. L. & O'CONNOR, G. A. 1979. *Soil chemistry*. Wiley-Interscience Publications, John Wiley.

CARTER, D., HEILMAN, T. & GONZALEZ, J. 1965. Ethylene glycol monoethyl ether for determining surface area of silicate minerals. *Soil Science Journal*, **100**, 356–361.

CHHABRA, R., PLEYSIER, J. & CREMERS, A. 1975. The measurement of the cation exchange capacity and exchangeable cations in soil: a new method. *Proceedings of the International Clay Conference, Illinois, USA*, 439–448.

CROOKS, V. E. & QUIGLEY, R. M. 1984. Saline leachate migration through clay: a comparative laboratory and field investigation. *Canadian Geotechnical Journal*, **21**, 349–362.

ELLIOTT, H. A. & LIBERATI, M. R. 1981. Properties affecting retention of heavy metals from wastes applied to north-eastern US soils. *In*: HUANG, C. P. (ed.) Industrial waste. Proc. 11th Mid-Atlantic Conference, Newark, DE., Ann Arbor Science, 95–104.

——, —— & HUANG, C. P. 1986. Competitive adsorption of heavy metals by soils. *Journal Environmental Quality*, **15**, 214–219.

FURTH, R. 1931. Diffusion. *In*: AURBACH, F. & HART, W. (eds) *Handbook Physik. Techn. Mechanik*, **7**, 635.

HARTER, D. 1983. Effect of soil pH and adsorption of lead, copper, zinc and nickel. *Soil Science Society American Journal*, **47**, 47–61.

MOHAMED, A. M. O., YONG, R. N., TAN, B. K., FARKAS, A. & CURTIS, L. W. 1994. Geo-environmental assessment of a micaceous soil for its potential use as an engineered clay barrier. *Geotechnical Testing Journal*, GTJODJ, **17**(3), 291–304.

STUMM, W. & MORGAN, J. J. 1981. *Aquatic chemistry*. 2nd edn. John Wiley & Sons, New York.

YONG, R. N. & PHADUNGCHEWIT, Y. 1993. pH influence on selectivity and retention of heavy metals in some clay soils. *Canadian Geotechnical Journal*, **30**, 821–833.

—— & WARITH, M. 1990. Contaminant migration effect on dispersion coefficients. *In*: HODDINOTT, K. B. & LAMB, R. O. (eds) *Physico-chemical aspects of soil and related materials, ASTM STP 1095*. American Society for testing and Materials, 69–80.

——, —— & BOONSINSUK, P. 1986. Migration of leachate solution through clay linear and substrate. *In*: LORENZEN, D. *et al.* (eds) *Hazardous and Industrial Solid Waste Testing and Disposal: Sixth Volume, ASTM STP 933*, American Society for Testing and Materials, 208–225

——, MOHAMED, A. M. O. & SAMANI, H. V. 1990a. Adsorption/desorption in multicomponent contaminant transport in clay barriers. CSCE, 1990 Annual Conference, Hamilton, Ontario, 689–706.

——, WARKENTIN, B. P., PHADUNGCHEWIT, Y. & GALVEZ, R. 1990b. Buffer capacity and lead retention in some clay materials. *Water, Air, and Soil Pollution*, **53**, 53–67.

——, MOHAMED, A. M. O. & WARKENTIN, B. P. 1992a. *Principles of contaminant transport in soils*. Elsevier, Amsterdam.

——, TAN, B. K. & MOHAMED, A. M. O. 1992b. Assessment of chemical buffering capability of a micaceous soil. *International Conference on the Implications of Ground Chemistry/Microbiology for Constructio*. University of Bristol, UK, 1–12.

An assessment of permeability of clay liners: two case histories

N. J. Langdon, M. J. Al Hussaini, P. J. Walden & C. M. Sangha

Department of Civil Engineering, University of Portsmouth, Burnaby Building, Burnaby Road, Portsmouth PO1 3QL, UK

Abstract. Two sites, each utilizing natural clay from the same geological formation but from sources in different depositional basins, are compared together with their method of site control and their test results. From a series of laboratory test results the geotechnical differences in the material are compared and tentative trend lines for various parameters drawn. Conclusions from these two sites and these soils suggest that precise determinations of permeability would not have been easily predicted from relationships in published information. However, there remains a major need for a method of determining permeability appropriate to the site conditions and limitations pertaining to a landfill site. The concept of the falling head test 'failing safe' and determining a rough threshold level at $10^{-9}\,m\,s^{-1}$ can be used but an on-site water absorption approach or infiltrometer may be more readily adaptable to determine permeability rapidly in quantities and also cost-effectively.

This paper describes two sites in southern England that are currently being developed as landfill sites and utilizing imported clay material to form the landfill liner. Although the material comes from geologically the same formation, results clearly demonstrate some significant differences in the Hampshire and London depositional basins.

The first site is an existing gravel pit/landfill site situated at Netley, Southampton, which is being expanded and modernized to cater for increased demand and the changes in recent legislation. The geology of the landfill site precludes the use of an on-site natural clay liner and the liner uses material from another landfill site some 30 km along the M27 at Southleigh near Emsworth and close to the Hampshire/West Sussex border. Both sites are owned by Leigh Environmental Limited. The excavation of London Clay from Southleigh creates void space and provides the material for the construction of waste disposal cells at Netley. The London Clay contains varying amounts of silt, siltstone and occasionally sand, which requires the adoption of rigorous site control and fast, cost-effective monitoring and soil testing.

The second site is situated at Gerrards Cross, Buckinghamshire, and has been operational for more than 20 years. Leigh Environmental Limited have been operating the site since June 1991, the previous owner being The Gerrards Cross Sand and Gravel Company. The void space has been created by the extraction of sand and gravel which still occurs on a small scale. The capacity of the site is some $4\times10^6\,m^3$ and has an anticipated life of 15 years. half the site has already been restored, with the remainder to be lined with clay from a variety of sources but consisting of London Clay. This part of the site is to be used for either difficult or special waste or for domestic, industrial and inert waste. Extraction of landfill gas to an existing on-site 3.6 MW power station is carried out by Greenlands.

Site locations

The Netley site is situated to the southeast of Southampton between Old Netley and Burseldon and is some 3 km from the M27. The area of the site is approximately 36 ha and the containment area forms some $170\,000\,m^2$ of the site. The eastern part of the site has for some years served as an 'attenuate and disperse' site taking commercial, industrial and domestic dry waste from Southampton and Fareham.

The site at Gerrards Cross lies within Wapsey's Wood which is approximately 3 km west of the town. It is bounded to the south by the M40 and the A40 London to Oxford road to the north. The site is a multi-user site with batching and ready-mix plant as well as landfilling. The total area of the site is about 150 ha.

Geology

The Netley site is shown by the 1:50000 British Geological Survey Sheet (Sheet 315) to be underlain by River Terrace Gravels associated with the Rivers Test and Itchen, and the Marsh Farm and the Earnley Sands belonging to the Bracklesham Group. These strata are described in detail by Edwards & Freshney (1987a). The Terrace Gravels have long been extracted commercially

From BENTLEY, S. P. (ed.) *Engineering Geology of Waste Disposal*,
Geological Society Engineering Geology Special Publication No. 11, pp 291–297

in the area and the Earnley Sand, typically some 15 m or so thick, comprises glauconitic clayey and silty fine-grained sand. It is being extracted at Netley by Hall Aggregates (South Coast) Ltd for Building sand. Neither these strata nor the laminated clayey and sandy soils of the Marsh Farm Formation are suitable for the construction of containment cells for domestic refuse disposal. Small quantities of the most clayey facies have been used for capping the existing landfill where permeabilities are shown to be acceptable.

The source of liner material is a landfill site at Southleigh some 2 km north of Emsworth near Havant in Hampshire. A site investigation of this site showed it to be underlain by London Clay and more particularly the stratum referred to as the Bognor Member. This is described as a glauconitic bioturbated or cross-bedded fine- to medium-grained sand with nodules of shelly sandstone by Edwards & Freshney (1987a, b). Borehole logs indicate the material to be a generally stiff dark-grey clayey Silt or silty Clay, the former being very typical of material ascribed to the London Clay Formation in the Hampshire Basin which also contains significant sequences of sands and silty sands (Edwards & Freshney 1987a, b). Experience of the formation in the London Basin and indeed the vast majority of information in published literature (Cripps & Taylor 1981) would lead the engineer to believe that this formation almost without exception is an ideal material for landfill lining. Typically plasticity indices for London Clay are quoted as between 40% and 65% and clay content in the range 40% to 70%.

Local experience of the Hampshire Basin strata shows that the material can be variable and often occurs as a sandy silty clay or clayey silt. Indeed some exposures of the formation are recorded as at least one-third sand and laminated sands and silts. It is not uncommon to find plasticity indices dropping to 20% or 30% and with correspondingly lower clay contents in the range 20% to 30% in the recognizably clayey fraction. Preliminary tests from Southleigh suggested that with good site control and monitoring the material would provide satisfactory material for use at Netley to give a maximum permeability of $10^{-9}\,\mathrm{m\,s^{-1}}$ when compacted.

The site at Gerrards Cross is underlain by a thin sequence of the Reading Formation which in turn overlies the Upper Chalk, and it lies at the very edge of the London Basin to the northwest of London. These strata are overlain by Glacial Gravel which is worked for the construction industry. Source material for the liner at Gerrards Cross was London Clay from a number of locations in the London Basin, in particular the London Ring Main and sites near Hillingdon, northwest London. The precise geology of each source is not known but the material conformed to type for the material in the nothwest of the London Basin. This has necessitated rigorous site control and testing.

Site operations

Netley

Material was excavated from two borrow pits at Southleigh to a depth of some 10 m, which allowed adequate side slopes to be maintained within the constraints of the site boundaries. This left approximately 50 m of the remaining London Clay Formation and the Reasing Formation shielding the Upper Chalk aquifer. Not only did the borrow pits provide material for use at Netley but also useful void space for controlled disposal of inert fill at Southleigh. Careful selection of material avoided incorporation of siltstone fragments in the clay fill for Netley although some hand-picking had to be done occasionally. An average volume of 1200 $\mathrm{m^3}$ of clay was transported daily 30 km down the A27/M27 during the filling season. A total of 90 000 $\mathrm{m^3}$ was placed in the first year to construct the first three containment cells at Netley. Completion of the project will result in approximately 16 such cells. Each cell was approximately 1 ha in size. The basal liner was 1 m thick with side slopes at the perimeter of 2 m thick. Excavation and transportation was restricted to days on which the clay could be placed and compacted since no stockpiling was allowed. Compaction was to Department of Transport specifications involving the use of a sheepsfoot roller of 4000 $\mathrm{kg\,m^{-1}}$ width capacity and eight passes on lifts of no more than 250 mm. A smooth vibrating roller was used at the end of shifts to roll to closure and ensure as weather-resistant a formation as possible. On occasions, water was added by bowser prior to placing to assist with compaction. Rapid moisture content determinations were made on site by using a microwave and a balance accurate to 0.01 g. However, the decision to add water to the clay was mostly based on physical and visual observation and was found to be regularly necessary during the dry periods of the last two summers. Batters and intermediate cell walls were overfilled and trimmed to side slopes of 1.5 : 1, which, although steeper than conventionally used in earthworks, is required to maximize available volume and is justified by the short period between construction and filling with waste. Regular sampling of the liner material and completed works was undertaken.

Gerrards Cross

A clay stockpile of some 70 000 $\mathrm{m^3}$ of material from a number of sources was built up over a two-year period in anticipation of lining operations. Further clay, mainly from the London Ring Main, was imported during filling operations. The containment cells were built to have a basal liner thickness of 2 m with bunds 3 m thick tapering to 2 m at the top. At the periphery of the tip the external bunds were trimmed on a 1 : 3 slope to ensure a minimum of 1 m of clay liner. Work over 1992 amounted to 105 000 m^3 to construct two cells. The clay was compacted by a sheepsfoot roller of some 12 t, in eight passes on lifts of 250 mm. A smooth vibratory roller was used to roll to closure.

Due to the variation of sources, constant monitoring of the fill was required to prevent contamination of the liner by unsuitable material. A number of operators were employed to act as pickers and control staff.

Testing programme

Netley

Hampshire County Council required a minimum of two permeability tests, two moisture content tests and four *in situ* dry density tests at regular intervals on the recompacted clay. In practice, the number of moisture content tests was increased to four, with two 100-mm open-tube/core-cutter samples and two U38-mm samples taken for assessment of density, and a falling head permeability test on the 100 mm samples.

In addition to these tests, a series of particle size distribution tests were also done to confirm the proportion of clay in the liner material, which was crucial. A typical grading curve is shown Fig. 1. Bulk disturbed samples of clay were taken for B. S. 1377 2.5-kg rammer compaction tests and comparisons made with *in situ* density tests. Where the liner material failed to comply with the permeability requirements, and furthermore the particle size distribution tests indicated low fractions of clay, the liner material was removed and new material compacted and tested. A average of more than one permeability test every 200 m^3 was achieved in the first season with typically moisture content, PSD and density determinations every 1000 m^3.

Gerrards Cross

Buckinghamshire County Council required a minimum of one sample per 5000 m^3 of imported material to be tested for plasticity index, optimum and natural moisture content, PSD and coefficient of permeability. It also required *in situ* clay to be tested every 5000 m^3 for permeability and for moisture content and dry density every 700 m^3. In practice, the testing regime produced one permeability determination every 3000 m^3 or less, and one density or moisture content determination every 400 m^3. Most moisture content or dry-density determinations were done by the site engineer using 50-mm samples with 100-mm core cutter samples taken for permeability testing using the triaxial equipment.

During construction of the two containment cells, all *in situ* samples were found to comply with the specified coefficient of permeability of 10^{-9} m s^{-1}. This was due to the very strict monitoring of material. Furthermore, the two to three week turn-round of permeability results meant any non-compliance would result in an expensive reconstruction of the completed works.

A PSD grading curve of the typical source material is shown in Fig. 2.

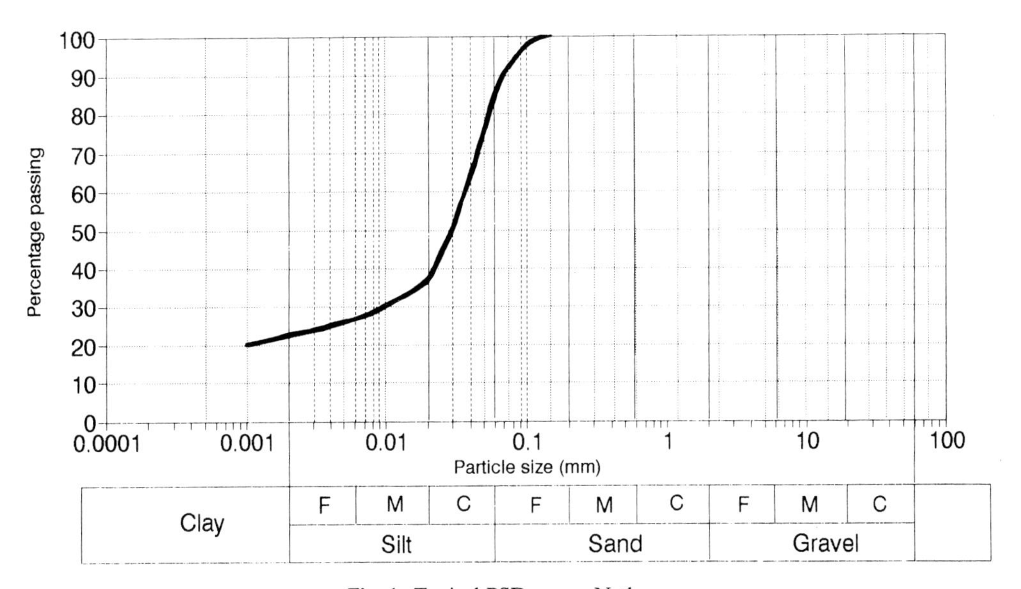

Fig. 1. Typical PSD curve, Netley.

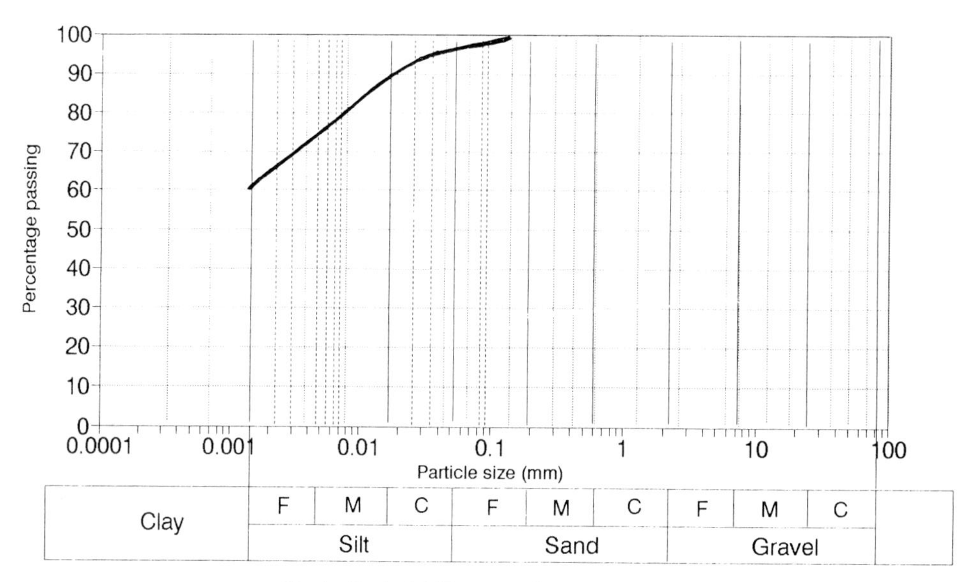

Fig. 2. Typical PSD curve, Gerrards Cross.

Verification of permeability specifications

The taking of the maximum coefficient of permeability for clay to be used for landfill liners as $10^{-9}\,\mathrm{m\,s^{-1}}$ (NRA 1989) has given the industry a number of headaches in suitably verifying this critical parameter. The triaxial cell permeability test has found increasing favour with regulatory authorities for its determination. It is well recognized that the inherent variability of soils gives particular difficulty in obtaining repeatable results. In addition, such permeability tests, as carried out in the triaxial cell, take time to saturate and test, use relatively expensive equipment which is often not available in any quantities in many laboratories, and are subject to variations by individual laboratories. Very often the turn-round of results is not appropriate for the rate of construction during a busy summer earthworks programme.

Enquiries during early 1992 showed a threefold variation of cost between laboratories. In addition, sample diameters varied, as did the aspect ratios for the test specimen, which varied from 2:1 to 1:1 and 1:2. There was a total unwillingness on the part of some laboratories to commit themselves to testing the two or three samples a week that only a single modest earthworks programme can generate. In addition, hydraulic gradients in excess of 50 were reported in some tests and it remains unclear, at least to the authors, as to what situations in landfill containment cells may give rise to these sorts of hydraulic gradients. It is not surprising that a wide range of laboratories are used with great variation in results. Parkinson (1991) has suggested up to three orders of magnitude difference in the permeability determined being due to variability of specimen diameter. This scale effect, particularly with fissured silty clays, has long been recognized as influencing shear strength (Marsland 1973) and it is logical to extent this concept of fissure/sample size ratio to permeability.

Alternatives to monitoring the permeability of the soil directly have been adopted to give the required rapidity of testing, albeit these tests are then referred back to a series of permeability tests. It was this approach that was adopted for both Netley and Gerrards Cross. Although control was essentially by monitoring clay content and density, other indirect approaches are used, as described by Barsby (1991) and Murray *et al.* (1992), the latter putting up a strong and attractive case for the use of the MCV.

The plasticity chart of available results from both Netley and Gerrards Cross (Fig. 3) confirms the essentially more silty nature of the London Clay in the Hampshire Basin compared to the material described as the same formation in the London Basin. Whilst there are fewer results from Netley, the observation from site and a local knowledge give greater certainty to this result. The 'A Line' clearly separates the two source materials.

Particle size distribution tests confirm the general trend of greater silt content in the Hampshire Basin and that permeabilities of $10^{-9}\,\mathrm{m\,s^{-1}}$ are achievable although with less of a margin than with the London Basin derived material, as shown by Fig. 4. The scatter of results makes the drawing of firm conclusions for this particular batch of soil difficult and a generality of a minimum clay content of about 25% would seem an appropriate minimum threshold to achieve a target permeability of 10^{-9} m s^{-1} based on these results. However, clay contents as low as 20% have the potential to achieve the desired permeability.

Murray *et al.* (1992) discuss the use of a series of Glacial Tills used as landfill liners and present a series of figures

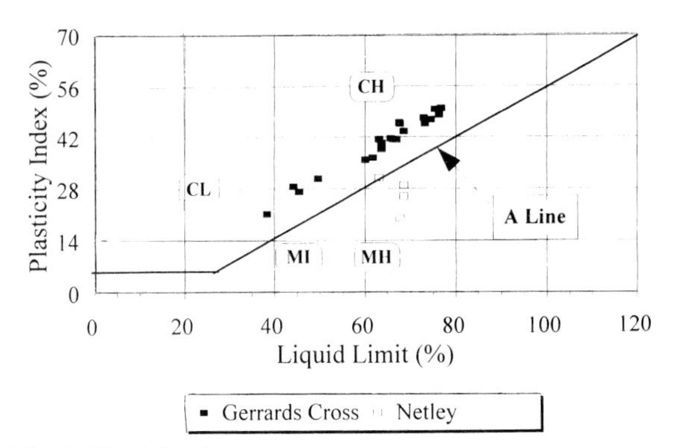

Fig. 3. Plasticity chart.

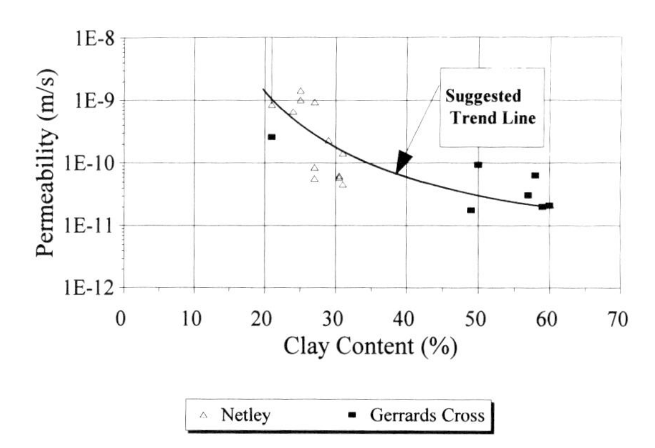

Fig. 4. Relationship between permeability and clay content.

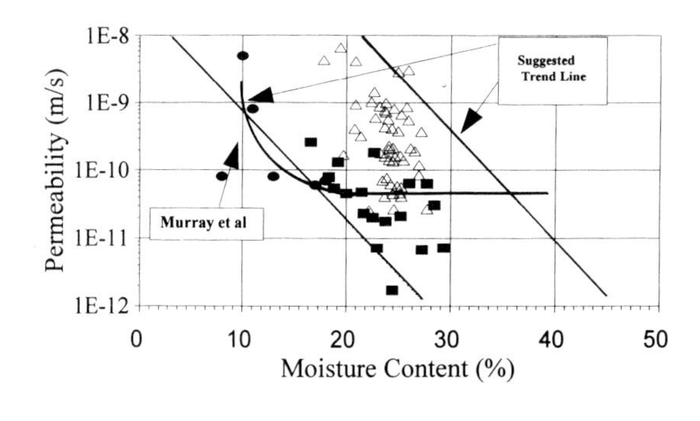

Fig. 5. Relationship between permeability and moisture content.

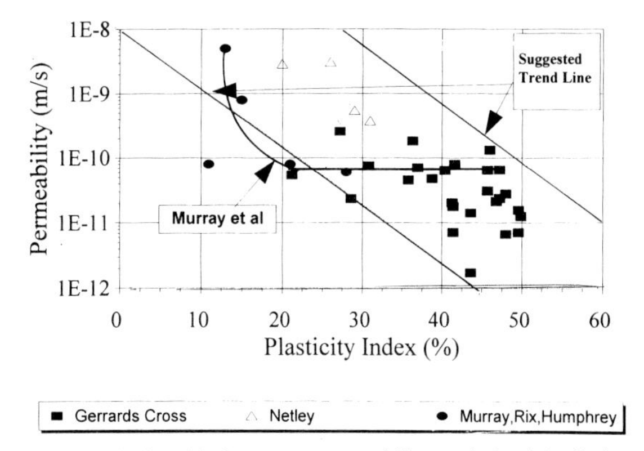

Fig. 6. Relationship between permeability and plasticity index.

that seem to show an asymptotic relationship between permeability, moisture content and plasticity index for the Glacial Tills used (see Figs 5 and 6). The results from Netley and Gerrards Cross are perhaps too scattered to allow interpretation in this manner and a broad band or pair of 'trend lines' certainly seems more justifiable in this case. Furthermore, they do not show the inverted curve relationship between permeability and moisture content reported by Barsby (1991) for results from two London Clay sites and by Harrop-Williams (1985).

Whilst logic suggests this sort of relationship (i.e. an inverse of the density relationship) may be valid for a material which has intra-particulate water transmission, such as sand, the authors remain sceptical that this is valid for a material that is transmitting water through the remains of a fissure structure and the new discontinuities induced by placement and compaction. No amount of compaction representative of site conditions will destroy all the randomly orientated fissures in a material such as London Clay nor will it destroy them at a uniform rate with moisture content. The general scatter of results from both Netley and Gerrards Cross seems to obscure any form of inverted curve relationship for the London Clay although a greater population of tests may allow some of the wilder results to be discounted.

Whatever the relationship selected there remains the need to interpret a trend for the particular site and soil formation which is adequately backed up by actual determinations of permeability in a manner consistent with and appropriate for the construction process.

Alternative test procedures

Whilst the triaxial test cell is favoured at present for the determination of permeability, the falling head permeameter test done on 100 mm samples has been used for far longer. Utilizing far lower hydraulic heads, it uses inexpensive and far simpler equipment and has the advantage of 'failing safe'. That is to say, if the water does not penetrate the test specimen it will pass down the annulus between the specimen and ring and give a higher than actual reading of the permeability, thus giving a 'safe' reading. It is also less prone to obstruction in water lines and other problems not always easily identified in the use of triaxial cells.

Nevertheless the use of a laboratory test requiring various sizes of undisturbed samples to be taken from site has long been seen to have major disadvantages. Many current *in situ* permeability tests are not suitable for testing clay layers which are typically 1.0 m or 2.0 m thick since they are either borehole or trial-pit orientated. Neither approach is satisfactory for testing liners *in situ* given that the liner is destroyed in the main part by the procedure and 'making good' will result in an inherent weakness in the homogeneity of the liner. Day & Daniel (1985) report two orders of magnitude variation of field permeability when compared to triaxial and consolidation cell laboratory tests on a trial clay liner in the USA. The *in situ* ring infiltrometer is the only test to predict accurately the permeability of the liner as calculated from a full-scale ponding test. There is little information on these sorts of test in the UK although there has been work reported in both the USA and Japan on such tests.

For some years the ISAT water absorption test has been used for determination of rate of flow into concrete, a material with a typical permeability of 10^{-9} m s^{-1}, as a British Standard. Some trial tests of this equipment have been done on samples of saturated kaolin and London Clay from Southleigh as a separate pilot investigation in the Geotechnics Laboratory in the School of Civil Engineering at the University of Portsmouth and also on site. In essence, the equipment becomes a site portable falling-head test which is capable of giving results in a few hours. Whilst accepting that permeability results are being compared with results from a test designed to measure water absorption of concrete, there is some correlation between properties that are seeking to quantify flow of water into and through a material. Figure 7, although based on few results, suggests that a correlation might be possible which could allow routine testing using a version of the ISAT equipment backed by regular check tests by more conventional determinations of permeability. Greater numbers of tests with a quicker turn-round are certainly possible if this approach is developed.

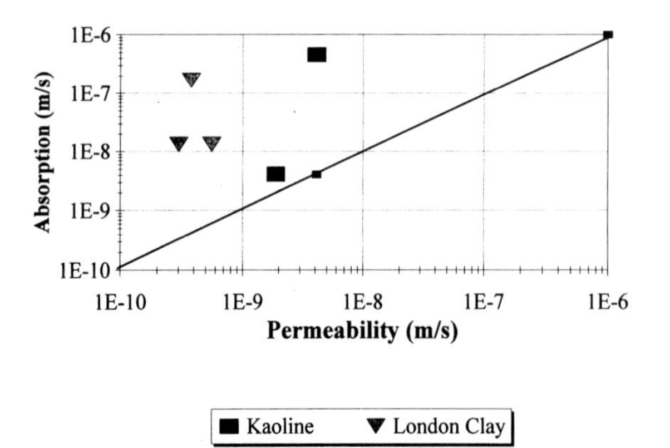

Fig. 7. Comparison of water absorption result with permeability.

Conclusions

The two sites demonstrate the variability of material from the same geological sequences but from differing depositional basins. This variation is often ignored by the bulk of published geotechnical work and can get overlooked by those unfamiliar with local variations.

The many results have not been susceptible to interpretation in a manner chosen by others although other factors which the authors are unaware of may be influencing one or all sets of results. However, for both sites the minimum of 20–25% clay content is an appropriate working threshold.

Enquiries around commercial testing organizations suggested, certainly in 1992, that there was remarkable variation in the manner in which triaxial permeability tests were being carried out, despite B.S.1377: Part 6 (Anon. 1990).

There remains a need for a test to determine *in situ* and *en masse* the permeability of landfill liners. The infiltrometer investigated in the USA and Japan or in the form of the ISAT test would appear to have potential for the UK landfill control testing requirements.

Acknowledgements. The authors are grateful to Leigh Environmental Limited for permission to present the test results and to L. Attril and R. Chown for their contribution to the investigation of the absorption test for clays.

References

ANON. 1990. B.S.1377: Part 6: *Soils for civil engineering purposes. Consolidation and permeability tests in hydraulic cells and with pore pressure measurement.*

BARSBY, R. 1991. The lining of landfill sites with natural clay materials. Two case histories. *The Planning and Engineering of Landfills, Conference*, Midlands Geotechnical Society, University of Birmingham, 193–197.

CRIPPS, J. C. & TAYLOR, R. K. 1981. The engineering properties of mudrocks. *Quarterly Journal of Engineering Geology*, **14**, 325–347.

DAY, S. R. & DANIEL, D. E. 1985. Hydraulic conductivity of two prototype clay liners. *American Society of Civil Engineers Proceedings*, **8**, (August), 957–970.

EDWARDS, R. A. & FRESHNEY, E. C. 1987a. *Geology of the county around Southampton, Memoir for 1:50 000 geological sheet 315.* HMSO, London.

—— & —— 1987b. Lithostratigraphical Classification of the Hampshire Basin Palaeogene Deposits (Reading Formation to Headon Formation) Tertiary Research, Leiden, February 1987.

HARROP-WILLIAMS, K. 1985. Clay liner permeability: evaluation and variation. *American Society of Civil Engineers Proceedings*, **8** (August), 1211–1225.

MARSLAND, A. 1973. Large in situ tests to measure the properties of stiff fissured clays. CP/1/73 Building Research Establishment Current paper.

MURRAY, E. J., RIX, D. W. & HUMPHREY, R. D. 1992. Clay linings to landfill sites. *Quarterly Journal of Engineering Geology*, **25**, 371–377.

NRA 1989. *Earthworks to landfill sites*. National Rivers Authority North West Region.

PARKINSON, C. D. 1991. The permeability of landfill liners to leachate. *The Planning and Engineering of Landfills, Conference*, Midlands Geotechnical Society, University of Birmingham, 147–152.

The containment properties of glacial tills: a case study from Hardwick Airfield, Norfolk

J. M. Gray

Department of Geography, Queen Mary & Westfield College, University of London, Mile End Road, London E1 4NS, UK

Abstract. Although of relatively low permeability, glacial tills may have deficiencies that make them less suitable as natural leachate containment materials. These deficiencies include sand lenses, fissures, and weathering zones. Geotechnical investigations of proposed unlined containment sites on tills must fully assess the impact of these and other deficiencies. This paper examines the proposals for a household and commercial waste containment site on Lowestoft Till at Hardwick Airfield, Norfolk, and concludes that the site deficiencies had not been adequately investigated or recognized. This led to difficulties for the consultants at a public inquiry into the proposals.

Glacial till is arguably the most important engineering geology material in the UK. It occurs as the surface or near-surface geology over the majority of the country north of the Thames (Eyles *et al.* 1983, fig. 9.6). The major exceptions include highland areas where bedrock rather than glacial drift predominates. However, in most lowland areas till thicknesses of several metres occur at shallow depths, and till properties in these areas have major implications for many engineering applications including foundation design, slope stability and hydrogeology.

Glacial till is a very variable material. Flint (1971, p. 154) described it as 'possibly more variable than any other sediment known by a single name'. However, normally it has a bimodal particle size distribution, with a matrix often dominated by silt and clay. Most of the variations in lithology and character are caused by the variable bedrock lithologies in the up-ice direction and the processes/position of debris entrainment, transport and deposition.

Because of its widespread, near-surface occurrence and its typically relatively low permeability, till is potentially important as a containment material and is attracting increasing attention in this regard (Daly 1992). The lowest permeabilities would be expected to occur on clay-rich subglacial lodgement tills, while granular meltout tills or sandstone-dominated deformation tills would be expected to have much higher permeabilities (Younger 1992). However, as described below, even clay-rich lodgement tills may be more permeable than is often assumed.

Permeability of clay-rich lodgement tills

Lloyd (1983, p. 353) states that clay-rich tills 'may provide excellent containing ground for landfill sites' ($k = 10^{-8}$ to 10^{-11} m s^{-1}), but as Lloyd points out, certain problems may be encountered.

Sand bodies. Lenses of sand and gravel are common in lodgement tills due to subglacial fluvial activity. They are frequently interconnected, can be laterally extensive, and are 'critical' in controlling till permeability (Eyles 1983; Lloyd 1983).

Fissures. Fissuring at scales varying from metres to centimetres are also very common in till and result from several processes including shearing during glaciation, unloading during deglaciation, drying and freezing (Boulton & Paul 1976). Numerous studies over the last 25 years from various parts of the world have demonstrated that fissures are responsible for increasing the permeability of tills (Williams & Farvolden 1969; Kazi & Knill 1973; Grisak & Cherry 1975; Grisak *et al.* 1976; McKinley *et al.* 1978; Lloyd 1983; Sharp 1984; Keller *et al.* 1986, 1988; Hendry 1988; Cherry 1989; Fredericia 1990; Haldorsen & Kruger 1990; Ruland *et al.* 1991). Fredericia (1990) reviewed much of the existing data and found that due to fissuring, the field permeabilities in till are typically one to two orders of magnitude higher than the laboratory permeabilities. 'This means that the laboratory measurements only represent the hydraulic conductivity of the material between this kind of fractures' (Fredericia, 1990, p.127). Subsequent work in Denmark has confirmed these conclusions (J. Fredericia, pers. comm., 1992). In order to obtain an accurate measure of the bulk permeability of fissured till, McKinley *et al.* (1978) recommended that a sample size 20 times the fissure spacing should be used for laboratory analysis.

Weathering. According to Sladen & Wrigley (1983, p. 190), 'till at the surface weathers chemically through

From Bentley, S. P. (ed.) *Engineering Geology of Waste Disposal*,
Geological Society Engineering Geology Special Publication No. 11, pp 299–307

oxidation, hydration, leaching of soluble materials—mainly carbonates, and by mechanical disintegration of particles, changes in till structure by fracture and in some cases, by downward movement of very fine material.' They have developed a four-zone weathering scheme for lodgement tills. Lloyd (1983) states that weathering zones 'cause leachate problems in landfill sites' since they increase permeabilities albeit by variable amounts.

Since these three potentially problematic characteristics of lodgement tills are fairly well known, any proposal to construct an unlined landfill containment site on lodgement till would be expected to assess whether there are likely to be problems at the site in question.

The proposed landfill site at Hardwick Airfield, Norfolk

In May 1991, Norfolk County Council applied for deemed planning consent for a $1.5 \times 10^6\,m^3$ household and commercial waste disposal site at the disused US World War II Airfield at Hardwick, about 18 km south of Norwich in South Norfolk (Fig. 1). The life of the site was to be 22–30 years at a fill rate of 50 000–70 000 tonnes per annum.

The proposal involved excavating a pit 2–4 m deep into the Lowestoft Till (a lodgement till of the Anglian Glaciation). In their geotechnical report prepared prior to the planning application (Norfolk County Council, 1990, p. 1), the County Council's consultants (W. S. Atkins Environment) argued that 'the substantial thickness of clay across the site is of sufficiently low permeability for groundwater protection without reworking'. Only where deficiencies became apparent on the floor of the site would 'some over-excavation and recompaction of clay' occur, though the consultants did propose a reworking of the perimeter wall of the excavation because sand lenses were said to occur 'in the upper 2 m of the till'.

It was proposed to develop the site on a cellular basis, each cell being excavated to a depth of 2 m in the centre of the site and 4 m at the periphery to provide a gradient

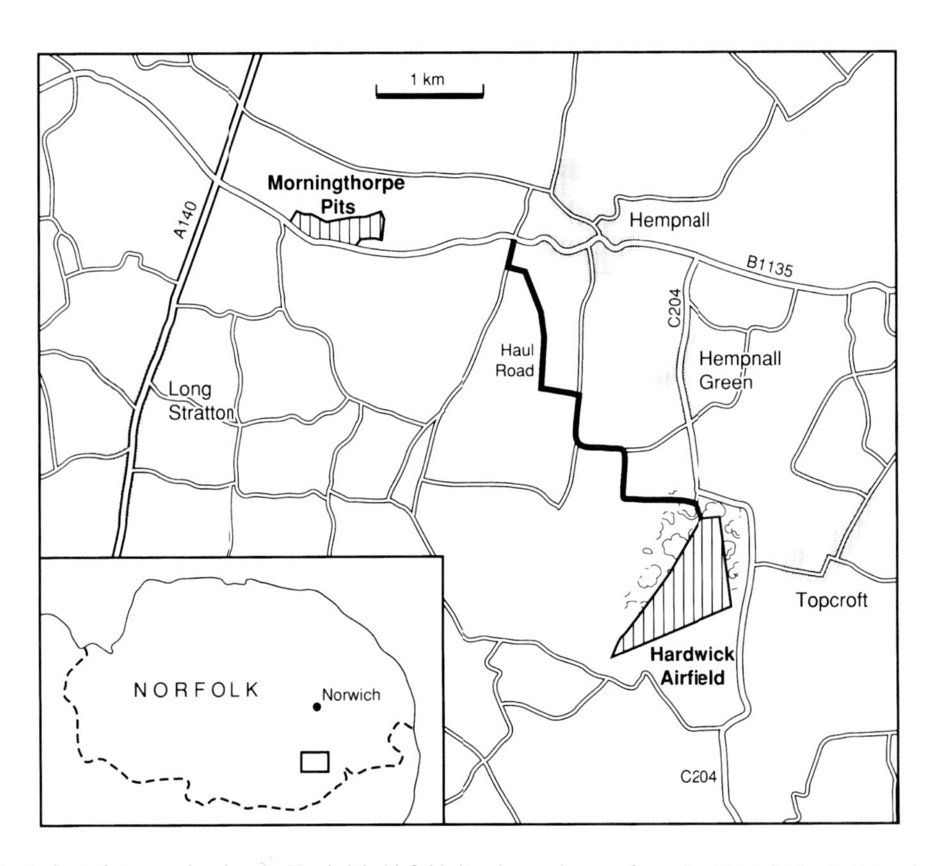

Fig. 1. Location map showing the Hardwick Airfield site, the road access from the A140 via the B1135 and C204 through Hempnall and Hempnall Green, and the proposed alternative haul road access.

for leachate drainage to a peripheral collection system. The excavated material would be used to provide a low permeability cap to the tip, over which a topsoil covering would be laid to restore the site to eventual agricultural use.

The site was to be a 'landraise' (as opposed to landfill) one, comprising a low hill rising to 10 m above the surrounding till plain. the County Council's policy is to establish landfill sites within each District Council waste collection area and it was argued that the landraising approach was necessary because of the shortage of disused mineral workings suitable for landfilling in south Norfolk.

Opposition to the proposals came from the South Norfolk District Council, eight parish councils around the proposed site, a local recycling company, and many individual farmers and residents. The grounds for objection included traffic, ecological, landscape, pollution and residential amenity issues. The planning application was 'called in' by the Secretary of State for the Environment in January 1992 and a three-week Public Inquiry was held in January/February 1993.

This paper is based on a comprehensive geological assessment of the site investigation data, consultants' reports, environmental statements and numerous other documents. As a result of this work and a failure to convince the County Council and its consultants that they had underestimated the geotechnical deficiencies of the site, the author submitted his own proofs of evidence to the Public Inquiry, attended the whole of the Inquiry, and cross-examined all nine County Council witnesses including the County's consultant geologist.

Assessment of the borehole and trial pit data

As part of the Hardwick Airfield site investigation carried out in December 1989, 13 boreholes were sunk (three to a depth of at least 20 m, 10 to a depth of about 7.5 m), and 18 trial pits were dug to a 5 m depth. Subsequently (May 1992), a further eight trial pits were excavated to a 4 m depth. Particle size tests on the till showed its average composition to be about 45% clay, 30% silt, 20% sand and 5% gravel.

The site investigation information as well as a number of boreholes indicate that Hardwick Airfield is underlain by 15–20 m of Lowestoft Till overlying a complex series of clays, sands and gravels, which in turn overlie the chalk aquifer at a depth of about 60–80 m.

An assessment of the trial pit and borehole data indicated that there are two main deficiencies in the till which could significantly affect its properties as a natural leachate containment system.

First, about 20% of the boreholes and trial pits indicate that sand layers and lenses occur below the 2 m depth mentioned in the geotechnical reports prepared for the County Council, though they appear to be mainly restricted to the upper 6 m of the till. They can be expected to cause relatively high horizontal permeabilities.

Secondly, all 26 trial pits describe the till as 'fissured', yet there is no mention in the planning application or environmental statements of the potential influence of fissuring on till permeability. According to the County Council's consultants (G. Raybould, pers. comm., 1991), 'fissures were not observable in the sides of the trial pits at Hardwick, and only became evident when hand samples of the clay were broken open, appearing as smooth surfaces of a few centimetres extent. It is unlikely that this feature significantly affects the overall transmissivity of the clay.' However, four of the eight trial pits excavated in May 1992 describe fine decayed roots, iron staining and/or silt within the fissures, indicating that they have been hydraulically active in the past and suggesting that leachate may move preferentially through them. Furthermore, the fissures below the floor of the site could be expected to open slightly when the pit is excavated.

Assessment of the permeability data

A limited number of permeability measurements were made as part of the site investigation. These tests, which were carried out by sub-contractors, involved four measurements of laboratory permeability (falling head permeameter), five of field permeability (constant head test through installed piezometers), and several laboratory permeability tests on till reworked at different compaction efforts in the presence of water.

In his Proof of Evidence to the Public Inquiry (Norfolk County Council 1993, p. 14), the County's consultant geologist was still maintaining that the field and laboratory tests on undisturbed material indicated that in general the site would meet the permeability specification for the site of 1×10^{-9} m s^{-1} for at least 1 m below floor level. 'However, where this requirement is not met, or where visual evidence of silt and/or sand layers suggests the possibility of higher horizontal permeabilities, the clay will be excavated and recompacted to a depth of at least one metre.'

The permeability measurements for the Hardwick site are shown in Fig. 2. It will be noted that the four laboratory permeability measurements cluster around the permeability specification (Fig. 2(B)), and if accurate and representative, would indicate that the till, in undisturbed laboratory samples, only just meets the permeability requirements.

Of the five measurements of permeability in the field, three exceed the permeability specification, and the highest is 350 times above the specification and above

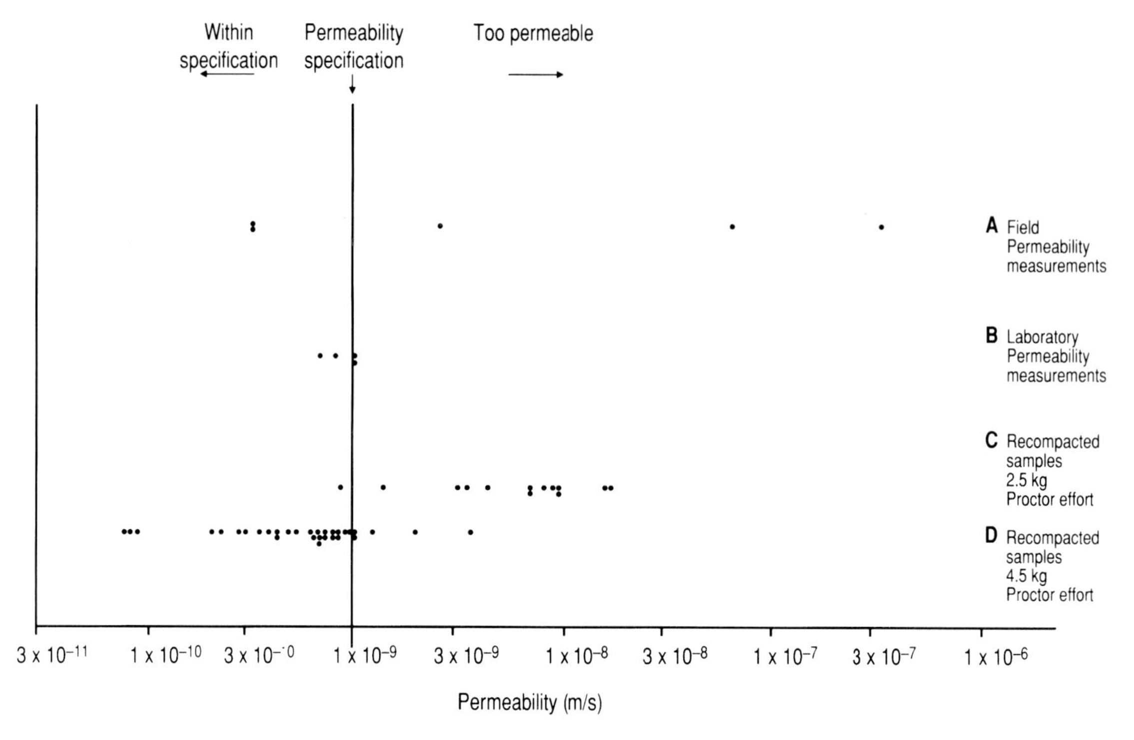

Fig. 2. Field, undisturbed laboratory, and recompacted laboratory permeability measurements on till from Hardwick Airfield.

the laboratory values (Fig. 2(A)). In his Proof of Evidence to the Public Inquiry (Norfolk County Council 1993, p. 9), the County Council's consultant geologist argued that 'The higher permeability results from some of the field tests probably result from the presence of sandy and silty layers in the clay'. However this is not borne out by the stratigraphies of the five boreholes where field permeability measurements were made (Fig. 3). The highest field permeability was measured in borehole 28, yet the logged stratigraphy specifically states that no sand pockets were observed in the vertical range of the permeability measurement. The second highest permeability (borehole 24) could be explained by the sand pockets or possibly the very chalk rich till mentioned in the borehole stratigraphy. However, the third highest field permeability (borehole 29) was taken over the smallest vertical range (0.7 m) and the stratigraphic log indicates very stiff till (Fig. 3). The two field permeability measurements that do meet the specification (boreholes 30 and 33) come from depths of about 6–7 m, well below the planned floor of the tip. It must be stressed that the number of measurements is rather low in number and that the field permeabilities have generally been made over too high a vertical range, but the conclusion from this analysis is that the till itself in its upper part appears to be more permeable than the specification.

This is also illustrated in Fig. 4(A) which plots permeability against depth of both laboratory and field permeabilities. There is a slight decrease in laboratory permeabilities with depth possibly due to weathering. However, the field permeabilities tell a different story, the two measurements that include levels within the range of the floor of the site and 1 m below it (2–5 m) giving permeabilities much higher than the specification. This diagram suggests that laboratory and field permeability measurements might converge at depths of about 7–8 m, but that above this, field permeabilities diverge from laboratory ones by two orders of magnitude or more (Fig. 4(B)).

The spatial distribution of permeability measurements over the site is shown in Fig. 5. In fact, only part of site was selected for landfilling because of the need to conserve the ecological value of Runway Plantation and the requirement for extensive landscaping along the southern section. Figure 5 illustrates the poor relationship between the sites where permeability was tested and the area subsequently selected for tipping, and clearly a better spatial design of permeability measurements would have been appropriate. Only three of the nine permeability measurements fall within the proposed tip area. Furthermore, Fig. 6 shows that only one permeability measurement was made that includes

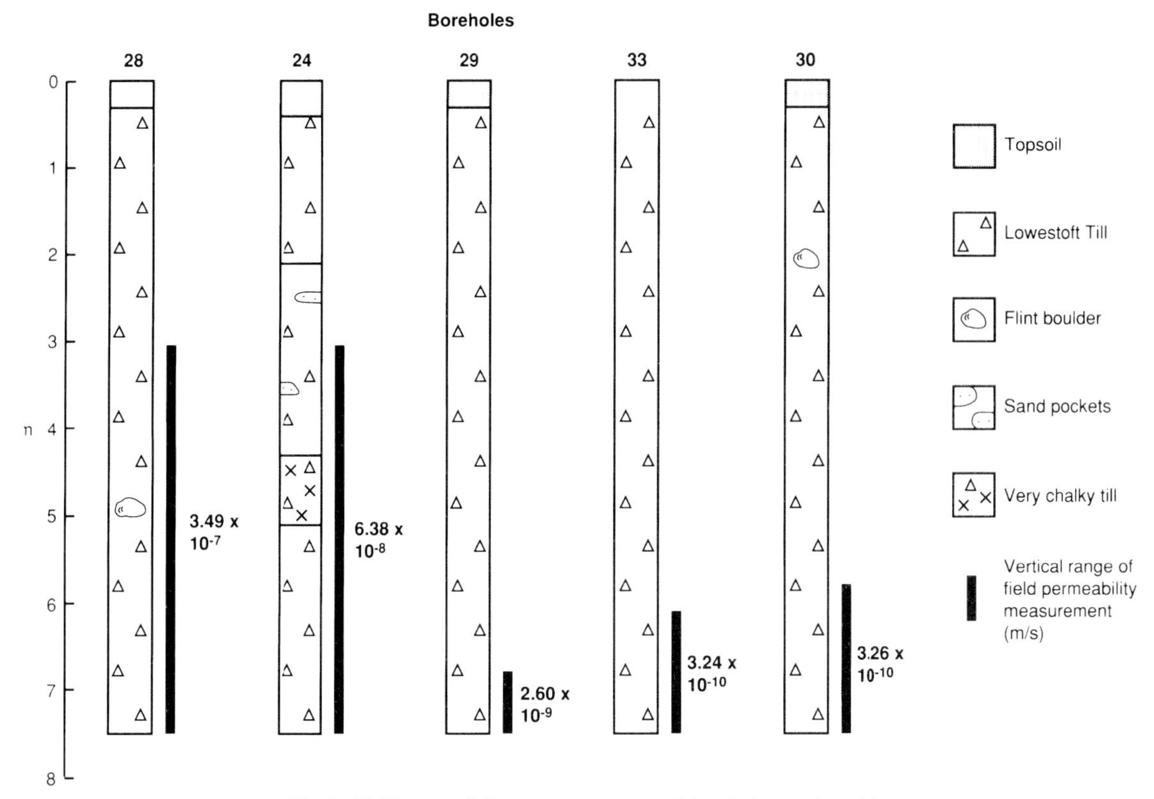

Fig. 3. Field permeability measurements and borehole stratigraphies.

material within the critical 1 m below the floor of the tip, and this is the highest permeability measurement, 350 times above the specification. At the Public Inquiry the author argued that there was no evidence to demonstrate that the till comprising the 1 m depth below the floor of the pit would meet the permeability requirement under operational conditions, and that the evidence that did exist indicated the opposite.

This difference between the field and laboratory permeabilities in the upper few metres of the till probably reflects the combined influence of the horizontal till structure, the presence of sand lenses, and the role of the fissures. Contrary to the view expressed by the County Council's consultant geologist, these results indicate that the site will generally *not* perform to the specified permeability, and it is suggested there is a strong possibility that leachate will escape the containment system and move horizontally towards surface water courses and wells of the many domestic and agricultural abstracters around the site. Under cross-examination at the Public Inquiry, the County's consultant geologist admitted that his Proof of Evidence was incorrect in stating that the till would generally meet the permeability specification and agreed that there was a strong possibility that the entire floor of the site would have to be engineered.

The NRA had also recognized a problem with the till permeability. After initially objecting to the proposals and requiring a reworking of the till to a 1 m depth across the entire floor of the site, this approach was later modified to one in which the floor of each cell would be tested on an agreed grid pattern. 'Should the initial testing of the prepared base show that our requirements are met across the entire base area of the cell, then the base would, of course, not require to be reworked.... If any part of the cell base does not meet our permeability requirement, then the entire cell base will be-reworked' (D. Taylor, pers. comm., 1992). A crucial issue concerns how the permeability is to be measured in these tests. The NRA indicated that the permeability 'is likely to be assessed in the laboratory using the falling head method' (D. Taylor, pers. comm., 1992). If so, the laboratory permeability measurements so far made indicate that the till will just about satisfy the permeability requirements (Fig. 2(B)). However, as explained above, laboratory permeabilities are likely to underestimate seriously the permeability of the till under field conditions, and the use of laboratory permeabilities was therefore queried at

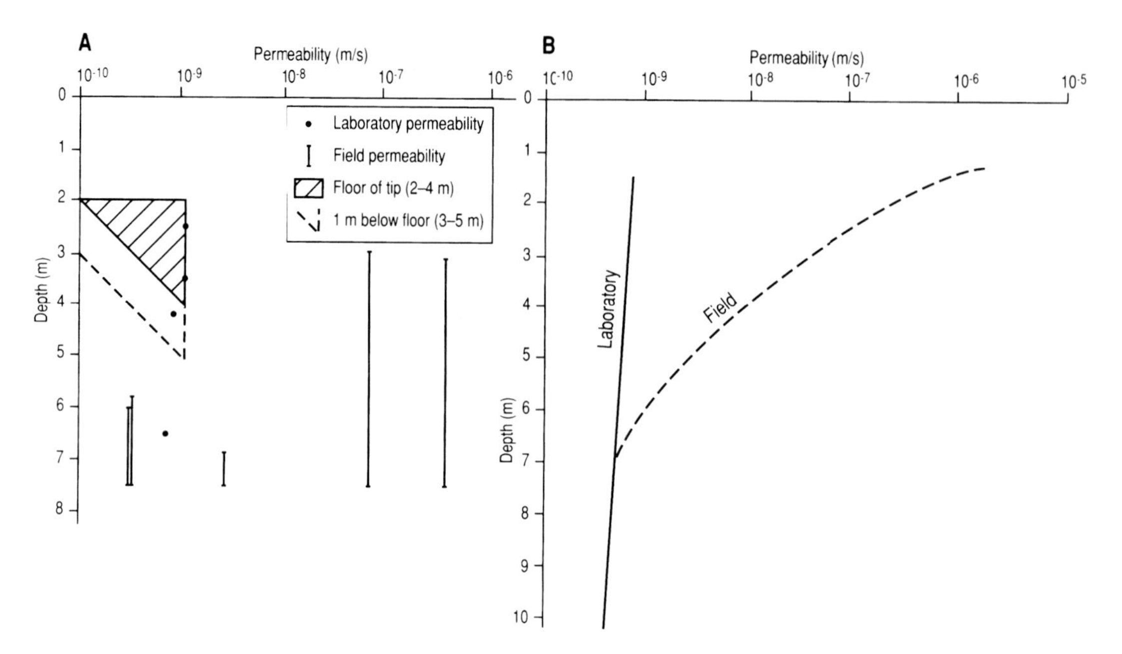

Fig. 4. (a) Depth versus permeability plot for field and undisturbed laboratory samples at Hardwick Airfield. Also shown is the floor depth of the pit (2–4 m) and the 1 m layer below the floor throughout which the permeability specification must apply. (b) Suggested generalized relationship between depth and permeability in tills. Note the divergence between undisturbed laboratory and field measurements in the upper few metres probably due to surface fissuring, sand lenses and till structure.

the Public Inquiry. The author argued that the limited permeability data available suggested that any field testing of the base would demonstrate that the entire floor would have to be reworked and that therefore the cost of the testing could be saved. It was argued that one of the main aims of a site investigation should be to establish what engineering measures will be required and their likely cost, and that the stance agreed between the County's consultants and the NRA did not conform to this normal practice.

The question then arises as to whether the till can be reworked to meet the NRA specification. A series of laboratory tests was undertaken by the consultants, involving reworking till samples at a range of moisture contents and recompacting using a 2.5 kg rammer (B. S. 1377). Only 1 of the 13 tests achieved the permeability specification using the falling head test (Fig. 2(C)). A second series of tests was therefore carried out on samples taken from the May 1992 trial pits using moisture contents of 14–18% and the heavier 4.5 kg rammer (B. S. 1377). The majority of the resultant tests did meet the permeability specification (Fig. 2(D)) and the County's consultants argued that the till 'can be recompacted as necessary to give permeabilities well below 1×10^{-9} m s^{-1}, provided appropriate site plant is used . . . and due control is exercised over moisture contents'. However, the District Council's consultant landfill engineer, in his Proof of Evidence to the Public Inquiry (South Norfolk District Council 1993), argued that 'it is highly unlikely that the compactive effort achieved in the laboratory can be replicated in the field during placement of the clay liner'. He therefore argued for the installation of a composite liner at the site, involving both a reworking of the till and a HDPE membrane. The costs were estimated at £7–8 m^{-2} at 1992 prices, or about £2.5 million for the 32 ha site.

Other considerations

The previous landfill site for South Norfolk was a disused gravel pit at Morningthorpe (Fig. 1) where infilling was completed in 1990. Two other disused gravel workings at Morningthorpe were considered as alternatives to the Hardwick Airfield site at Stage 1 of the investigation; Hardwick was chosen because of the site geology, cost, capacity and ecological sensitivity, though it was recognized that Morningthorpe had better traffic access. The cost of lining Hardwick Airfield clearly altered one of these factors, while other factors such as ecology were challenged at the Public Inquiry.

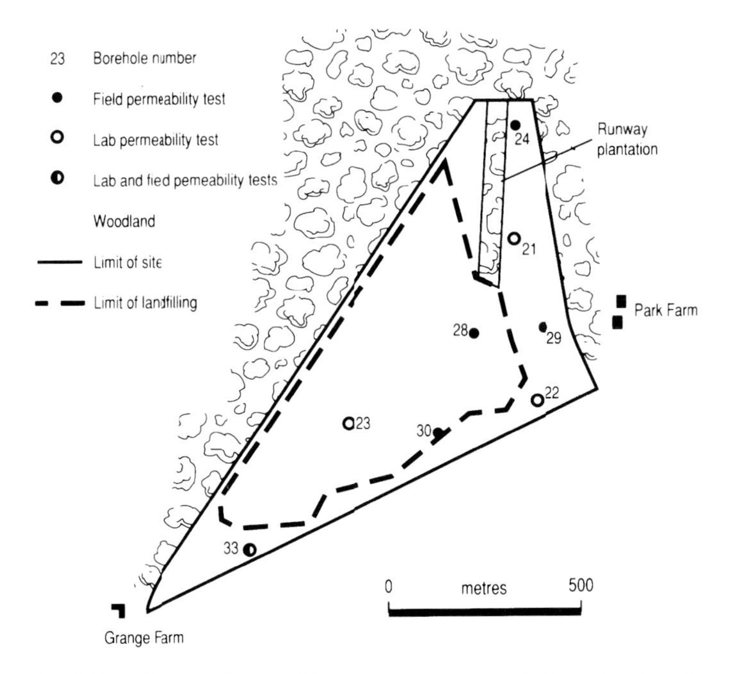

Fig. 5. Distribution of permeability measurements in relation to the landfill area.

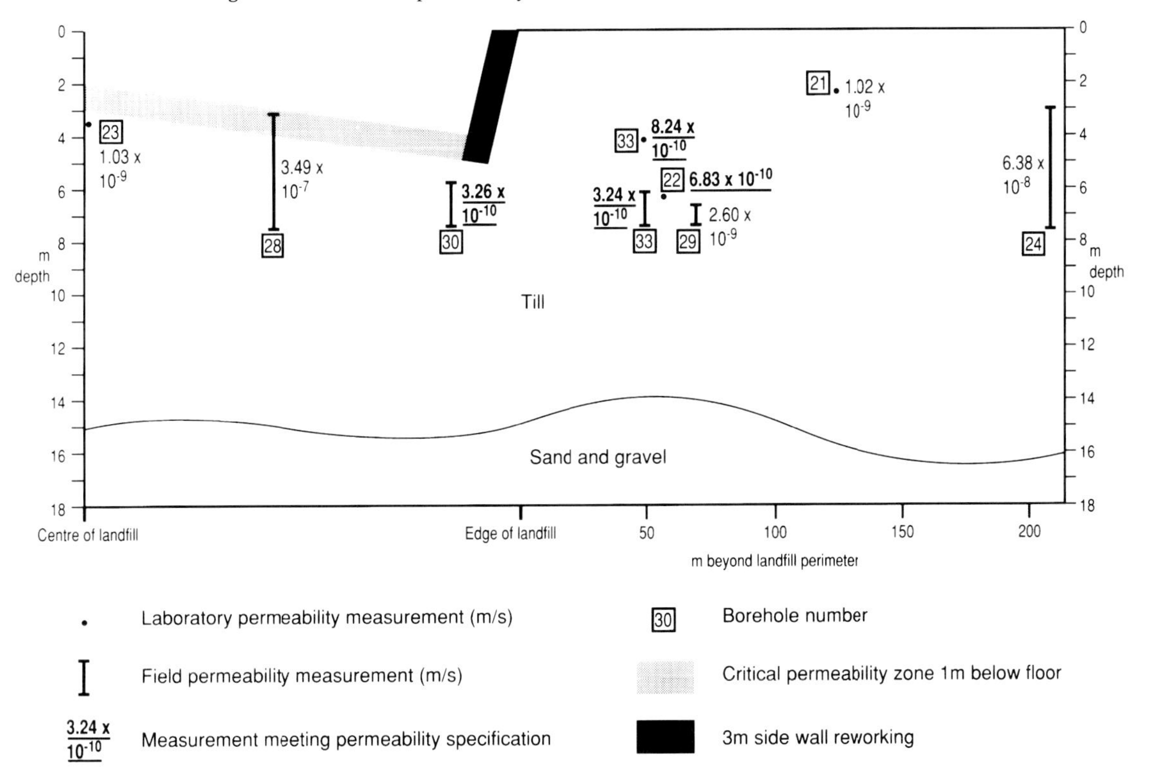

Fig. 6. Field and laboratory permeability measurements plotted on a composite cross-section from the centre of the pit (left) to 210 m beyond the landfill perimeter (right).

The public outcry over the traffic impact of the proposals led the County Council to investigate the option of a 3.3 km long haul road to the site (Fig. 1) through an area of High Landscape Quality identified in the South Norfolk Rural Area Local Plan. The cost of this road was estimated at £1.75 million. Objectors to the Hardwick Airfield site argued that because over 70% of South Norfolk is underlain by Lowestoft Till, landraise siting could be highly 'footloose' and there should be no need for a site that involves either the cost or impact of a haul road or the routing of lorries via the B1135 and C204 through Hempnall and Hempnall Green (Fig. 1).

The County Council also commissioned its consultants to investigate the presence of an alternative landraise site in the vicinity of the A140, the major highway through South Norfolk. Unfortunately this study misidentified the geology of 5 of the 13 alternative sites (Gray 1993), and was heavily criticized for being subjective in its support for the Hardwick Airfield site.

There were also objections to the landraise landform. Even though the gradients were relatively low (maximum 1 : 23), the site itself is on a very flat till plain, and a 10 m high hill would be noticeable. The County Council argued that the existing woodland on two sides, together with new planting on the third, would effectively screen the feature (Fig. 5).

Result of the Public Inquiry

The Inspector's Report and Secretary of State for the Environment's decision refusing permission for the site and haul road were published in August 1993. The Inspector (DOE 1993, p. 111) concluded that a composite liner 'would be necessary' and 'in my opinion . . . the evidence is clear that the original Norfolk County Council assumption about the geological advantage of the Hardwick site was misplaced and that the cost of providing a fully engineered containment was not included in the site evaluation exercise'.

On the other issues raised at the Public Inquiry, the Inspector argued that alternative sites were available, but did not accept the landform impact argument because of the screening from existing or new planting. He was very critical of the proposed access routes and did not believe that the use of either the existing highway or the construction of a haul road was acceptable. The Inspector also supported some of the ecological, historical and agricultural objections to the site, concluding that 'It is the combination of these harmful features which, in my opinion, would be contrary to the objectives of Development Plan policies . . . and add up to a significant objection to the application' (DOE 1993, p. 115).

Conclusions

Although tills, particularly clay-rich lodgement tills, have relatively low permeabilities, there are several potential deficiencies that can raise their permeabilities, viz. the presence of sand lenses, fissures and/or weathered zones. Geotechnical investigations of landfill containment sites seeking to exploit the natural containment properties of tills must assess the possible impact of these and other deficiencies if the proposals are to withstand the scrutiny of a public inquiry. In the case of Hardwick Airfield, the Public Inquiry revealed a number of inadequacies in the site investigation design and the consultant's reports:

- significance of fissuring in the till not recognized;
- significance of discrepancies between laboratory and field permeabilities of the till not adequately recognized;
- presence of sand lenses not fully recognized;
- low number of permeability measurements;
- poor spatial and vertical distribution of permeability measurements in relation to the planned tipping area and floor depth;
- significant errors in identifying the Quaternary geology of alternative sites.

It is argued that this is a further example of inadequate site investigation (Instruction of Civil Engineers 1991) and a 'failure to appreciate the geomorphological and Quaternary features of sites' (Hutchinson & Wilson 1993).

Acknowledgements. I am very grateful to H. M. Haslam (Chief Planning Officer) and D. Willis (Senior Planning Officer) of South Norfolk District Council, and S. Ralph, J. D. Brigham and D. T. Beadle of Norfolk County Council for providing several reports on which this paper is based. A. Street (MRM Partnership), J. H. Atkinson (City University) and A. Leach (Engineering Geology Ltd) clarified several of the geotechnical issues but the opinions expressed in this paper are those of the author. E. Oliver kindly drew the figures.

References

BOULTON, G. S. & PAUL, M. A. 1976. The influence of genetic processes on some geotechnical properties of glacial tills. *Quarterly Journal of Engineering Geology*, **9**, 159–194.

CHERRY, J. A. 1989. Hydrogeologic containment behaviour in fractured and unfractured clayey deposits in Canada. *Proceedings of the International Symposium on contaminant transport in groundwater*, Stuttgart, Germany, April 1989.

DALY, D. 1992. Quaternary deposits and groundwater pollution. *In:* GRAY, J. M. (ed.) *Applications of Quaternary Research.* Quaternary Proceedings, vol. 2. Quaternary Research Association, Cambridge, 79–89.

DOE, 1993. *Hardwick landfill site and haul road: inspector's report.* File no. E1/X2600, E1/X2600/3/2,Y/DN/5066, Department of the Environment, London.

EYLES, N. 1983. Glacial geology: a landsystems approach. *In*: EYLES, N. (ed.) *Glacial geology: an introduction for engineers and earth scientists*. Pergamon Press, Oxford, 1–18.

——, DEARMAN, W. R. & DOUGLAS, T. D. 1983. The distribution of glacial landsystems in Britain and North America. *In*: EYLES, N. (ed.) *Glacial geology: an introduction for engineers and earth scientists*. Pergamon Press, Oxford, 213–228.

FLINT, R. F. 1971. *Glacial and Quaternary Geology*. Wiley, New York.

FREDERICIA, J. 1990. Saturated hydraulic conductivity of clayey tills and the role of fractures. *Nordic Hydrology*, **21**, 119–132.

GRAY, J. M. 1993. Quaternary geology and waste disposal in South Norfolk, England. *Quaternary Science Reviews*, **12**, 899–912.

GRISAK, G. E. & CHERRY, J. A. 1975. Hydrogeological characteristics of response of fractured till and clay containing a shallow aquifer. *Canadian Geotechnical Journal*, **12**, 23–43.

——, ——, VONHOF, J. A. & BLUMELE, B. 1976. Hydrogeological and hydrochemical properties of fractured till in the interior plain region. *In*: LEGGET, R. F. (ed.) *Glacial till*. Royal Society of Canada, Special Publication 12, 304–335.

HALDORSEN, S. & KRUGER, J. 1990. Till genesis and hydrogeological properties. *Nordic Hydrology*, **21**, 81–94.

HENDRY, M. J. 1988. Hydrogeology of clay till in a prairie region of Canada. *Ground Water*, **26**, 607–614.

HUTCHINSON, J. N. & WILSON, D. D. 1993. The influence of education and training in engineering geology on site investigation practice in Britain. *Geoscientist*, **6**(3), 23–24.

INSTITUTION OF CIVIL ENGINEERS 1991. *Inadequate Site Investigation*. Institution of Civil Engineers, London.

KAZI, A. & KNILL, J. L. 1973. Fissuring in glacial clays and tills on the Norfolk coast, United Kingdom. *Engineering Geology*, **7**, 35–48.

KELLER, C. K., KAMP, G. van der & CHERRY, J. A. 1986. Fracture permeability and groundwater flow in clayey till near Saskatoon, Saskatchewan. *Canadian Geotechnical Journal*, **23**, 229–240.

——, —— & —— 1988. Hydrogeology of two Saskatchewan tills. 1. Fractures, bulk permeability and spatial variability of downward flow. *Journal of Hydrogeology*, **101**, 97–121.

LLOYD, J. W. 1983. Hydrogeology in glaciated terrains. *In*: EYLES, N. (ed.) *Glacial geology: an introduction for engineers and earth scientists*. Pergamon Press, Oxford, 349–368.

MCKINLEY, D. G., MCGOWN, A., RADWAN, A. M. & HOSAIN, D. 1978. Representative sampling and testing in fissured lodgement tills. *In*: *The engineering behaviour of glacial materials*. Geoabstracts, Norwich, 129–140.

NORFOLK COUNTY COUNCIL 1990. *Replacement landfill site in South Norfolk: Stage 2—site evaluation*. W. S. Atkins Environment, Epsom.

—— 1993. *Hardwick landfill site with and without haul road. Public inquiry proof of evidence on geology and engineering design by John Garth Raybould*. W. S. Atkins Environment, Epsom.

RULAND, W. W., CHERRY, J. A. & FEENSTRA, S. 1991. The depth of fractures and active groundwater flow in a clayey till plain in Southwestern Ontario. *Ground Water*, **29**, 405–417.

SHARP, J. M. 1984. Hydrogeologic characteristics of shallow glacial drift aquifers in dissected till plains (North-Central Missouri). *Ground Water*, **22**, 683–689.

SLADEN, J. A. & WRIGLEY, W. 1983. Geotechnical properties of lodgement till—a review. *In*: EYLES, N. (ed.) *Glacial geology: an introduction for engineers and earth scientists*. Pergamon Press, Oxford, 184–212.

SOUTH NORFOLK DISTRICT COUNCIL 1993. *Hardwick Airfield, Norfolk: Proof of Evidence of Andrew Street*. MRM Partnership, Bristol.

WILLIAMS, R. E. & FARVOLDEN, R. N. 1969. The influence of joints on the movement of groundwater through glacial till. *Journal of Hydrology*, **5**, 163–170.

YOUNGER, P. L. 1992. Quaternary geology and hydrogeology: the value of an interdisciplinary approach. *Geoscientist*, **5**(2), 24–27.

Suction-controlled oedometer tests in montmorillonite clay: preliminary results

M. V. Villar & P. L. Martín

Técnicas Geológicas, CIEMAT, Avda. Complutense 22, 28040 Madrid

Abstract. A Spanish montmorillonite clay is being characterized in its unsaturated state for use as a sealing material in high-level radioactive waste repositories. The 'water content versus suction' curve has been determined and oedometer tests are being performed under suction pressures, ranging from 0.1 to 100 MPa, controlled by air pressure or sulphuric acid solutions. The methodology of the tests is described and the first results are presented.

The Spanish concept for the disposal of high-level radioactive wastes foresees the construction of tunnels and shafts in deep geological formations of granite, clay or salt (ENRESA 1994). In granite sites an engineered clay barrier would be placed between the canister and the rock to avoid the release of radionuclides to the environment, retardate the entrance of water to the canister walls and protect the canister structure of external stresses.

To achieve these goals, the material used in the construction of the barrier must have certain physico-chemical and mechanical characteristics, i.e. high swelling pressure, low hydraulic conductivity, sufficient thermal conductivity, mechanical stiffness and plasticity, a high adsorption rate of radionuclides and good compaction properties.

Smectite clays have been proved to satisfy these requirements (Yong *et al.* 1986; Pusch & Börgesson 1989), and are being studied in CIEMAT as a potential backfilling and sealing material for radioactive waste repositories.

The state of the clay in the barrier will probably be unsaturated while the heat source of the wastes remains active. For this reason, the mechanical characterization of the unsaturated clay is being performed. As the stress–strain behaviour of partially saturated soils depends on their pore water pressure (i.e. on suction, $s = u_a - u_w$, where s is suction and u_a and u_w are air and water pore pressure respectively), suction-controlled laboratory tests have been designed and are now being carried out. The relation between clay water content and suction has also been determined.

Materials

A montmorillonite clay from Almería (southeast Spain) was chosen from among other Spanish bentonites after a selection process. It contains low quantities of biotite, plagioclase and quartz (less than 5%), fragments of volcanic rocks and probably colloidal silica. It was formed by the hydrothermal alteration of volcanic rocks.

Some physico-chemical characteristics of this clay (Pérez del Villar *et al.* 1991; Rivas *et al.* 1991; Villar & Rivas 1994) are shown below:

- specific surface (internal + external): $570\,m^2\,g^{-1}$
- liquid limit: 213%
- CEC: 88 meq per 100 g
- water content at laboratory conditions: 10–13%
- specific weight: $2.78\,g\,cm^{-3}$
- saturated swelling pressure at $\rho d = 1.60\,g\,cm^{-3}$: 5 MPa
- saturated hydraulic conductivity at $\rho d = 1.60\,g\,cm^{-3}$: $2.10^{-13}\,m\,s^{-1}$
- thermal conductivity at $\rho d = 1.60\,g\,cm^{-3}$ and $w = 11\%$: $0.77\,W\,mK^{-1}$

Methods of suction control

Both for the determination of the relation between water content and suction and for the oedometer tests, the suction is controlled by two different methods, i.e. air pressure (Escario 1969) and control of the relative humidity in a vacuum-closed atmosphere by sulphuric acid solutions (Biarez *et al.* 1988).

The theoretical concept of both methods is different. If suction changes are given by the difference between pore air pressure and pore water pressure ($s = u_a - u_w$), by applying air pressure to the sample we introduce an increase in u_a, while u_w remains equal to the atmospheric pressure. Thus, suction varies in the same quantity that nitrogen pressure does. In the second method, the relative humidity of the atmosphere in which the sample is placed controls u_w and u_a, as there is no external exchange.

From BENTLEY, S. P. (ed.) *Engineering Geology of Waste Disposal*,
Geological Society Engineering Geology Special Publication No. 11, pp 309–312

Air pressure method

In the air pressure method, the compacted sample is placed in a stainless steel pressure cell (Fig. 1(a)) in contact with a dialysis membrane which is permeable to water but not to air (Escario & Sáez 1973). The membrane is in contact with free water at atmospheric pressure. The pressure in the cell is immediately raised by injecting nitrogen to the desired value, which increases the air pressure in the pores of the clay. This new pressure situation forces the clay to exchange water through the membrane until equilibrium is reached again.

To remove the air bubbles that diffuse through the membrane, a pump is located between the cell and the water deposit, in order to circulate water in contact with the membrane, as was suggested by Bishop & Donald (1961). In the oedometer cells, a peristaltic pump is used for this purpose, and the water deposit is a buret in which water taken or given by the sample can be directly measured. Over the water surface in the buret an oil layer avoids evaporation.

Because of the mechanical limitations of the cell, this method is fitted only for suction pressures below 14 MPa.

Sulphuric acid solutions method

The samples are introduced in a closed atmosphere, a vacuum desiccator for the determination of the water content versus suction curve, or in a hermetic cell shown in Fig. 1(b) for the oedometer tests (Esteban 1990), in which a known relative humidity is imposed by a sulphuric acid solution of a given density. The pressure in the pores of the sample is related to the water activity of the solution by Kelvin's law. The clay exchanges water until it reaches a thermodynamic equilibrium with the vapour pressure of the solution.

The relation between the activity of the solution and the weight percentage of H_2SO_4 used to prepare it, is reflected in experimental tables. This relation depends on temperature, so it is necessary to keep it constant during the tests (20 °C). There is also an experimental relation between acid weight percentage in the solution and solution density. The water transfer between clay and atmosphere makes the density of the solution vary during the stabilization period. For this reason, the solution densities are checked with picnometers before and after the experiment, in order to know the actual suction pressure to which the sample has been submitted.

Suction pressures from 13 to 300 MPa can be obtained with this method.

Test performed

For the determination of the relation between water content of the clay and suction, samples of different

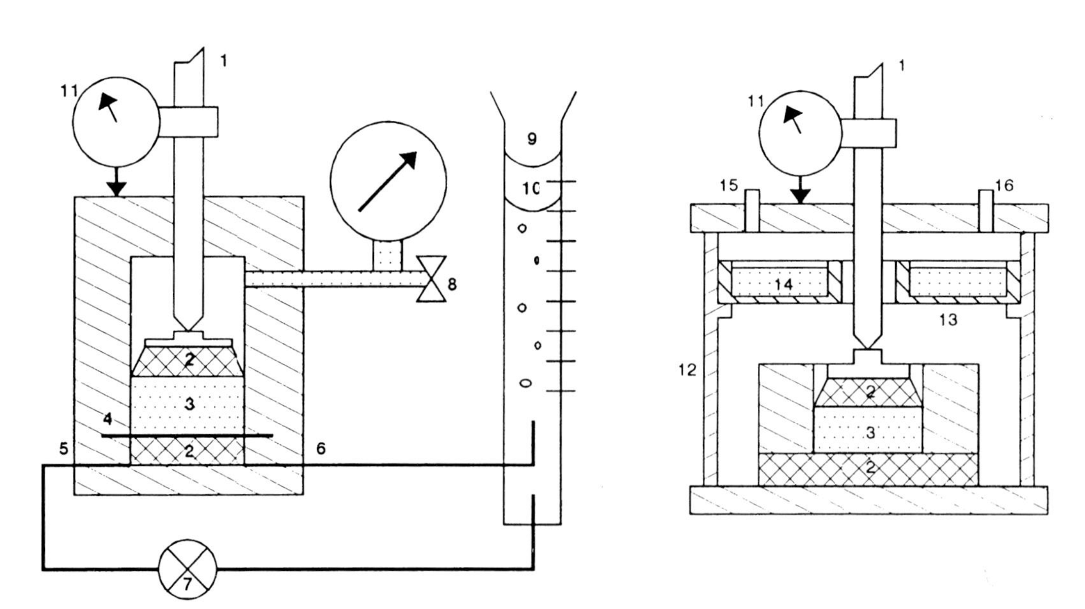

Fig. 1. Schematic representation of the suction-controlled oedometer cells (a) with nitrogen pressure, and (b) with sulphuric acid solution. Note: 1, loading ram; 2, porous stone; 3, sample; 4, semipermeable membrane; 5, water inlet; 6, water outlet; 7, peristaltic pump; 8, nitrogen supply; 9, buret; 10, coloured oil layer; 11, deformation gauge; 12, metacrilate wall; 13, glass deposit; 14, H_2SO_4 solution; 15, vacumm valve; 16, H_2SO_4 inlet.

initial dry densities (1.55–1.95 g cm^{-3}) and water contents (natural, 15 and 20%) have been uniaxially compacted, their diameter being 3.03 cm and their height 1.23 cm. They are introduced either in the pressure cells described above (one sample in each cell) or in a vacuum desiccator with a sulphuric acid solution (5 to 12 samples inside each desiccator, to avoid a major change in solution density after stabilization). The water content of the samples is checked every week by weighting them, taking as the stabilization criterium a weekly oscillation of the water content less than 1%. As the samples can change their volume freely during the experiment, the dimensions of the sample are measured at the end of the stabilization process. With these final values we are able to plot the 'water content versus suction' curve.

The oedometer tests are performed in oedometer apparatus, in which the standard cell has been replaced by a pressure cell or by a hermetic sulphuric acid solution cell (Fig. 1). The clay has been uniaxially compacted under a pressure of 20 MPa to a dry density of 1.60 g cm^{-3} with the stabilization water content (around 12%), the height of the sample being 0.6 cm and the diameter 4.95 cm. The initial void ratio is 0.737 and the saturation index 0.45.

The set of oedometer tests reported here consisted of a desiccation–hydration cycle performed at different vertical loads for each test, starting with a suction pressure of 0.1 MPa in nitrogen cells and 13 MPa in sulphuric acid cells. To desiccate the sample, suction is increased, by subsequent steps, up to 14 MPa in the nitrogen cells and up to 100 MPa in the sulphuric solution cells. Hydration is achieved by decreasing suction gradually down to the former values. The sample remains in each step until no deformation is registered, maintaining the vertical load constant. As the stabilization process is very long, the duration of each step has been fixed to 20 days, a period that we consider enough to achieve the main percentage of total deformation in most of the cases. The vertical loads applied range from 0.05 to 1.5 MPa.

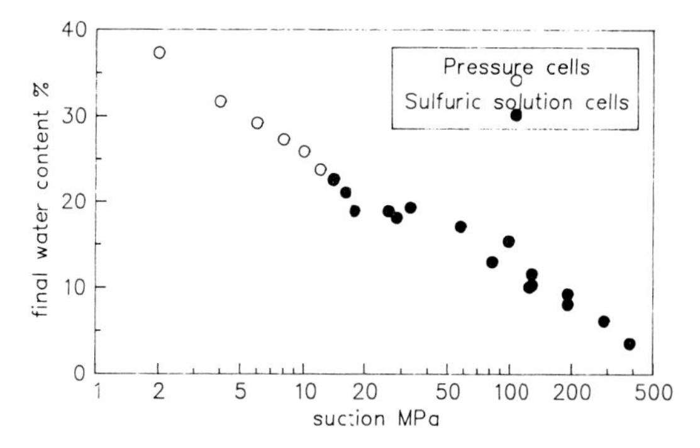

Fig. 2. Equilibrium water content versus suction for montmorillonite of initial dry density 1.65 g cm^{-3} and initial water content around 11%.

Results

A 'water content versus suction curve' has been determined for different initial water contents and dry

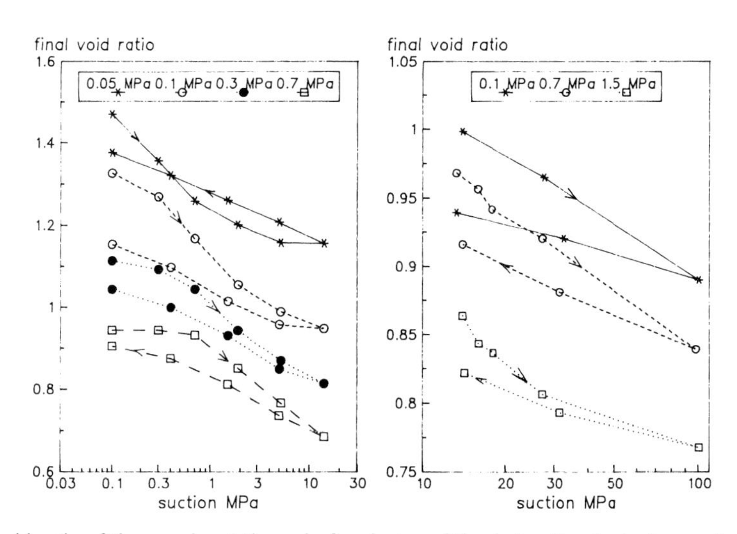

Fig. 3. The void ratio of the samples at the end of each step of the desiccation–hydration cycles of the suction-controlled oedometer tests performed in (a) nitrogen cells, and (b) sulphuric acid solution cells. The vertical load applied in each test is indicated inside the figure.

densities at 20 °C. There is a logarithmic relationship between water content at equilibrium and suction. The water content versus suction curve in Fig. 2 is for the montmorillonite with an initial dry density of 1.65 g cm^{-3} and a water content of around 11%, with values obtained both in nitrogen cells and sulphuric acid solution desiccators. The good correlation between both methods is evident. From these curves we obtain the suction pressure value for each water content, i.e. the suction for which the initial water content in the clay does not change during the experiment. For a dry density of 1.6 g cm^{-3}, these values are: 133 MPa for the water content at room conditions (11%), 61 MPa for 15% and 23 MPa for 20% water content.

The major percentage of water exchange is acomplished in the first week, but a good stabilization in water content values is not achieved until one to two months later.

The oedometer tests are still taking place. The void ratios attained by the samples at the end of each step of the desiccation–hydration cycle are plotted in Fig. 3. During the first step all the samples swelled because the initial vertical loads and suction pressures were not enough to counteract the high swelling pressure of montmorillonite. The void ratios are lower in the case of tests in sulphuric solution cells, as the samples have been submitted to higher suction pressures, and they have suffered greater initial consolidation. For the small range of suction pressures (nitrogen cells), a small increment in suction produces a high deformation. As the suction range is higher (sulphuric solution cells) the effect of suction changes on deformation is less evident, as if the sample stiffness were higher.

Acknowledgements. This work is being performed in the context of the CEC Contract FI2W/0102 and is partially supported by ENRESA. We are grateful to R. Campos (CEDEX, Madrid) for the performance of the tests. We thank J. Sáez (CEDEX) for his advice on suction-control techniques and E. Alonso's research group (DIT-UPC, Barcelona) for their suggestions on the development of the oedometer tests.

References

BIAREZ, J., FLEUREAU, J-M., ZERHOUNI, M-I. & SOEPANDJI, B. S. 1988. Variations de volume des sols argileux lors de cycles de drainage-humidification. *Revue Française de Géotechnique*, **41**, 63–71.

BISHOP, A. W. & DONALD, I. B. 1961. The experimental study of partially saturated soils in the triaxial apparatus. *Proceedings 5th International Conference Soil Mechanics*, **1**, 13–21.

ENRESA 1994. Almacenamiento geológico profundo de residuos radiactivos de alta actividad (AGP). Conceptos preliminares de referencia. *Publicación Técnica* 7/94, Madrid.

ESCARIO, V. 1969. Swelling of soils in contact with water at a negative pressure. *Proceedings 2nd International Conference Expansive Clay Soils*, A&M University, Texas, 207–217.

—— & SÁEZ, J. 1973. Measurement of the properties of swelling and collapsing soils under controlled suction. *Proceedings 3rd International Conference Expansive Soils*, Haifa, **2**, 195–200.

ESTEBAN, F. 1990. Caracterización experimental de la expansividad de una roca evaporítica. Identificación de los mecanismos de hinchamiento. Tesis Doctoral (PhD thesis), Universidad de Cantabria, Santander.

PÉREZ DEL VILLAR, L., DE LA CRUZ, B. & CÓZAR, J. S. 1991. Estudio mineralógico, geoquímico y de alterabilidad de las arcillas de Serrata de Níjar (Almería) y del Cerro del Monte (Toledo). División de Técnicas Geológicas, CIEMAT, Internal Report, Madrid.

PUSCH, R. & BÖRGESSON, L. 1989. Bentonite sealing of rock excavations. *OECD-NEA/CEC Workshop Sealing of Radioactive Waste Repositories*, Braunschweig.

RIVAS, P., VILLAR, M.V., CAMPOS, R., PELAYO, M., MARTÍN, P. L., GÓMEZ, P., TURRERO, M. J., HERNÁNDEZ, A. I., CÓZAR, J. S. & MINGARRO, E. 1991. Caracterización de materiales de relleno y sellado para almacenamiento de residuos radioactivos: bentonitas españolas. División Técnicas Geológicas, CIEMAT, Internal Report, Madrid.

VILLAR, M. V. & RIVAS, P. 1994. Hydraulic properties of montmorillonite–quartz and saponite–quartz mixtures. *Applied Clay Science*, **9**, 1–9.

YONG, R. N., BOONSINSUK, P. & WONG, G. 1986. Formulation of backfill material for a nuclear fuel waste disposal vault. *Canadian Geotechnical Journal*, **23**, 216–228.

Research into the mobility of a gas phase within a porous network

U. Boltze & M. H. de Freitas

Engineering Geology Section, Imperial College of Science, Technology and Medicine,
Prince Consort Road, London SW7 3BP, UK

Abstract. This paper presents work conducted at Imperial College on the mobility of a gas phase through a partially saturated porous medium. The research is based on the observation of the behaviour of a discrete gas phase in place rather than on gas in solution. The experiments were designed to ensure that chemical reactions and the nucleation of the gas phase were kept to a minimum if not prevented.

Glass micromodels of porous media were used to allow visual analysis of the movement of the gas phase in a partially saturated porous medium under conditions of rapidly changing head. Both conditions of heavy localized rainfall or rapid increase in atmospheric pressure, and rapidly changing water levels are simulated.

Images of gas bubble positions, sizes and frequencies under given potentials are analysed using a Geographical Information System (GIS) to enable the bubble population to be quantified in time and space.

From these studies conclusions concerning the behaviour of the gas phase under these conditions have become apparent. These have significant implications for the monitoring of gas emissions and movement of gas *in situ*.

Interest in the subject was triggered by reports of methane explosions at Abbeystead (Great Britain Factory Inspectorate 1985, Jefferis & Wood 1990) and Loscoe (Williams & Aitkenhead 1981; Tankard 1987) and the seeming lack of understanding of the processes leading to the release of gas at ground level.

At depth gas appears in solution, but as it approaches groundlevel it comes out of solution due to supersaturation and occurs in the form of gas bubbles, and near the surface it is released to the atmosphere.

In this research movement of gas in the form of bubbles is investigated since it seems to be a greater source for concern than diffusion and it has been proven to be a natural transport mechanism that operates near the surface as, for example, in the movement of radon gas (Kristiansson & Malmqvist 1982) and in the escape mechanism for air in irrigated soils (Powers 1934). Various ideas about the factors influencing gas emission are put forward in the literature, the main two ideas being rapid changes in atmospheric pressure (Kovach 1945; Kraner *et al.* 1964; Clements & Wilkening 1974; McOmber *et al.* 1982; Tankard 1987; Young 1990; Massmann & Farrier 1992) and a pumping effect of a moving water-table (Lapalla & Thompson 1983; Bishop *et al.* 1990). The research therefore concentrates on these two mechanisms, namely, the change in atmospheric pressure and the change in water-table elevation. The importance of the first is well known, if not entirely understood, whereas the importance of the second is largely unknown.

To investigate the mobility of the gas phase and especially the role of the water-table in this, field data from gas-emitting sites were studied. It was soon found that neither the quantity nor the quality of the data is sufficient to allow any analysis or prediction to be made, and the need for laboratory experiments to define the most important parameters governing gas movement was seen. Since observing the processes taking place in soil columns is indirect and very speculative, it was decided to use two-dimensional (2D) glass micromodels which allow direct observation of the processes taking place inside the model.

Use of 2D micromodels

2D glass micromodels

These models can be manufactured to an effective size of up to 10.5 cm by 5.5 cm by etching any porous network into one of two glass plates: the procedure is described in Dawe (1990) and Gray (1990). This plate is then sintered to a second glass plate. It is the size of the sinter oven used which dictates the size of the model produced. Access to the porous network is through two inlet channels at either end of the model (see Fig. 1).

Advantages of 2D micromodels for the research

The advantages in using these glass micromodels are as follows:

From Bentley, S. P. (ed.) *Engineering Geology of Waste Disposal*,
Geological Society Engineering Geology Special Publication No. 11, pp 313–322

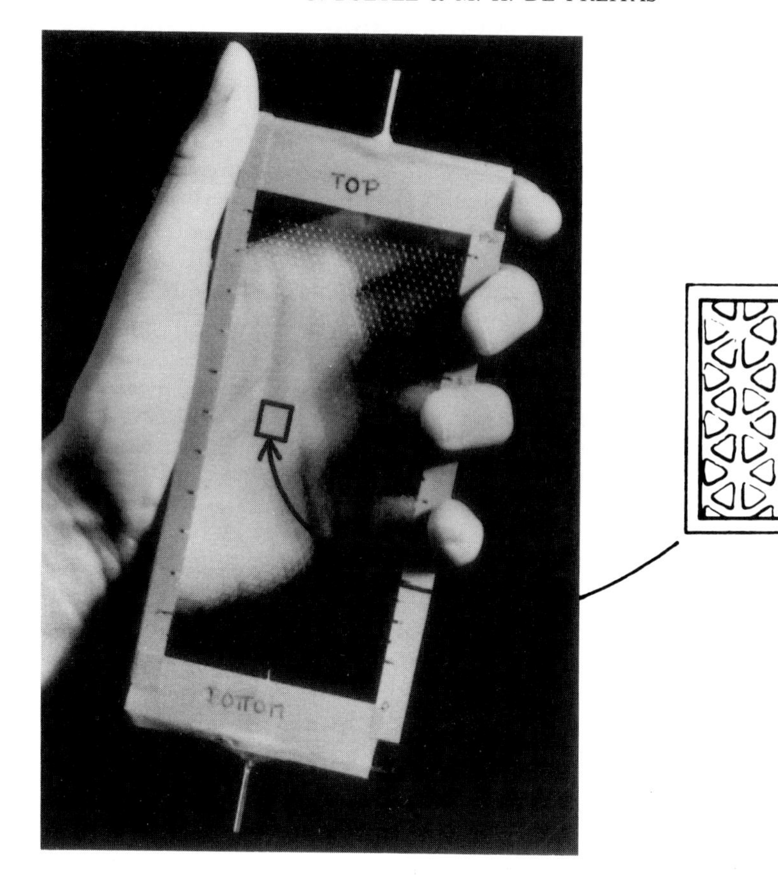

Fig. 1. Small hexagonal model (SHM).

- Any porous network that can be thought of in two dimensions can be modelled.
- Since the models are produced in glass, they are inert to chemical attack. The solubility of glass is very low so that the fluids introduced into the system will not change their physical parameters, like viscosity or surface tension, due to chemical interaction with the model.
- The models are transparent, therefore the processes inside can be viewed with a microscope camera allowing recording of the microscopic phenomena on video. This can be used to study the micromechanics of the gas mobility in a fluid–solid system.

Experiments

Set-up

To investigate the mobility of the gas phase through the 2D micromodels under controlled conditions, the test arrangement shown in Fig. 2 was assembled.

The set-up is a simple system which allows the elevation head and pressure head acting on the system to be known precisely. For the starting position the total head of water at the top of the model is exactly the same as at the bottom of the model. The hydraulic regime is therefore well controlled.

Only air and water are used in the system for economic and safety reasons. A condition for testing was that the volume of gas in the model was kept constant for each test cycle, therefore the gas is not generated in place but introduced in a controlled way.

The rig which houses the model (Fig. 3) consists of a motor-driven lift system for lifting the overflow chambers which are attached to either end of the micromodel. A reservoir is used to keep a constant water level in one of the overflows. A light source illuminates the model from behind, while a movable microscope camera connected to a video recorder allows the processes inside the model to be recorded.

Experimental procedure

For each micromodel three cycles of experiments are conducted as shown in Fig. 4. Each cycle investigates the

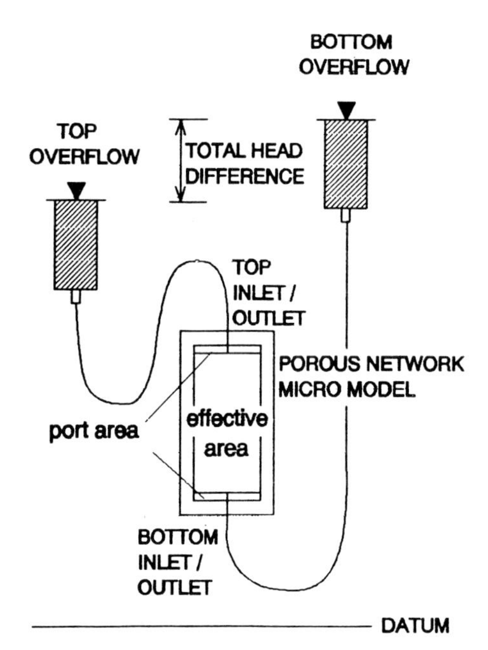

Fig. 2. The micromodel and its boundaries.

influence of the speed of water level rise (imbibition) on gas encapsulation. At the beginning of the first cycle the assumption that the volume of the gas bubble can be calculated, within experimental error limits, by multiplying the area measured by the depth of the pore channel is checked.

Furthermore, during the first and second cycle, the mobility of the gas phase is investigated. The first cycle simulates a sudden increase of atmospheric pressure or sudden, localized rainfall. The second cycle simulates the rise and fall of the water-table.

The detailed experimental procedure described below was chosen to achieve controlled seeding of the model with bubbles, movement of the gas phase under a given difference in head and a test of experimental reproducibility. The experimental procedure for one experimental run as shown in Fig. 4 is as follows.

The glass micromodel is saturated under vacuum with distilled, de-aired water starting from dry conditions. At the beginning of each cycle three falling head permeability tests are done. This gives a permeability value for that specific model and is used as a check for subsequent experiments that no changes have happened to the model. The still fully saturated model is then attached to the rig with both waterfilled overflows at 40 cm above datum (T1 on Fig. 4).

The model and its top overflow chamber are then drained to below the bottom inlet by lowering the bottom overflow with the winch under at a speed of approximately 42 m h^{-1} to 30 cm below datum (T2 on Fig. 4). After drainage has been achieved, the bottom overflow is lifted back to the starting level (40 cm above datum = T3) at a given speed (first cycle: 0.7 m h^{-1}, second cycle: 6.2 m h^{-1}, third cycle: 50.4 m h^{-1}). This procedure simulates the rise of a water level into a pre-wetted porous medium, thereby allowing encapsulation of gas bubbles, and is here used as a controlled way of seeding the model with gas bubbles. The set-up is left like this until the top overflow is full of water, the water level in the overflow chamber attached to the bottom of the model being constantly replenished from the reservoir higher up. The situation in the model at this starting position (T3) is recorded on video tape.

During the first cycle both overflows are lifted at the same time at the same speed, first to 90 cm above datum (T4) and then to 140 cm above datum (T5). For both levels the situation of the gas phase within the model is recorded. The reaction of the gas bubbles to increased total pressure allows validation of the assumption that the volume of the gas bubbles can be calculated by multiplying the area measured for each bubble by the depth of the pore channel.

To test the influence of the speed at which a pressure gradient is applied to the system, the top overflow is lifted consecutively at three different speeds (3.5, 7.2 and 55.4 m h^{-1}) to 140 cm above datum (giving a total head difference across the model of 1 m) and lowered at a speed of 60 m h^{-1} to the starting level (40 cm) (T6 to T11). The situation in the model is recorded on video tape after each raising and lowering of the overflow chamber.

At the end of the first cycle (T11), the model is dried, resaturated under vacuum, tested for its permeability, attached to the rig (T12), drained (T13) and seeded with gas bubbles (T14).

During the second cycle (T14 to T20) the overflow chamber attached to the bottom is lifted at three different speeds (3.5, 7.2 and 55.4 m h^{-1}) to 140 cm above datum (giving a total head difference across the model of 1 m) to simulate different rates of rise of the water-table. Between the lifts the overflow is every time lowered back to the starting level (40 cm above datum) at a speed of 60 m h^{-1}.

At the end of this second cycle (T20), the model is again dried, resaturated under vacuum, tested for its permeability, attached to the rig (T21), drained (T22) and seeded with gas bubbles (T23).

The third cycle consists of only one recording, namely the amount of gas encapsulated during seeding (T23).

Analytical procedure

Determining model parameters

To determine model parameters such as grain size, grain size distribution and porosity, the templates that were

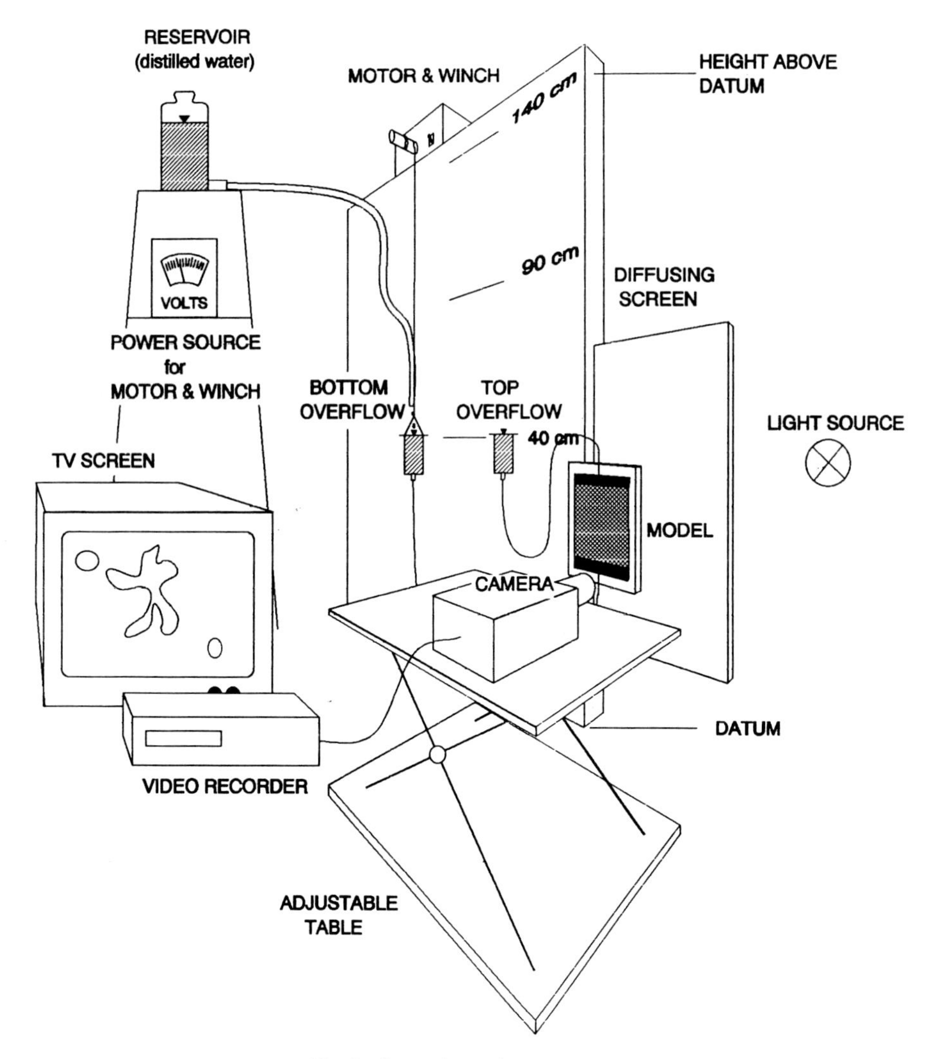

Fig. 3. General experimental set-up.

used for manufacturing the glass models are scanned (using SCANGAL) into the Geographical Information System (GIS) IDRISI in which the areas occupied by each single grain and by the pores can be measured. Knowing the area occupied by grains and pores, it is possible to calculate a porosity based on the ratio of these areas.

Analysis of video images

The GIS IDRISI is further used as the main tool for analysing the video images. After enhancing the images in the image analysis package KHOROS and printing the enhanced images to produce an enlarged replica of the model area, 115 cm long (including port areas), the gas bubbles are digitized into IDRISI, in which their size is measured allowing a bubble size distribution curve (Fig. 5) to be plotted in exactly the same way as a particle size distribution curve.

The image can further be converted into ASCII format giving a string of numbers which, after further manipulation, gives the saturation of the model, a gas phase distribution curve and the centre of gravity for the area under the curve (Fig. 6).

Analysis of bubble mobility

The distance between the centres of gravity of the area under the gas phase distribution curve of two experiments can be used as a measure of the mobility of gas

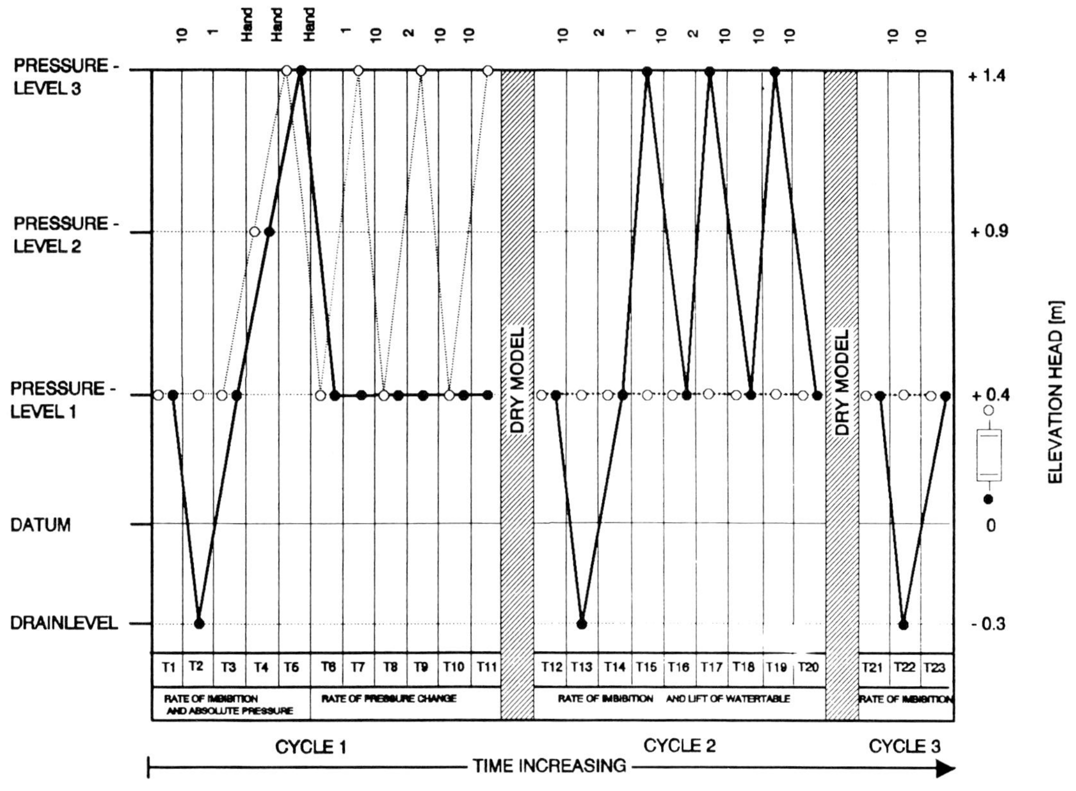

Fig. 4. Experimental procedure (Because the elevation head of the model is kept constant, the changes of total head are the changes in pressure head).

bubbles. A requirement for this method to be applicable is that only the distribution of the gas present changes while the total amount of gas present does not change. This assumes that no gas is either added or extracted during the test. Because the model and the leads are transparent any violations of this requirement can be identified.

A mobility factor is defined as being the ratio of the distance between two centres of gravity (ΔL) measured in the direction of maximum potential gradient through the model and the time it takes to apply the change in pressure gradient (potential) ($t_{dynamic}$). This definition has been chosen because the time over which bubble movement is observed is different from the time in which the change in pressure takes place. Because of possible inertia in the system the reaction of the bubbles might not be instantaneous, but the link between the two is not known. A distinction is therefore made between the time for observation, $t_{experiment}$, and the time for inducing change, $t_{dynamic}$. Hence a water level will change over a period $t_{dynamic}$ but its effects will be operating beyond that period, as can be see in Fig. 7.

Results

The following results were obtained from the Small Hexagonal Model (SHM), one of four micromodels being studied (Fig. 1). It has grains with an area of $0.85\,mm^2$. The pore nodes have a diameter of 0.7 mm, the pore throats are 1.19 mm long and 0.2 mm wide. Thin sections show that the pores are 0.07 mm deep. The effective size of the model is 10.5 cm by 5.5 cm by 0.07 mm. The hydraulic conductivity measured for this micromodel is $4.34 \times 10^{-4}\,m\,s^{-1}$, the intrinsic permeability being $4.2 \times 10^{-11}\,m^2$ (43 darcys). The porosity of

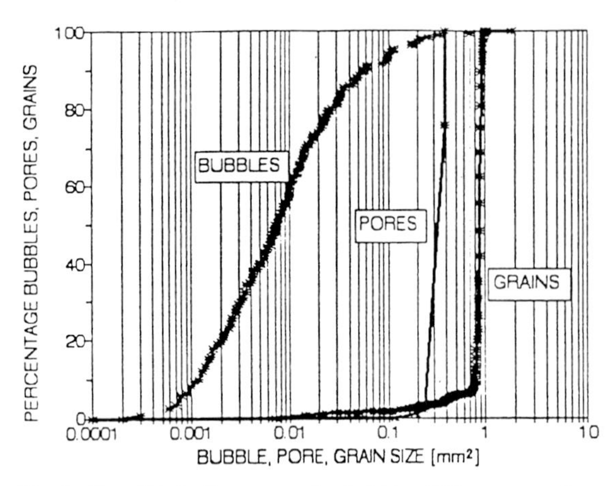

Fig. 5. Size distribution curves for bubbles (T3), pores and grains of a model.

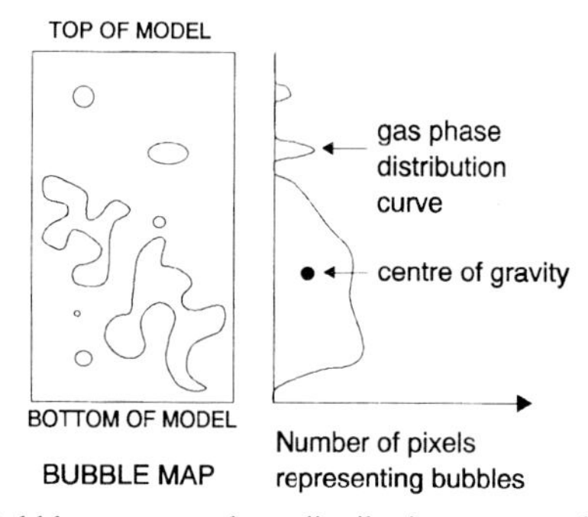

Fig. 6. Bubble map, gas phase distribution curve and centre of gravity for a micromodel.

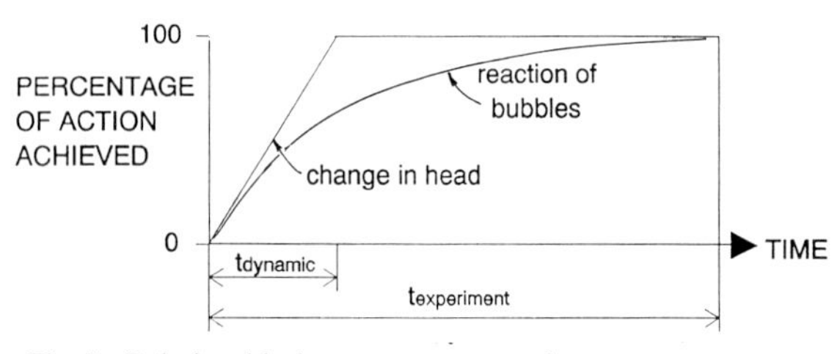

Fig. 7. Relationship between $t_{dynamic}$ and $t_{experiment}$.

the model, calculated as described earlier, was found to be 54.35%.The relative permeability and the flow rate through the model will have changed continuously, but unpredictably, with time due to changes in the position of the gas bubbles which can be regarded as barriers to flow.

Figure 8 shows the results from tests T3 to T23 which follow the procedure illustrated in Fig. 4. The procedure can be seen in row 1 where the fluid pressure at the upper and lower boundary of the model are recorded with time (note the gaps on the time axis). For details, refer to the description given for Fig. 4.

Rows 2 and 3 illustrate the changes in temperature and atmospheric pressure which occurred during the experiments. These affect the tests as the models are not enclosed in a pressure and temperature controlled chamber.

Row 4 records the time, $t_{dynamic}$, during which heads were changed, i.e. heads were raised or lowered over a relatively short period of time while the changes in gas phase in the model were observed over a period following such change in head (see Fig. 7). The rate of change of total head during such periods is shown in row 5.

Thus rows 1 to 5 illustrate the test progress and the environmental changes that occurred over that period. Rows 6 to 10 illustrate the results of such changes.

Row 6 records the saturation of the model with time and the extent to which the saturation with time can be explained by temperature and pressure applied during the period of testing. It shows that saturation is not unduly affected by the experimental procedure and that the changes in pressure head recorded in row 1 have essentially been shunting the same amount of gas up and down the model.

Row 7 shows the number of bubbles in the model. A sudden increase in the number of bubbles (T7) occurred when a pressure gradient was applied for the first time. Further applications of pressure gradients have far less influence. The first application of a pressure gradient results in splitting-up of bubbles, while any further changes in pressure gradients lead to no further change in bubble numbers or even result in the coalescence of bubbles.

Row 8 records the centre of gravity of the area beneath the gas phase distribution curve as described above and, especially in cycle 2, shows how this is related to the cyclic movement of head (row 1).

Row 9 illustrates the difference in the centres of gravity which are shown in row 8. Here negative values indicate a movement of the centre of gravity towards the bottom of the model while positive values indicate movement towards the top of the model. For the case of a pressure radient inducing upward flow (cycle 2, row 9), the values of L are positive as long as the pressure is acting upward, i.e. as long as the gradient is sustained. Once the upward acting pressure is reduced to zero, the gas phase does not carry on its upward movement but reacts with relaxation, resulting in slightly negative values for ΔL. These values do not necessarily indicate gas flow, but are due to surface tension effects and the energy balance.

Row 10 shows the mobility as defined above and illustrates how the distribution of the gas phase varies with changes of head, and the very clear difference that exists between the mobility of the gas phase for upward flow and for downward flow.

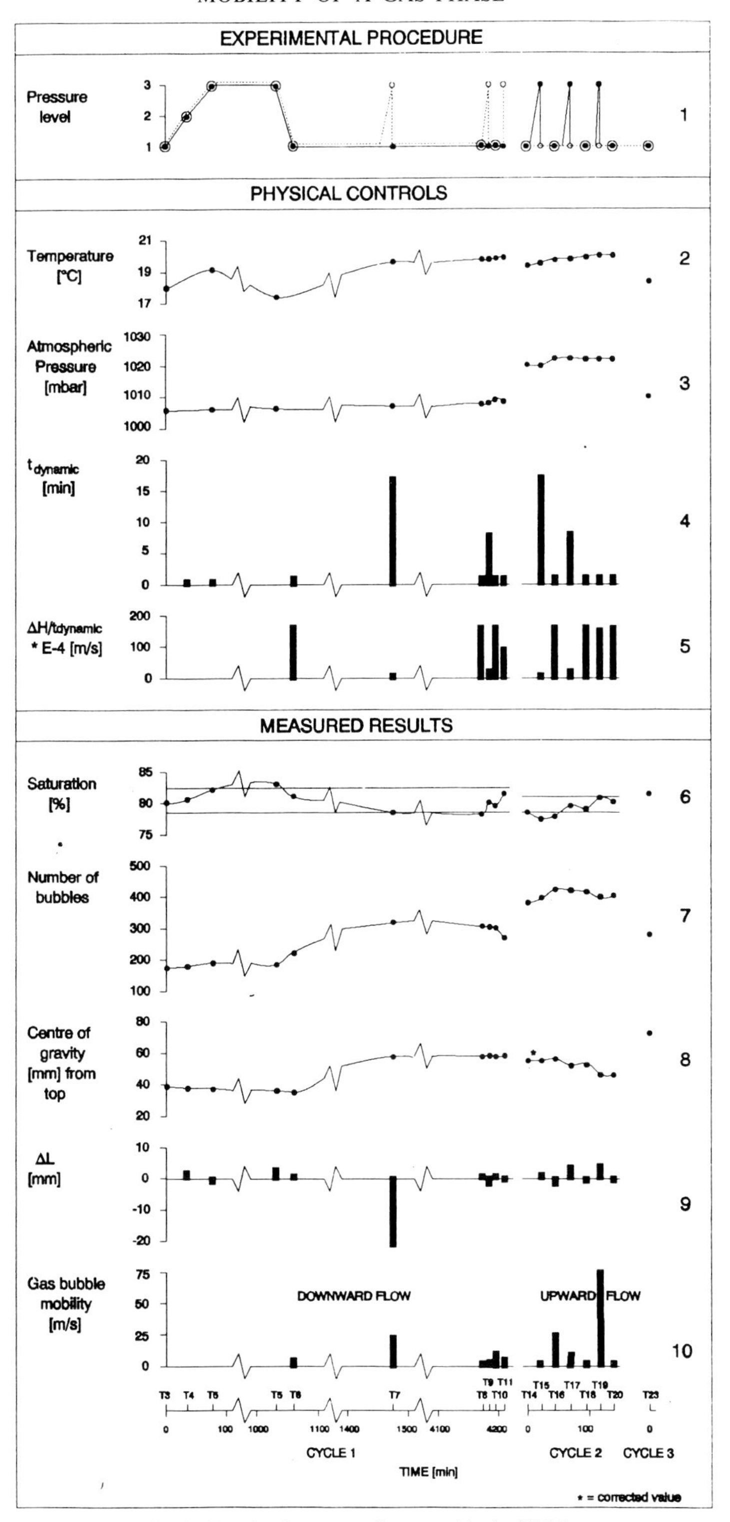

Fig. 8. Results from experiments with the SHM.

Interpretation

Four aspects of the test described will be presented, namely the measuring accuracy of the system, the changes that have been seen in bubble position, shape, number and size, and the influence of the rates of change of head on both encapsulation of gas, and the mobility of the encapsulated gas phase.

Measuring accuracy

Two images were taken of the same situation T5 (Fig. 8). The time difference between the two images was 953 min, (15 h 53 min, i.e. overnight). During this period the temperature fell by 1.8 °C which explains the change in saturation between the two images. The bubble number changed by eight bubbles from 189 to 181, which might be due to overlooking very small bubbles, coalescence of bubbles or due to the solubility of very small bubbles. From this it can be assumed that the error involved in identifying every bubble correctly is at most 8 in 189, i.e. 4%. Putting this into perspective, this change in bubble number is insignificant, compared with the increase of bubble number by 104 for the maximum change encountered (T6 to T7 in Fig. 8).

The difference in the position of the centre of gravity of the two images is 2.6 mm which, when used as the maximum error margin, would give 0.2% (2.6 mm in 1150 mm, including port areas).

Major changes in bubble position, shape, number and size

Most of the changes in bubble position, shape, number and size are taking place during the first application of a pressure gradient inducing flow through the model. Inspection of the images confirms that during cycle 1 the change happens between T7 and T8 and during cycle 2 the change happens between T14 and T15.

Influence of the rate of imbibition on gas encapsulation

Figure 9 illustrates the images for stages T3, T14 and T23, which represent the gas encapsulation process for different velocities at which the head of water leading to encapsulation was applied ($\Delta\phi/T_{\text{dynamic}}$).

When the images showing gas encapsulation (Fig. 9) are compared, the following facts become apparent. At the lowest velocity of imbibition (0.7 m h^{-1}), the gas phase is entrapped as large accumulations. The gas is mainly found near the top of the model. Considerable areas of the model are completely free of gas. At a medium velocity of imbibition (6.2 m h^{-1}), the entrapped gas is distributed much more evenly as smaller accumulations throughout the model. At the highest velocity of imbibition (50.4 m h^{-1}), the gas is mainly entrapped in the bottom part of the model, with the largest accumulations of gas lowest down. Again, larger areas appear throughout the model which are virtually free of gas.

If all three images are compared in more detail, it becomes obvious that some bubbles appear in all of the

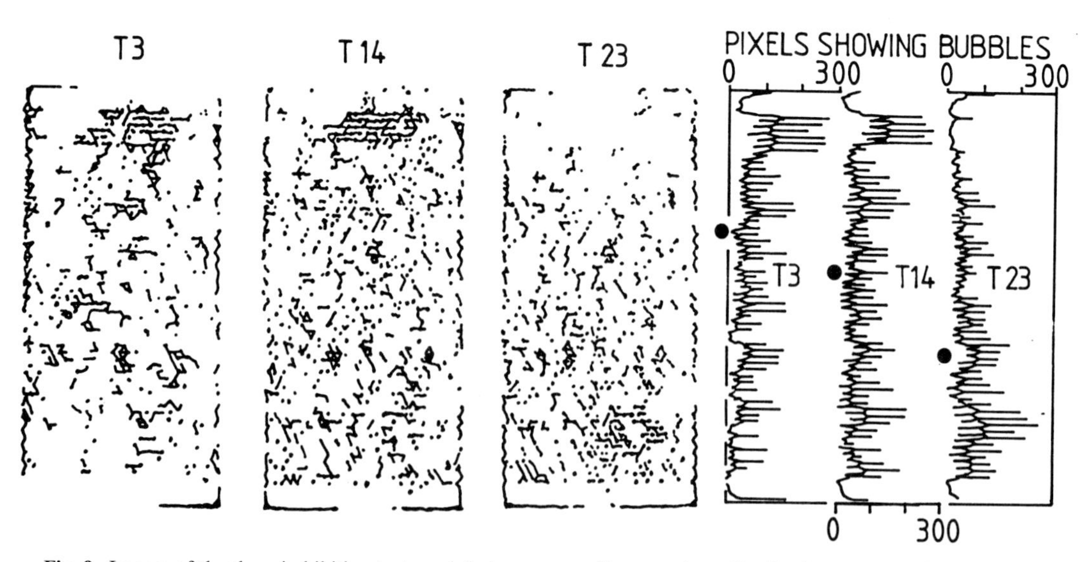

Fig. 9. Images of the three imbibition tests and their corresponding gas phase distribution curves and centres of gravity (indicated by •).

images in the same position, having virtually the same shape and volume. This indicates that the process of gas encapsulation is not entirely random but follows in part a repeatable pattern that causes certain pores to be occupied on repeated occasions.

The positions of the centres of gravity of the gas phase distribution curves confirm the impression gained by comparing the images for the imbibition case (T3, T14, T23), namely with increased speed of imbibition the centre of gravity moves further towards the bottom of the model.

Influence of the rate of pressure change on the mobility of gas

Figure 10 records the change in the position of the centre of gravity as a function of the rate at which the pressure head in the model was changed with time. Two basic cases are illustrated: upward and downward flow.

In downward flow any gas phase present will have to respond to the hydrostatic pressure under which the bubble exists. Hence there is a progressive decline in the migration with an increase in the rate of change of head. This has interesting implications for the movement of the gas phase under conditions of variable rates of infiltration or groundwater recharge. In upward flow quite a different response seems to occur in which there is no decrease in the mobility of the gas phase which carries on moving to the atmosphere.

Both these curves are shown on the right-hand side of the graph which is the area associated with the raising of head, whether it may be for generating either upward or downward flow. The left-hand side of the graph is associated with the lowering of head to a situation where there is no difference in head; this happens after each period of raising of head (see Fig. 4).

Conclusions

A number of important implications for field practice have emerged to date from this work and are listed below.

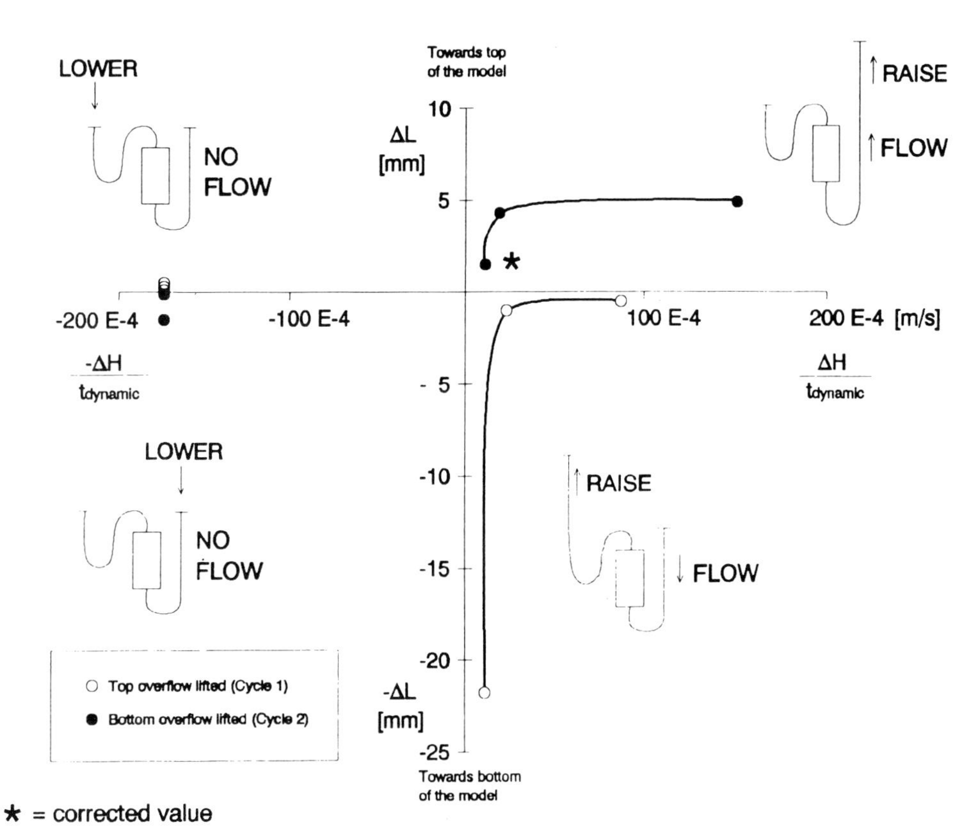

Fig. 10. Relationship between the distance between the centres of gravity and the speed with which the corresponding pressure head was applied (* = value affected by boundary conditions whose influence has been removed).

- The 2D glass micromodels can be made to simulate porous networks of given particle and pore size dimension, grading and conductivity. They are therefore an exceedingly useful tool for studying the movement of gas in a particular soil. They are particularly useful in assessing the sensitivity of the gas movement to changes in the potential which causes movement as well as to changes in the natural environment of temperature and pressure.
- They therefore permit a study of the repeatability of a material's performance to two-phase flow and facilitate a statistical study of the phenomena. This allows the confidence in design parameters used to predict gas migration under given circumstances to be assessed.
- The work does suggest, even though this has yet to be confirmed, that the rates at which driving forces change has a profound effect on the movement of the gas phase. This is most clearly seen in upward flow and has major implications for the frequency of monitoring of sites. It could suggest that continuous monitoring may be what is really required rather than intermittent monitoring.
- The research also suggests that bubble movement is related to the movement that has occurred before. The gas phase, once on the move, has a tendency to move into a stable position. If the potential applied to move it to this position stops, the gas phase locks up. It cannot then be remobilized by the same potential but will require a much higher potential to regain mobility. The difference between dynamic and static friction will be responsible for this effect.
- The mobility of the gas phase, under increasing rates of change of head, decreases in downward flow and increases in upward flow.
- The pattern of encapsulation of a gas phase under rising water levels is dependent on the speed of the rise. The slower the rise of water, the higher up in the soil does the gas encapsulation occur, i.e. the slower the water level rises, the more gas will get pushed towards the soil surface.
- Further investigation is required into the effect of the time elapsed between changes. Both physical and chemical potentials are involved and chemical potentials leading to diffusion, for example, will be time dependent.

Acknowledgement. This work is funded by the Directorate-General for Science, Research and Development of the Commission of the European Communities under the STEP programme.

References

BISHOP, P. K., BURSTON, M. W., LERNER, D. N. & EASTWOOD, P. R. 1990. Soil gas surveying of chlorinated solvents in relation to groundwater pollution studies. *Quarterly Journal of Engineering Geology*, **23**, 255–265.

CLEMENTS, W. E. & WILKENING, M. H. 1974. Atmospheric pressure effects on ^{222}Rn transport across the earth–air interface. *Journal of Geophysical Research*, **79**(33), 5025–5029.

DAWE, R. A. 1990. Reservoir physics at the pore scale. *In*: ALA *et al.* (eds) *Seventy-five Years of Progress in Oilfield Science and Technology*. Balkema, Rotterdam, 177–194.

GRAY, J. D. 1990. *Low interfacial tension (<1 mN/m) studies of multiphase fluid behaviour in porous media.* PhD thesis, Imperial College, London.

GREAT BRITAIN FACTORY INSPECTORATE 1985. *The Abbeystead Explosion.* Health and Safety Excecutive, HMSO, London.

JEFFERIS, S. A. & WOOD, A. M. 1990. Ground conditions at Abbeystead. *Proceedings of the Institution of Civil Engineers Part I. Design and Construction*, **88**, 721–725.

KOVACH, E. M. 1945. Meteorological influences upon the radon content of soil gas. *EOS Transactions of the American Geophysical Union*, **26**, 241–249.

KRANER, H. W., SCHROEDER, G. L. & EVANS, R. D. 1964. Measurements of the atmospheric variables on radon 222 flux and soil–gas concentrations. *In*: ADAMS, J. A. S. & LOWDER, W. M. (eds) *The Natural Radiation Environment*, University of Chicago Press, Chicago, Illinois, 191–215.

KRISTIANSSON, K. & MALMQVIST, L. 1982. Evidence for nondiffusive transport of ^{222}Rn in the ground and a new physical model for the transport. *Geophysics*, **47**(10), 1444–1452.

LAPALLA, E. G. & THOMPSON, G. M. 1983. Detection of groundwater contamination by shallow soil gas sampling in the vadose zone. *In*: *Proceedings of the NWWA/EPA Conference on Characterization and Monitoring of the Vadose (Unsaturated) Zone*, Las Vegas, Nevada. National Water Well Association, 659–679.

MASSMANN, J. & FARRIER, D. F. 1992. Effects of atmospheric pressure on gas transport in the vadose zone. *Water Resources Research*, **28**(3), 777–791.

MCOMBER, R. M., MOORE, C. A. & BEATTY, B. W. 1982. Field evaluation of methane migration predictions. *Canadian Geotechnical Journal*, **19**(3), 239–249.

POWERS, W. L. 1934. Soil–water movement as affected by confined air. *Journal of Agricultural Research*, **49**(12), 1125–1134.

TANKARD, J. 1987. Pressure drop has tragic consequences. *Weather*, **42**(9), 297.

WILLIAMS, G. M. & AITKENHEAD, N. 1991. Lessons from Loscoe. *Quarterly Journal of Engineering Geology*, **24**(2), 191–207.

YOUNG, A. 1990. Volumetric changes in landfill gas flux in response to variations in atmospheric pressure. *Waste Management & Research*, **8**(5), 379–385.

Programs used:

IDRISI 1990. Version 3.2., Clarke University, Graduate School of Geography, Worcester, MA 016610, USA. Sourcecode by R. Eastman.

KHOROS 1991. Version 0, release 1.0. University of New Mexico. Source code by M. Young & C. Williams.

SCANGAL 1988. Scanning Gallery Plus, version A.03.00. Hewlett-Packard Co. & Microsoft Corporation.

SECTION 5

STANDARDS OF LANDFILL ENGINEERING

Waste disposal, regulatory policy and potential health threats

Raymond N. Yong

Geotechnical Research Centre, McGill University, 817 Sherbrooke St W, Montreal, PQ, Canada, H3A 2K6

Abstract. The issues with respect to development of a safe liner-barrier system for waste containment facilities in the land disposal of wastes are examined in relation to the need for minimization of potential health and environmental threats. The influence of regulatory policies or attitudes, e.g. the 'Compliance (Mandated) Requirement' and the 'Performance Requirements' attitudes, on the requirements for barrier system design and performance is seen to be very significant. Ensuring protection of public health and the environment is central to the construction and performance requirements and criteria of the barrier system. Regulatory control and measures designed to fulfil minimization of public health threats must address the basic question of 'what constitutes a health threat', especially since one needs to ensure proper and secure barrier performance for 250–500 years or more. The performance of substrate material and engineered landfill barriers as competent liner systems, particularly with respect to substrate control on waste leachate containment and transport, must be well understood if proper protection of public health and the environment is to be obtained.

The term 'waste disposal' in the North American context generally refers to the discharge of waste forms into the atmosphere, receiving waters and land. Land disposal of waste is by far the most common form of waste disposal practised, not only in North America, but also in most other countries. In contrast to the North American perception of the term 'waste disposal', European Economic Community Directive 78/319, Article 1c, considers disposal to include both discharge of waste and the transformational operations necessary for recovery, re-use and recycle. The transformational operations identified in the EEC directive have been considered as somewhat separate items by the US-EPA context, and in fact occupy a prominent position in the 'National Policy' articulated in Section 1003 of RCRA 1976 together with the Hazardous and Solid Waste Amendments, 1984 (HSWA 1984), and more specifically with section 224(a) of HSWA 1984:

(1) the generator of the hazardous waste has a program in place to reduce the volume or quantity and toxicity of such waste to the degree determined by the generator to be economically practicable; and
(2) the proposed method of treatment, storage, or disposal is that practicable method currently available to the generator which minimizes the present and future threat to human health and the environment.

The safe management of waste containment systems, be it waste landfill facilities or systems for management or control of waste (leachate) contamination of ground (i.e. ground pollution), requires the construction of containment facilities in accordance with design specifications or expected performance criteria. The importance of available (published) criteria, guidelines, etc., as targets for control of the design, management and performance of the system is self-evident. In general, these criteria, standards, etc., are issued by various agencies and professional bodies, except that in the case of problems associated with the environment it is generally the government that is supposed to issue the guidelines, standards, etc., since protection of public health from environmental threats is the responsibility of the government. In this particular instance, the environmental threats are posed by the disposal of waste in the ground, and other problems associated with landfarming, illicit dumping, underground storage tanks, etc. The problem at hand relates to the extent of the participatory (active) role the government wishes to exercise in the issuance and enforcement of standards, etc., since this bears directly on the burden of responsibility for protection of public health that needs to be assumed by the government. For this presentation, it will be assumed that, by and large, if waste disposal facilities are constructed to meet the issued standards (i.e. standards designed to protect public health), these facilities should function adequately, and that protection of public health is obtained. This presumes that in the determination of standards and measures required for public health protection, one has considered not only what constitutes a health threat, but also the various human toxicological indices, phytotoxicological information, toxicological information for soil organisms and

From BENTLEY, S. P. (ed.) *Engineering Geology of Waste Disposal*,
Geological Society Engineering Geology Special Publication No. 11, pp 325–339

processes, and pathways to the various receptors. For this presentation, one assumes that if all the various pieces of information have been considered (as in the preceding) in the issuance of measures required for public health protection, the threats to the environment have also been correspondingly addressed.

As evident from the above, the issuance of standards and/or criteria for environmentally safe management of waste requires one to consider not only the nature of the health threats posed by the waste material, and the problem of acceptable daily intake (ADI) or its equivalent of the toxicant, but also the technical requirements for the waste barrier system. In land disposal of waste, the particular health protection issues that are considered as important (other than the nature of the waste) are those which are concerned with waste leachate propagation (transport) in the substrate, the toxicity of the leachate, and exposure routes (pathways). Considering water as the primary carrier for contaminants in the substrate, the barrier system which separates the waste material and/or waste leachates in a landfill from the natural substrate material and groundwater, constitutes the technical element which lends itself to control, either by technical specifications developed by the appropriate professional bodies, or through regulatory requirements and specifications. In the case of ground contamination occurring from spills and leaking underground storage tanks for example, where constructed liner-barrier systems are not necessarily present or obviously absent, the ground substrate must perform as a contaminant barrier system if the contaminants (plume) are to be controlled and groundwater protection is to be achieved. Under such circumstances, control is exercised through regulations governing illicit dumping and storage tank specifications.

The influence of the regulatory attitudes (policies and/or philosophies) in respect to the degree of involvement in development of guidelines, standards, etc., governing land disposal facilities for waste, cannot be overstated. This includes not only the perception of what constitutes a hazardous waste, but also whether one adopts drinking water standards for groundwater quality assessment, or develops appropriate groundwater quality standards. If the latter case is chosen, a new set of biostatics will be needed to provide for determination of ADI values, etc. One needs to note that in any event, there is a requirement to understand how measurements of groundwater quality and soil pore-water quality in the saturated and vadose zones (soil quality?) should be interpreted or used as controls for measures taken to protect or minimize public health threats. In regard to soil quality, the water that is held by the soil particle system (i.e. immobile water, water in the Stern layer (inner and outer Helmholtz planes), etc.) poses a problem that is particularly interesting inasmuch as one needs to be knowledgeable about contaminant hydrogeochemistry.

Regulatory attitudes and landfill performance

The regulatory attitude adopted by the government agency responsible for waste management and disposal (e.g. Ministry of Environment, Pollution Control Board, Environmental Management Bureau, Department of Environment, etc.) plays a very important part in the determination of the nature, role and effectiveness of the contaminant barrier. The various types of regulatory attitudes in evidence in the many countries surveyed can be broadly grouped into four types:

(1) *laissez-faire* or 'yet-to-be-crystallized',
(2) command, control and rectify (2CR),
(3) performance, monitor and rectify (PMR), and
(4) a combination-type which consists of a mixture of 2CR and PMR.

Of the above four types of attitudes, attitudes (2) and (3) will be discussed since attitude (1) does not merit consideration, and attitude (4) will be an outcome of both (2) and (3). Thus, for example, in the 2CR attitude which is essentially the US (EPA) attitude towards land disposal of wastes, a double liner and leachate collection system is mandated as a requirement for the liner barrier system (Section 3004(o) of RCRA, 1976), together with a groundwater monitoring system. In addition, an approved leak detection system (i.e. approved by the Administrator) capable of detecting leaks at the earliest practicable time, is also specified as part of the mandated requirement. Under Section 3008(h) of RCRA, 1976, detection of a release of hazardous waste into the environment from a facility authorized to operate . . . will require corrective action as deemed by the Administrator, necessary to protect human health or the environment.

The availability of the 2CR type of mandated requirements has been very favourably considered by many, inasmuch as this provides one with definite targets or objectives, and presumably relieves one from the responsibility of determination of 'design against future or potential health threats'. As noted previously, if one adopts the attitude that if the facility is constructed according to the RCRA 1976 mandated requirements (i.e. constructs the disposal facility according to rules), and that if the facility performs according to plan, one is assured of the safe protection of public health (and the environment). It would follow, therefore, that the fundamental question of What constitutes safe protection? will be adequately addressed. Whether this is true has yet to be fully tested.

The difference between the PMR approach and the 2CR attitude is in the requirements for the constructed facility. The PMR attitude requires the constructed facility to perform according to certain prescribed standards, particularly with respect to control of the waste leachate plume. By and large, the sets of leachate plume control adopted by the PMR type of attitude are in respect to the rate of transport of target contaminants in the substrate, e.g. allowable or acceptable concentrations of contaminants in the generated leachate at specified distance and time limits from the waste source. In essence, this type of attitude is a form of quality control on the constructed facility. If contaminants are detected in concentrations exceeding limit (trigger) values, at distances and time intervals previously specified as trigger values, this will constitute failure, and the facility will obviously need to be rectified.

Regulatory attitudes and technology development

Because the rectify aspect (R) of the 2CR attitude is by and large very costly, and could perhaps incur specific legal penalties, development of technology and research has been towards better and safer barrier liners and covers. The bulk of work performed has been towards procedures, materials, technology, durability, etc., that would ensure long-term safe encapsulation of the waste in the landfill. The analogy of the leakproof garbage bag has often been used to describe the end product. Emphasis in research and technology development has been directed towards more capable landfill barrier systems, i.e. systems that would satisfy the various sets of criteria implicit in the command and control guidelines. The need for development of a further understanding and sensible/rational utilization of ground substrate and soil barrier systems to provide physical and chemical buffer capabilities (for waste containment) has not been high on the list of priority items under the 2CR regime.

One would logically assume that ground substrate and soil barrier system research would be flourishing under the PMR attitude, but the record shows that, although some good research and technology developments are currently underway, these are not necessarily being driven by forces originating from a PMR attitude.

Regulation of hazardous waste

The different ways of categorizing waste materials are: (1) by the medium in which they are released, e.g. air, water or on/in land; (2) in accordance with their physical characteristics, e.g. gas, liquid or solid; (3) by their origin or generating source, e.g. mining, municipal, industrial, etc.; and (4) by the type of risk they pose, e.g. hazardous, non-hazardous, toxic or non-toxic, radioactive, etc. Whereas the first three categories, or methods of categorization, can be very useful, they require further sub-characterization if one wishes to determine whether the waste materials pose a health and environmental threat. Category (4) is the favoured method for categorizing waste materials (by many governments) inasmuch as this reflects the regulatory attitudes of governmental waste management authorities. The term 'regulated' wastes arising from the category (4) type of classification thus immediately identifies a hazardous waste material, and automatically triggers a tracking response sequence. The fundamental philosophy or underlying requirement in the tracking scheme is one which seeks to protect the public, i.e. minimize or avoid potential risk to public health (and the environment). The 2CR and PMR attitudes towards land disposal of waste are good examples of the final disposal end of the tracking spectrum.

The characterization of municipal solid waste (MSW) as a non-hazardous waste by most regulatory agencies, in essence placing it as a non-regulated waste, belies the fact that MSW leachates contain many toxic substances that pose dangerous threats to human health and the environment. Evidence from field studies of leachates from MSW sites show that dissolution of various kinds of household wastes such as drinks cans, transistor batteries, natural and synthetic cloth, leather products, wood, paper, etc., produce contaminants in the leachate such as uncombined metals, Fe, Ca and Mg bicarbonates, oxides of Sn, Zn and Cu, aldehydes, ketones, organic acids, sulphates, phosphates, carbonates and

Table 1. *Typical hazardous substances in waste streams*

Industry	*Arsenic*	*Heavy metals*	*Chlorinated hydrocarbons*	*Mercury*	*Cyanides*	*Selenium*	*Misc. organics*
Chemical	—	×	×	×	—	—	×
Electrical & electronic	—	×	×	×	×	×	—
Electroplating, metal industry	×	×	—	×	×	—	×
Leather	—	×	—	—	—	—	×
Mining, metallurgy	×	×	—	×	×	×	—
Paint and dye	—	×	—	×	×	×	×
Pesticide	×	—	×	×	×	—	×
Pharmaceutical	×	—	—	×	—	—	×
Pulp and paper	—	—	—	×	—	—	×
Municipal solid waste (MSW)	×	×	×	×	×	×	×

phenols, to name a few. Table 1 shows that there is proper justification in considering MSW as a regulated waste.

Determination of health and environmental threats

The health and environmental threats posed by storage and disposal of hazardous wastes and materials in the ground via landfills, landfarms and underground storage tanks, have prompted many regulatory agencies to develop legislation or recommended guidelines and/or standards requiring some proper means for neutralization and detoxification of the waste materials before discharge/disposal. It is not only the waste materials themselves that pose the threats, but also the many other discharges from the waste materials (leachates, volatiles, etc.). An example of this can be seen in the various pathways for waste materials (waste pile) emanating from a landfill disposal site shown in Fig. 1. In the schematic shown, the various forms and ways in which the waste pile can impact on humans will be dictated by the pathways taken. It is not difficult to perceive that if regulatory agencies are to assume the responsibility for eliminating or minimizing the threats to humans, these same agencies must have sufficient toxicological, phytotoxicological, and epidemiological information relative to the impact of these toxic substances and their respective pathways to the various receptors. Present evidence shows that insufficient information exists to permit one to render the necessary sets of judgements with any degree of certainty. The problem is not restricted to waste materials (as can be seen in the bar chart shown in Fig. 2), but also with respect to various kinds of substances assessed for health hazard threats.

The information shown in Fig. 2 indicates that only a very small percentage of the substances can be completely assessed *vis-à-vis* health hazard threat. In the simple case of cosmetic ingredients shown as the second grouping of bars in Fig. 2, the US National Academy of Sciences found that in a study of 3410 substances in the 100 chemical compounds sampled, it was only possible to complete health hazard assessment for 2% of the substances. Of the remaining 98%, no toxicity information was available for 56% of the substances. The problems arising relate to information availability for the vast number of substances, and the very systematic and tedious requirements in assessment of health effects.

To determine what waste material constitutes a health threat, one needs to revisit the definition of a waste material. Common acceptance of waste is generally confined to solid waste, primarily because of the frequent references to hazardous solid waste (HSW) and municipal solid waste (MSW). Where necessary, one refers to liquid waste streams to distinguish between solid and non-solid wastes. It would be useful to look at the term solid waste contained in the US Resource Conservation Recovery Act (1976) (RCRA 1976). This is defined in Section 1004(27) of RCRA (1976), as:

> any garbage, refuse, sludge from waste treatment plant, water supply treatment plant or air pollution control facility, and other discarded material including

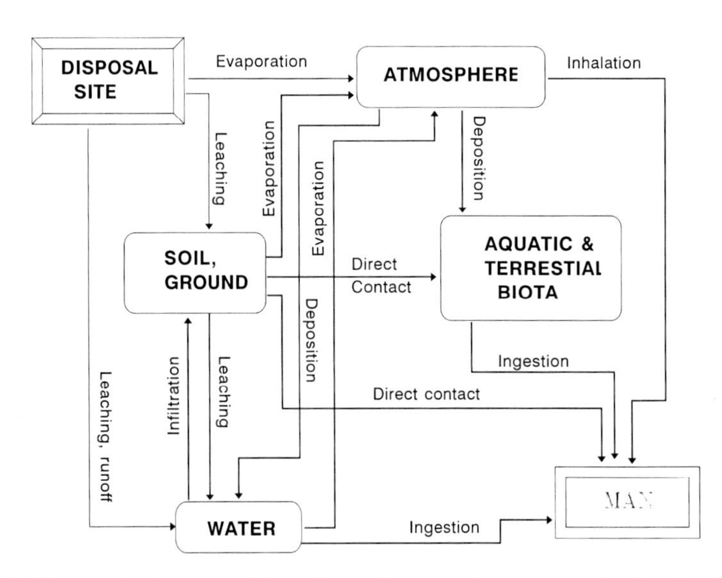

Fig. 1. Schematic diagram showing some of the pathways for contaminants amanating from landfill disposal site and how they can impact on humans.

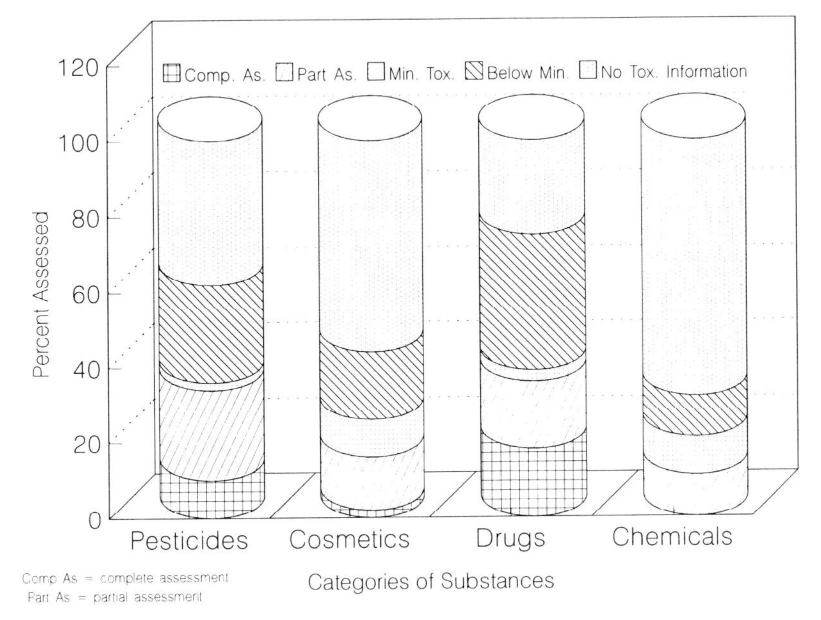

Fig. 2. Health hazard assessment of some selected categories of substances. Number of substances in each category: pesticides, 3550; cosmetics, 3410; drugs, 1815; chemicals, 48 523. (Adapted from Anon. 1994).

> solid, liquid, semi-solid, or contained gaseous materials resulting from industrial, commercial, mining and agricultural activities, and from community activities, but does not include solid or dissolved material in domestic sewage or irrigation return flows.

If one uses this definition, it is immediately obvious that one includes almost all kinds of industry and consumer waste discharge (solid, semi-solid, semi-liquid and liquid) for management under criteria and standards established for handling of regulated wastes. The record shows that not all regulatory agencies (in many other countries) have similar perceptions of the constituents of solid waste, resulting thereby in the development of guidelines and regulations which at times may appear to be confusing and overlapping.

2CR and PMR implications in the classic landfill problem

The diagram given in Fig. 3 portrays the classic landfill problem, i.e. containment of a wastepile and the subsequent development of a contaminant leachate plume. Under the 2CR type of regulatory attitude, the contaminant plume would not be present since detection of such a plume would require one to re-engineer the barrier, with correction procedures and requirements to be specified by the Administrator (RCRA 1976). A not uncommon procedure is to require removal of the wastepile and replacement of the liner system—a very costly procedure which is obviously one which the developer of the landfill facility would like to avoid.

Under the PMR type of attitude, the regulatory agency (or permitting agency) places limits and controls on the concentrations of the leachate plume at specific spatial and temporal intervals to meet standards deemed by the agency as being safe (thereby assuring minimization of potential threat to public health and the environment). The burden of responsibility assumed by the regulatory agency may or may not extend to the monitoring system needed for detection of the chemistry (concentration and type) of the pollutant in the substrate. The problem that needs to be addressed in regard to monitoring strategy and protocols relates to who pays for the monitoring system, and who decides on the monitoring strategy. The strategy for placement of monitoring wells depends on whether the monitoring system is placed by a regulatory/permitting agency (via direct subcontract) or directly by the constructor of the facility. Other than monitoring systems for research purposes designed to study groundwater quality and contaminant movement, at least three different types of monitoring strategies can be cited: (1) ambient-trend systems designed to measure quality (chemistry and specific concentrations) of the groundwater in relation to standards (e.g. drinking water standards or groundwater standards) at various locations—a general approach adopted to assess groundwater quality;

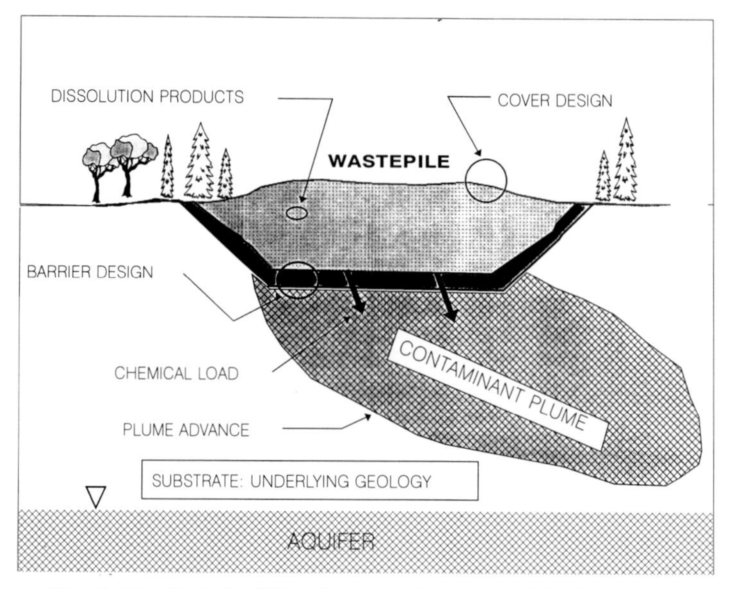

Fig. 3. The classic landfill problem showing generated leachate plume.

(2) source monitoring systems designed to measure groundwater quality in relation to contaminant source—a general PMR approach; and (3) a case-preparation type of monitoring system where attention is paid to data accumulation for enforcement actions. The number, type and placement of monitoring wells, such as that shown in the typical scheme in Fig. 4 will obviously depend on which one of the above is chosen.

It needs to be noted that whereas monitoring wells will provide the samples needed for tests on groundwater quality, the question of soil quality is not addressed. The problem that needs to be resolved is whether a knowledge of soil quality is necessary for the PMR regulatory requirements, i.e. is soil quality an important and necessary piece of information for decision-making with respect to determination of the

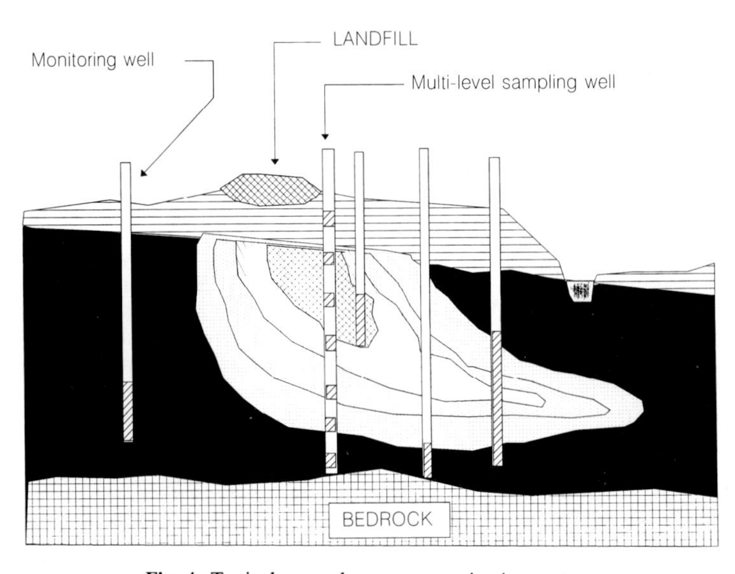

Fig. 4. Typical general-purpose monitoring system.

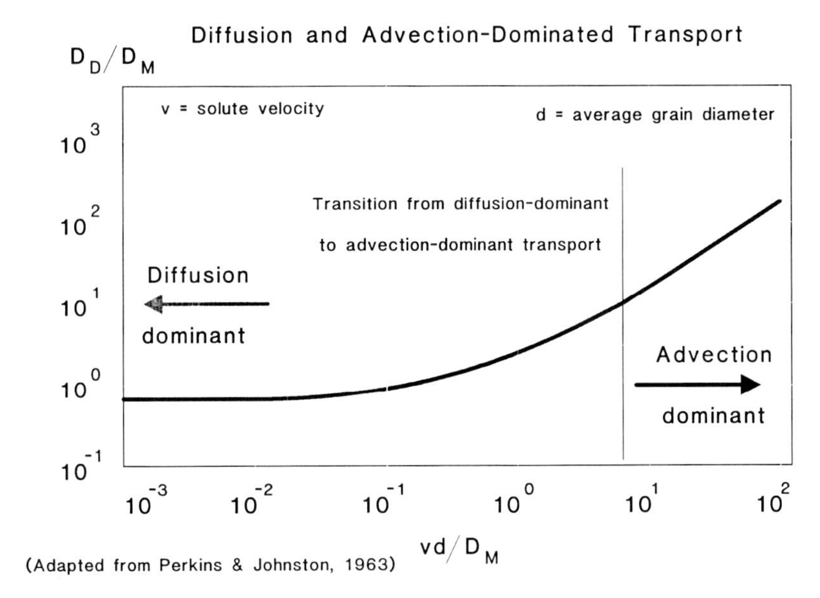

Fig. 5. Relationship between Peclet number P_e and D/D_0.

threats to public health posed by contaminant plumes in the substrate. As will be seen in the discussion of results given in Fig. 5 and in the examination of groundwater and soil qualities, one needs to be concerned with the mobility of the toxicants, and the extractability and presence of the toxic ions/solutes in soil pore water. If the waste containment facility is designed for best optimum conditions, the facility will generally be located in the vadose zone. Accordingly, the contaminant plume shown in Fig. 3 will not be sampled by monitoring well systems designed to test the groundwater, as opposed to the classic diagram in Fig. 4 which recognizes that the monitoring wells have been designed to sample below the water-table.

Landfilling and waste management

From the preceding discussions, one will observe that there are at least six important parts of landfilling of waste that need to be considered in landfilling waste disposal management:

- the waste product and leachates generated,
- the siting, engineering/design, substrate and underlying geological controls,
- the physics, chemistry and biology of contaminant transport in and through the substrate (contaminant transport processes in soil substrate), i.e. adsorption, interaction, transformation and attenuation characteristics,
- contaminant transport prediction,
- monitoring and measurements, and
- assessment of operations and monitoring measurements to satisfy regulatory and other safety standards.

As stated previously, the 2CR regulatory climate mandating and specifying liner and cover design details reduces (to a certain extent) the need to evaluate the underlying geology insofar as chemical buffering capability is concerned. The general requirement of a limiting maximum permeability value is often considered sufficient to satisfy the 2CR attitude, provided that the appropriate monitoring scheme is introduced.

In the design and construction of the landfill facility to satisfy the PMR attitude, the importance and role of analytical modelling cannot be overstated, since one does not have any means to examine the 50-year or 100-year performance prediction of a proposed landfill design in any real sense except through modelling predictions. The importance of development of a model which truly represents the problem at hand is obvious. The problem that presents itself is: How can we distinguish between a good and a bad model? That is, how does one know which model can better predict the landfill performance? The calibration and validation opportunities are generally not available, and also the chemical load, initial and boundary conditions over the entire landfill (underlying geology) are not usually properly represented.

To provide a proper analytical model of the transport of contaminants through soil, both physical and chemical mass transport processes should be included in the model. By and large, models currently available consist of the essential elements which try to portray the relationships established through the Fickian process. Considering that an analytical model consists of a set of equations which describe the physical/chemical process being studied, the particular concern in regard to the modelling of contaminant transport in a clay substrate is how well, i.e. how accurately, the physical/chemical problem is characterized. Apart from the assumptions dealing with initial conditions and loading conditions, the critical assumptions or relationships dealing with the fate of the contaminants as a result of physico-chemical interactions with (and in) the soil-water system are central to the success of prediction of contaminant accumulation, attenuation and transport through the soil substrate. It is important to note at this juncture that contaminant transport in the underlying rock constitutes another set of considerations, which are outside the scope of this presentation. Modelling of contaminant transport in underlying rocks can become very complicated (and challenging) because of the presence of fissures and cracks.

The general procedure for development of the sets of differential equations describing contaminant transport in soils combines the equation of motion (flow equation for the water) with the convective–dispersion equation for mass transport of the contaminants. This is, by and large, the mass transport model which considers both quantity and quality of the groundwater flow. Assessment of the capability of the various models to properly represent the conditions at hand needs to include consideration of the many processes that influence mass transport, e.g.

- physical: advection, hydrodynamic dispersion, molecular diffusion, and density stratification;
- physico-chemical and chemical: mechanisms of accumulation (physisorption, chemisorption, complexation, desorption, speciation, precipitation, etc.), oxidation–reduction reactions, immiscible phase partitioning;
- biological: biotransformation, co-metabolism, biotransformation and microbial population dynamics.

If groundwater resources are to be protected, prudent engineering practice requires the siting of waste landfills in unsaturated ground, some distance above the water-table. Hence, generated waste leachates will be transported through the vadose zone before reaching the groundwater table, requiring one to study contaminant transport both in the vadose and saturated zones. Whereas the greatest attention has been on pollutant transport in the saturated zone, transport through the vadose zone is indeed a very significant problem that has yet to be fully addressed.

The development of the relationships for transport in the saturated zone, in terms of Fickian relationships or irreversible thermodynamics, has been well covered by many researchers (see Yong *et al.* 1992). A sample of the various equations, written to represent the one-dimensional case (for simplicity) is given as follows:

$$\frac{\partial c}{\partial t} = D\frac{\partial^2 c}{\partial x^2}$$

$$\frac{\partial c}{\partial t} + v_x\frac{\partial c}{\partial x} = D\frac{\partial^2 c}{\partial x^2} \tag{1}$$

$$\frac{\partial c}{\partial t} + v_x\frac{\partial c}{\partial x} = D\frac{\partial^2 c}{\partial x^2} \pm \frac{\rho_s}{n}\frac{\partial S}{\partial t}$$

where c is the concentration of the particular contaminant species being considered, v_x is the average advective velocity, ρ_s is the dry density of soil, n is the porosity of soil, S is the mass of a particular contaminant species adsorbed per unit mass of dry soil, x and t are spatial and time variables, and D is the diffusion-dispersion coefficient.

The first relationship in the above sets of equations (Equation (1)) is the familiar diffusion relationship. The second relationship is the diffusion–advection equation, and the third relationship represents the adsorption–desorption 'add-on'—via the last term on the right-hand side of the diffusion–advection equation. The points of interest are: (a) 'when is the diffusion process dominant?' or conversely, 'when does advective flow dominate—to the point where diffusion flow of solutes is considered vanishingly small and insignificant?'; (b) the coefficient D; and (c) the adsorption term S.

Defining the ratio vd/D_o as the Peclet number P_e, the diagram in Fig. 5 shows the region where diffusion of solutes in water (in the soil) is moving faster than advective flow movement of solutes, i.e. the movement of solutes is diffusion-controlled ($P_e < 1$). In the figure, v denotes pore water velocity, and d is the average diameter of the soil particles. For P_e values much greater than 1, advection controls solute movement. If one makes the necessary sets of calculations for substrate materials consisting of clays, with typical permeability coefficients (i.e. k values) of 10^{-7} cm s^{-1} or less, the Peclet numbers obtained will be less than 0.1, showing that the transport of pollutants in the clay substrate will be dominantly diffusive.

The combination of non-uniform and non-steady-state leachate composition and loading at the boundary of the landfill (into the substrate) makes for very uncertain conditions in the diffusion transport of solutes (pollutants) in the soil substrate. In the absence of significant advective flow, i.e. diffusion of solutes, the processes which control diffusion transport, in combination with the very uncertain leachate composition and loading condition and various adsorption mechanisms will result in diffusion transport that is not readily modelled with a constant diffusion coefficient. In the face of these observations, the challenge facing the analysts is to continue using constant D values, without compromising the validity of the model chosen to represent the problem at hand. There is sufficient capability in modelling procedures to provide one with the opportunity of using variable D values, even in the steady-state diffusion models, by assuming for example that the D coefficient is a function of concentration of solutes ($D = f(\text{conc.})$), or a more realistic (complex?) relationship such as: $D = f(x, \text{conc. adsorp.})$.

Contaminants and health threats

A common procedure for the assessment of health threats posed by contaminants in the groundwater is to use 'drinking water' guidelines. Some of the more common inorganic contaminants required for identification in determination or assessment of water quality, and also commonly associated with concerns as threats to human health include Al, Sb, As, Asbestos, Ba, Be, Cd, Cl, Cr, Co, Cu, Cn, Fl, I, Fe, Pb, Mg, Mn, Hg, Mo,

Table 2. *Health effects in respect to deficiency and/or excess of ingestion of some essential inorganics (adapted from Lappenbusch 1988)*

Inorganic	*Deficiency*	*Toxic effects*
Chromium, Cr	Atherosclerosis	Tubular necrosis of the kidney; Cr^{+6} penetrates cell membranes, reduced to Cr^{+3}, and is mutagenic
Copper, Cu	Anemia, loss of pigment, reduced growth, loss of arterial elasticity	Disorder of copper metabolism; hepatic cirrhosis
Fluoride, Fl	Possible osteoporosis	Dental fluorosis (mottling); possible osteosclerosis
Iron, Fe	Decreased hemesynthesis, anemia	Gastrointestinal irritation; heart disease
Magnesium, Mg	Electrolyte imbalance of calcium and potassium	Muscle weakness
Phosphorous, P	Weakness, bone pain, rickets	Kidney/liver damage; gastrointestinal irritation
Potassium, K	Hypokalemia, muscle weakness, hypertension, fatigue	Diarrhea; nephrotoxicity
Selenium, Se	Myopathies, possible liver injury	Growth inhibition, liver damage, central nervous system and gastrointestinal disturbances, dermatitis
Zinc, Zn	Reduced appetite and growth	Irritability, nausea, immune response reduction

Table 3. *Average daily intake of some inorganics in typical North American adults compared with typical dosages given in common dietary supplements*

Constituent	*Daily intake (mg day^{-1})* typical average values*	*RSC %, drinking water**	*RSC % food**	*Typical dosages in vitamin pills (mg day^{-1})*	*Acceptable daily intake, (ADI)* (mg day^{-1})*
Potassium	3750	<1	>99	32	5600
Calcium	420	24	76	530	800
Sodium	5660	<1	>99		2200
Phosphorous	1500	<1	>99	400	800
Chloride	8440	<1	>99		5450?
Magnesium	375	3	97	100	350
Zinc	13	3	97	22.5	15
Iron	19.5	3	97	12	18
Chromium	0.115	13	87	0.027	0.34
Fluoride	3	~50	~50		4.2
Copper	1.7	12	88	2	1.3
Lead	0.57	57	42		
Molybdenum	0.35	3	97	0.027	0.2
Selenium	0.19	8	92	0.027	0.07
Cadmium	0.28	13	87		

RSC, relative source contribution. Percentages given are approximate.
Note: Most of these are toxic substances and should not permit display of any acceptable daily intake.
* Adapted from Lappenbusch (1988).

nitrite, nitrate, P, K, Se, Ag, Na, sulphate, Ti, V and Zn. Many of these are naturally occurring substances in the ground and also in the daily food intake of humans. Table 2 shows that whereas many of these inorganic contaminants pose risks to human health, they are also essential in the health maintenance of humans.

The information given in Table 2 shows that the questions 'How little is too little?' and 'How much is too much?' are not easily answered. The few simple examples cited in Table 3, using information given as typical dosages in dietary supplements (column 5, vitamin pills) is compared with the average daily intake obtained from some representative values given by Lappenbusch (1988). Most of the ingredients in the dietary supplements are obtained as chelates and/or compounds, e.g. potassium is provided in the form of potassium chloride, zinc is given as zinc chloride or zinc sulphate, etc., complicating the picture with respect to the amount of intake of chloride and sulphates. One should note that the acceptable daily intake (ADI) or recommended daily allowance (RDA) are continuously being updated, as more information and studies become available.

The corresponding health risks arising from exposure to synthetic and volatile organic chemicals (SOCs and VOCs) can be seen in Table 4, where the type of health risk is determined on the basis of exposure level and duration of exposure (acute and chronic effects). In general, one classifies a substance as acutely toxic if it produces a lethal or sub-lethal effect within a short time-frame, whereas a chronic toxic substance is viewed as one which requires a relatively longer period of time to manifest itself, e.g. carcinogenic, mutagenic, teratogenic effects.

As shown in Table 4, in terms of statistically verifiable effects, extrapolation from tests with animals is required in many instances. It is useful to note that the classification of toxic substances with regard to the level of toxicity differs between different countries, not only with respect to the terminology used, but also with respect to the criteria used to distinguish between the various levels of toxicity. Thus, for example, if one compares the classification which denotes the most *acute* toxic level, where threshold substances (toxicants) can be defined, the Committee on Transport of Dangerous Goods (CTDG) within the Economic and Social Council (ECOSOC) of the UN uses the general term 'poisonous' and the classification of 'severe risk' to denote the most toxic level. The EU, on the other hand, uses the term 'very toxic', whereas the US uses 'highly toxic' to denote the comparable level of toxicity. ECOSOC uses three levels (severe risk, serious risk, low risk), as does the EU (very toxic, toxic, harmful), whereas the US uses two levels (highly toxic, toxic) for categorization.

The dosages used to determine the most acute toxic level category (severe risk, very toxic or highly toxic) differ. Examples are shown in Table 5.

For the chronic toxic levels, one should note that there are no easily defined thresholds or threshold substances. Nevertheless, whereas the EC uses two categories (toxic and harmful), where 'toxic' refers to serious damage (clear functional change of toxicological significance) likely to be caused by repeated and prolonged exposure. The US, in turn considers specific target organ effects due to hepatotoxins, neurotoxins, blood toxins, etc., whilst the

Table 4. *Acute and chronic effects of some hazardous wastes on human health**

	Acute effects						*Chronic effects*		
Waste type	*Nervous system*	*Gastro system*	*Neuro system*	*Respiratory*	*Skin*	*Death*	*Carcino-genic*	*Muta-genic*	*Terato-genic*
Halogenated organic pesticide	H		H	H		H	A	A	A
Methyl bromide			H						H
Halogenated organic phenoxy herbicide					H		A	A	A
2-4-D	H								
Organophorous pesticide	H		H	H		H	A	A	A
Organonitrogen herbicide		H				H	A	A	A
Carbamate insecticide	H		H	H		H			
Dimethyldithio-carbonate fungicide				H					
Aluminum phosphide		H							
Polychlorinated byphenyls		A			H		A		A
Cyanide wastes	H		H	H		H			
Halogenated organics	H		H		H		H	H	
Non-halogenated volatile organics			H	H			A	A	

H, A = statistically verifiable effects on humans and animals respectively.
* Adapted from Governor's Office of Appropriate Technology, Toxic Waste Assessment Group, California (1981).

Table 5. *Dosages used to determine acute toxic levels*

	ECOSOC severe risk	*EC very toxic*	*U.S highly toxic*
Oral lethal dose (OLD_{50})	<5 mg kg^{-1}	<25 mg kg^{-1}	<50 mg kg^{-1}
Dermal lethal dose (DLD_{50})	<40 mg kg^{-1}	<50 mg kg^{-1}	<200 mg kg^{-1}
Inhalation lethal concentration (ILC_{50})	<0.5 mg l^{-1}	<0.5 mg l^{-1} 4 h^{-1}	<0.2 mg l^{-1} 1 h^{-1}

ECOSOC provides no specific categorization. It is clear that much work remains to be done to arrive at more definitive indices, especially if the pathways to the biological receptors are to be reckoned with. The problem of determination of threshold levels becomes further complicated if one considers both groundwater and soil qualities as being significant health threat agents. In addition to water consumption in the general procedures used for health risk assessment, one needs to consider soil ingestion, soil covering or physical contact, and ingestion of contaminated produce.

An anecdotal recounting of the harmful effects of saccharin given in the 11 May 1992 issue of *Newsweek* magazine illustrates some of the problems of determination of what constitutes a 'lethal' dosage. In that particular report, more than 700 rats and 100 mice were fed high doses of a non-caloric saccharin substance, equivalent to human consumption of 1000 packs of sweetener a day. The rats developed malignant bladder tumours whereas the mice did not. The conclusion reached was that the saccharin developed a specific reaction with a rat urine protein not present in mice (or in humans).

Absolute to relative scales

As noted in the preceding discussion, one of the principal requirements in both the 2CR and PMR attitudes is the specification or articulation of the limiting target values (trigger levels) for pollutants, such as OLD_{50}, DLD_{50} and ILC_{50}. One has the option of specifying the target values without regard to the particular situation, as for example in general standards, criteria or guidelines, or one can specify the values with respect to specific situations and circumstances. In the former instance, one adopts the absolute scale and in the latter case, one adopts the relative scale. The absolute scale provides one with goals and targets for design/construction, monitoring and assessment of pollution, and is essential in the PMR regulatory position. In the 2CR case, target values assume importance in site remediation programmes and assessment of level of pollution for contaminated sites. They assume a different role in newly constructed landfill disposal facilities, i.e. the detection of a toxicant (pollutant) outside the facility which is traceable to the facility will trigger requirements for immediate total re-engineering of the facility. Under such circumstances, the limiting target values may be different from the trigger levels. With the absolute scale, one can establish identifiable goals, which under the 2CR attitude provides for a very simple set of rules for environmental waste management.

In the PMR attitude which has been adopted in certain countries, site and situation specificities are considered in the articulation of the target values, i.e. a form of risk management has been introduced wherein the specified target values will vary in accordance to some previously established acceptable risk. In essence, this identifies the relative scale with respect to target values, for which one needs to either develop or adopt a procedure for determination of acceptable risk. Of the many models which exist in decision-making directed towards determination of acceptable risk, two common fundamental characteristics exist: toxicity assessment and exposure assessment. In toxicity assessment, the main sets of problems relate to definition of acceptable concentrations of toxicants, the outcome of which will generally be expressed in terms of quantitative indices such as NOAEL (no observed adverse effects level) and LOAEL (low observed adverse effect level). As might be deduced from the preceding discussion, determination of what constitutes an adverse effect and the level of adversity would be most challenging. The fact remains that one needs a proper assessment of not only the health threat, but the severity of the threat and the manner in which such a threat is posed or delivered. This requires determination of the most likely exposure path. Restricting oneself to the substrate and the delivery manner posed by contaminant plumes, the problem translates itself into determination of the bioavailability of the pollutants.

Bioavailability

'Soil pollution' as a term has been used in soil science to denote a soil where the quality of the soil has been degraded by the presence of contaminants. Considering the contaminants as pollutants, or more specifically as toxicants, one recognizes that the concentration of toxicants in the substrate is a variable quantity, dependent not only on the spatial location (from the pollution source) but also on how the toxicant is held within the substrate system. To illustrate the problem of bioavailability, i.e. what or how much of the toxicant is available as a threat, the results of the study by Yong & Galvez-Cloutier (1993) on the various forms of a lead (Pb) contaminant in a pure (clay) soil-water system are used. In Fig. 6, the various Pb forms in the soil-water system are shown as a function of the local equilibrium pH of the total soil-water system and initial concentration of the Pb contaminant in the soil-water. Note that pH_0 represents the zero point charge (zpc) of the clay (kaolinite) in the presence of the Pb contaminant in the soil-water, and that the initial Pb concentrations in the soil-water are those values obtained at equilibrium pH 0.5.

At least three significant items can be noted from Fig. 6 in relation to the problem of determination of bioavailability of the Pb as a toxicant.

(1) The amount of Pb adsorbed by the soil particles between pH 2 and about pH 7.5 represents Pb uptake by the soil as specifically and non-specifically adsorbed Pb. The amount of Pb specifically adsorbed, and thus by inference not bioavailable, cannot be exactly quantified. Studies on the blank clay suspension (i.e. using distilled water) show a zpc of about 4, whereas the zpc shift from 4 to about of 4.5 in the presence of the Pb in the soil-water, as shown in Fig. 6, can be interpreted as the result of specific adsorption of Pb, in accordance with other studies reported by Pyman *et al.* (1979) and Uehara (1981).

(2) The various precipitated forms of Pb can be presumed to partly bioavailable, assuming utilization of groundwater as the health threat. The various oxides and hydrous oxides of Pb cannot be totally identified, thus rendering it difficult to determine the amount and degree of soil particle attachments. One assumes that some precipitated Pb will be available in solution as free-phase precipitated Pb.

(3) The Pb remaining in solution is assumed to be the amount of Pb freely available in the groundwater. Thus, with the non-specifically adsorbed Pb (as in (1) above), and the free-phase precipitated Pb (as in (2) above), the total amount of bioavailable Pb will be somewhat less than the amount (concentration) of Pb originally introduced into the soil-water system.

The preceding three items testify to the difficulties in rendering a proper determination of measured toxicant concentration and its relationship to bioavailability, particularly if one recognizes that the local equilibrium pH environment is neither stagnant nor uniform throughout the substrate underlying a waste landfill. In a multi-toxicant soil-water system, the problem of determination of bioavailability of individual toxicant species/concentration becomes considerably magnified because of the competition between species, and also because of the various ligand attachments. The obvious fall-out from this set of observations relates not only to determination of bioavailability, but also to problems of monitoring, sampling and testing for ground contamination for determination of potential health threats.

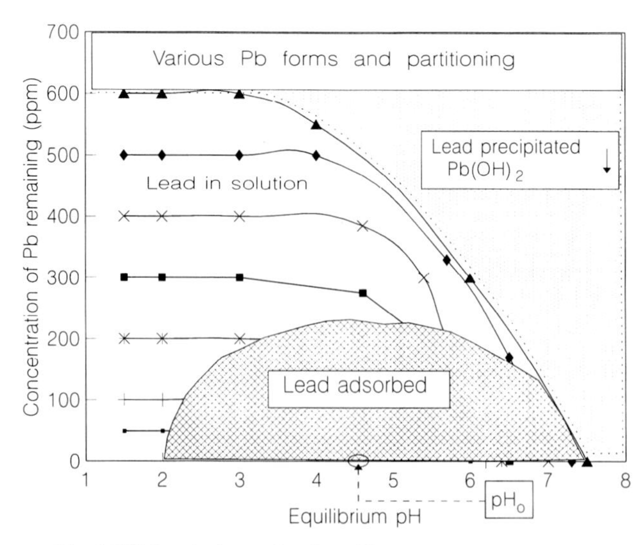

Fig. 6. Different forms of lead (Pb) in relation to local equilibrium pH (from Yong & Galvez-Cloutier 1993).

Soil and groundwater qualities

Considering the soil substrate as an environmental compartment, one needs to recognize that contaminant transport processes involve interaction between the contaminants in the fluid phase of soil, i.e. soil pore water in the vadose and saturated zones. The term 'groundwater' is commonly used to refer to water that occurs in the permeable saturated strata of rock and soil granular media (gravels and sands), 'carrying' agent (groundwater, i.e. soil pore water) and the soil solids. It is not always clear that when one refers to soil quality this means both soil solids quality and groundwater quality, or for that matter what 'soil quality' means. Because of the various sorption mechanisms developed between contaminants and soil solid constituents, and because some of these can develop retention bonding relationships that are more or less permanent, there is a need to distinguish between extractable and non-extractable contaminant ions/solutes. Since chemical mass transfer can be a significant factor in the control of the environmental mobility (transport) of contaminants in the soil substrate, one needs to recognize that the determination of the quality of the soil pore water (groundwater) via testing of extracted groundwater will be confined to those solutes that are extractable with the pore water. For simplicity in discussion, one generally uses the terms 'non-extractable' and 'extractable' ions/solutes to distinguish between those ions that are 'held' to the soil solids and those remaining in the pore water that are 'mobile'. One recognizes, however, that the aggressivity (efficiency?) of the extraction method or process will undoubtedly influence the quantities measured. Leaching processes and their efficiencies are as much affected by the immediate chemical environment, temperature and other biological and physical agents. Thus, it becomes possible to measure different concentrations of a target species in the groundwater for the same contaminated patch, dependent on any or all of the preceding effects.

The results of selective sequential extraction analyses shown in Fig. 7 (from Yong *et al.* 1992) for an assessment of the amount of lead retained by the various soil fractions for an illitic soil provide an example of the problem of extractability of the lead contaminant. The sequence of extraction of the Pb was as follows: (1) extraction of the exchangeable Pb using KNO_3; (2) extraction of Pb retained by the carbonates using a sodium acetate reagent (NaOAc) buffered to a pH of 5 by acetic acid; (3) extraction of the Pb retained by precipitation as hydroxides and/or adsorbed on the oxides or amorphous hydroxides of the soil using hydroxylamine hydrochloride; (4) extraction of the Pb retained by the soil-organic fraction using H_2O_2. The amount of Pb retained in the phase identified as 'residual' in Fig. 7 was removed by digestion with strong acids.

With the information gained in extracting the contaminant Pb from the illitic soil, one not only obtains an appreciation of the amount of effort needed to extract the Pb from the various soil fractions, but also the amount of Pb that will be released by the soil

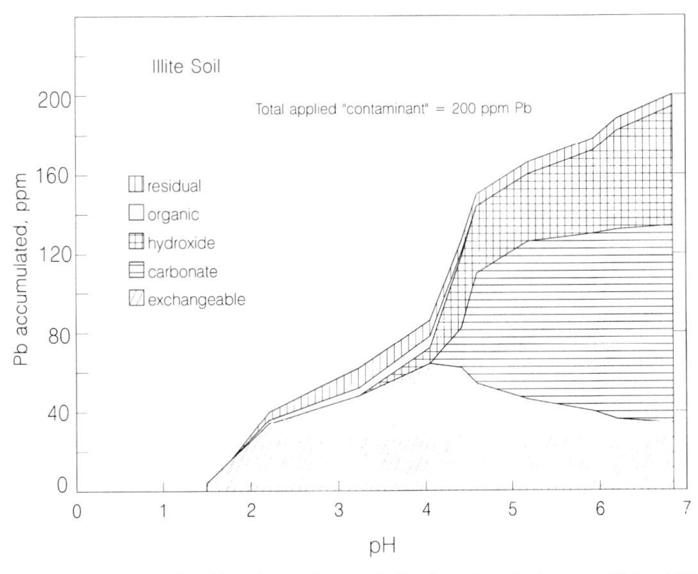

Fig. 7. Amounts of Pb retained by the various soil fractions in relation to pH fro illitic soil.

(i.e. available in the soil pore water) under conditions that may or may not exist in the natural field situation. One could infer from the results shown in Fig. 7 that the Pb obtained through standard monitoring and extraction techniques would not necessarily reflect the actual amount of Pb in the total soil sample.

Considering the term 'groundwater' to include soil pore water, and in the broader sense the water contained in perched and fossil aquifers, the question of groundwater quality, or more specifically the problem of determination or preservation of the quality of groundwater, requires one to seek, or to establish the various rules concerning groundwater use. This is performed either in conjunction with land use planning where direct use of immediate underlying groundwater is involved, or with present (ongoing) groundwater use. A convenient compromise is to group groundwater into five broad categories based upon the manner in which it can be used: (1) sole-source drinking water requiring no treatment, (2) drinking water requiring some necessary treatment, (3) drinking water requiring extensive treatment, implying thereby that the treatment is still cost-effective in comparison with delivered (and treated) pipe-water from a surface reservoir source, (4) water for agricultural use, i.e. assuming no treatment of the groundwater and that contaminants present within the water do not pose a health threat via plant uptake (i.e. no plant contamination), and (5) untreated groundwater for industrial use, assuming that one is assured that no pathways to biological receptors exist within the 'use' procedure. One can make the case that one does not often distinguish between categories (1)–(3), inasmuch as common practice generally requires treatment (and testing) of groundwater before actual use as drinking water. However, for the other categories, the determination of the separation between categories (3) and (4), i.e. treated groundwater as opposed to agricultural water use, poses one of the more interesting issues.

Whereas determination of ground (soil substrate) contamination requires assessment of contamination of both soil solids and groundwater, it is not apparent that this is always reflected in the statement of procedures, guidelines or requirements articulated by many regulatory agencies. A study of various guidelines which refer to 'trigger levels', 'action levels', 'threshold levels', 'background levels', etc. of target species for contaminated soils, do not indicate how such target species are to be measured. One can assume that this generally implies levels or concentrations of extractable target species, inasmuch as total extraction or removal of the target species of contaminants can often involve complete soil destruction methods. The problem of standardization of contaminant extraction techniques from soil solids (soil fractions) remains to be fully addressed, together with techniques for measuring the concentration of contaminants extracted from the soil fractions, as might be deduced from the results portrayed in Fig. 7. The question that needs to be addressed concerns soil solids contamination (i.e. contaminants specifically adsorbed onto soil solids' surfaces), as opposed to groundwater contamination, and whether this poses another significant factor that contributes to a proper assessment of environmental health threats. To fully appreciate the problem, one needs to study exposure pathways and associated health threats. Two simple scenarios can be posed to demonstrate the problem: (1) ingestion of a contaminated soil, representing a direct exposure pathway, and (2) plant contamination, assuming that the plant uptake forces are sufficient to detach the non-extractable contaminant ions.

One can argue that the question of soil or groundwater quality assessment is not particularly important in the case of the 2CR regulatory climate, if the position taken is one that considers any detectable target species level above background to be a failure of the waste containment facility and hence to be corrected. However, the same cannot be said for facilities constructed under the PMR regulatory attitude, since assessment of the capable performance of the waste containment facility is generally based on established trigger levels or action levels. How these trigger and action levels are determined and how they relate to adequate assurance of protection of public health can be a very interesting set of issues.

The criteria used to determine ground (site) contamination can be used to provide information on the various concentration levels of target species which classify under threshold, trigger, action, etc. The meaning of any or all of the terms may or may not be common to regulatory agencies in various countries or regions. By and large, these terms should be considered in relation to land use and also with respect to site specificities. The procedure adopted by some of the Provinces in Canada, for example, considers three broad categories of ground contamination in relation to land use.

Concluding remarks

The design and construction of land disposal facilities for management of waste to meet regulatory requirements often require informed interaction between regulators and the constructors of the facility, if protection of public health is to be assured. The difficulties in determination of what constitutes proper safeguards and how these are to be incorporated in the constructed facility stem not only from an inadequate understanding of the total waste containment issues, but also because present knowledge of the many aspects of the problem is somewhat limited. Regulatory attitudes exert considerable influence on the design of the constructed facilities, particularly with respect to barriers and covers, and the monitoring of the short-term and long-term performance of the facility. It is not uncommon to observe the use of 2CR type criteria and guidelines in the construction of disposal facilities in situations where the clear mandate is dictated by the requirements of the PMR. The use of contaminant transport models differs considerably, depending on whether one needs to satisfy the 2CR or the PMR scenario.

The assessment of failure of the facility to function in a manner designed to ensure protection of public health requires one to properly appreciate what constitutes a health threat. However, it is not immediately clear that the controls needed to establish safe protection of public health are well founded, or sufficiently diligent. A good working knowledge of the various interactions occurring in the substrate during contaminant transport is required if one seeks to assess the fate of contaminants. In addition, it would contribute well to the determination of the quality of information concerning sampling, monitoring and testing for evaluation of contaminant presence and concentration.

References

ANON. 1984. *Toxicity testing strategies to determine needs and progress.* US National Academy of Sciences, National Research Council. National Academy Press, Washington, DC.

LAPPENBUSCH, W. L. 1988. *Contaminated waste sites, property and your health.* Lappenbusch Environmental Health Inc., Virginia.

PERKINS, T. K. & JOHNSTON, O. C. 1963. Reviews of diffusion and dispersion in porous media. *Journal of the Society of Petroleum Engineering*, **17**, 70–83.

PYMAN, M. A. F., BOWDEN, J. W. & POSNER, A. M. 1979. The movement of titration curves int he presence of specific adsorption. *Australian Journal of Soil Research*, **17**, 191–195.

UEHARA, G. 1981. *The mineralogy, chemistry and physics of tropical soils with variable charge clays.* Westview Tropical Agricultural Series, No. 4, 137–150.

YONG, R. N. & GALVEZ-CLOUTIER, R. 1993. pH control on lead accumulation mechanisms in kaolinite–lead contaminant interaction. *Proc. Environment et Géotechnique*, Paris, 309–316.

——, MOHAMED, A. M. O. & WARKENTIN, B. P. 1992. *Principles of contaminant transport in soils.* Elsevier Science Publishers, Netherlands.

Risk assessment: where are we, and where are we going?

Michael Jefferies, David Hall, John Hinchliff & Morag Aiken

Golder Associates (UK) Ltd, Landmere Lane, Edwalton, Nottingham NG12 4DG, UK

Abstract. This paper was presented as the keynote address to a subsequent session on 'Investigation, hazard assessment and remediation of existing landfills'. As such, the paper is written as an overview to introduce risk assessment, and hopefully encourage and convince engineering geologists to adopt the techniques.

The paper outlines how modern risk-based management techniques are applied within the engineering geology of waste storage and disposal. Both current usage ('where are we ?') and future developments ('where are we going ?') are discussed. References and a discussion of useful software is provided to assist those starting out.

Identification and control of risk is important in the 1990s. Risk is the natural consequence of uncertainty and affects virtually every aspect of business today. Sponsor companies, regulators, consumers and the public are all affected. Consider the following issues.

Projects have to be sponsored, financed, designed, constructed and operated in ever changing conditions. Sponsors cannot control all the influences on their business and things may not turn out as planned. Geological conditions are uncertain yet fixed-price design-build is increasingly the contractual basis for a project. Customers, and the public in general, demand higher standards of care and quality while also expecting lower costs. Governments change standards (environmental ones in particular) which compound with the legal doctrine of strict liability to give financial liability. And, the courts in the English-speaking world are beginning to transfer some aspects of corporate decision-making from the civil to the criminal arena if things go wrong.

Although the above issues might suggest a bleak outlook for business, this is not a fair view. We are simply witnessing the global pressures for companies to be as scientific in their management as they are in product development. And, the required tools are available. These tools are commonly called 'risk assessment', 'decision analysis' and 'risk management''. They are united by a common basis in the mathematics of probability theory using the Bayesian approach, and together they represent appropriate methods to manage the business pressures of the 1990s.

Before proceeding, however, it is appropriate to state who are the 'we' in the paper's title. Originally, the intention was to provide an industry consensus on geologically related aspects of risk issues. However, it became apparent that although several companies are active on this issue within the UK, there is by no means an accepted standard view as can be ascertained by reading the papers submitted to three geologically related 'risk' conferences that will have taken place within a year (Limit state design in geotechnical engineering, Copenhagen, May 1993; Risk and reliability in geotechnical engineering, London, November 1993; Risk assessment in the extractive industries, Exeter, March 1994). A state-of-the-art is premature. The 'we', therefore, is the opinion of approximately a third of the senior staff of Golder Associates UK who all routinely work with the techniques described in the environmental, landfill, and nuclear waste fields. We are a sample of what the leading geoscience/environmental companies are doing.

What do we mean by 'risk'?

Risk is an imprecise word in common usage. The Oxford Dictionary defines the noun risk as: hazard; chance of bad consequences; exposure to mischance. In fact each of these definitions is a distinct concept. Risk invokes ideas of potential loss, but in truth can also include gain (specifically in contracting). The generally accepted and precise definitions of various words used in risk studies are set out on Table 1. Note in particular that risk combines the probability of an outcome with its consequence, summed (or integrated) over all possible outcomes.

Within geological engineering or environmental management, our experience is that when a 'risk assessment' is requested, one or more of six issues might be meant. These issues are:

- environmental effects (health/toxicity)
- financial factors (liability and costing)
- uncertainty in performance (under normal conditions)
- hazard identification
- performance under disruptive events

From BENTLEY, S. P. (ed.) *Engineering Geology of Waste Disposal*,
Geological Society Engineering Geology Special Publication No. 11, pp 341–359

- consequences of alternative decisions (particularly downside outcomes)

Environmental risk often arises in the context of contaminated sites or a pollution incident. Broadly, some contaminant has been inadvertently introduced to the environment and a question arises as to what risk this poses. Risk here denotes the probability of damaging the environment or human health. There are two risk issues here: what is the uncertainty in the processes involved and how certain is an undesirable outcome; and, the outcome may be expressed as a risk to life as, for example, causing no more than a one in a million increase in an individual contracting cancer because of the pollution. Regulators and the public may be interested in both.

Financial risk arises out of uncertainty in what must be done. Uncertainty arises pervasively in ground engineering since actual conditions are only revealed during excavation; design and costing is based on the spot data provided by a few boreholes. Uncertainty arises pervasively in environmental contamination because both the extent and nature of the contaminants may have been unrecorded, inferred from incomplete information, or estimated from a spot test in a few places. Both situations can lead to substantial liabilities or cost over-runs. Often, conventional contingency margins are quite inadequate. Although financial uncertainty arises in any contracting situation, we have found that the issue of financial liability becomes most important during negotiations over land which is suspected to have industrial contamination. Equally, all competitive tendering is an exercise in financial risk: the tender price is the balance of likelihood of winning against potential income or profit.

Table 1. *Definition of terms used in risk/reliability assessments*

Damage	Loss of value or fitness for purpose. Damage includes personal harm, property loss, environmental contamination, and financial loss (direct or opportunity)
Failure	an event in which an artefact or project is damaged
Accident	An unplanned failure
Disruptive event	an event which changes a component or part of a system; generally only modelled for situations leading to damage; commonly used in the context of an anticipated accident
Hazard	A set of conditions with the potential for initiating an accident
Danger	Exposure or probability [$P(a)$] of an accident
Pure risk	The combined effect of the chance of an accident and the damage caused by that accident; commonly taken as the product so that risk = danger × damage. Occurrence of pure risk accidents always lowers value
Speculative risk	The product of the probability of an outcome and the consequences of the outcome. Unlike a pure risk, a speculative risk may have positive outcomes
Safety	Freedom from unacceptable risk
Reliability	Probability that an accident will not occur: [$1-P(a)$]

Performance assessments are estimates of likely outcome given uncertainty. In fact, both the two previous examples of risk issues are handled by carrying out a performance estimate to understand how our confidence in the parameter of interest (contaminant concentration and cost, respectively) is influenced by uncertainty in the data and the situation. Performance assessments are often carried out as part of permitting for many projects as regulators have moved to health-risk criteria for project acceptability. Although this health-risk approach was started in the nuclear field, it is now frequently found in the environmental arena. In conventional civil engineering, performance assessments are being used to select or identify particular construction techniques when there is the potential to damage adjacent structures, and for assessment of design criteria for high-cost projects such as offshore platforms.

Hazards are conditions that could potentially disrupt the normal operation of a system leading to an unplanned disruption. Hazard assessments are carried out both to determine those factors that should be included in a complete performance assessment as well as in their own right to identify factors that management could (should) control. In particular, the offshore industry tends to view risk management as equivalent to hazard control. Hazard assessments have also been used in tunnelling to identify appropriate management strategies.

Performance under disruptive events is the combination of a normal performance assessment with the hazard assessment to estimate system performance even in the face of unplanned events. To date, this approach seems largely restricted to the nuclear industry, and in a few cases, waste containment.

Decision analysis provides a rational and documentable basis for selecting between options. It has widespread application. Selection of alternative remedial treatments is one application of decision analysis while a related variation has been to evaluate which site investigations or data collection can meaningfully contribute to project success.

Although the above topics may appear quite different at first glance, they are united because the same approach is used to express uncertainty and its consequences. The approach is Bayesian probability theory.

Some ideas: uncertainty and Bayesian concepts

Although some risk studies have been carried out qualitatively using words such as high, medium, low or class a,b,c to rank risk, real progress and understanding requires the language of mathematics. However, only elementary probability theory is required and computer software automates many of the calculations. The fundamental concept is that we use a probability, which is simply a dimensionless real number p such that $0 \leq p \leq 1$, to represent uncertainty.

Uncertainty is the lack of knowledge. It is a neutral term and does not imply incompetence, lack of education, or lack of diligence. For example, the value of the Dow Jones index in three months time is uncertain; no amount of studies or other research can tell us what the value of the index will be. We are uncertain because the value can only be established when we get to that time in the future. However, what we can do is express our judgement about the future value in terms of probability as in, for example, we expect the Dow Jones index to have a 90% chance (probability $p = 0.9$) that it will be greater than 3100 in three months time.

By considering a range of values and their corresponding probability we define a probability distribution expressing our uncertainty in an outcome based on our present knowledge. For a continuous variable such as cost, two graphical representations of probability distribution are commonly used: the probability density function (PDF) and the cumulative density function (CDF). A PDF shows the distribution of probability on the value space, which makes the most likely value and skewness in the distribution obvious. However, the vertical scale depends on the plotting interval; the requirement is that the area beneath the PDF must equal unity. For this reason, it is more convenient to use a CDF which is simply the integral of the PDF; a CDF always has the probability scale from 0 to 1 and the probability that a variable is greater (or less) than a selected value can be seen by inspection. The two forms of presenting a continuous probability distribution are illustrated in Fig. 1 using infiltration rate into a landfill as the notional uncertainty.

Sometimes the uncertainty is in terms of several different scenarios may occur rather than a continuous range of values. In the case of alternative scenarios a 'tree' representation is used, as illustrated in Fig. 2. The chance event is shown by the node with the branches leading from the node denoting the alternative outcomes. The probability of each branch is written on the branch and the sum of the branch probabilities at the node must add to unity.

The above view of how we can express uncertainty in terms of probability is called a *Bayesian* approach. Bayesian approaches differ in concept from a traditional statistical view. In a traditional statistical view probability is seen as expressing the relative frequency of an outcome (assessed by observing repeated trials). In the Bayesian approach, probability expresses our current judgement of likely outcomes and we do not need repeated occurrence data to estimate the probabilities, although such data can be used if available. The key attribute of the Bayesian approach is that probabilities quantify judgement. Quantification both clarifies communication and allows systematic examination of difficult problems where we may have interlinked and uncertain cause–effect relationships and different people giving judgements.

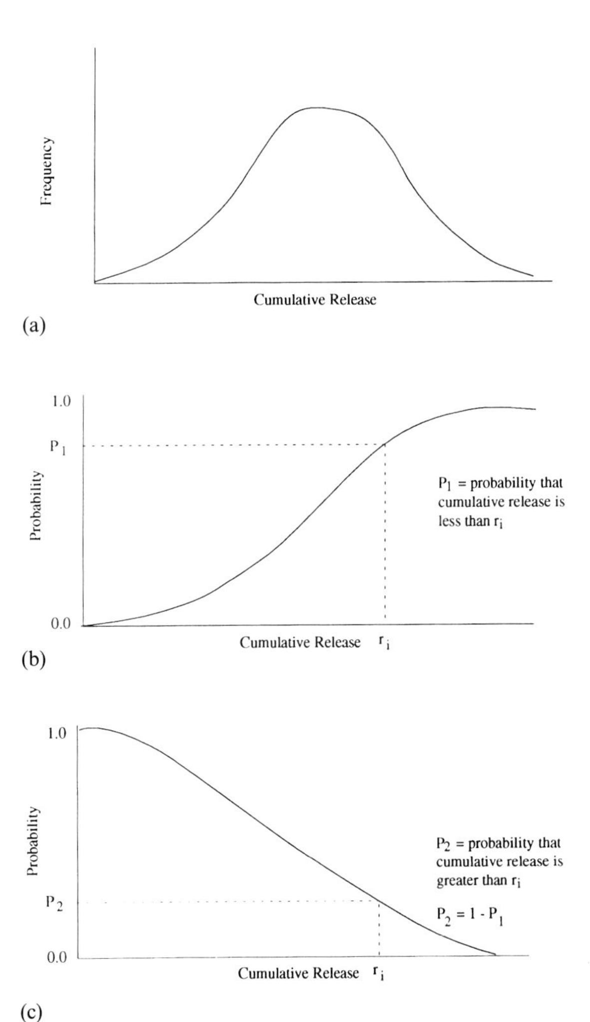

Fig. 1. Alternative representations of continuous uncertainty. (a) Probability density function. (b) Cumulative distribution function. (c) Complementary cumulative distribution function.

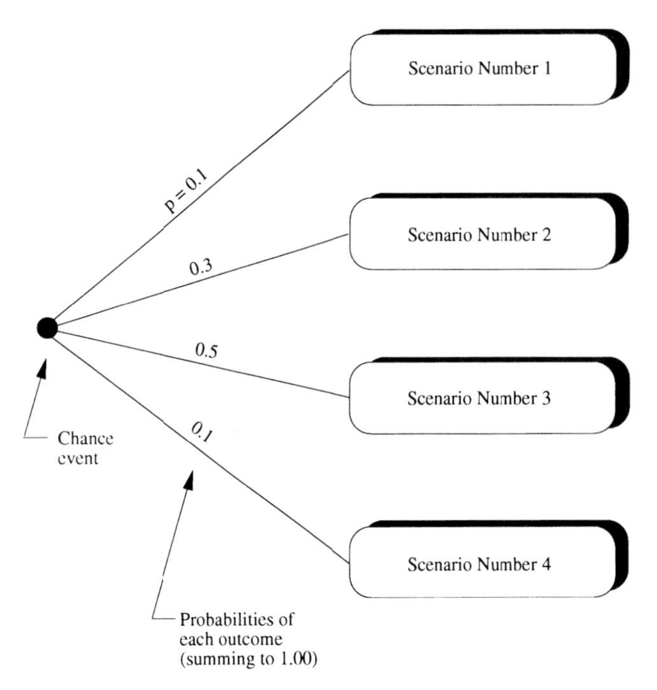

Fig. 2. Alternative representation of uncertainty with 'tree' form.

We emphasize that there is no magic in a Bayesian approach. We simply use the formalism of mathematics to quantify judgement and allow for the manipulation of the various judgements in a problem according to the rules of probability theory. Clarity follows from the use of numbers as opposed to mere words, provided that the numbers are a realistic reflection of a situation.

The use of mathematics can throw a chill into people. This is unwarranted for risk-based applications in engineering geology as most problems are easily implemented on personal computers using commercial software. Problems involving continuous distributions and relatively few trees are easily modelled using the Monte-Carlo method (or a related variant).

Such calculations can be carried out within the familiar spreadsheet environment using add-ins*. For more complicated problems, such as those involving radioactive decay or strategic simulation, we use our own custom software running on RISC workstations.

*We use two such add-ins: @RISK from Pallisade Software (tel. 607-277-8000); and, Crystal Ball from Decisioneering (tel. 303-292-2291). Tree-type analysis, and decision analysis in particular, is easily handled with SUPERTREE from Strategic Decisions Group (tel. 415-854-9000) or DATA from Tree Age Software (tel. 617-536-2128).

Importance of Bayes theorem

The theorem

Bayes theorem gives a formula for estimating the probability of an original event given the observation of an outcome. The formula and its proof are given in all standard statistical texts which discuss probability. What is important in the present context is to understand that this formula is used to underpin risk in two ways: (i) defining the power of an indicator to resolve a situation in decision analysis; and (ii) providing a way to update judgmental probabilities as new data become available.

Indicator performance

Typically, experience allows us to assess the performance of how well a test works. But what we need to know is the probability of contamination given a test result. Bayes theorem allow us to compute this reversed probability from our experience with the test and hence anticipate in terms of probability what decision we might come to if we carried out a test. This is most easily understood by considering an example.

Suppose there is a concern for a potential contaminant plume from a municipal landfill. It is proposed to put down several monitoring wells and test for ammonia (a common and easily detected component of municipal waste degradation). Experience tell us that if ammonia is present in the groundwater at concentrations above current UK and EEC drinking water regulatory limits ($0.5\,mg\,l^{-1}$) then we will detect it 95% of the time using standard protocols. Reasons for not detecting ammonia include dipping strata so the plume is not where expected and the boreholes did not extend deep enough; fractured rock with preferential flowpaths; very localized discharge; and sampling errors. On the other hand, ammonia is not infrequently present in groundwater from causes other than landfills, for example farming. Our experience is that there is a 10% chance of detecting ammonia above the regulatory limit even if there is no contaminant plume (a false positive). A tree representation of this experience is shown in Fig. 3.

Now, consider a situation in which a landfill may be contaminating the groundwater and it has been proposed to install monitoring wells to resolve responsibility. Based on various site-specific factors, we judge that there is a relatively small chance that the landfill is actually contaminating the groundwater, a probability $p = 0.15$ capturing our judgement. The probabilities of the various outcomes of the monitoring wells are shown in Fig. 4(a). We then use Bayes formula to give the probability that the ground actually is contaminated, given a borehole which indicates ammonia above the regulatory limit; this is also shown in tree form in Fig. 4(b).

In this landfill example a positive test result has only a slightly better than even chance of showing the landfill actually is contaminating the groundwater despite the apparently high success rate (95%) of the test. The low probability of contamination combines with the likelihood of a false positive to substantially reduce the value of the test to resolve the issue.

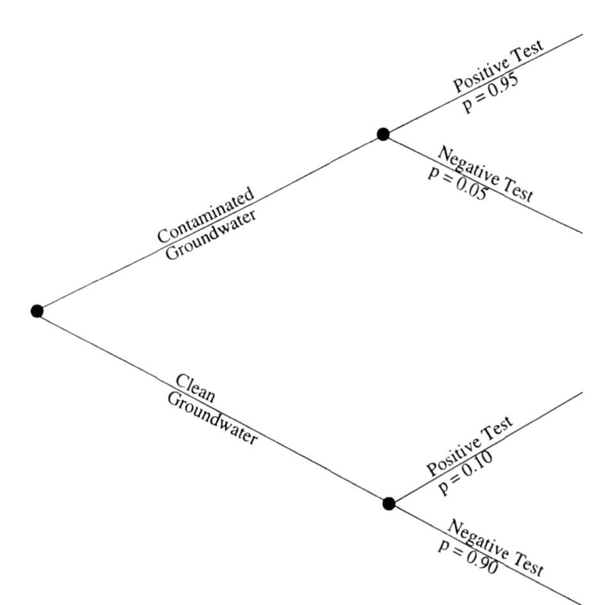

Fig. 3. Summary of experience with ammonia test on borehole water sample to detect municipal landfill contamination.

Bayes theorem used in this manner contributes to the decision analysis as to whether further testing is warranted, a situation that occurs frequently in engineering geology.

Updating with new information

Many projects involve an anticipated reduction in uncertainty as the project proceeds and information becomes available. Again, this is most easily understood by considering an example.

Suppose an oil company was concerned about how many of their underground storage tanks at retail petrol stations were leaking petroleum products to the groundwater. Initially, there is no knowledge so that the proportion leaking could range from none to all, represented by a uniform distribution between 0 and 1, but this is felt to be far too uncertain for making financial provision to fix the potential problem. An investigation is mounted and after several months a small fraction of the petrol stations have been investigated, actually 27, with the result that two were found to be leaking. What is the uncertainty in the true fraction of tanks that are leaking?

In this case a uniform distribution is our prior estimate. Computation of posterior distributions from continuous prior distributions using Bayes theorem is difficult and tedious if done by hand. However, it is simple with modern software (we use our in-house programs for this). In the case of the petrol tanks example, the prior and updated (posterior) distributions are illustrated in Fig. 5. Note that both the expected value and the shape of the distribution change during the updating.

The important point to recognize is that updating using Bayes theorem is not an arbitrary process: it has the substance of a formal mathematical basis. Original uncertainty is not ignored and the new information is included.

What does a risk study involve?

The two common threads in the risk issues outlined above are these. First, there may be uncertainty about an existing situation and its evolution in time. Second, there may be a need to express the consequences of alternative events which occur either by chance or by decision. In terms of the risk procedures used, it does not matter whether the issue involved is money, health risk, structural settlement, time delay or contaminant concentration. The general procedure in all cases is the

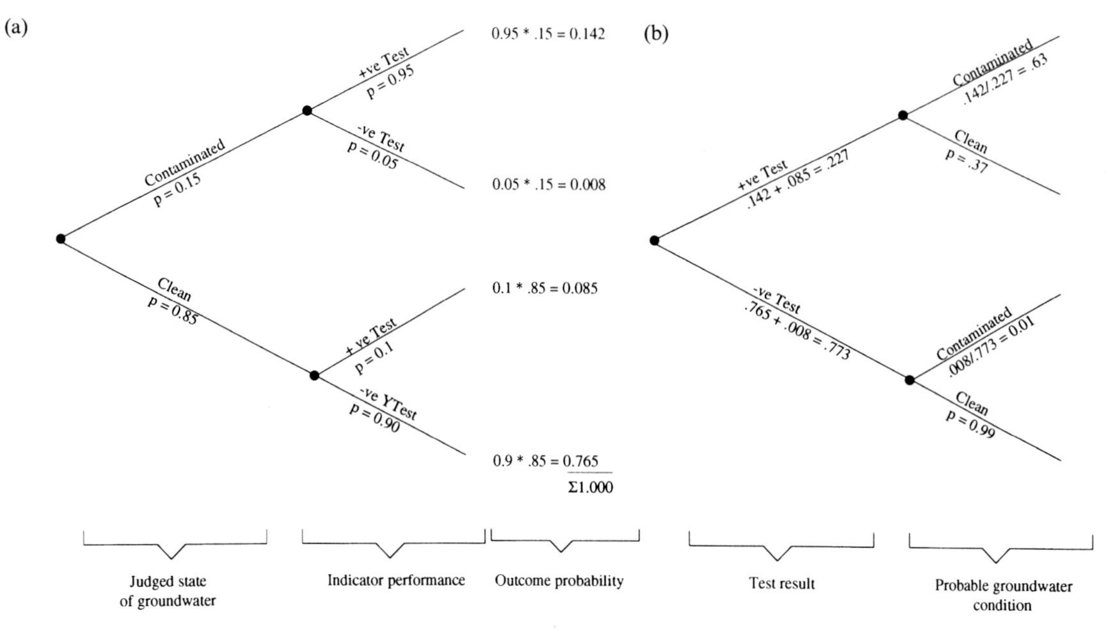

Fig. 4. Computing resolving power of a test with Bayes theorem. (a) Test outcome given pollution. (b) Chance of pollution given test result (situation (a) reversed by Bayes theorem).

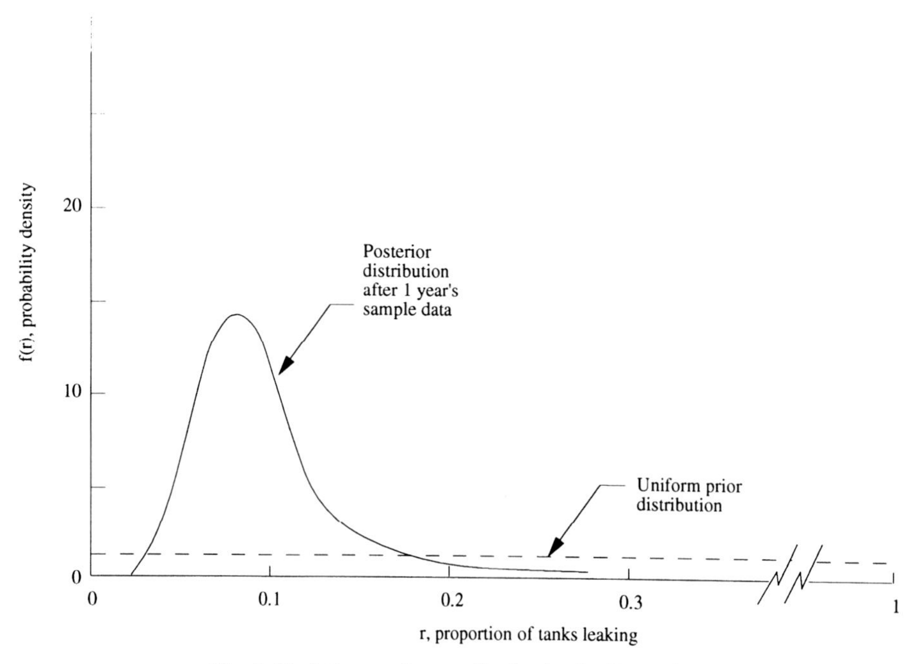

Fig. 5. Updating continuous distribution by Bayes theorem.

same, individual studies only differing in the degree of thoroughness, defensibility and extent of analysis.

The kernel of any risk study is the *performance assessment* (PA). A performance assessment describes the possible future behaviours of a system, allowing for uncertainties in knowledge about the system being modelled, including uncertainties about the processes, and uncertainties in the parameters controlling those processes. The overall performance is presented in terms of uncertainty in one value parameter, common examples of which are: cost, health risk, and project duration.

The system description is the problem statement, and consists of the various components of the system (i.e. design elements, site conditions, etc.), their functions and interrelationships, and how they will be developed with time (i.e. development strategy). The major site factors and their uncertainties are identified. Potential 'failure modes' (ways in which the system will perform in an unexpected and undesirable manner, such as accidents) and alternative assumptions are evaluated.

The backbone of the system is the *model* which is the set of rules relating input to output; commonly, this will be one or more equations. The equations will require numerical input at various places, the numerical inputs being called the system *parameters*. The model expresses not only what may normally occur, but also the hazards which are treated as chance events. Figure 6 shows the schematic breakdown (influence diagram) of a general system for a PA model.

The concept that a result can be given by an equation (such as cost $= ay + bx$) is familiar to everyone. A model with uncertainty is conceptually identical to a single equation except that instead of a single value for each parameter there is a range of values represented by a probability distribution. And, the result is no longer a single value but is also a range of values represented by a probability distribution. The calculation of the distribution of results ('propagation of uncertainty') can be carried out in several ways. However, for most people the availability of inexpensive software and personal computers makes it a simple choice: use the Monte-Carlo method (or one of the related variants).

In general terms, performance assessments can be broadly divided into two categories: those based on a 'top-down' approach, and those based on a 'bottom-up' approach.

Bottom-up (or process level) approaches to performance assessment attempt from the outset to model the various controlling processes in detail. The emphasis is on understanding and explaining the processes in great detail in order to eventually describe the behaviour of the entire system. Under a bottom up approach, a large amount of time and money can be inadvertently spent studying processes and parameters which are of little or no importance to the overall goal.

Top down approaches to performance assessment start from a desired performance statement and identify those factors which contribute to performance. Each factor in turn can be broken into greater detail. The

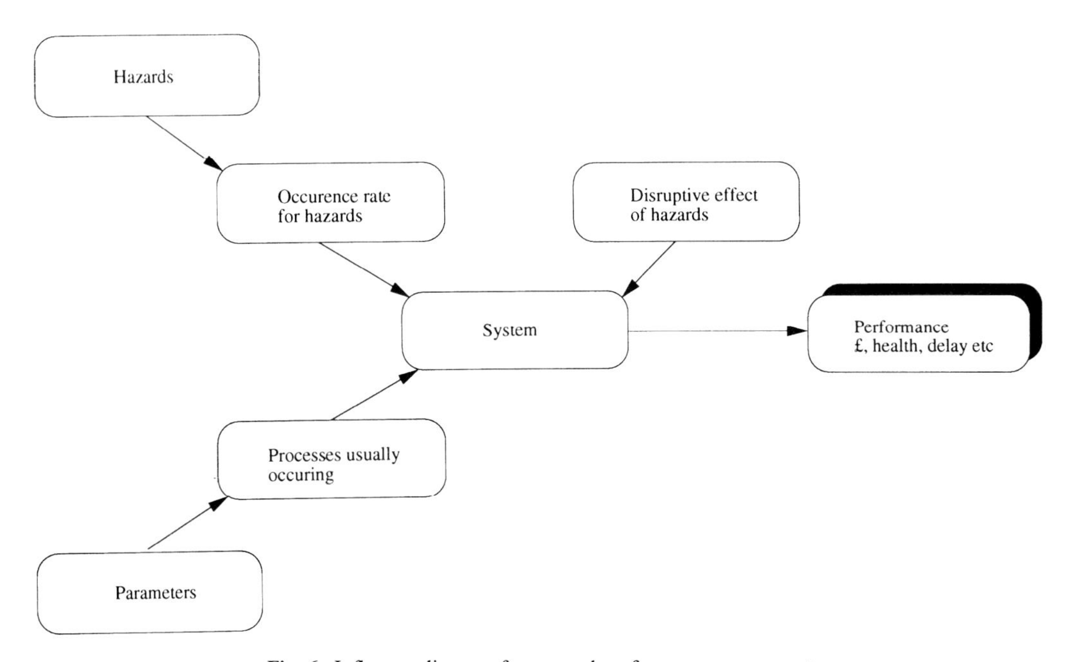

Fig. 6. Influence diagram for general performance assessment.

controlling processes may initially be represented by approximate (high-level) models and parameters. In general, these high-level parameters will take the form of subjective assessments (of probability distributions) from qualified experts. The process of obtaining subjective assessments from experts (i.e. eliciting expert opinion) is therefore critical to the successful application of a top-down performance assessment approach. Note that a top-down model does not have to be simplistic. Whereas a simplistic model might completely ignore a key process, a well-designed top-down model approximates the process while explicitly incorporating the resulting uncertainty introduced by the approximation.

A performance assessment will only be useful if it includes all meaningful contributions to uncertainty and all relevant processes. Because a top-down approach starts with the desired goal and adds detail, it naturally leads itself to completeness. The key to ensuring completeness is essentially the same open-minded attitude as used in brainstorming. A technique called influence diagrams gives an easily understood graphic framework to ensuring completeness.

An influence diagram is shown in Fig. 7. This diagram is the first few items of a situation in which we wish to estimate the health risk for someone living downgradient of a landfill and drawing their drinking water from a well. The diagram consists of ovals and arrows. An oval represents an uncertainty and the arrows represent an influence on the uncertainty. Thus, in Fig. 7, the goal is health risk and working downwards from this goal we conceive that health risk would be influenced by ingested contaminants. Exploring the issue further, the effect of ingested contaminants will depend upon their dose-effect action while the ingested amount will depend upon both the contamination in the groundwater drawn up by the well and the dietary habits.

Drawing influence diagrams is an excellent way of bringing issues into focus. It is an iterative process involving three steps:

- determine an uncertainty you would like resolved (write it on the influence diagram and draw an oval around it);
- ask whether there is an uncertainty that would help resolve the identified uncertainties, and draw an arrow showing how this uncertainty would be resolved by the new information;
- keep repeating the above two steps until a consensus is reached that all important aspects of the system have been identified.

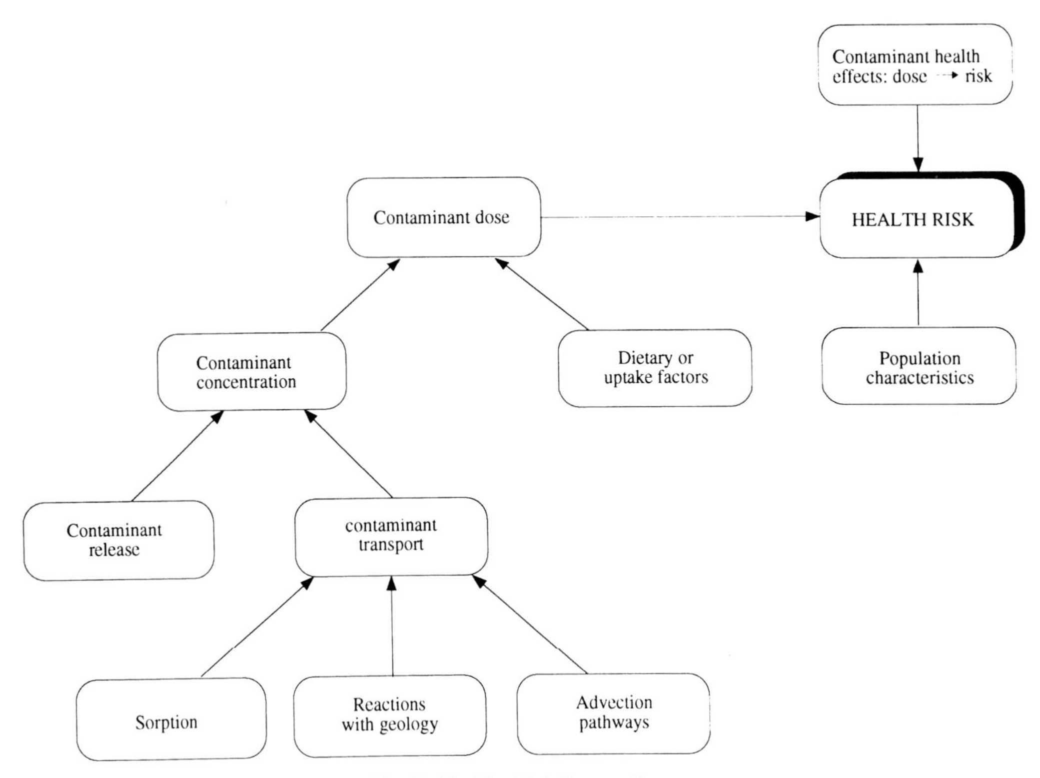

Fig. 7. Health risk influence diagram.

The key in using the technique is both to keep pushing the boundaries of what processes influence the system and to keep in mind one simple question. When anything is explained to a five-year-old child, they invariably respond with a further question: 'But why mummy?'. Essentially, we all need to remember this attitude when constructing influence diagrams, as it is this attitude that gives us the assurance the diagram will be complete and hence that the PA will be meaningful.

Once the influence diagram has been developed, we have defined the system structure for PA. Also, the influence diagram gives a basis for getting the various judgements about the various parameter and process uncertainties.

How is judgement and experience represented?

A model or idealization underlies all our thoughts about business or technical problems. A model can range from a series of cause–effect rules deduced from years in business (for example, if I drop the product price to increase market share this will be matched by competitor A while competitor B will respond by offering longer warranties) to the more familiar equations of engineering and science. The backbone of the model is the *system* which is the set of rules relating input to output; commonly, this will be one or more equations. The equations will require numerical input at various places, the numerical inputs being called the system parameters. Uncertainty usually exists in both the system and the *parameters*. Judgement/experience is used to quantify these uncertainties.

System uncertainty arises in idealizing reality in two ways. Either a simplified representation is used approximating reality out of conscious choice, or true conceptual uncertainty arises when different groups of experts offer different theories/equations for the same process. In our experience, system uncertainty is commonly ignored by scientifically trained people; there is an unwarranted confidence in the precision of our representations of nature and people tend to assume that (their) current theories are all-powerful, despite the evidence that theories evolve to either better or completely different ones with time. We incorporate judgement about system uncertainty in two ways. In the case of conscious approximation, we introduce a random variable to modify the output, the extent of the modification being treated as simply another parameter uncertainty and elicited as described below. In the case of true conceptual uncertainty, we include the alternative theories in the modelled system giving each a probability that it is correct; in assigning probabilities to competing theories we usually judge that all experts are equal unless there is compelling evidence to the contrary (i.e. if there are three competing theories, then each is assigned a probability of one in three).

Parameter uncertainty is well understood by most people with a scientific education, at least in concept, and as its name implies is the uncertainty in the input values to the system. Parameter uncertainty is quantified using expert judgement through the process of elicitation (also called encoding). Commonly, a probability is determined for a particular value of the parameter in question, say x, by comparing the probability that x will be exceeded with an alternative bet on a probability wheel. The probability wheel (Fig. 8(a)), is divided into two sectors (usually blue and orange), the relative size of which can be adjusted. The expert being encoded is asked: 'would you rather bet that the value of the parameter is less than x or that the pointer will land in the blue region when the wheel is spun ?'. The blue/orange ratio is adjusted until the expert is indifferent as to whether they would rather bet on the parameter or the wheel; the corresponding probability is then read from the scale on the probability wheel. Repeating the process for several different values of x leads to a collection of points which are joined by a smooth curve to from a CDF, as illustrated in Fig. 8(b).

(a)

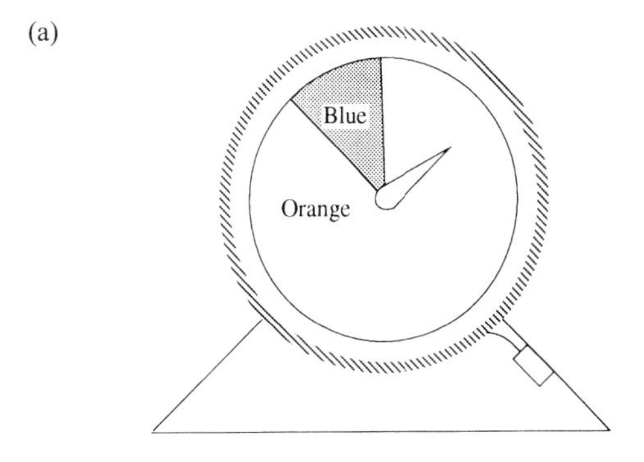

(b)

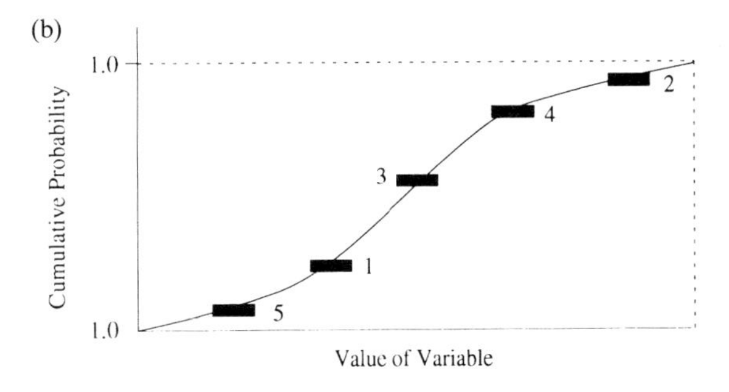

Fig. 8. Quantifying judgement. (a) Random chance indicator ('probability wheel') used during encoding of judgement. (b) CDF encoded from judgements at five different probabilities by drawing smooth curve through the elicited values.

There are five techniques for assisting the coding of judgement into a probability:

- *decomposition*: estimating probabilities of simple events is much easier than for a complex process with several contributory factors;
- *consistency*: if the probabilities of a multiple scenario do not sum to unity then something has been missed and/or the probabilities are incorrect;
- *ranking*: relative likelihoods are usually easier to express than absolute probabilities ('A is twice as likely as B'), and are subsequently easily converted into probability;
- *limits*: start from limits of complete uncertainty or certainty and iterate to a best judgement;
- *normalizing*: use case history data of some related experience to develop a first estimate and then assess the extent that the situation at hand is 'better' or 'worse' than the reference experience.

We have found that in many instances where there is relatively little information, or where we are making a first-pass through the problem, the elicitation process can be conveniently expedited by using a triangular distribution. Three questions are posed: what is the minimum credible value, what is the maximum credible value and what is the best guess? The three answers are sufficient to define a triangular distribution (Fig. 9). Use of simple triangular distributions works particularly well when the system has been broken down into many elements, each of which is readily understood by a knowledgeable group.

Human biases can distort the elicited probabilities, and these possible distortions must be recognized and dealt with. Common biases are motivational (the person has a personal stake in the action selected), availability (more available information unduly influences the judgement), representativeness (overall experience is undervalued in comparison to the most recent piece of information), and hidden assumptions. There is a wide literature on biases and how they can be avoided in the elicitation. Two good papers are Tversky & Kahneman (1973), and Spetzler & von Holstein (1975).

When the elicitation process is first encountered by people with a scientific training there tends to be incredulity that such a 'soft' process as elicitation can actually result in hard numbers for subsequent mathematical use. Proof of adequacy depends on subsequent experience, but this is difficult because we are predicting probabilistically and replicate trials are usually impossible. However, encoded probabilities have been tracked in the commercial business and reported results show that encoded probabilities correspond to the actual outcome frequencies (Sander 1969; Kabus 1976; Balthasar *et al.* 1978). And, the process yields results which are felt to be reasonable by the people involved.

Judgement is also incorporated in choosing between alternatives. Most people and companies are what is called 'risk adverse'. While average values are a reasonable basis for decisions of low consequence, as the stake increases in comparison to the wealth of the risk taker then the risk taker generally chooses to give up some of the expected gain for greater certainty in the outcome. This risk aversion requires elicitation from the decision-makers in a similar manner to elicitation of parameter uncertainty. In most instances, human risk-adverse behaviour can be approximated with an exponential function characterized by a single parameter called the risk tolerance.

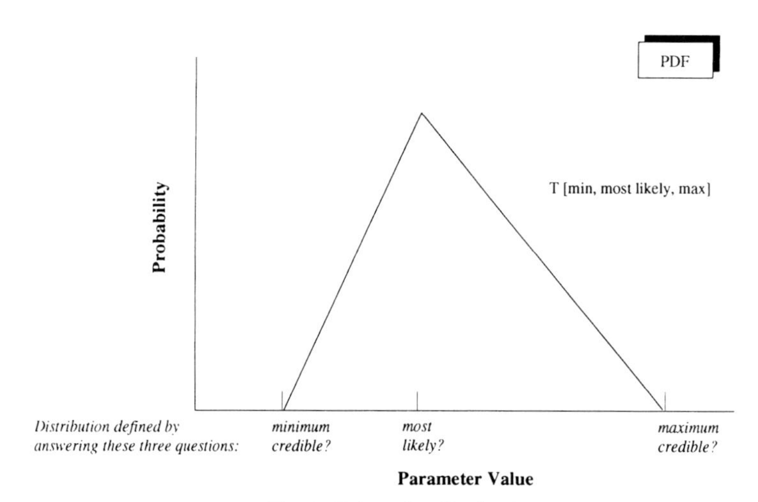

Fig. 9. Triangular distribution.

An example of performance assessment

So far we have set out the framework for the various ideas and techniques used in risk studies. We will now work through an example to crystallize how these ideas are used.

The example application is a risk assessment for hazardous waste landfill, illustrated in Fig. 10. Risk here means health risk to the affected group by the landfill.

The scenario sketched in Fig. 10 requires many parameters and mechanisms to represent the situation. However, the elements of performance assessment can be seen by considering just one of the processes involved. We illustrate PA by considering one element of the groundwater contamination pathway, specifically the leachate leakage through the composite liner system.

The performance of a clay liner with respect to leakage of leachate is a function of its thickness, permeability and the hydraulic gradient developed by the accumulation of leachate. Giroud *et al.* (1992) suggest that leakage (q) through defects in a composite liner depends on the hole size (a), shape, hydraulic head (h), permeability of subgrade (k) and the degree of contact between the membrane and the clay subgrade. Two equations are proposed:

Good contact: $q = 0.21h^{0.9}a^{0.1}k^{0.74}$

Poor contact: $q = 1.15h^{0.9}a^{0.1}k^{0.74}$

At present, the choice of one or the other equation cannot be defended as we do not know the contact conditions. We have *model uncertainty*. In this instance, if we examine the two equations we see that they are of the same form and only differ in the first coefficient. Although each equation could be assigned a probability of correctness of $p = 0.5$, it was thought that the uncertainty was related to uncertainty in the calibration constant rather than in the nature of the equations themselves. Hence, it would be appropriate to use a single equation with a continuous distribution for the calibration constant. As we have no information to suggest a most likely value for this constant, a uniform distribution is appropriate. Hence we use $C_d = U[0.21, 1.15]$.

The head driving the leachate through the defects in the liner is a function of both the rainfall infiltration through the top and the internal drainage details. With time, drainage blanket clogging or drainage pipe blocking may be important. In terms of the composite liner system, not only is the defect rate in the synthetic membrane important, but so is the actual size of the defects. And, the bentonite-enriched sand underlying the synthetic liner will control the seepage getting through the synthetic liner; the leachate may also react with the bentonite (which depends upon its swell potential for its effectiveness). The influence diagram for these processes is presented in Fig. 11.

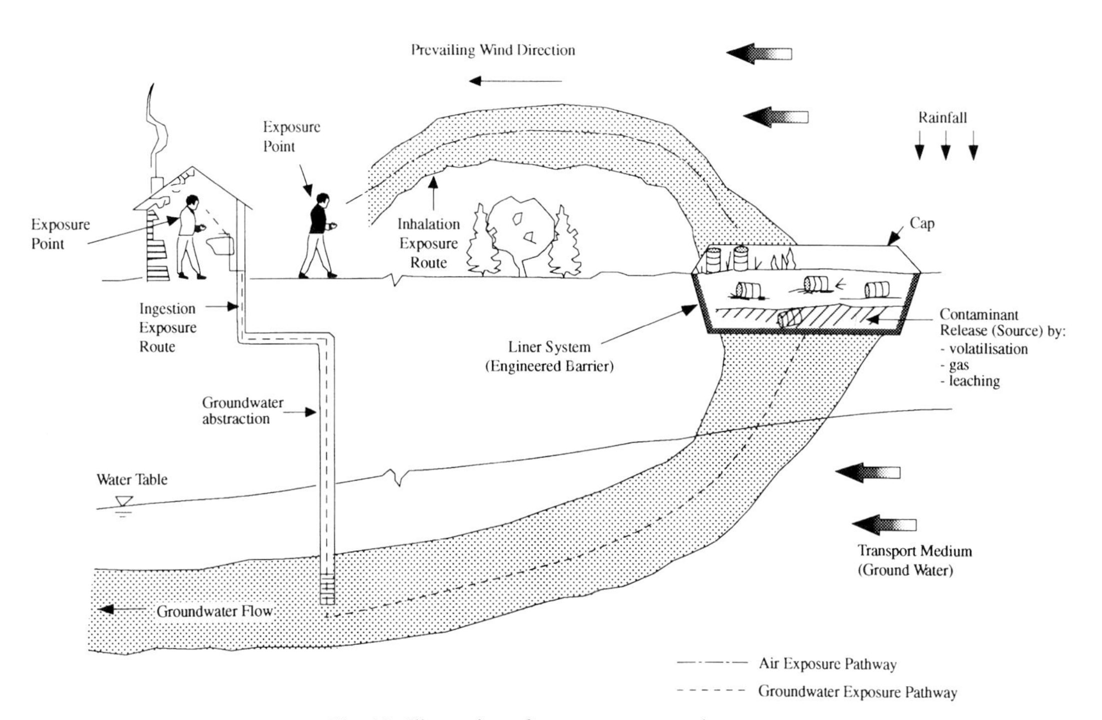

Fig. 10. Illustration of some exposure pathways.

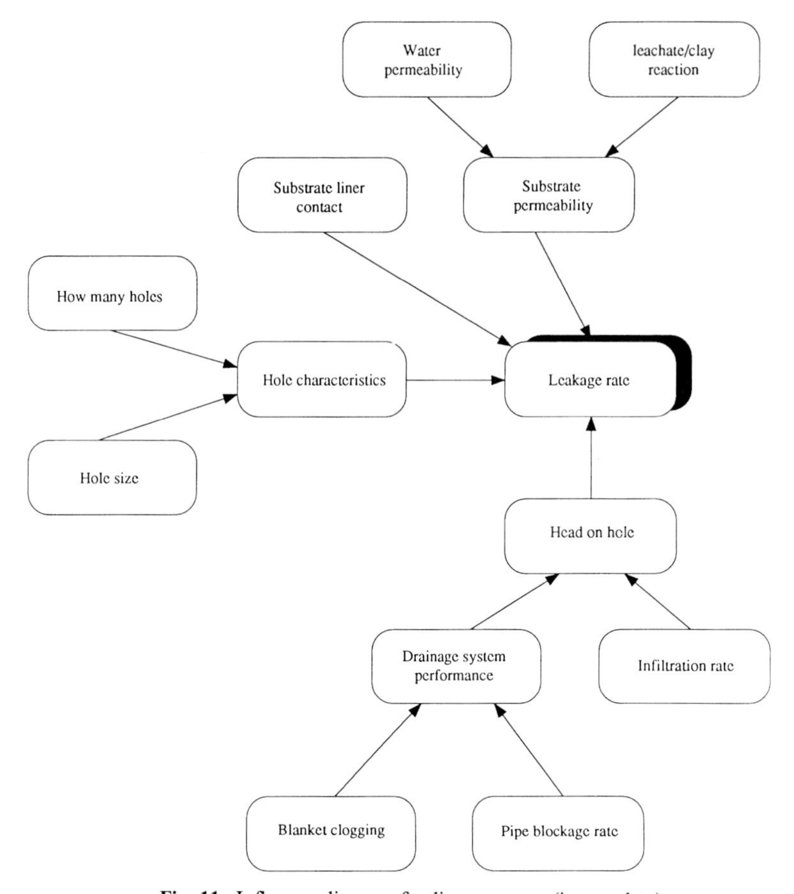

Fig. 11. Influence diagram for liner seepage (incomplete).

Analytical equations have been developed to calculate hydraulic heads within the drainage blanket for a variety of geometric configurations of landfill base and drainage blanket combinations (i.e. single slope, corrugated, herringbone pipe system, no pipe system, etc.). This hydraulic head across the liner varies with position. As defects in the liner system are considered to be random, it follows that the driving head across the defects would also be random from the range of possible values. Therefore the driving head was computed by sampling 1000 random locations.

The formulation of the database for defining hole sizes and frequencies for membrane defects has been developed from in-house records of all construction quality assurance (CQA) work over the past four years plus partial records of CQA projects undertaken worldwide. An allowance was made for the ratio of defects located against those not located and for additional membrane damage caused during early operations. Due to the uncertainties associated with making such predictions the model uses a wide distribution with a maximum number of defects typically 5–15 times the most likely value. The distributions for number and size of defects is presented in Table 2 (see Bonaparte & Gross 1990; Feeney & Maxson 1993).

Internal checks are made to ensure that leakage cannot exceed leachate production from individual areas of the landfill under extreme conditions.

We are now in a position to realize results. The uncertainties and processes identified in the influence diagram have been developed into a model and parameters with the various uncertainties specified. The overall uncertainty in leachate escape is now computed using the Monte-Carlo method, and the result is shown in Fig. 12. This figure is not general but is specific to one site; many of the input parameters relate to local conditions and the site design.

In a full performance assessment, this leachate production rate would be passed to the groundwater flowpath part of the simulation.

Table 2. *Summary of geomembrane defect rates*

Size	*Distribution of number*	*Distribution of size*
Pin holes 0.1–$5\,mm^2$	triangular $T[0,5,25]$ per hectare	uniform $U[0.1,5]$ mm^2
Small holes 5–$100\,mm^2$	triangular $T[0,2,5]$ per hectare	uniform $U[5,100]$ mm^2
Large holes 100–$10\,000\,mm^2$	triangular $T[0,0.15,2]$ per hectare	uniform $U[100,10\,000]$ mm^2

What about risk management?

No human activity is risk-free. Risk management requires determining whether assessed risks are acceptable, and if they are not then what actions have the best chance of producing the desired outcome.

In terms of risk management, the assessment and analysis procedures provide the information on which to select which actions to take. Proactive, rather than a reactive, use of risk studies minimizes unpleasant surprises and allows successful projects within modern time, budget and performance constraints. Optimum solutions can be selected that are neither too much (i.e. unnecessarily conservative and thus expensive) nor too little (i.e. insufficient and thus expensive in the long run). This process is consistently aimed at the ultimate consequences of the project (the 'big picture').

There are two distinct aspects to risk management. In one situation, we expect the project to be acceptable within its risk basis, but we are not sufficiently confident: risk management amounts to reducing uncertainty in the performance estimate. In an alternative situation, the project leads to risks that are greater than its risk basis: risk management requires risk reduction.

Risk management, then, requires a value system against which decisions are to be made (the risk basis), determination of possible actions (for uncertainty reduction or risk control), and techniques to select the situation/actions leading to the best chance of desired outcome under the value system. We will consider these issues in turn.

Value system

Risk occurs in terms of either monetary value or in terms of potential loss of life (usually taken to be human, although recent trends might argue for a wider group of organisms). Despite efforts to correlate societal death-rate acceptance with money to form a single value system, it is desirable (and possibly essential) to use separate health and money value systems.

The practical reason to separate the monetary and health value systems is very clear in the United States. If people are killed on a project or with a product and it is found during the discovery stage of the damages trial that corporate management made decisions based on $/life, then any reasonably skilled lawyer is going to show that management was callous to the public apparently regardless of the value used in the $/life factor; corresponding jury awards for damages will be large. Now although it could be argued that the UK legal system is significantly different from that of the US, the trend to charging companies and their decision-makers with corporate manslaughter, possibly leading to prison sentences for executives, suggests that the US situation of separating health and money values should be carried through in the UK. Basically, the difficulty is

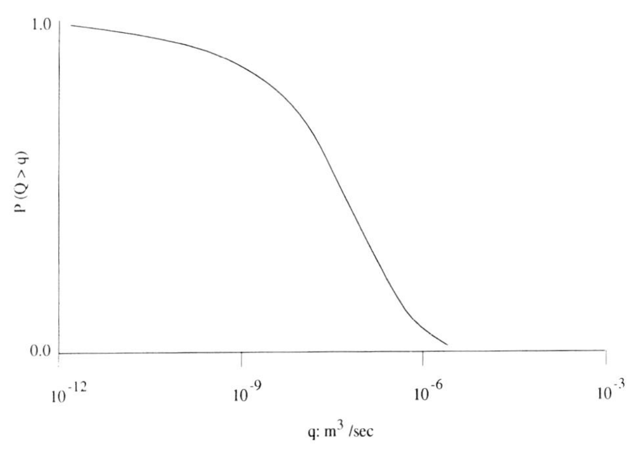

Fig. 12. Distribution of landfill leachate discharge to groundwater.

arguing that the company has the right to impute a value on human life. On the other hand, it is rather easy to discuss comparative health risks without getting into a moral or legal mire. Where both health and money are involved, the correct action is to optimize the financial scenario using the health risk as a must-satisfy constraint (i.e. run two sets of calculations in parallel).

Risk management decisions not involving health risk can use a straightforward monetary value optimization; all costs and benefits are presented in money terms which reduces multiple time-varying uncertain results to a single time-varying uncertain result. However, money now is worth more than money in the future so that some time correction must be applied to correct for this. There are various money value systems that might be used, but it has been found that the net present value (NPV) of cashflow leads to the most consistent results. The choice is however, entirely that of the sponsor company as there are no societal constraints. It is usual to use greatest expected NPV of cashflow as the basis of monetary decisions, adjusted for sponsor-company risk adverse behaviour.

When using a health risk as the value system the average incremental mortality rate of the affected population is commonly used as the value criteria. This mortality rate is the expected additional deaths per year divided by the total affected population. Other health risk measures can be used such as average loss of life expectancy, for example.

There is significant data and precedent on health risks and what society finds acceptable, although the bulk of the literature is in the social sciences rather than the sources familiar to engineering geologists. However, the findings of the various studies are summarized in three documents which amount to professional guidance notes for the UK. These three are:

- *Risk—analysis, perception, and management*, published by the Royal Society;
- *The tolerability of risk from nuclear power stations*, by the Health and Safety Executive and available from HMSO;
- *Guidelines on risk issues*, published by the Engineering Council.

Acceptable risk is inferred by comparing accident statistics from activities that the public freely undertakes, and imputing that the computed risk is acceptable since the activities occur. An example is driving a car. Driving a car exposes the driver to a chance of death, yet virtually all adults in western society own a car or want one; hence, the average death rate from driving a car is acceptable. This form of analysis has identified three trends:

- people accept greater risk voluntarily than they will allow to be imposed on them;
- people expect progressively greater safety as the number of deaths in a single accident increases;
- societal, cultural and political attitudes all influence people's acceptance of risk.

The first two points are readily derived by basic numeric processing of accident data to calculate actuarial risks for different activities, but such processing throws no light on why society should have these attitudes. It is the third point that gets to understanding what people find acceptable by looking at psychology and motivations. But, this third point gets those trained in engineering geology into unfamiliar areas. And, it is this third point that drives public policy which leads the regulatory criteria for project development. Arguably, these perception issues are at least as important as the actuarial risk.

The initial direction of studies into societal view on risk was to define some numerical risk acceptance value (or relationship). Thus guiding principle has effectively been replaced, at least in the UK following the Layfield enquiry into the Sizewell B nuclear power station, with the concept of *tolerability*. The Health and Safety Executive (1988, p. 1) have defined tolerability as:

> willingness to live a risk to secure certain benefits and in the confidence that it is being properly controlled. To tolerate a risk means that we do not regard it as negligible or something that we can ignore, but rather as something we need to keep under review and reduce still further if and as we can.

The use of tolerability to guide public policy implicitly bridges technical and societal considerations, but does so by acknowledging that people must be granted a role in making decisions about acceptability since it is they who must live with the risks. Clearly such an attitude puts an onus on project sponsors to communicate the benefits of their proposed project.

Even though tolerability appears to lead to rather open-ended values for risk management, the situation is not quite as undefined as might be thought. The HSE has summarized the various actuarial studies with some judgement about tolerability to develop a societal risk tolerability guide (HSE 1992) shown as Fig. 13.

As the UK moves to risk tolerability as the value-criterion, there remains in the world a rather wider use of the older risk acceptability concept. Commonly, an annual mortality risk of 10^{-6} is used as defining a risk that should be acceptable. Wilson (1979) gave data on common activities with an equal 10^{-6} risk, reproduced here as Table 3. As can be seen, this table would indeed suggest the 10^{-6} criterion be acceptable on the grounds that this level of risk is negligible. But, comparison with Fig. 13 shows that while indeed 10^{-6} is negligible on an individual basis, society does not regard it as negligible if upwards of 1000 people might be affected. And 1000 people is certainly in the realms of possibility in large-scale industrial accidents as, for example, Bhopal and Chernobyl.

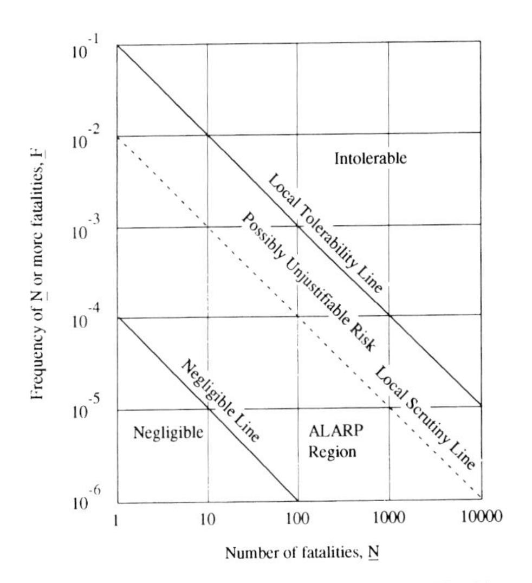

Fig. 13. Inferred societal risk tolerance (source: Health and Safety Executive 1992).

Uncertainty reduction

When an uncertainty is quantified it is expressed as a probability. For example, an uncertainty as to whether a project will be completed within the planned budget would be quantified by a statement such as 'There is only a 20% chance of completing this project within the budget'. Probabilistic performance assessment, discussed above, propagates the user's uncertainties about the elements of a project to calculate uncertainties in its outcomes.

Uncertainty is reduced by obtaining additional information, for example by more boreholes. We may want (need) to obtain more information before committing to a project. Presently, the various nuclear waste disposal programs are all gathering information to reduce uncertainty (i.e. increase confidence) that geologic disposal of nuclear waste will indeed be as safe as thought.

The power of Bayesian performance assessment is that it allows us to compute the effect of additional information before that information becomes available. In the case of an uncertain parameter, more information always reduces uncertainty. How more information moves the mean is not known but we do know that more information always reduces the standard deviation. Figure 14 illustrates the point. Alternative outcomes of the reduced uncertainty are propagated through the performance assessment to identify those parameters where more information will significantly change our understanding of the overall uncertainty.

Alternatively, we can carry out some diagnostic test (for example, install some monitoring wells). Given prior experience with the diagnostic test in other situations, Bayes theorem allows us to compute how the results of the additional testing will resolve the performance in question. This then allows the computation of revised distributions of outcomes if we undertake the tests. Only diagnostic tests that will potentially lead to a change in decision are worthwhile; in fact, the worth of a test can be calculated using the expected net worth with and without the information gained from the test.

Table 3. *Risks which increase the chance of death by one in one million*

Activity	*Cause of death*
Smoking 1.4 cigarettes	Cancer, heart disease
Drinking 1/2 litre of wine	Cirrhosis of the liver
Living 2 days in New York or Boston	Air pollution
Living 2 months in Denver on vacation from NY	Cancer caused by cosmic radiation
Living 2 months in average stone or brick building	Cancer caused by cosmic radioactivity
Travelling 6 minutes by canoe	Accident
Travelling 10 miles by bicycle	Accident
Travelling 300 miles by car	Accident
Flying 1000 miles by jet	Accident
Flying 8000 miles by jet	Cancer caused by cosmic radiation
One chest x-ray	Cancer caused by radiation
Eating 40 tablespoons of peanut butter	Liver cancer caused by aflatoxin B
Drinking Miami drinking water for 1 year	Cancer caused by chloroform
Drinking 30 12-oz cans of diet soda	Cancer caused by saccharin
Eating 100 charcoal-broiled steaks	Cancer from benzopyrene

Risk control

Referring back to the overall contributors to risk (Fig. 2), we see that there are three contributors: hazard (what might go wrong), danger (how likely is it to go wrong), and consequences (what is the effect of the accident). Risk can be controlled by controlling one or more of these contributors. The usefulness of breaking risk control actions into the different contributing classes is primarily as an aid to getting all alternatives identified: an incomplete list of possible actions may lead to wrong decisions. Not all types of actions are possible in all circumstances, there being a difference between commercial and safety-related problems. Some ideas in each class of action are as follows.

Hazard control. In some instances it may be sufficient to manage risk through managing the hazard directly, a common approach in industrial safety. Identification of fault trees and their likelihood provides a defensible basis for subsequent actions. Blockley (1992) outlines

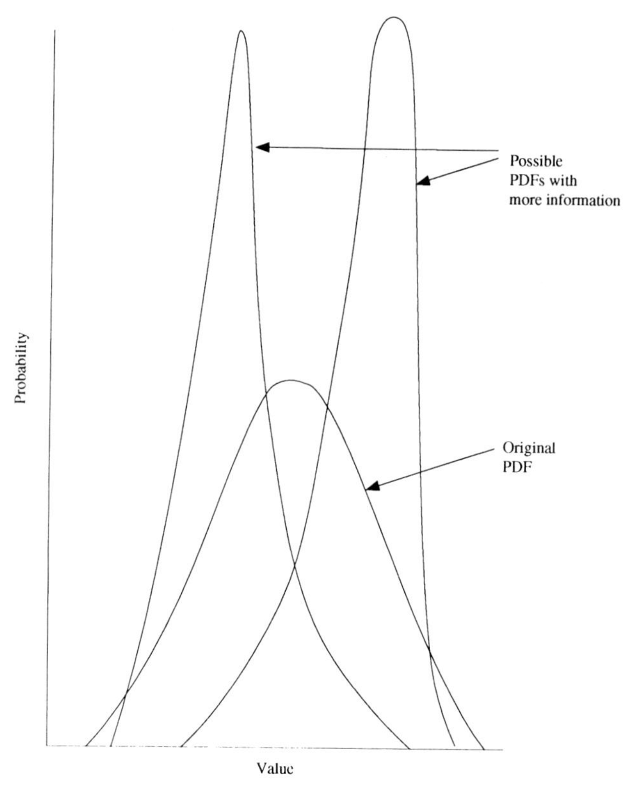

Fig. 14. Effect of new information on uncertainty during a strategic simulation.

the merits of this approach and it clearly can be applied where hazard control is possible; slope remedial measures in a rockfall or debris slide prone areas are examples in the geological context.

Danger control. The chance that a particular hazard may occur (danger) can often be changed by engineering. For example, a liner system for a landfill constructed under 'construction quality assurance' (CQA) procedures has about a 100-fold better performance than the same physical liner system constructed without CQA. In nuclear waste disposal, considerable engineering goes into the source containers: the Swedish copper canister system has better than a million-year mean time to failure in the proposed disposal configuration, so greatly reducing the risk even though the hazard (plutonium) is unchanged.

Consequence control. In other instances, identification of the consequences of various accidents allows actions to change the consequences. For example, tailings embankments constructed by the upstream method are very vulnerable to liquefaction during earthquakes, leading to a flow slide. The consequences of this hazard can be mitigated by moving people and buildings out of the potential slide area; this course of action has been pursued when the cost of treating the embankment exceeds the value of the land and asset that might be inundated in a flowslide.

Decision analysis and strategy simulation

The purpose of decision analysis is to select the alternatives with the greatest chance of producing the desired outcome. As noted earlier, for commercial decisions the assumption is that the rational action is to maximize the NPV of the certain equivalent money. Note that as decisions are made in the face of uncertainty, there is no assurance that the desired outcome will be realized in any situation.

The extension of decision analysis of a single choice into a strategy is perhaps the most valuable contribution of Bayesian risk procedures to modern management.

What differentiates a strategy from a plan is the explicit incorporation of contingency plans in a strategy. In other words, a strategy will set out a plan as to what is to be done if all goes well, but the strategy will also set out what is to be done if all does not go well. Thus, a strategy incorporates planned responses to alternative outcomes of uncertain aspects of the main plan. Performance assessment modelling allows for uncertainties in all aspects of the system being modelled: its actual make up, the processes that it will undergo, and the external events that may impinge upon it. The same modelling techniques lend themselves equally well to strategic simulation. Typically, the decision to carry out a contingency plan would be represented by a parameter, which might be a function of certain events that could occur, or of project status at a certain time.

The Monte-Carlo method is used to propagate the input uncertainties through the strategic project model to the outputs. The performance of a strategy is simulated under a large number of possible, equally likely combinations of the uncertain factors. The statistics of the set of calculated project consequences then provide the measures of the success of the strategy. Simulation of alternative strategies allows identification of the strategy with the best chance of producing the desired outcome.

A goal is a special type of parameter used in strategic modelling. A goal can be achieved by completing one or more of the elements in a project, and achieving a goal can be used to trigger further project actions. After completing a strategic analysis, one of the outputs is the probability distribution of when (and if) each project goal was achieved. Events are used to define sudden impacts on a project, such as storms or other natural occurrences, lawsuits, accidents, etc. The probability of an event and of its consequences can be affected by the project elements or achieved goals. For example, a project element might consist of the construction of a dyke that would prevent the project from floods: the probability of the 'flood' events would drop to zero once the dyke was constructed (goal achieved).

Where are we now in the UK?

Golder Associates is an international company and our experience with the variety of risk studies developed in the US and Canada, initially in connection with nuclear waste. The range of work found in the US, which was the basis of the earlier part of the paper, has not been encountered in the UK as yet.

Within the UK, over the last 18 months or so, risk studies in terms of relative numbers have been as follows:

- financial liability assessments mainly for landfills, but also including contaminated sites: averaging one site per week;
- landfill performance assessment for use in planning appeal or as part of an environmental impact study: four projects;
- health risk assessments for proposed geologic disposal of nuclear waste: three projects;
- mineral reserve estimates for asset transfer: several sites;
- tunnel construction hazard assessment: one project.
- tunnel delay and cost over-run assessment: one project.

The North Sea oil industry routinely uses hazard assessment and control so, although not environmentally related, does represent another area of risk studies within current UK practice. Clearly by way of comparison with the possible range of risk techniques, the UK does not appear to be making serious use of the available technology. Perhaps it is a reflection of the times, but it is the financial liability and 'due diligence' studies that have been numerically dominant. However, in terms of effort it is the nuclear work that has used the about two-thirds of our total effort in risk studies. Broadly, when comparing the UK with our US operations, we suggest the UK is about a decade behind the US; the sole exception is societal risk tolerance discussions where the UK is much more advanced.

Where are we going?

We expect that the use of risk techniques in the UK will parallel the US with a lag of at least five to ten years. This view is based on both the specific relative similarity of what we are presently doing in the UK with earlier experience in the US as well as the well-known tendency of the UK to follow the US in many things. Thus, we expect to see the following emerge.

Probabilistic health risk criteria may be formulated for landfills and contaminated land. The UK already has probabilistic health criteria for nuclear-related facilities and it is reasonable to expect these approaches to find themselves in other public policy areas. Thus we expect performance assessments of landfill designs to become routine, and for contaminated site remediation be designed on a rational health risk-basis.

Decision analysis will be used to evaluate the cost-effectiveness of site investigation. We have a trend in geotechnics under the sponsorship of the ICE to place more emphasis on site investigation. However, it is by no means clear that this emphasis will give the desired end result. Thus, we anticipate a counter movement from project sponsors to verify the reasonableness of large site investigation expenditures through uncertainty/decision analysis; this will be particularly required under environmental monitoring scenarios.

The move to design-build contracting as well as the move to eliminate 'Clause 12' will require careful understanding of financial risk by contractors if they are to remain in business in the longer term. Both these new contractual trends are moving financial risk from sponsor to contractor, and this must be correctly priced for the contractor to remain in business in the longer term.

The estimation of financial liability for contaminated land will continue and possibly expand.

There will be a wider use of risk procedures in conventional civil engineering. We have begun to observe this trend with tunnelling and expect it to spread across the industry. Perhaps the greatest benefit will come through probabilistic scheduling, although the contribution of hazard assessments to health and safety should not be underrated.

We will begin to see strategic simulation of large development projects. In the US, Golder Associates have just completed a simulation of alternative strategies for the WIPP (Waste Isolation Pilot Plant) facility. As Europe moves towards development of several nuclear waste facilities, we think it is possible that strategic simulations will be used to drive the programs. We are slightly reticent in this prediction because of European parochialism which may prevent adoption of ideas and software originating in the US despite GATT.

Final thoughts

Perhaps the most important benefits of a risk study are part of the process itself. The formalism of the process promotes unambiguous communication across a corporate culture and provides decision-makers with a clear understanding of the set of knowledge (including uncertainty), alternatives, and values brought into the decision or performance estimate.

In the case of performance estimates, the goal is to understand not just the best-guess scenario, but how likely and by how much this best guess might be wrong. This aspect is often important in the construction industry with its propensity to large cost overruns. For example in environmental liability assessments it is not unusual to find a 1 in 10 chance that costs might be more than twice the best guess. While this may not matter for small projects in comparison to the capital of the sponsor company, it could be very unwise if a downside outcome would break the sponsor company.

An advantage of risk-based procedures is that they are explicit and thus easily defended. This is especially important when dealing with the public and regulators, for example when estimating the confidence that health risk would meet regulatory criteria. The adoption in the UK of the concept of risk tolerance rather than acceptability squarely places an onus on project sponsors to explain the benefits of a proposed project. In the case of waste management projects, not permitting the project does not eliminate the waste; risk assessments should therefore consider not only the proposed project but also the status quo if the project does not proceed.

Implicit in the concept of risk tolerance is that the comfort margin demanded by the public depends on their confidence in not only the project proponents but also the regulators. Presently, there is widespread cynicism about the political establishment and their values. The effect of this cynicism on the public's risk tolerance can be ameliorated if industry is seen to be conducting itself in the most ethical manner possible. Risk-type studies, if clearly explained on a comparative basis, are one element in building trust. But, equally important is openness. The Swedish nuclear waste program is a remarkable example of how to conduct a technologically driven project and build public trust. The UK would benefit by the waste disposal industry adopting many aspects of the Swedish approach. Obviously self-serving advertising campaigns are not sufficient: what appears to work is systematic consideration of risk and active uncertainty reduction carried out in a public form. Proprietary studies are an anathema to the public.

Turning to the risk procedures themselves, these will have to become part of normal company operations as the current commercial pressures to transfer risk and operate on smaller margins increase. There appears to be every sign that these commercial pressures are intensifying, not decreasing and there is little hope that the drive to greater efficiency will moderate as the worlds economies come out of recession. Companies who do not understand their risk exposure are betting their continued existence on luck.

To conclude, in our experience, initial scepticism about using the risk methods described in this paper gives way to enthusiasm once an end result is seen. Subsequently, the sceptic becomes an evangelist for the methods. Based on the acceptance within our company, we think the approaches outlined in this paper will become pervasive within the next decade. You might as well adopt them now as later.

References

ANON. 1993. *Guidelines on Risk Issues.* The Engineering Council.

BALTHASAR, H. U., BOSCHI, R. A. A. & MENKE, M. M. 1978. Calling the shots in R&D. *Harvard Business Review*, May-June.

BLOCKLEY, D. 1992. *Engineering Safety.* McGraw-Hill.

BONAPARTE, R. & GROSS, B. A. 1990. *Field behaviour of double liner system*, ASCE Geotechnical Special Publication No. 26.

FEENEY, M. T. & MAXSON, A. E. 1993. *Field performance of double liner system in Landfills.* Geosynthetics 1993, Vancouver.

GIROUD, J. P., BADU-TWENEBOAH, K. & BONAPARTE, R. 1992. Rate of leakage through a composite liner due to geomembrane defects. *Geotextiles and Geomembranes*, **11**, 1–28.

—— & BONAPARTE, R. 1989a. Leakage through liners constructed with Geomembranes. Part I Geomembrane Liners, *Geotextiles and Geomembranes*, **8**(1).

——&——1989b. Leakage through liners constructed with Geomembranes. Part II Composite Liners. *Geotextiles and Geomembranes*. **8**(2).

HEALTH AND SAFETY EXECUTIVE 1988. *The tolerability of risk from nuclear power stations*. HMSO, London.

KABUS, I. 1976. You can bank on uncertainty. *Harvard Business Review*, May–June.

ROYAL SOCIETY SAFETY GROUP 1992. *Risk—Analysis, perception and managements*. Royal Society.

SANDER, W. E. 1969. The value of subjective forecasts by R&D project managers. *IEEE Transactions on Engineering Management*, 35–43.

SPETZLER, C. S. & VON HOLSTEIN, C-A. S. 1975. Probability encoding in decision analysis. *Management Science*, **22**.

TVERSKY, A. & KAHNEMAN, D. 1973. Availability: a heuristic for judging frequency and probability. *Cognitive Psychology*, **4**, 207–232.

WILSON, R. 1979. Analyzing the risks in daily life. *Technology Review*, February 1979.

The protection of groundwaters from the effects of waste disposal

P. A. Hart[1] & I. Davey[2]

[1] NRA Anglian Region, Peterborough, Cambs, UK
[2] NRA Thames Region, Kings Meadow House, Kings Meadow Road, Reading, Berks RG1 8DQ, UK

Abstract. Powers available to the NRA are summarized. Definitions of waste and groundwater are given. The scale of the problem is described by reference to the legacies of historical 'dilute and attenuate' landfills, mainly occupying disused surface mineral excavations.

An explanation is given of an interpretation of the National Rivers Authority *Policy and Practice for the Protection of Groundwater* in context of waste disposal. Special emphasis is placed on the protection of groundwaters by development control. Reference is made to work regarding best practice for containment and management of leachates generated at landfills.

Scenarios are described which are then used to assess the likely impact of the policy on the future of waste disposal.

It is concluded that the NRA's policy provides a framework which is being used to promote the practical protection of vital natural groundwater resources.

Disposal and storage of waste in a safe manner is important. This has to be considered within the legislative framework for environmental protection. Controlled waters (including groundwaters) are protected by the National Rivers Authority (NRA) whose powers and duties in England and Wales are explained in the Water Resources Act 1991. Also NRA has indirect powers in connection with the control of waste disposal to land where it may cause pollution of water resources, under the Environmental Protection Act 1990, and principally the Waste Management Regulations 1994, which are enforced by Waste Regulation Authorities. As far as these UK authorities are concerned, the 1994 Regulations are the yardstick during the waste management consultation process.

Support of these powers arose from a meeting of European Community Environment Ministers at The Hague in November 1991, resulting in the 'Hague Declaration', which recognized the need for management and protection of groundwater on a sustainable basis by preventing its overexploitation and pollution.

What is groundwater?

A single definitive explanation of 'groundwater' has been delayed because of the contradictory definitions arising from the Groundwater Directive and the Water Resources Act 1991. The former regards 'groundwater' as the liquid in the zone below the uppermost level of the water-table, whilst the latter includes all water contained in underground strata. Under the 1994 Waste Management Regulations, Section 15 (12), the Directive definition is upheld, and this is now used in context of waste management sites, which are potential point sources of pollution.

In consideration of diffuse pollution, NRA may use the definition of groundwater given in the Water Resources Act 1991, which includes all water below ground level.

The question of when does groundwater become polluted may be answered by using the 'fit for usage' philosophy as the basis for an assessment. So, for example, should a potable groundwater resource become affected as a result of leachate leaking from a landfill site, the raw water at an abstraction borehole intended for potable usage would be deemed 'fit for usage', provided no treatment (other than disinfection) would be necessary prior to consumption. Determination of natural background quality, and historical usage of the land and the aquifer are thus important baseline considerations in the determination of liability.

What is 'waste'?

To give a full explanation of the definition of waste is beyond the scope of this paper, but it is useful to refer to the meanings of 'waste'. The EPA '90 Section 75 defines waste as any substance (excluding an explosive substance as defined by the Explosives Act 1875):

- which constitutes a scrap material or an effluent or other unwanted surplus substance arising from the application of a process
- or article which requires to be disposed of as being broken, worn out, contaminated or otherwise spoiled.

From BENTLEY, S. P. (ed.) *Engineering Geology of Waste Disposal*,
Geological Society Engineering Geology Special Publication No. 11, pp 361–366

The 1994 Waste Management Regulations adopt the same definition as in the EC Directive on Waste, 75/442/EEC, as amended by 91/156/EEC and 91/692/EEC. 'Directive waste' basically means any substance which the producer or person in possession of it discards, or intends or is required to discard.

NRA is concerned about all wastes which have, or can, give rise to pollution of controlled waters. Broadly these wastes may be classified, according to current UK legislation, into 'controlled wastes' (household, commercial, industrial; inert, and putrescible; special wastes, and hazardous wastes), and 'non-controlled wastes' (agricultural, mining, quarrying). For example, in connection with EC Groundwater Directive (80/68/EC), NRA is charged with setting up of a national register of List I and II (Dangerous) substances which may be contained in some controlled and non-controlled wastes.

Impact of waste disposal and storage on groundwaters

Importance of groundwaters

Groundwater is important. In England and Wales, it provides on average around 35% of public water supplies, is the sole source of drinking water for thousands of private users, supports agriculture by spray irrigation, and is utilized by a wide range of industries, from cooling waters in power generation to drinking in soft drinks manufacture and brewing.

Effect on groundwater quality

NRA has encountered demonstrable contamination of both surface and groundwaters throughout England and Wales from historical landfills (e.g. Palmer & Young 1991; Lewin 1992; Heathcote 1993), waste transfer stations, and uncontrolled waste disposal. In connection with the latter, land uses include chemical manufacturing, smelting, heavy engineering, dry cleaning, tanning, military, mining and agriculture (e.g. Lawrence & Foster 1991; Lerner & Tellam 1992; Lloyd 1992; Kershaw & Clews 1993; Fellingham *et al.* 1993).

Case history of a site in the Vale of St Albans

Gravel extraction began in the 1930s and continued to the late 1970s, eventually covering an area of about 75 ha to a maximum depth of 12 m. Perched groundwater occurs locally within gravels and may seep into the site. Groundwater in the underlying Chalk lies at about 15 m below ground level, flowing towards a public supply borehole (commissioned in 1950) situated 750 m to the south.

Putrescible wastes have been tipped throughout the site, disposal being first consented in 1954. Limited quantities of liquids have also been deposited, including slaughterhouse wastes, and there is evidence which indicates the depositing of solvents.

First attempts at aquifer protection were rudimentary, with quarry reject fines being deposited at the base of the site. Later the requirement was for compaction of 1.5 m of clayey materials across the base. Recent investigation suggests this activity may have been haphazard. The most recent worked area, filled with commercial and industrial waste from 1990, was lined with 2 m of clay compacted to a permeability of less than $1 \times 10^{-9}\,m\,s^{-1}$.

Clear signs of contamination to the public supply source were evident by 1975 with indications that the process may have begun much earlier. Ammonia concentrations, undetected previously, rose to $2\,mg\,l^{-1}$ (EC Maximum Acceptable Concentration (MAC) $0.5\,mg\,l^{-1}$) and chloride concentration doubled to $50\,mg\,l^{-1}$ (EC MAC $400\,mg\,l^{-1}$).

In 1989 the site operators applied to dome the surface using commercial and industrial wastes, incorporating a low-permeability capping in the site restoration. Site investigation revealed potential for further pollution to groundwater. Up to 4 m of leachate was found, with ammonia concentrations of $300\,mg\,l^{-1}$ in wastes more than 20 years old and $600\,mg\,l^{-1}$ in wastes of 15 years. Monitoring of Chalk groundwater down-gradient of the site showed variable pollution levels with ammonia concentrations in one perimeter borehole varying between 2.2 and $111\,mg\,l^{-1}$. This may be attributed to pulses of leachate entering groundwater and moving along discrete fissure systems.

The scheme was also to provide for leachate management. On this basis and in view of the proposed capping improvements, the NRA raised no objections. However, it was recognized that the consequent reductions in leachate percolation to groundwater would be at the expense of extending the site's polluting life.

The planning application was turned down on appeal on the grounds that the proposed end use, a golf course, was inappropriate development in the Green Belt. Nevertheless, the site remains a risk to groundwater quality and improved capping is desirable.

Groundwater protection policy for England and Wales

The favoured methodology for the protection of groundwater quality is based on the concept of practical risk management.

By launching its *Policy and Practice for the Protection of Groundwater* (NRA 1992), NRA established a technical framework for the practical protection of groundwaters. This framework is applied in carrying out of its own powers. It can also be used in connection with indirect powers to seek to influence policies and decisions of others, for example, in response to consultations under planning legislation.

Protection of groundwaters is being sought by way of the following.

(1) Classification of groundwater vulnerability (resource protection of future or potential groundwater supplies):

- major and minor aquifer { high vulnerability / intermediate vulnerability / low vulnerability }
- non-aquifer

This classification is the basis for the subdivision of area on the NRA's Groundwater Vulnerability 1:100 000 scale map series (view at NRA offices, local planning authority offices, and available from HMSO).

(2) Definition of source protection zones (source protection around existing abstractions):

- Inner Zone (Zone I): 50 day travel-time in groundwater to abstraction point
- Outer Zone (Zone II): 400 day travel-time in groundwater to abstraction point
- Catchment Zone (Zone III): water will eventually find its way to the point of abstraction

Maps showing these zones are available for consultation at NRA offices.

(3) Eight policy statements in relation to groups of activities posing a threat to groundwaters, five being particularly relevant here.

C waste disposal to land
D contaminated land
E the application of liquid effluents, sludges and slurries to land
F discharges to underground strata
H additional activities (e.g. the production, storage and use of chemicals, storage of farm wastes, animal burial sites, sewage works, oil and petroleum storage, major infrastructure developments)

Whilst the prime objective is to encourage the siting of new landfill sites away from major aquifers, according to the NRA's policy in general terms the acceptability of any new landfill proposal to the NRA can be summarized as follows.

(1) In the Inner Zone (I), NRA will normally object to potentially polluting activities.
(2) In the Outer Zone (II), NRA will normally object, unless it is satisfied that the waste materials do not contain significant biodegradeable or other potentially polluting matter, and the site will have acceptable operational safeguards.
(3) In the Catchment Zone (III) and on major, minor and non-aquifer, landfill sites with potentially polluting wastes may be acceptable, subject to satisfactory containment measures, and are considered on a case-by-case basis.

Depositing of polluting waste below the water-table is unacceptable to the NRA.

Sites which may be acceptable for a particular category of waste will be assessed by the NRA which will base its assessment on one or a combination of other factors: the sites will be considered on a case-by-case basis, for instance taking into account the depth and nature of the unsaturated zone (see Mather 1989); engineered containment and operational safeguards will be considered (NRA 1992, 1995; Seymour 1992; Walker 1993).

Groundwater protection policy in action

Many areas have been subjected to land uses which are now recognized as conflicting, that is arising between activities associated with commercial developments and requirements for aquifer protection. Many landfill sites in the past were located near to public water supply points, and there is evidence of localized groundwater contamination and pollution around landfill sites and of impact on public water supplies, giving rise to breaches of EC Groundwater Directive. Location of sites for waste disposal, waste handling and waste storage, and further developments at existing licensed sites, are now influenced by the *Policy and Practice for the Protection of Groundwater*. As examples, several areas for which the policy has been implemented are presented below.

South Essex

The industrialized and urbanized area between Purfleet and Stanley-le-Hope which forms the north bank of the Thames Estuary in south Essex (Fig. 1), has many demands on use of land.

A summary description of the stratigraphy of the area is recent river alluvium (bordering the Thames) and gravels, Palaeogene London Clay, Lower London Tertiaries, and Cretaceous Chalk (Fig. 1). Structurally these strata dip at low angles to the south, steepening beneath the estuary.

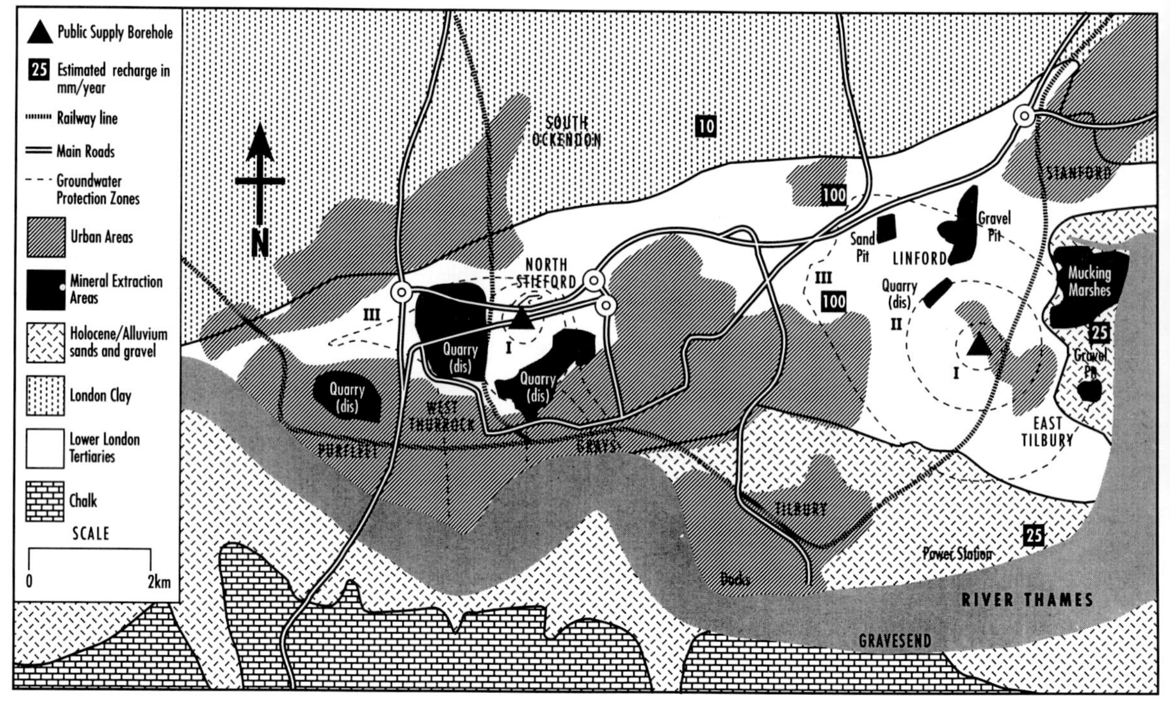

Fig. 1. Protection of groundwater in South Essex (simplified).

Chalk groundwaters have been abstracted at two locations for public supply, which constitute strategic sources for the area.

Groundwater flow in the multilayer aquifer system is generally southwards. Groundwater within the gravels and Lower London Tertiaries represented by the Thanet Sand leak to the Chalk aquifer below, and as far as the public water supply sources are concerned, provide recharge particularly in the area to the north of the sources. Transmissivity, recharge distributions and localized groundwater flow directions are complicated by discrete karstic systems beneath the Palaeogene cover, and subjacent karst-related collapse phenomena. Recharge is also supplemented by a component 'spilling off' the edge of the London Clay outcrop.

Cretaceous and Palaeogene strata have been quarried at numerous locations in the area, leaving void spaces, several of which have been historically utilized for disposal of a variety of wastes using the dilute and disperse philosophy. Utilization of the abandoned mineral workings for future landfilling is an option that has to be carefully considered if groundwater quality is not to deteriorate. An assessment of whether or not a site within this area is suitable for landfill use, is possible based on an understanding of the hydrogeology and the establishment of Source Protection Zones around the public water supply sources (Fig. 1). NRA's stance on planning applications for new sites and new licence applications may thus be determined according to the locality in question and the nature of the proposed waste.

The Vale of St Albans

In the Vale of St Albans, landfill sites have frequently been developed in abandoned sand and gravel pits near to Chalk-fed public water-supply boreholes. There is evidence of uncontained landfill sites having caused both localized groundwater pollution and contaminated some public water supplies.

The Vale is a broad shallow valley originally formed by a precursor to the River Thames, having extensive deposits of glacial till (clayey), sands and gravels, which lie on weathered Upper Chalk. Sands and gravels, of high quality, up to 20 m thick, occur as variably distributed interlayers with the tills. The weathered Chalk in places is 'putty' like with a clayey consistency.

Perched groundwater in the sands and gravels supports small abstractions, such as for watering cattle. Where clay restricts infiltration or springs occur, in some areas wetlands have developed.

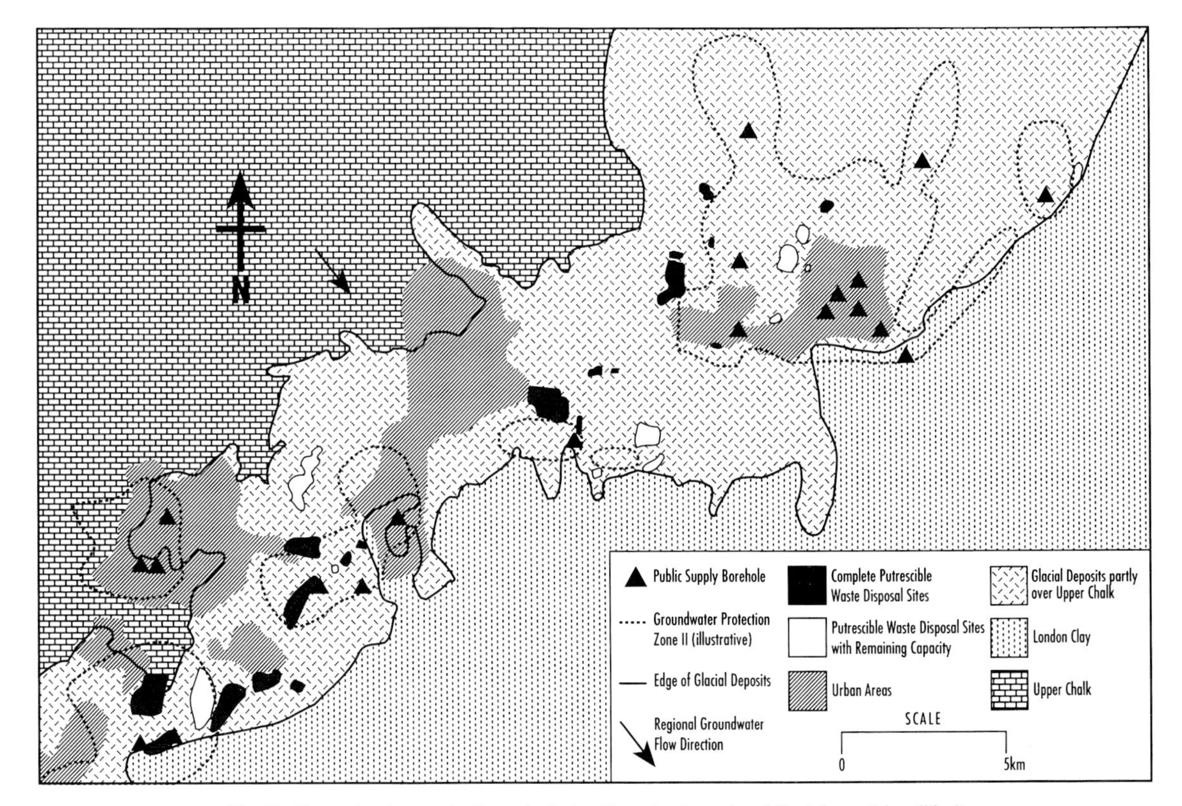

Fig. 2. Groundwater protection at abstractions in the vale of St Albans (simplified).

In the Chalk, regional groundwater flow is to the southeast towards the London Basin, saturated conditions occurring at depths up to 40 m below ground with fluctuations in levels of up to 10 m following recharge. Locally confined conditions occur beneath boulder clay and/or 'putty' chalk.

Micro-karstification, particularly below river valleys, produces preferential flow zones both for groundwater recharge and contaminant migration. This was demonstrated in the 1930s when the Mimmshall Brook near North Mymms was dosed with fluorescein. The tracer appeared in groundwater three days later in springs and boreholes to the south of Ware, a distance of 16 km. Flow had occurred northeastwards, perpendicular to the apparent regional flow direction.

The Chalk aquifer provides high-quality drinking water meeting most of the demand in this part of Hertfordshire and, via the New River, a proportion of London's requirement. Base flows to rivers are sustained via Chalk springs. However, rainfall is low and annual recharge amounts to about 230 mm. Consequently combined with the low effective porosity (2%), abstractions draw on large volumes of aquifer. Virtually the whole area is classified as the source catchment, a large area of this being classified as the outer source protection zone (Fig. 2), which has considerable influence on its future for waste disposal.

Concluding remarks

It should be noted that two of the reasons for the locations of urban and industrial developments are the existence of potable water supplies and useful mineral deposits. Waste is generated as a result of anthropogenic activities, and as far as possible needs to be minimized, and disposed of in a safe and practicable manner. Mineral deposits for construction, manufacturing and energy production, are wasting assets and their importance is great. Groundwater resources, however, must be regarded as sustainable and therefore new sites, if needed, must be encouraged which are not going to threaten the quality or quantity of existing or future drinking water supplies. The NRA's policy and its application is considered to be a practical way forward and a major contributor to the sustainability of these supplies. In addition, new sites may be required to be designed according to the engineered containment principle. Acceptable design principles are described in a series of pollution prevention guidelines, published by NRA (NRA 1995).

Groundwater is out of sight, but not out of mind. Prevention is better than cure.

Note. The views expressed are those of the authors, and not necessarily those of the NRA.

References

FELLINGHAM, L. R., ATYEO, P. Y. & JEFFERIES, N. L. 1993. The investigation and remediation of the groundwater pollution at Harwell Laboratory. Groundwater Pollution Conference, Documentation E0136, IBC Technical Services Ltd (unpublished).

HEATHCOTE, J. A. 1993. Mayton Wood Landfill-Groundwater Pollution from an attenuate and disperse site. Groundwater Pollution Conference, Documentation E0136, IBC Technical Services Ltd (unpublished).

KERSHAW, M. & CLEWS, J. E. 1993. Investigation of a solvent problem incident in North East England. Groundwater Pollution Conference, Documentation E0136, IBC Technical Services Ltd (unpublished).

LAWRENCE, A. R. & FOSTER, S. S. D. 1991. The legacy of aquifer pollution by industrial chemicals: technical appraisal and policy implications. *Quarterly Journal of Engineering Geology, London*, **24**, 231–239.

LERNER D. N. & TELLAM, J. H. 1991. The protection of urban groundwater from pollution. *Journal of the Institution of Water & Environmental Management*, **6**(1), 28–37.

LEWIN, K. 1992. The fate of leachate and landfill gas in the Sherwood Sandstone. Groundwater Pollution Conference, Documentation E7600, IBC Technical Services Ltd (unpublished).

LLOYD, J. W., 1992. Urban pollution—a case history of the Birmingham Triassic Sandstone aquifer. Groundwater Pollution Conference, Documentation E7600, IBC Technical Services Ltd (unpublished).

MATHER, J. D. 1989. The attenuation of the organic component of landfill leachate in the unsaturated zone: a review. *Quarterly Journal of Engineering Geology, London*, **22**, 241–246.

NRA, 1992. *Policy and Practice for the Protection of Groundwater*. HMSO, London.

——, 1995. *Pollution Prevention Manual* (in press).

PALMER, C. & YOUNG, P. J. 1991. Protecting water resources from the effects of landfill sites: Foxhall Landfill Site. *Journal of the Institution of Water & Environmental Management*, **5**(6), 682–696.

SEYMOUR, K. J. 1992. Landfill lining for leachate containment. *Journal of the Institution of Water & Environmental Management*, **6** (August), 389–396.

WALKER T. 1993. Managing leachate. Conference on Landfilling of Wastes, Harwell, May 1993 (unpublished).

Development of a lined landfill site adjacent to a major potable supply river

J. P. Apted,[1] M. Philpott[2] & S. W. Gibbs[3]

[1] Acer Geotechnics, Acer House, Medawar Road, The Surrey Research Park, Guildford, Surrey GU2 5AR, UK
[2] Shanks and McEwan, Woodside House, Church Road, Woburn Sands, Milton Keynes, Bucks MK17 8TA, UK
[3] Robinson Fletcher Consultants Limited, Pier House, Wallgate, Wigan, UK (formerly of Acer Environmental)

Abstract. The site at Pen-Y-Bont is based on a deep clay pit in Ruabon Marl and is on a piece of land almost completely surrounded by a loop in the River Dee, a major source of potable water. The proposals for developing a landfill at the site required two stages of investigation with particular regard being given to permeabilities and groundwater flow. A strategy for landfilling was developed that took into account the need to protect the Dee with particular emphasis being given to leachate control. In addition, the requirements for landfill liners in terms of current guidance were also considered.

The Shanks and McEwan Group purchased the site at Pen-Y-Bont in the late 1980s. As a deep worked-out brick pit in argillaceous rocks, it appeared ideally suited to provide a containment site for landfill. However, as the proposals for the site were developed, a number of issues had to be addressed that lead to a radical review of the approach to landfilling.

The site

Pen-Y-Bont is a small village in the County of Clwyd around 10 km east of Llangollen and 35 km northwest of Shrewsbury (Fig. 1). The site is to the south of the village, on a high spur almost completely surrounded by a large loop in the river Dee (Fig. 2). Immediately to the south of the site is the small hamlet of Pentre. The River Dee is a major source of potable water.

The spur is surrounded by a flood plain of variable width, at its maximum around 100 m wide, at its narrowest nearly non-existent. Over a period of many years a large clay pit had been excavated into the side of the spur, with its base well below the level of the flood plain. At its deepest point the pit is around 50 m deep. As well as extraction of mudstone rocks for brickmaking, the site has been used for the dumping of colliery spoil and coal washings as well as the overburden from above the clay minerals.

Geology

The site is underlain by the Ruabon Marl of the Upper Coal Measures. The Ruabon Marl consists of silty mudstone and subordinate bands of siltstone and sandstone. It is underlain by the Middle Coal Measures. While a number of coal seams are known to be present within the Middle Coal Measures, the available information indicates that it is not likely that any such seams exist beneath the site. A major fault cuts across one corner of the site, as well as passing beneath the river. This is shown on Fig. 2.

On the higher ground, the Coal Measures rocks were mantled by cohesive Glacial Till. A typical section through the site is shown in Fig. 3.

Hydrogeology

The hydrogeology of the site is dominated by the Ruabon Marl. As a mudstone this was considered to be an aquiclude, effectively excluding significant groundwater penetration and transmission. Both the overlying superficial deposits and underlying Middle Coal Measures were considered to be significantly more permeable than the Ruabon Marl, but to be effectively separated by the Marl.

As the investigations of the site proceeded, and the landfill proposals were developed, the permeability of the Marl became a key issue.

Site investigations

Two major investigations of the site were undertaken. These were both planned and supervised by Geotechical staff from Acer Geotechnics. The first was a preliminary investigation of the site to establish the general characteristics of the ground and to confirm the feasibility of the proposals. The second stage was used to provide

From Bentley, S. P. (ed.) *Engineering Geology of Waste Disposal*,
Geological Society Engineering Geology Special Publication No. 11, pp 367–372

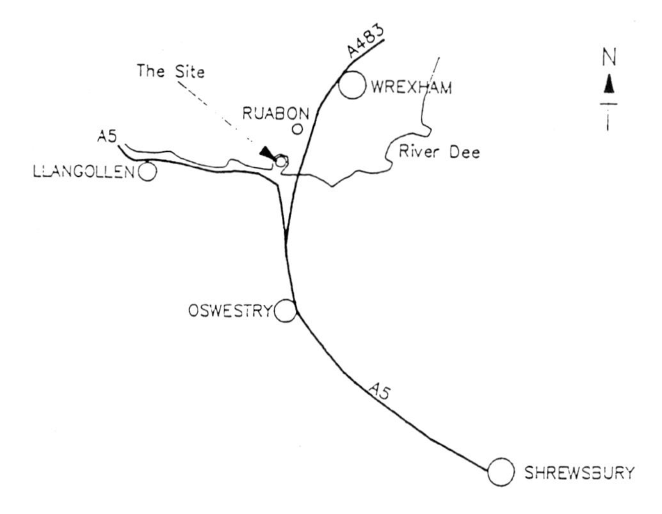

Fig. 1. Site location plan (not to scale).

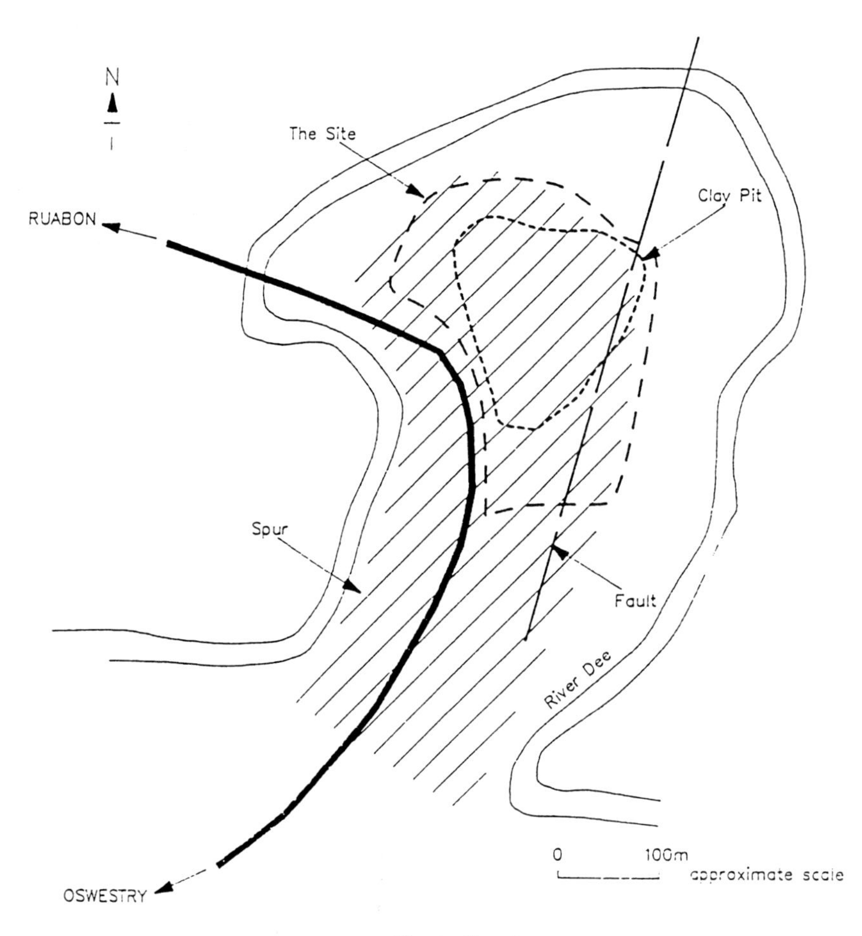

Fig. 2. Site plan.

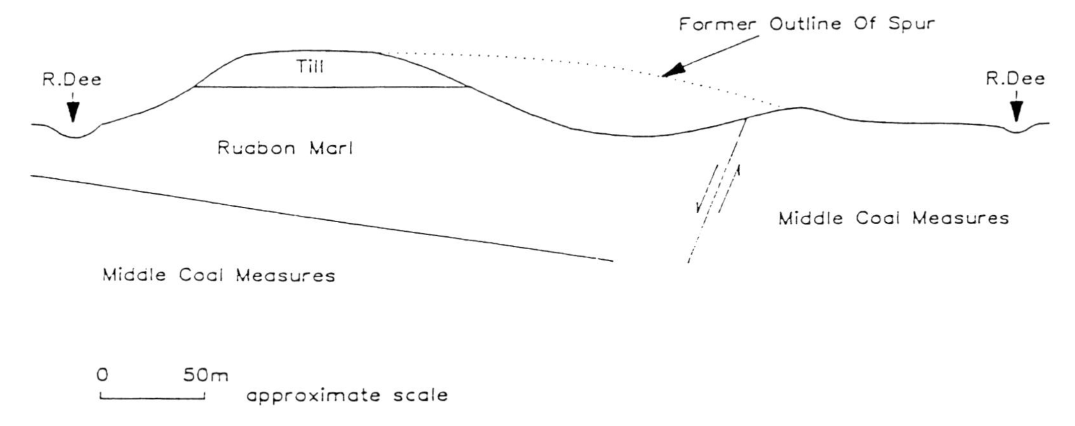

Fig. 3. Typical section.

additional information on the ground conditions, and in addition provide information on earthworks suitability, permeability, and groundwater levels.

The basic details of the investigations are given in Table 1. In both cases the investigations comprised a combination of cable tool boreholes and rotary cored holes.

A key feature of the investigations were the *in situ* permeability tests; the numbers of which are given in Table 2.

Ground conditions

While it is not possible here to go into all the details of ground conditions some of the principal characteristics are given in Table 3.

Establishment of initial proposals—the key problems

In the beginning the concept was simple. The basic scenario was that a deep clay pit provides the ideal means of establishing a containment landfill.

The development of the proposals followed on from the initial environment studies carried out in 1989 and 1990. The geotechnical studies were part of this process and were based on 'desk study' information; the results of on-site inspections, and the results of the 1989 site investigation.

It was considered that the landfill would effectively be seated into a 'bowl' of Ruabon Marl and this, by itself, would form the containment barrier. The landfill would be shaped to follow the side of the spur the clay pit had been cut into. Thus the sides of the landfill would rest against both Ruabon Marl and the various overburden materials. Leachate would be allowed to build up to the level of the top of the *in situ* Ruabon Marl bowl, approximately coincident with the level of the flood plain. Overflow, when it occurred, would be controlled via a properly constructed drainage system and passed through a treatment plant.

While these proposals appeared to correspond to established good practice and published guidelines, there were features which promoted significant further discussion, both internally within the professional team, and externally with the NRA. In addition, the progress with preparing the planning application had enabled discussion with local residents together with their professional advisers to take place.

In terms of the engineering geology of the site the key feature that needed to be addressed in more detail was the threat of leachate pollution to the River Dee. The

Table 1. *Investigation details*

Stage	*Year*	*Number of boreholes*	*Approximate depth range (m)*
Stage I	1989	13	13–50
Stage II	1991	21	5–36

Table 2. *Number of permeability tests*

Stratum	*Falling head tests*	*Packer tests*
Natural superficial deposits	6	—
Waste tip materials	9	—
Ruabon Marl	3	25
Middle Coal Measures	3	10
Fault zone	—	3

Table 3. *Characteristics of the ground conditions*

Strata	*Description*	*Permeabilities* ($m\,s^{-1}$)		
		Maximum	*Minimum*	*Mean*
Made ground	(i) Reworked glacial till (ii) Coal Washings (iii) Colliery Spoil	1×10^{-6}	2×10^{-7}	6×10^{-7}
Terrace deposits	Clays and gravels	4×10^{-6}	4×10^{-7}	2×10^{-6}
Ruabon Marl	Red brown and grey green mudstones with subordinate siltstone and sandstone bands	2×10^{-7}	0	0.9×10^{-7}
Middle Coal Measures	Interbedded mudstones, siltstone, sandstones and coal	2×10^{-6}	8×10^{-8}	5×10^{-7}

River Dee is the major source of potable water for significant areas of north Wales and the northwest of England. As such it was considered that any scheme must avoid putting the river under any significant threat of leachate pollution.

Development of proposals

The targets

Following intensive re-examination of the proposals and the philosophy of approach it was still considered that the establishment of a landfill was appropriate, and that it should still take the form of a clearly defined containment site.

As part of developing these proposals the professional team, that is both the consultant's (Acer) and the client's staff, set down a number of questions which it was considered that any scheme should address.

- Can the site be effectively developed as a containment site?
- Can leachate egress from the site be controlled to prevent a pollution hazard to the Dee?
- If all leachate control fails can it be shown that there is not a significant risk of leachate egress from the site producing a pollution hazard to the River Dee?
- Is the nature of the site and the ground conditions such that the required earthworks can be successfully and reliably implemented?

The problems

Clearly while it was considered that the Ruabon Marl has a low permeability, it was not below the value of $1 \times 10^{-9}\,m\,s^{-1}$ recommended by Waste Management Paper No. 26 (Department of the Environment 1986). However, that reference gave no clear guidance on thickness requirements.

Other guidance (NRA 1989; Commission of the European Communities 1991) provided much clearer requirements on the thickness of strata or barriers in relation to the formation of containment sites. Notwithstanding these guidelines it was not easy to extract clear criteria from the guidelines that could be consistently and logically enacted.

However, it was clear that the key issue was the development of a methodology that would ensure the protection of the Dee.

This was carried out by establishing the following:

- the hazard,
- the risk,
- the control and/or avoidance of the risk,
- the nature of the works to control and/or avoid the risk,
- a means of assessing and demonstrating the adequacy of the proposals.

The proposals

The threat to the Dee was clear. With a leachate build-up within the site, it was feasible that an outflow gradient between the landfill and the Dee could be established. In particular, the following were considered particularly critical:

- while the Marl had a low permeability, it did not have, in mass terms, permeabilities that would be considered negligible in terms of transmission potential;
- the deposit abutting the Marl had a significant permeability;
- the fault crossing the eastern edge of the site could provide a significant conduit for leachate.

Having reviewed the above, a radical change to the landfill philosophy was made. This was that the leachate would no longer be allowed to build up within the 'bowl' of the landfill, but instead would be controlled by pumping and maintained well below the level of the Dee. Thus a groundwater gradient into the site would be created. This is shown in Fig. 4.

As part of the development philosophy a fault tree diagram was established to target critical areas of the operation and identify what action would be available if a failure of the system occurred.

It was noted that if everything failed, and no monitoring occurred, leachate would eventually build up, to a point where it could theoretically leak from the site to the Dee.

An assessment was made of the probability of leachate in given quantities reaching the Dee within given travel-times.

Initially the proposals included lining the 'bowl' of the landfill with recompacted Ruabon Marl. However, during the planning enquiry this was further developed to include lining the full height of the sides of the landfill, even where this was well above the level of the flood plain. A combination of HDPE membrane and recompacted marl was proposed.

The proposed EU directive on the landfill of waste

A particular area of concern in developing the landfill proposals was the requirement for lining. The last published draft of the proposed landfill directive contains a requirement that:

'The non-saturated geological formations constituting the substratum of the landfill base and sides shall satisfy the following permeability and thickness requirements.

Maximum values of the permeability coefficient, K ($m\,s^{-1}$), for a substratum thickness of 3 m measured under conditions of water saturation was defined as follows:

Landfill for hazardous waste: $K = 1.0 \times 10^{-9}$ m s
Landfill for municipal and non-hazardous wastes and for other compatible wastes: $K = 1.0 \times 10^{-9}$ m s
Landfill for inert waste: $K =$ no limit value'
(Commission of the European Communities 1991).

The provision goes on to state that equivalent conditions may be met by engineering measures if not achieved naturally, and that a standard permeability testing method is to be developed and approved.

The proposed Pen-Y-Bont landfill site would appear to conflict with the draft directive, which seems to imply that if there are no non-saturated strata in the base of the site then landfill is unacceptable. (An alternative interpretation, but one which the authors do not believe is intended, is that if the strata are saturated then no thickness or permeability criteria apply!)

The fundamental concerns must be how much leachate will pass through the strata and whether that quantity is unacceptable in the receiving water. The former is a function of leachate head, thickness and permeability of the strata and the latter of the quality, quantity and use of the receiving water.

The issue is raised as to what is meant by 'equivalent conditions'. The authors believe that the test of equivalence must be the amount of leachate passing through a unit area in a unit time. This might be achieved by having a thinner layer of less permeable material, or by adding a flexible membrane liner to a more permeable material. In fact, there is an unwritten preference for a composite liner consisting of 1 m of clay of permeability less than or equal to $1 \times 10^{-9}\,m\,s^{-1}$ and an HDPE membrane.

As equivalence is a function of head of leachate, the target equivalent is unspecified in the directive, which does not state a permissible head, the UK norm is to use a 1 m head of leachate for assessment purposes.

It is the authors' view that saturation of the strata is not important in the assessment, except in as much as permeability of unsaturated clays is difficult to measure.

In the case of the Pen-Y-Bont site the strata are saturated. However, flow is into the site, so that pollution of surface or groundwaters will not occur. If for any reason the leachate controls should ever fail then the flow through the liner and the many metres of marl strata will be less than that through a 3 m layer at $1 \times 10^{-9}\,m\,s^{-1}$. The fact that the strata are saturated is unimportant. The authors consider it a suitable site, and the Secretary of State of Wales has agreed.

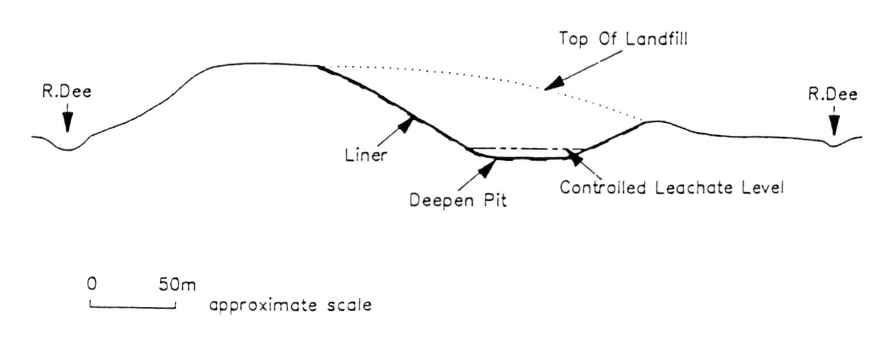

Fig. 4. Proposals—schematic.

Clearly the total flow of leachate from a site is a function of area; under the same leachate head and with the same liner specification, a 5 ha site will leak five times the quantity of leachate than a 1 ha site. The authors also contend that the characteristics of the receiving water must be taken into account. Indeed, although a 1 m, 1×10^{-9} m s^{-1} liner has become the norm in the UK (NRA 1989), Waste Management Paper No. 26 (Department of the Environment 1986) does not state such a requirement and the NAWDAC landfill guidelines (NAWDAC 1991) adopt the same approach as is advocated above by the authors as can be seen from the following extracts:

> One of the most important design considerations is how many factors, including the leachate within the site, are to be managed. This will depend upon the predicted leachate quality and the degree of protection that the surrounding environment requires. For example, if there is an aquifer capable of yielding economic quantities of usable water below or in the vicinity of the site it may be necessary to maintain the minimum practically achievable head of leachate over the liner system. However, if the groundwater quality is poor and is unlikely to be used and/or has a piezometric surface above the base of the site it may be necessary to permit the leachate to be maintained at a greater head.
>
> The basic requirement of a natural clay liner is that it maintains a low permeability. Limits on the liner specification may be set dependent upon the hydrogeological situation and risk analysis in consultation with the appropriate authorities.
>
> The standard adopted by the North West Region of the National Rivers Authority in 'Earthworks on Landfill Sites' (1989) is that the permeability of a clay liner must not exceed 1×10^{-9} m s^{-1}; that the minimum liner thickness is one metre and that leachate head should be kept below 1000 mm. Other limits may be appropriate depending on site specific conditions.

In the case of Pen-Y-Bont it is felt that the water in the River Dee and in any permeable strata nearby requires protection, and that the water in the Ruabon Marl immediately surrounding the site does not. The proposed scheme developed an appropriate philosophy to protect the identified targets, and did not restrict itself to following a rigid 'recipe' approach.

Conclusion

A sensitive site for landfilling has been addressed by reference to the nature of the site, the risk to the environment and the ability to adequately control the risk. By considering all aspects of the project an operational philosophy was developed that took into account the constraints that applied, and the possible variation in the ground.

The need to line the site was clear but the ability to define unequivocal standards to adopt was much more difficult.

Acknowledgements. A great number of people participated in developing the proposals and assisting the authors. The following individuals are thanked: Acer, R. Hoare and K. Longworth; Shanks and McEwan, D. Greedy.

References

COMMISSION OF THE EUROPEAN COMMUNITIES 1991. *Proposal for a council directive on the landfill of waste.*

DEPARTMENT OF THE ENVIRONMENT 1986. *Landfilling wastes.* Waste Management Paper No. 26, HMSO, London.

NRA 1989. *Earthworks on landfill sites.* National Rivers Authority, North West Unit.

NAWDAC 1991. *Guidelines on the use of landfill liners.* NAWDAC Landfill Committee, September.

Engineering geological and legal aspects involving proposals for a large waste disposal facility in The Netherlands

Pieter Michiel Maurenbrecher

Delft University of Technology, Faculty of Mining and Petroleum Engineering, PO Box 5028, 2600 GA Delft, the Netherlands

Abstract. In the densely populated area of the Netherlands proposals for a waste disposal tip can involve drawn-out legal proceedings before permission is granted. A case history is given in which the author acted as advisor for one of the objectors to the scheme and hence gives insight into shortcomings in the quite extensive environmental laws concerning, amongst other items, waste disposal facilities. The licensing authorities for such a scheme are the municipal and provincial authorities who, in this instance, are also supporters of the scheme. Such authorities, hence, do not maintain an unbiased role. High-court arbitration procedures are also outlined and the role of an expert witness. To date, the proposers of the scheme have not been successful in being granted a licence, in part, due to the proposer trying to precede new environmental laws before their ratification by Parliament.

The Wet Algemene Bepalingen Milieu-hygiëne (The General Conditions Environmental Hygiene Act) requires that new projects be subject to a MER or Milieu-effect rapportage (environmental impact report). The findings of a MER are advisory. The only legal requirement is the production of such a report including investigating various scenarios which represent a parametric-type study to see what influences a project could have on the environment. One mandatory requirement is the 'nulalternatief' or zero-option; how would one solve the waste disposal problem if the site remains as it is or what would be the effects if no action is undertaken. The law is, however, not very clear as to the reporting extent and detail. Adaptations in the law can in the course of time result in new demands during the planning of a project. The project would have to meet the update demands until actual permits/concessions have been granted. Various statutes can influence a project. Depending on the type of project, many statutes may have to be considered in order to carry out a satisfactory MER. A listing of Dutch statutes that could influence various waste disposal projects is given in Fig. 1. A further list of the environmental statutes are given in the references (Anon 1986).

The effectiveness of such statutes can only be realized with experience resulting from case histories. These examples may not cover every aspect and are usually influenced by the interests of the proposers of a scheme and their objectors. Both make use of specific legal elements to suit their case. The courts can be regarded as a third party as they may invoke a further set of legal elements neither party had used but the court requires, for further information or actions based on a ruling. The ruling used in this case history touched on the procedures adopted by the proposers to introduce, prematurely, in the court's opinion, the scheme to avoid the more stringent environmental laws expected to be introduced.

Waste dump scheme

The province of Utrecht is not only one of the smaller provinces of the Netherlands but it forms the inland apex of the Randstad in the Netherlands. Its central position makes it the main cross-roads for east–west and north–south routes for all the major means of transportation. This position causes it to be a popular location for industry. The area has seen a large expansion in industrial estates and urbanization, especially the new towns of Nieuwengein and Houten. Such growth cannot occur without a parallel growth in waste. Options for waste disposal sites are limited by the province's small land area, its dense population and an area of scenic beauty; the ice-pushed ridges of the uplands covering much of the eastern half of the province. As existing waste dumps are nearing their capacity the province is examining future sites for waste disposal. One such area is the large meadow/polder fields situated in an area wedged in by the Amsterdam–Rhine canal and a major new relief motorway taking traffic from the south of the Netherlands to the northeastern part of the City of Utrecht and major centres such as Hilversum, Amersfoort and those in the northern half of the Netherlands. (The older motorway

From Bentley, S. P. (ed.) *Engineering Geology of Waste Disposal*,
Geological Society Engineering Geology Special Publication No. 11, pp 373–377

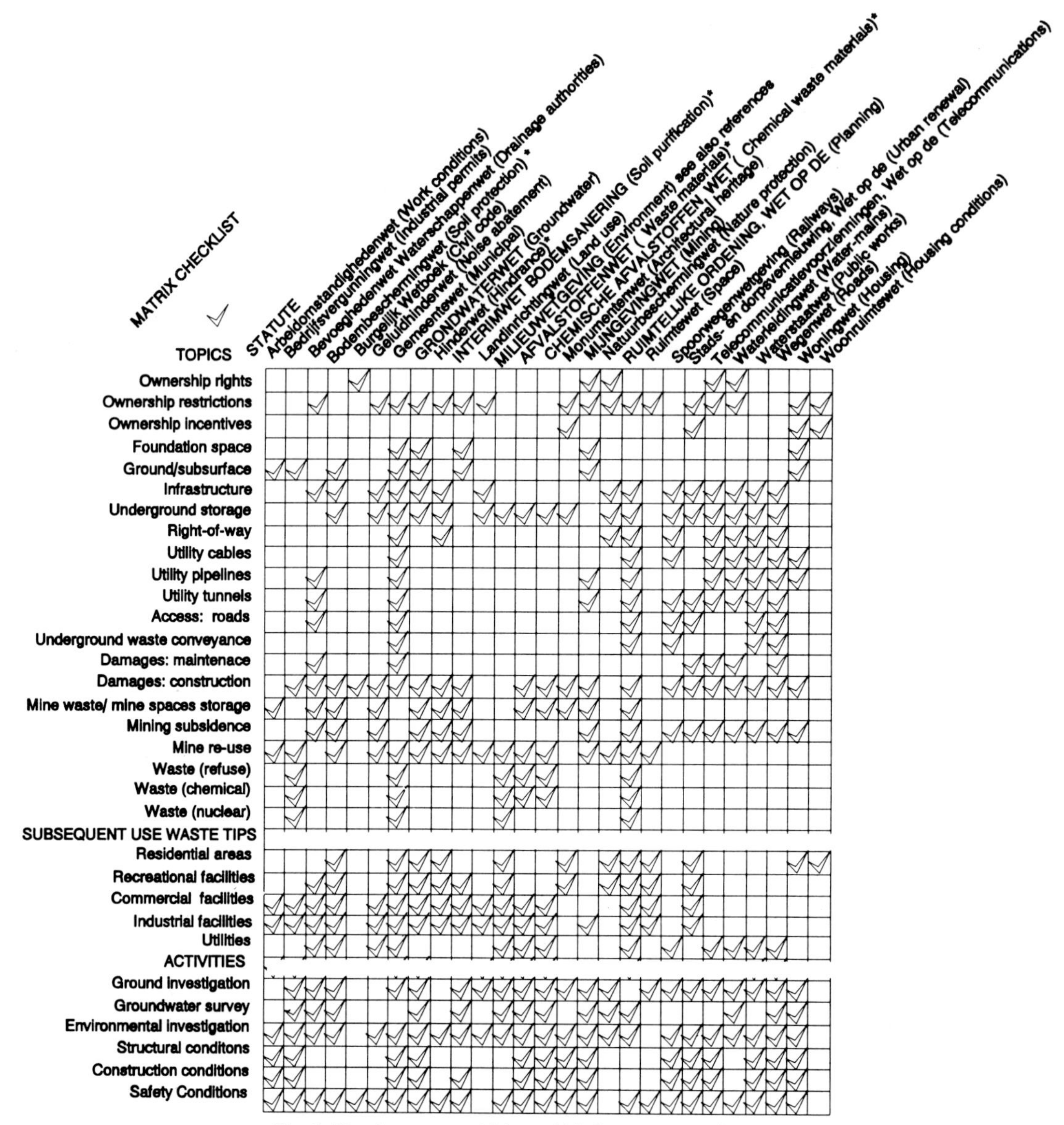

Fig. 1. Dutch statutes which could influence waste disposal.

heads towards Amsterdam and the western flank of the city of Utrecht.)

The motorway makes little contribution to the inhabitants living nearby and hence a situation arose which made their area more suitable for less attractive development projects such as waste disposal. The quiet back-water would now be intruded by a second environmentally deleterious project, the first having been the motorway, with its noise, visual obstruction and contaminants.

Role of engineering geology in support of objectors

The scheme met with a vociferous and (since the motorway project) seasoned local residents' committee. A second objecting party is the Amsterdam water company (NV Rijn-Kennermerland WRK) which pumps water from the nearby Rhine tributary, the Lek, through a pipeline to the dune infiltration plants near Haarlem. The Lek water during low flow periods drops in quality due to the increasing concentrations of existing pollutants. During such periods the water company can obtain groundwater from aquifers situated beneath the site of the proposed refuse dump. The water company appears to have drawn up its objections through experts in their employ.

The chairman of the residents' committee, Prof. de Vries of the veterinary department of the University of Utrecht, did not have to enquire far to obtain experts to assist his committee in drawing up objections to the scheme. Through the Institute of Earth Sciences at the University of Utrecht, the Engineering Geology Section at Delft was asked to review the scheme's proposals, designs and investigation documentation. Through a relatively informal request, Engineering Geology at TU Delft became more involved with environmental problems. This was not the first time; earlier work had been carried out by Salters & Verhoef (1980) and Maugenest & Schokking (1992) who investigated the possibilities of nuclear and chemical waste disposal respectively. The latter study assists government ministries in preparing background information for 'notas', i.e. parliamentary responses to possible questions raised on these matters.

Initial communications

The type of information provided by the objectors for critical review and comment consisted of the following:

- Site investigation report 1986 and 1989
  Report on refuse waste disposal:
  Section 1: Request permit in accordance with the refuse-waste disposal act (Afvalstoffenwet),
  Section 2: Environment-effect note (in accordance with the refuse-waste disposal act),
- Public announcement by the Province of Utrecht of the proposal for a refuse disposal area in accordance with general conditions of the Environmental, Hygiene and Refuse Disposal Act.

The strategy was to comment on the reports on engineering geological aspects concerning the geotechnical and hydrological information, the analysis and design aspects involving geotechnical and hydro(geo)logical models, and on monitoring systems. Hence little consideration was given to challenge shortcomings in the statutes. It was hoped that the proposers and/or arbitrators or legal advisors engaged by the parties concerned in the undertaking would be able to underscore such reviews as relevant, irrelevant, in conflict with, or, in need of appending to the various statutes which could have a bearing on the granting of a permit.

Comments generally concerned the lack of proper coverage of the site investigation. Initially results from the motorway investigation were used with only a cursory additional investigative work in the form of cone testing on the site of the refuse disposal. The soil mechanics analyses for the proposed retention embankments were very preliminary. For example, factors of safety were calculated but no indication was given of the extent of deformations that would have to take place before the shear strength of the very weak clays lying underneath the site would be mobilized. Other aspects of the site investigation report were emphasized, such as one location not being tested because of the 'poor ground conditions'. This statement served to doubt the low shear strengths used in the stability calculations. The low design factor of safety of 1.1 was also challenged in terms of a storm recurrence cycle and that the value was too low for an 'important structure retaining possible contamination accumulations'. Consolidation predictions, especially in connection with the existence of peat deposits and the soft clays, combined with little in the way of laboratory testing, were criticized as being too speculative. The geological 'pancake' interpretation was challenged on the basis of insufficient investigation and the type of environment (close to a meandering river) as well as the existence of a continuous clay layer acting as a seal against any percolate leakage.

The design report was mainly challenged on the section concerning monitoring. Doubt was expressed as to the number of monitoring devices 'for such an extensive site' and for the frequency of monitoring (the report stated yearly measurements).

The Provinces in the Netherlands are, since the last revision of the Water Act, responsible for the water management in their jurisdiction. The report concluded on the basis of groundwater modelling they had performed, the aquifers would not be affected by leakage of percolate, should it occur. The opposers of the scheme requested further information:

- What were the input data for the model?
- Which modelling method was used?
- On which information was the deeper geology of the site based or interpreted?

The fact that the Province did the modelling appears to suggest a conflict of interest with respect to the Amsterdam Water Company as the Province is one of the proposers for the waste disposal facility.

A public hearing evening 'inspraak avond' was organized by the municipality to discuss the pros and cons of such a scheme. It is not clear how many of the objections were passed on to the developers and to what extent they reacted to them. Subsequent to the meeting, further investigation and analyses were carried out with regard to the earth structures. It appears that a portion of the critical comment has been heeded.

Second round communication

The public hearing, however, did not resolve the issue between the objections of the opposers to the scheme and the whole matter was then referred to the high court in the Hague known as the Raad van State (the Council of State). The main purpose of the afdeling voor geschillen van bestuur (Department for Disputes of Governmental Administration Institutions) is to arbitrate in cases where objections are made towards decisions made by government and semi-government institutions such as is the case with the Province of Utrecht and the municipality of Nieuwengein. Three judges, specialized in aspects of environmental and planning law, presided during the meeting, at which were submitted the dossiers of both the proposers and opposers for the project. The parties can also present arguments verbally during the meetings. They were requested by the judges not to repeat what was already stated in the dossiers. The hearing allows the lawyers representing the parties concerned the opportunity to solicit clarification on possible conflicts, inconsistencies or comprehension of the law and to add new facts or oversights. Conversely, the judges can solicit clarification on various aspects of the submissions.

Depending on the content of the dossiers it can be several months before advice or a ruling is given.

In preparation for the above hearing, the latest additional information was critically examined. Despite the extra geotechnical investigation by geophysical resistivity methods and extra analysis of the possible behaviour of the retention embankments, further criticism was submitted. Comments on the geotechnical matters from two 'experts' were submitted. One remained neutral by stating that the investigation geophysical survey interpretations appeared in order. Indeed, the quality of the investigative work could not be challenged. What did seem obvious was that the developer appeared to try and limit the costs of such work as the principal comment remained the lack of good coverage of the site. Increasingly such work demands giving some indication as to the confidence levels of the interpretations and the risks associated with analyses work. This was submitted as the main deficiency in the site investigative work. Such techniques are a requirement in standard codes of practice (draft form) in the Netherlands.

The developers or authorities had not reacted to comments on monitoring or the aquifer modelling. It may be that the latter aspect may not have been an objection of the Amsterdam Water Company and hence would have little further bearing on objections from the Residents' Committee.

During the court proceedings in May 1991, considerable time was spent discussing the necessity to carry out a proper environmental assessment report in accordance with The General Conditions Environmental Hygiene Act. The proposers of the scheme objected to having to compile such a report as they said such a requirement was introduced, by law, after the development of the scheme has been started. They also objected on the grounds that introducing such requirements would put the Province in a difficulties with regard to delays in meeting demands for near-future disposal capacity for the anticipated waste. They further argued that for each revision of the law such new demands could mean that a project be delayed continuously. Towards the end of extended proceedings (programmed from 9:45 am to 11:00 am but extending to 13:45 pm) an opportunity was given for only a minute's comment. The comment was restricted to emphasizing the lack of sufficient information, stating that a Municipality would normally not grant permits for a construction project on the density of information given for the waste dump site. Although the Amsterdam Water Company would have better reasons for opposing the scheme, the lack of deeper information was again stressed with regard to the groundwater modelling.

The opportunity to comment was first objected to by the proposers but they were over-ruled by the judges who stated that they could, had they wanted, provided expert witnesses on ground matters.

Subsequent developments and comments

The case has appeared twice more before the Council of State. The judges first ruled that an Environmental Impact Report should be carried out. On submission of the report for the second court hearing, the report appears not to have covered sufficient possibilities; nor to have looked at other potential sites for waste dumps. Hence a third hearing will be required before the judges can rule for or against the use of the site.

In retrospect, the Council of State obliged the Province to carry out a proper environmental impact report because as a government body the Province is party to the enactment of laws, especially concerning the environment. Hence when the project was started the law would have been in a draft/committee stage, including the requirement for a environmental impact report. The Province, as party to such a law, should have then already acted in the spirit of the law rather then circumvent it by initiating the waste dump scheme to save itself or its developer the more lengthy process the new law would cause, and lessen, thereby, the chances of obtaining a permit to develop their favoured site.

Author's note. The paper is a personal assessment and hence legal aspects and procedures need further verification.

References

ANON. Milieuwetgeving, 1986, 1988, 1990, etc. (Environmental Laws):

Wet algemene bepalingen milieuhygiëne (The General Conditions Environmental Hygiene Act)

Wet chemische afvalstoffen/Afvalstoffenwet III (Chemical Waste Materials/Waste Materials Act

Wet bodemsanering VIa/ bodembescherming VIb (Ground Purification Act/Ground Protection Act)

Wet op de invordering (Collection Act)

Wet inzake luchtveronreiniging (Air Pollution Act)

Wet olieverontreiniging zeewater (Sea Water Oil Pollution Act)

Wet verontreiniging oppervlaktewateren (Inland 'Surface Water' Pollution Act)

Wet veronreiniging zeewater (Sea Water Pollution Act)

Wet voorkomen verontreiniging door schepen (Prevention of Pollution by Ships Act)

Nederlandse staatswetten Schuurman & Jordens (Netherlands state statutes Schuurman & Jordens series) published by W. E. J. Tjeek Willink, Zwolle. (For further relevant statutes see Fig. 1.)

MAUGENEST, C. & SCHOKKING, F. 1992. Chemical wastes above or underground; a pre-evaluation study. *ICUSESS 1992 Proceedings 5th International Conference on Underground Space and Earth Sheltered Structures*. Delft University Press, 467–474.

SALTERS, V. J. M. & VERHOEF, P. N. W. 1980. *Geology and Nuclear Waste Disposal*. Proceedings of the Symposium and Colloquia held in October and November 1979, Institute of Earth Sciences, Utrecht, Geologica Ultraiectina Special Publication No. 1.

Appendix

A newspaper article appeared in the *NRC-Handelsblad* of 3 August 1993. Its translation from Dutch into English is presented below.

According to the water-supply company, a joint venture of the Amsterdam Municipality and the Province of North Holland, the waste tip threatens the quality of the drinking water.

The claimed amount is required for additional purification measures for the drinking water, which is earmarked for 2.5 million inhabitants of North Holland, amongst others in Amsterdam. The Provincial Council of Utrecht appears to approve the installation of the waste tip scheme in the polder of Klein Vuylcop, adjoining the Lek Canal. In September, a definite decision will be made, after opposition had earlier been submitted by the WRK. From the Lek Canal, Rhine water is abstracted, which is then transported to the dune area around Haarlem for further filtering.

The waste-disposal scheme, which must obtain a capacity of 5 million cubic metres, is proposed for the disposal of waste products, dredged sediments, sewerage sediments and contaminated ground. The WRK is concerned about leakages from the waste dump at the Lek Canal, pollution from infiltration basins through birds and waste material blown in by wind. The water supply company calls the waste tip area, which has to be filled in the next thirty years, 'a covered environmental time bomb'.

According to the representative of the WRK it is necessary to move the water intake to another location along the Lek Canal during the construction of the waste tip facility. At the same time the infiltration basins would have to be roofed over, as protection against the birds that will be attracted to the waste tip area.

The Amsterdam council member M. Witte-Buijserd from the environmental party 'Eco-200' has, in connection with the proposed waste tip facility, put questions to the City Executive Council. 'This is a typical example of a lapse in properly appraised environmental supervision', said Mrs Witte.

The Amsterdam municipal water-mains department views the Province's (Utrecht) proposals for the establishment of a waste tip facility with astonishment. 'We assumed it would be impossibie that the Province would take such a decision', claimed the acting director ir. A. J. Roebert, 'Naturally we require a waste facility. But it goes without saying that you do not establish one (waste tip) directly opposite the extraction point for drinking water.' If the claim of 104 million guilders (40 million Pounds Sterling) cannot be obtained from the Province, then, according to Roebert, the users' water rates for drinking water will have to rise by 10%.

In total, the Province has investigated eight locations for a possible waste tip area. An Environmental Impact Report (MER) has shown that the risks of a tip area in the polder Klein Vuylcop would be 'manageable'. The Province, this morning, could not be reached for comment. It had declared earlier that the waste tip would have, at most, 'minimal effects' on the quality of the water.

Landfill failure survey: a technical note

David Roche

Frank Graham Consulting Engineers, 22 Waterbeer Street, Guildhall Centre, Exeter, Devon EX4 3EH, UK

A survey of reported pollution incidents and other modes of failure affecting landfill waste disposal sites has been undertaken, and the key results are summarised here. A questionnaire survey was made of selected representative organizations in the waste management industry in England and Wales, including operators, regulators and specialists, and the results from which constitute a substantial database. The majority of the sites included in the survey have been active during the past 20 years. Currently active sites and recently commenced sites are included.

The Initial Survey comprised around 1000 sites. Of this sample, about one in six sites had suffered a significant pollution incident or failure which has required substantial remedial treatment (see Fig. 1(a)). Landfill gas migration was the most frequent form of problem, identified affecting nearly 1 in 10 sites. Surface water pollution incidents affected more than 1 in 20 of the sites and groundwater pollution more than 1 in 40 sites (see Fig. 2(a)).

For Stage 2, a more detailed questionnaire survey has been undertaken. The survey targeted over 200 selected representative members in the waste management industry. The survey now takes account of around 4000 landfill sites.

The findings of the more detailed survey essentially confirm the results of the initial survey, although with the much larger sample the trends are modified in certain respects.

Of the larger sample, about 1 in 20 sites have suffered a significant pollution incident or failure which has required substantial remedial treatment (see Fig. 1(b)). This indicates a lower incidence than the initial survey findings; however, only half of the initial sample responded in the second questionnaire survey. More interestingly, the data provided in the second survey by the operators correlate closely with the overall findings of the original survey. Regulators and client authorities embrace more sites but indicate a lesser frequency of problems than is actually revealed by the operators.

Overall it would appear that there is a fairly even subdivision between the main types of problem, involving groundwater, surface water and landfill gas (see Fig. 2(b)).

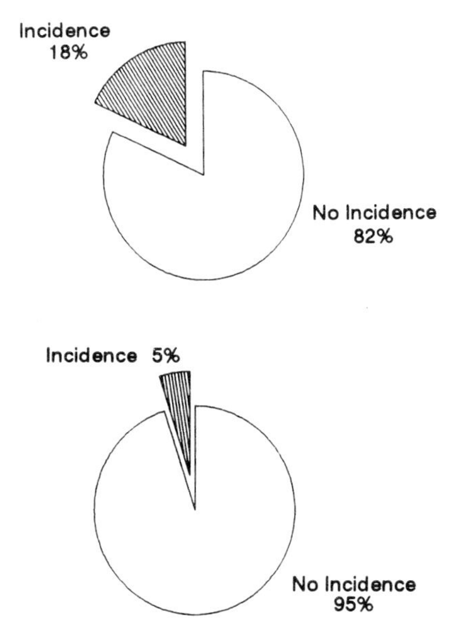

Fig. 1. Incidence of significant pollution or failure: (a) 1st survey; (b) 2nd survey.

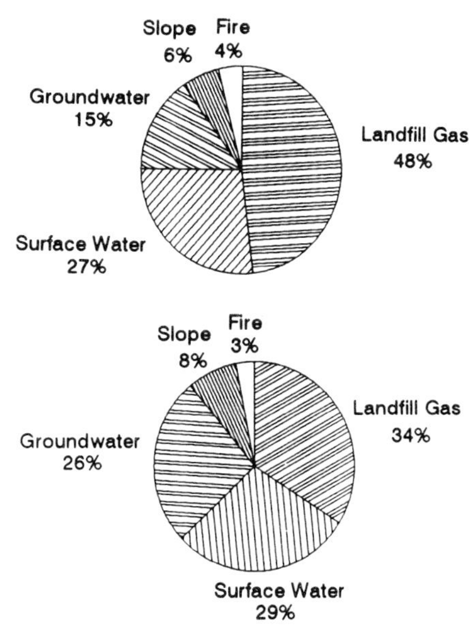

Fig. 2. Subdivision of failure types: (a) 1st survey; (b) 2nd survey.

From BENTLEY, S. P. (ed.) *Engineering Geology of Waste Disposal*, Geological Society Engineering Geology Special Publication No. 11, pp 379–380

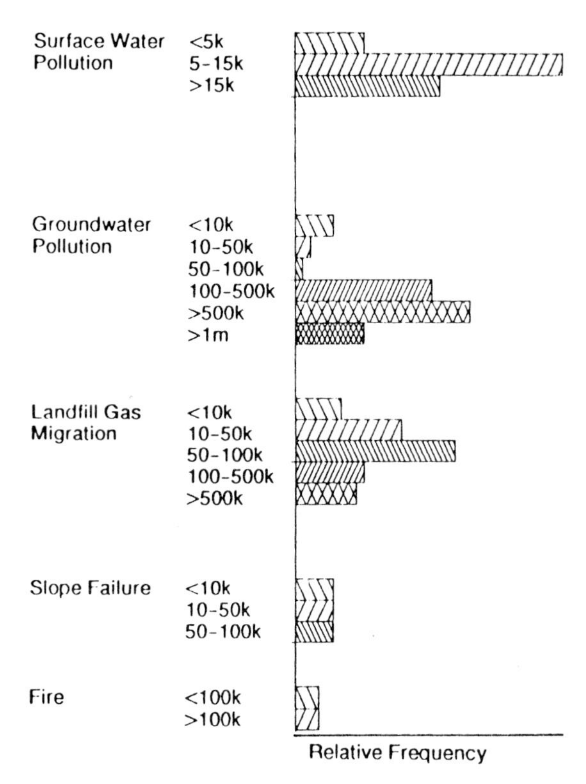

Fig. 3. Relative frequency and scale (£).

Groundwater pollution affects around 1 in 60 sites and the majority have required remediation treatment costing in excess of £100 000. Several incidents have been intimated in which very large-scale remedial costs in excess of £1 million have been necessary (see Fig. 3).

Surface water pollution incidents are also reported on around 1 in 60 sites but are generally less costly to remediate, typically ranging up to about £15 000.

Landfill gas migration also affects about 1 in 60 sites and typically requires control and remedial schemes costing in the order of £100 000 and sometimes in excess of £0.5 million (see Fig. 3).

Slope instability has been reported at about 1 in 200 sites, and fire problems at about 1 in 600 sites; both types of problems can incur substantial remedial costs (see Fig. 3).

The survey stems from the need to provide advice on the probabilistic risks of significant problems arising on landfill sites and the absence of documented historical data on such incidents. Although risk assessment requires evaluation of the exposure of targets to hazard sources, the results of the survey will assist the quantification of landfill failure risk assessment.

Some very interesting trends are indicated by those respondents who completed the second part of the questionnaire. Of the problem sites it would appear that almost a third are considered to be 'containment' sites, the remainder are presumed to be 'dilute and disperse'. Also, more than 5% were commenced after 1990. In addition, of the sites where remedial measures have been implemented, a substantial number will still require further remediation to resolve the problems. Also, over half of the problem sites have third-party habitable properties within 250 m of the site boundary.

The overall conclusion of the survey is that there is a very significant incidence of problem landfill sites in the United Kingdom. On this basis, more detailed survey and evaluation is needed to evaluate the problems and resource the solutions. A greater depth of information and scientific research should provide more reliable data and conclusions.

Acknowledgements. The following organizations are thanked for their provision of data used in the survey (see Fig. 4): Cambridgeshire CC WRA; Cumbria CC WRA; Devon CC WRA WDA; Gloucestershire CC WRA WDA; Surrey CC WRA; WIltshire CC WRA; East Waste; Haul Waste; The Waste Company; Wyvern Waste Services; Dorset CC WRA; East Sussex CC WDA; Greater Manchester WRA; Hampshire CC WDA; Hertfordshire CC WRA; Isle of Wight CC WDA; North Yorkshire CC WRA; Nottinghamshire CC WRA; Somerset CC WRA; South Yorkshire Waste Regulation Unit; Suffolk CC WDA WRA; Tyne & Wear WRA; NRA South West; London Borough of Croydon; Durham County Waste Management Co; Northumberland Waste Management; Waste Notts; Gloucestershire Sand & Gravel Co; Hales Waste Control New Soils Reclamation; N J Banks & Co; Skipaway; Ashact; Babtie Geotechnical; Bingham Cotterel; Coal Processing Consultants; Crouch Hogg Waterman; EAU; Golder Associates; Johnson Poole & Bloomer; Marcus Hodges Environment; Mott MacDonald; Aberconwy BC; Alyn & Deeside BC; Colwyn BC; Delyn BC; Ruddlan BC; Avon CC WRA; Buckinghamshire CC WRA; Cleveland CC WRA; Essex CC WRA; Hereford & Worcester CC WDA; Kent CC WRA; Warwickshire CC WDA WRA; County Environmental Services (Cornwall); Greenways Landfill; Springfield Disposal; Tarmac Econowaste; Waste Management.

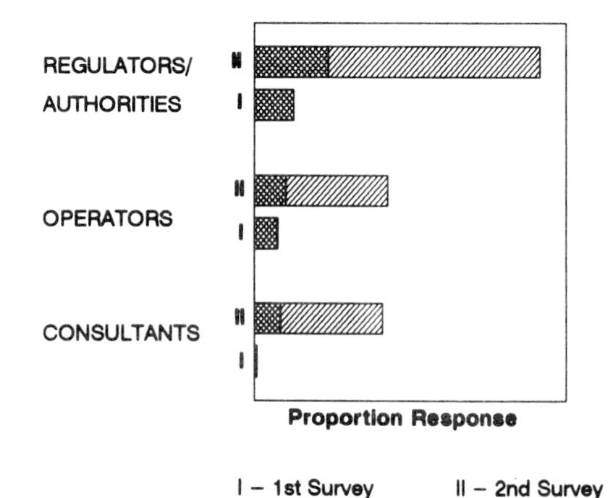

Fig. 4. Subdivision of respondents.

Educational issues in environmental geological engineering

D. G. Toll

School of Engineering, University of Durham, Durham, UK.

Abstract. The educational implications of introducing environmental issues into engineering geology courses are discussed. It is suggested that there is a need for specialist courses in environmental geological engineering specifically designed for engineering geologists who intend to work in the environmental field. Such courses should introduce the various issues related to air, water and ground pollution; noise and vibration. The courses also need to deal with environmental planning issues such as risk assessment, hazard evaluation and the appropriate legislation. However, it is argued that the main focus of these courses should be the disposal or treatment of waste and the investigation and remediation of contaminated land and groundwater.

The term 'environmental' tends to mean different things to different people. The perception of environmental issues for society at large concerns the 'global' environmental crises: acid rain, global warming (the greenhouse effect), sustainable development and conservation. It is these issues which are accorded greatest prominence within the education system, both at school (Brody 1994) and at university (Patterson 1993).

For geologists, the term 'environmental geology' is now well established. It covers such factors as natural hazards (seismic, volcanic, floods, landslides etc), land use and planning, use of natural resources, waste disposal and pollution. By comparison, the engineering community has traditionally focused on water resources and wastewater treatment (public health engineering), environmental impact due to construction, energy efficiency, as well as renewable energy and waste management as environmental concerns.

One way of considering these factors is to present them as a spectrum, ranging from those which can be considered as having local impact at one end, to those which are global in effect at the other. Figure 1 shows such a spectrum. The scale also represents 'achievability', i.e. the degree and extent to which human intervention is possible and desirable. One end of the scale represents issues where action is possible at the local level, i.e. a high degree of achievability. At the other end, humans are largely restricted to observing changes, and those factors which can be affected require international cooperation. We have yet to see significant impact from human intervention at this end of the spectrum.

At the global end of the spectrum are the issues which tend to be the preserve of the 'environmentalist'. These are the issues which are taught at primary and secondary school level (e.g. Brody 1994). In reality, however, humans can least directly affect these factors. We are largely restricted to observing the changes produced. Engineering Geologists can have little input at this level in a professional capacity; that which can be achieved falls largely into the preserve of politicians. One might argue that we should be encouraging more engineering geologists to enter the political arena. However, that is a wider issue than can be addressed in this paper.

In the mid-ground of the environmental spectrum fall issues such as landslide hazard and flood hazard, where some degree of mitigation can be achieved, or at least the effects are regional, and planning of land use can have an effect. At the local end of the scale are those factors which humans can control to a reasonable level, where action can be taken which will have an effect. It is at this end of the spectrum that engineering geologists can have the greatest input in their professional capacity.

Environmental issues can also be categorized according to subject specialisms. One has to be careful in doing this since one of the main aspects of environmentalism is its multi-disciplinary nature. However, there are some issues which most closely relate to particular disciplines. Figure 2 shows two sub-sets of 'environmental issues' which can be considered to be those relating to geologists (environmental geology) and those of concern to engineers (environmental engineering).

Defining environmental geological engineering

The overlap area between environmental geology and environmental engineering (shaded in Fig. 2) can be called *environmental geological engineering*. This is the area within which engineering geologists have the greatest role to play. It is, of course, closely allied to *environmental geotechnics*; the differentiation reflecting

From Bentley, S. P. (ed.) *Engineering Geology of Waste Disposal*,
Geological Society Engineering Geology Special Publication No. 11, pp 381–388

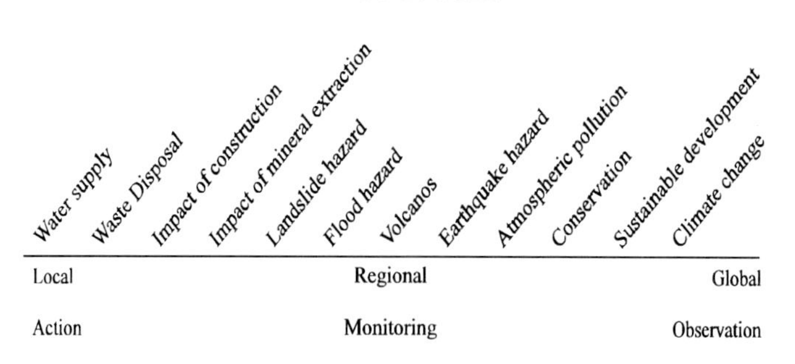

Fig. 1. The spectrum of environmental issues.

the divide (often vague or blurred) between engineering geologists and geotechnical engineers (Toll 1991; Atkinson 1993).

Carrier *et al.* (1989) attempted to classify the discipline which has become known as *environmental geotechnology* (they admitted that this was a struggle). Although they discussed some of the global environmental issues (climate change, extinction of species) they focused on problems associated with solid, liquid, hazardous, industrial and mining waste.

Sembenelli & Ueshita (1981) had previously defined the term *environmental geotechnics* as:

> '*a tool* that contains all the different branches of soil and rock mechanics. A tool shaped to investigate, to understand and decide, *used beforehand* to avoid geotechnical problems that might materialise and geotechnical activities which might later develop into problems. A tool aimed at minimising the entropy of a given process through geotechnical knowledge. A tool to be used together with other similar ones in a planning effort shared by many disciplines.'

Similarly, Hamel (1993) reviews a number of definitions of *environmental geotechnology*, and favours the rather abstract definition given by Caldwell & Hobbs (1987):

> 'a planning tool; a set of problems (and the appropriate solutions) in the relevant disciplines; an attitude of mind; or simply the science and art of the interaction of soils and rocks with their surroundings'.

Although Sembenelli & Ueshita (1981) give an abstract definition, they did attempt to classify the subject. They suggested the following categories:

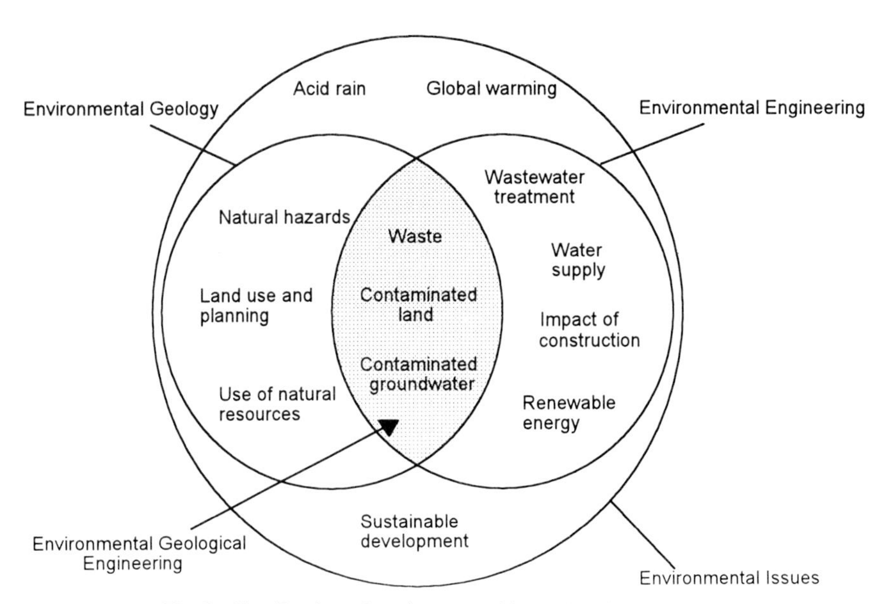

Fig. 2. Classification of environmental issues by discipline.

- Modifying physiochemical properties of materials (causing pollution, peptization, surficial degradation, contamination of surface waters, change of agricultural value of soils, heave)
- Reuse of wastes (Mining, metallurgical, industrial and urban wastes)
- Solids removal at the surface (Cuttings, borrow pits, opencast mines)
- Solids removal from underground (Mining, tunnelling, artificial caverns)
- Solids accumulation on the surface (Coal tips, mine waste piles, tailings dams, dredge dumps, reclamation fills)
- Fluids extraction from underground (Water pumping, oil draft, gas tapping, geothermal vapour trapping, brine extraction, coal gasification insitu)
- Fluids storage on the surface (reservoirs)
- Underground deposits (recharge, water storage, thermal energy storage, disposal of wastes, grouting)
- Towns (impact of urban development)

What is clear from such a classification is that many main-stream engineering geology activities (cuttings, tunnels, mines, reservoirs) are listed as part of environmental geotechnics. This emphasises the difficulty in separating out 'environmental issues' as a distinct discipline. In essence, everything we do has an environmental impact.

However, I believe there is now a general consensus on a sub-set of issues which fall under the banners of *environmental geological engineering* or *environmental geotechnology*. At the 8th Meeting of Teachers of Geotechnical Subjects (Toll, 1993) the following areas were considered to define the field:

- Environmental risk assessment and risk analysis
- Contaminated land
- Waste disposal (particularly landfill)
- Migration of pollutants and gases from contaminated and waste sites

Educational implications

The educational implications of the development in environmental issues were addressed at the 8th Meeting of Teachers of Geotechnical Subjects (Toll 1993). The discussion revolved around whether these subjects should be taught as part of undergraduate civil engineering degrees. The general feeling was that, while an awareness of the environmental impact of engineering work should be an integral part of an engineering education, such specialist topics were beyond the scope of undergraduate programmes. Such a degree of specialization should be left to postgraduate MSc courses.

While most undergraduate geology degrees have an environmental geology component, they would not specialize in the engineering geology aspects. Again, it would be expected that such a degree of specialization would be covered by postgraduate MSc courses.

Many graduate students specializing in engineering geology or geotechnical engineering now leave to work in the environmental field. MSc courses are adapting to this change in emphasis and most have incorporated an environmental component. In addition, specialist courses on environmental engineering, environmental geotechnics, waste management etc. are now available and are very popular. However, Morgenstern's (1991) comments are worth bearing in mind that, in environmental geotechnics, many academic programmes lag behind what is happening in industry. I believe that is less true than it was in 1991, but the comment may still be pertinent.

Gasoski (1990) drew attention to the fact that 'Training and education is a weak point of the environmental management structure. Most of the environmental authorities are inefficient due to the lack of trained manpower for specific tasks'. To meet this need in the environmental geological engineering field we need people who are educated in engineering geology but who have also been introduced to the concepts of environmentalism. There is a danger that general environmental courses may produce graduates who have wide awareness, but have limited specific skills to tackle real problems at a practical level.

As has been pointed out, environmental geological engineering is not a new discipline; it has always been part of engineering geology. What is different about the 'environmental component' is that it does have to breach institutional and interdisciplinary barriers. As is suggested by the rather abstract definitions given by Sembenelli & Ueshita (1981) and Caldwell & Hobbs (1987) it is vital that one thinks beyond the technical aspects of engineering geology and considers the sociopolitical implications, as well as the other technical aspects (chemical, biological etc.).

One might therefore argue that environmental issues should feature in all engineering geology (or geotechnical engineering) courses, and this is now, in general, true. However, the introduction of new material into a course inevitably squeezes something else out. The overall balance of a course must therefore be considered. In order to provide sufficient education for graduates to proceed into the environmental field may involve removing what might be considered core material for an engineering geologist. This would suggest that there should be separate courses catering for the different needs: The traditional engineering geology course should remain and environmental geological engineering courses should be provided for those intending to specialise in the environmental field.

The content of an engineering geology course can be considered as a 'spectrum' ranging from the 'hard' subjects (not necessarily in terms of difficulty!) dealing with rocks, through the 'soft' subjects dealing with soils, into the 'wet' end dealing with groundwater (Fig. 3). The environmental field tends to be an extension of the 'soft' and 'wet' end of the spectrum, and have less involvement with the 'hard' end. This is, of course, a generalization; for instance underground storage of nuclear waste will be closely related to rock behaviour (although it may be the groundwater flow through the rocks which is of greatest importance for environmental impact).

Figure 3 shows the topics which would normally be covered in a traditional engineering geology course. This covers the rock, soil and groundwater part of the subject spectrum and partially overlaps the environmental end. A suggested range of subjects for an environmental geological engineering course is also outlined. This starts from the environmental end, spans groundwater and soil and partially covers the rock end. It can be seen that there is a large area of overlap between the two areas. Both should include all of soil behaviour and groundwater and each should have some component of rock behaviour and environmental matters. It is simply a question of the degree of emphasis placed on each end of the spectrum.

Further, it is suggested that the environmental component of environmental geological engineering courses should introduce the issues surrounding air, water and ground pollution, as well as noise and vibration. It also needs to deal with environmental planning issues such as risk assessment, hazard evaluation and the appropriate legislation. However, the central subjects that need to be covered in detail are the disposal or treatment of waste and the investigation and remediation of contaminated land and groundwater.

A typical syllabus might therefore cover the following topics:

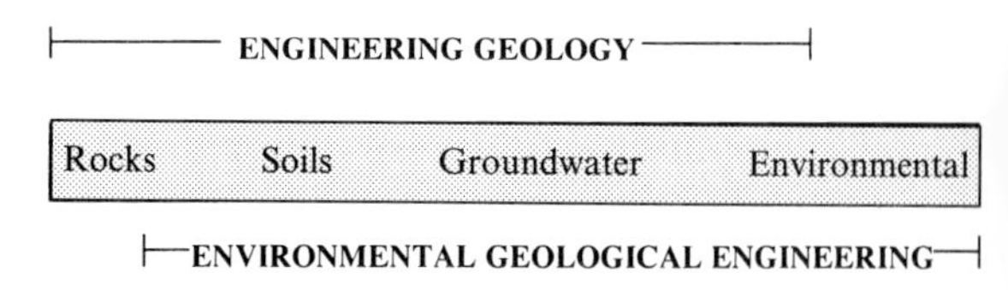

Fig. 3. The subject spectrum of engineering geology and environmental geological engineering.

Environmental planning and legislation

- Introduction to environmental concepts: air, water and ground pollution, noise and vibration
- Environmental assessment and impact statements, hazard evaluation, risk analysis.
- Environmental Legislation: Environment Agency, European Union directives, Environmental Protection Act, Integrated Pollution Control. BATNEEC[1] and BPEO[2].

Waste

- Types and sources of waste (domestic, industrial, hazardous). Methods of disposal or re-use.
- Landfill: Containment vs dilute and disperse philosophies. Site assessment and material suitability. Containment design. Lining materials. Quality assurance. Landfill gas.

Contaminated land

- Types and sources of contaminants. Soil chemistry.
- Site investigation. Site assessment.
- Prevention of contaminant movement.
- Remediation.

Contaminant transport

- Groundwater sampling and chemistry.
- Designation of groundwater protection zones. Aquifer testing.
- Saturated and unsaturated flow and transport. Flow of miscible and immiscible fluids. Heat transport in groundwater flow. Mathematical modelling of contaminant flow.
- Remediation of contaminated groundwater.

A detailed syllabus for a new MSc course in environmental geological engineering to be run at Durham University is given in the Appendices. The course will comprise six taught modules: rock behaviour, geotechnical engineering; engineering geology; soil engineering; water engineering; environmental engineering. The environmental engineering module is based on the typical syllabus set out above. Students will also undertake a dissertation on a chosen aspect of environmental geological engineering.

Conclusions

All aspects of engineering geology have an environmental impact. However, a general consensus appears to exist as to which topics are central to *environmental geological engineering* or *environmental geotechnics*. These are:

- Environmental risk assessment and risk analysis
- Contaminated land
- Waste disposal (particularly landfill)
- Migration of pollutants and gases from contaminated and waste sites.

These topics cannot be covered in sufficient detail in undergraduate courses and should form part of specialist MSc courses. However, it is argued that even

[1] Best available technology not entailing excessive cost.
[2] Best practicable environment option

to cover these topics fully within MSc courses in engineering geology would mean that core engineering geology material would have to be excluded. Therefore it is suggested that, in addition, environmental geological engineering courses should be provided for those engineering geologists who intend to specialise in the environmental field.

These environmental geological engineering courses should introduce the issues surrounding air, water and ground pollution, as well as noise and vibration. They also need to deal with environmental planning issues such as risk assessment, hazard evaluation and the appropriate legislation. However, the main focus of these courses should be the disposal or treatment of waste and the investigation and remediation of contaminated land and groundwater.

It is to be hoped that, as more such courses become available, we will develop a greater environmental awareness within the profession. If environmental skills are applied *beforehand* we should be able to avoid some of the environmental problems that currently confront us.

Acknowledgements. I would like to thank all my colleagues in the School of Engineering and the Department of Geological Sciences at Durham University (particularly Dr Steve Thomas) who have been involved in the planning of the new MSc course in environmental geological engineering. I also acknowledge the input of the external members of the course management committee, Dr Alan Common of Exploration Associates and Bruce Underwood of FWS Consultants. Many of the ideas put forward in the paper have emerged from discussions held while planning the course. Thanks also to Linda Thompson for a wider view on environmental education.

References

ATKINSON, J. H. 1993. Engineering geologists in the construction industry, *Geoscientist*, **3**, 3, 9–11.

BRODY, M. J. 1994. Student science knowledge related to ecological crises, *International Journal Science Education*, **16**, 4, 421–435.

CALDWELL, J. A. & HOBBS, B. T. 1987. Environmental geotechnology and tailings reclamation, *In*: *Geotechnical Practice for Waste Disposal '87*. ASCE, New York,362–376.

CARRIER, W. D., DE MELLO, L. G. & MOH, Z. H. 1989. Environmental impact in geotechnical engineering, *Proceedings 11th ICSMFE, Rio de Janeiro*. Balkema, Rotterdam, **4**, 2121–2156.

GASOWSKI, N. 1992. The greening of business, *Proceedings of Global Opportunities for Business and the Environment '90, Globe '92*, Vancouver, Canada, 93.

HAMEL, J. V. (1993) Environmental geotechnology in the United States: A consultant's perspective. *In*: CHOWDHURY, R. N. & SIVAKUMAR. M. (eds), *Environmental management, Geo-water and Engineering Aspects*, Balkema, Rotterdam, 37–48.

MORGENSTERN, N. R. 1991. The emergence of environmental geotechnics. *Presidential Address, 9th Asian Regional Conf. SMFE, Bangkok*.

PATTERSON, J. G. 1993. Preparing for the future—Will our university graduates be scientifically literate? *Geoscience Canada*, **20**, 4, 165–172.

SEMBENELLI, P. & UESHITA, K. 1981. Environmental Geotechnics—State-of-the-art report, *Proceedings 10th ICSMFE, Stockholm*. Balkema, Rotterdam, **4**, 335–394.

TOLL, D. G. 1991. Editorial. *Geotechnical and Geological Engineering*, **9**, 1, iii–iv.

—— 1993. Report on the 8th Meeting of Teachers of Geotechnical Subjects (MTGS), Dublin. *Ground Engineering*, **26**, 10, 22–24.

Index

References to figures are given in *italic* type; references to tables are in **bold** type.